T0406554

The Twin Sister Planets Venus and Earth

Robert J. Malcuit

The Twin Sister Planets Venus and Earth

Why are they so different?

A TREATISE ON HOW THE PROCESS OF GRAVITATIONAL CAPTURE OF A MAJOR SATELLITE CAN ALTER THE EVOLUTIONARY HISTORY OF TWO OUTSTANDING TERRESTRIAL PLANETS

Springer

Robert J. Malcuit
Geosciences Department
Denison University
Granville
Ohio
USA

ISBN 978-3-319-11387-6 ISBN 978-3-319-11388-3 (eBook)
DOI 10.1007/978-3-319-11388-3
Springer Cham Heidelberg New York Dordrecht London

Library of Congress Control Number: 2014950150

Printed on acid-free paper

Springer is part of Springer Science+Business Media (www.springer.com)

This book is dedicated to my wife, MARY ANN, for her multidimensional support, assistance, and counsel over the years which helped to bring this project to completion. In our years together we have had many great experiences in both well-travelled and remote areas on this amazing planet.

Preface

Earth is a very unusual planet and the Moon is a very unusual satellite. Our "twin sister" planet is also a very unusual planet. If Venus had a sizeable satellite in prograde orbit as Earth does, then life would be a bit simpler for the earth and planetary science community. Furthermore, if planet Venus had a transparent moderate density atmosphere, oceans of water, and continents with linear mountain belts, we would have far fewer questions about planet Venus. And if we could spot a few pyramids, whether trigonal, tetragonal, hexagonal, or octagonal, we would be very curious about the nature and history of the civilizations on our sister planet.

The characteristics of planet Venus, however, are much different than what we normally would predict for a "sister planet". The only attributes that Earth and Venus have in common are planetary mass, planetary density, and location at a similar distance from the Sun. So the main question of this book is: What happened to planet Earth to make it the **"paradise planet"** that we live on today? Or we can turn the question around and ask: "What happened to planet Venus to make it into the **"hades planet"** that it is today?

The purpose of this book is to present a scientific story for the development of the twin sister planets, Earth and Venus, that is consistent with our present knowledge of the Solar System and that is very testable by present and future earth and planetary scientists. This book has ten chapters of varying lengths and I think that it takes ten chapters to ask the questions and to tell the story. If only Earth, Venus, the Moon, and the Sun were involved, perhaps five chapters would be sufficient. But nearly the entire Solar System is involved in this scientific story of the twin sisters. Thus the narrative gets a bit lengthy and more complex than most of us would like to believe. This book represents something like the antithesis of the basic tenants of the test of Occam's Razor. A common statement of this test is that **"the simplest explanation for some phenomenon is more likely to be accurate than more complicated explanations."** I will admit that Occam's Razor works well for many problems in the world of science. But its usefulness in the earth sciences has some limitations. As a first example let us consider the Continental Drift/Plate Tectonics model. Although the model is simple in principle, it is very complex in detail and some 40 plus years after general acceptance, and a full century after conception of the idea, concerned scientists still cannot agree on when our modern style of plate tectonics began. As a second example let us briefly consider the model of Milankovitch for explaining the ice ages on Earth. When I took a "Glacial and Pleistocene Geology" course in 1965, there were perhaps three pages on the idea in one of the best textbooks on the subject. The conclusion was that the model had failed many scientific tests and the implication was that the idea was not worthy of serious scientific consideration. Then by the middle 1970s, the Milankovitch Model emerged for serious consideration and is now a very useful paradigm in the earth sciences. Is it a simple model? NO! Is it easy to test? NO! Does this work with the Milankovitch Model support the basic tenents of the test of Occam's Razor?

How does the concept of Occam's Razor fit in with the story of the twin sister planets? I can state that the story of the twin sister planets is "simple in principle but somewhat complex in detail". In my opinion, the complexities, including the probability of a favorable outcome for certain processes, outweigh the simplicities.

A major theme of this book is that our twin sister planets had very similar physical and chemical characteristics soon after they were formed about 4.6 billion years ago. They formed in the same region of space and, according to the most recent work on planetary orbit resonances, they have been "shepherding" each other, as many twin sisters do, in a way to make their sun-centered orbits mutually stable. As a result the orbits of Venus and Earth are by far the most stable orbits in the inner part of the Solar System. Their life-long planetary partners (the planetoids that eventually become their satellites in this story) were accreted from material much closer to the Sun but also ended up in fairly stable sun-centered orbits. According to the most recent calculations their orbits were, in their first several 100 million years, the second most stable orbits in the inner Solar System. These planetoids, Luna and Adonis, eventually get perturbed out of their orbits of origin into a Venus-like orbit (for Adonis) and an Earth-like orbit (for Luna). They then get gravitationally captured by these large terrestrial planets and this is where the life histories of these planet-satellite pairs diverge. Venus captures Adonis into a **retrograde** orbit (that is, the planet is rotating in one direction and the satellite is orbiting in the opposite direction). Earth captures Luna in a **prograde** direction (that is, the planet is rotating in the same direction as the satellite is orbiting the planet). The remainder of the story is "planetary history". The end result is that the twin sisters are now "polar opposites" for habitability.

Chapter 1 is an introduction to the characteristics of the twin sister planets. Chapter 2 is a brief discourse on the early history of the Sun and how and why this early history is so important to the formation of Luna and Adonis, in particular, but really for all planets, planetoids, and asteroids out to at least the vicinity of planets Jupiter and Saturn. Chapter 3 is a brief treatment of intellectual endeavors associated with developing explanations for the origin of the Earth-Moon system. Chapter 4 is an explanation of my model of gravitational capture of planetoid Luna by planet Earth to form the Earth-Moon system. Chapter 5 consists of a series of "vignettes" (brief to the point stories) on various topics in the planetary sciences. The first three vignettes involve an analysis of the patterns of lunar maria and mascons (mass concentrations) that I think is crucial for developing a more meaningful interpretation of the features of the Moon than we have at present. Another vignette is on a process for partially recycling a primitive crust inherited from a "cool early Earth" into the Earth's mantle by way of the very high amplitude rock tides caused by the process of gravitational capture. The final vignette in the chapter is on the subject of the origin of water on the twin sister planets, a somewhat neglected subject in the earth sciences but an important issue for our planet in particular. Chapter 6 is an explanation of my model for the retrograde capture of planetoid Adonis (a planetoid that is only one-half as massive as Luna) by Venus and the subsequent evolution of the system over time. A major feature of this model is that planet Venus has the possibility of being habitable for about 3 billion years before surface conditions deteriorate. Chapter 7 is a bit different in that it is more like "make believe" or "alternate reality" in that it presents a model for a retrograde capture of a lunar-mass satellite by Earth. The final result is an earth-like planet that has many of the features of planet Venus today: i.e., a "hades-like" scene, featuring a dense, corrosive atmosphere on a planet rotating slowly in the retrograde direction and with no satellite. In Chapter 8 some new concepts are introduced. The general theme of the chapter is that Planet Orbit—Lunar Orbit Resonances may be important for explaining some critical events in Earth History such as the beginning of the modern style of plate tectonics and the transition from a bacteria-algae based biological system to one featuring metazoan-type organisms. The main characters in the Planet Orbit—Lunar Orbit Resonances are "big brother" Jupiter for one resonance episode and "twin sister" Venus for three episodes. Chapters 9 and 10 pose some questions about how all these complex episodes in the historical development of Earth and Venus, and indeed in the history of the Solar System, relate to the habitability of our planet and for the possibility of finding other candidate planets for habitability.

Acknowledgements

This project has been something like a forty-year endeavor. I thank Dick Heimlich (Kent State University) for getting me acquainted with the primitive Earth by way of the Bighorn Mountain project. Introduction to the geological literature on the primitive Earth eventually led to some very fruitful interactions with Preston Cloud (University of California, Santa Barbara). Thanks to Tom Vogel (Michigan State University) for permitting me to do a somewhat "risky" Ph. D. project on Precambrian interactions of the Earth and Moon in contrast to a more normal Precambrian Geology project. Thanks to Tom Stoeckley (Physics-Astronomy Department, Michigan State University) for the use of an n-body computer code that my Denison University colleagues and I have been using, in modified form, over the years. I also acknowledge the very positive interactions and discussions with former graduate school colleagues at MSU: Gary Byerly (University of Louisiana, Baton Rouge) and Graham Ryder (now deceased but for many years at the Lunar and Planetary Science Institute, Houston) as well as to former MSU professor Bob Ehrlich. Special thanks to Ron Winters (Denison University Physics and Astronomy Department) for sticking with the project and for the physics and computer work and dedicated time working with physics and computer science students, David Mehringer, Wentao Chen, and Albert Liau, in the late 1980's and early 1990's without whom this project would not have progressed. I give special thanks to John Valley (University of Wisconsin, Madison) for developing the "Cool Early Earth" model and for encouragement to pursue a planetoid capture model for the origin of the Earth-Moon system. I also give thanks to Fred Singer for his pioneering work on the concepts of gravitational capture.

On the local front, I am grateful to Dave Selby, Cheryl Johnson, and Leslie Smith (Denison University ITS Group) for technical support and manuscript preparation, respectively. Thanks to Denison University for office space, computer equipment, and other support over the years since retirement. I note here that most of the critical concepts that made the world safe for capture studies occurred after my retirement (e.g., stable Vulcanoid planetoid orbits, a Cool Early Earth, recycled primitive crust on Earth). Special thanks to my retired colleague, Ken Bork, for his encouragement over several decades to pursue my ideas on the origin and evolution of the Earth-Moon system. Thanks to Tom Evans for his long-term interest in this project and for sharing his knowledge of the resources in the Harold Urey Collection in the Archives at the University of California, San Diego. And finally, thanks to my active departmental colleagues and generations of geoscience students for the many questions and comments that helped in the development of my ideas on gravitational capture of satellites by planets and thanks to Ron Doering, Springer, for his guidance in manuscript preparation. And on a very personal level, I am very grateful to my wife, Mary Ann, for her perusal of several renditions of each chapter and for her special advice and constructive criticism. The manuscript is much improved because of her careful attention to detail.

Contents

Chapter 1
Introduction

"We believe that life in the form of microbes or their equivalents is very common in the universe, perhaps more common than even Drake and Sagan envisioned. However, complex life—animals and higher plants—is likely to be far more rare than is commonly assumed. We combine these two predictions of the commonness of simple life and the rarity of complex life into what we will call the Rare Earth Hypothesis." From Peter D. Ward and Donald Brownlee, 2000, Rare Earth (Why complex life is uncommon in the Universe): Copernicus (An imprint of Springer-Verlag), p. xiv.

The above statement will resonant very well with the content of this book in that I am attempting to explain some of the unusual events that have acted on our planet, EARTH, and our sister planet, VENUS, over the course of Solar System history to make them polar opposites for habitability. A currently popular explanation for the unusual nature of planets Earth and Venus is that they were affected differently by giant impacting bodies early in their respective histories. Mackenzie (2003) expounds on the virtues of the "GIANT IMPACT MODEL" for Earth and recently Davies (2008) explained how a giant impact might relate to some of the outstanding features of Venus. I do *not* agree with these *ad hoc* explanations for the contrasting differences in the conditions of these sister planets. The explanation I am developing in this book is that both Earth and Venus captured planetoids from heliocentric orbits into planet-centered orbits early in their planetary lives. Earth captured a planetoid that I am calling LUNA [a name coined by Alfven and Arrhenius (1972) for our satellite before capture] into a prograde orbit about 3.95 billion years (Ga) ago and it is still with us at the present time. In contrast, Venus captured a sibling planetoid (a sibling to Luna that I have named Adonis that was about one-half as massive as Luna) into a retrograde orbit: i.e., the satellite orbits in the opposite direction to the rotation of the planet. To make a three billion year story a bit shorter, Adonis no longer exists! This satellite coalesced with Venus after despinning the planet to essentially zero rotation rate over a period of about 3 billion years. [Note:

Robert J. Malcuit, *The Twin Sister Planets Venus and Earth*,
DOI 10.1007/978-3-319-11388-3_1

The basic scenario of retrograde capture for a satellite for Venus was first proposed by Singer (1970) and I have simply added a few phases to his model.]

Let us keep in mind that a giant impact of a Mars-sized body on an Earth-like planet would be an unusual event. Likewise, capture of a moon-sized body from a heliocentric orbit is an unusual event! Indeed, gravitational capture of a one-half moon-mass planetoid from a heliocentric orbit into a venocentric orbit seems to be an additional complication, but as we will discover, this retrograde capture model for planet Venus explains many of the features of planet Venus very well.

1.1 The Scientific Method

The SCIENTIFIC METHOD is a procedure for testing new ideas in the natural sciences. It works for "big picture" ideas as well as for local field study problems. There are five basic steps to the process:

a. We start with a LIST OF OBSERVATIONS OF A NATURAL PHENOMENA (i.e., the facts to be explained by our hypothesis or model—a model simply being a somewhat more detailed explanation than an hypothesis)
b. We FORMULATE AN HYPOTHESIS (An hypothesis is simply an untested explanation)
c. We TEST THE HYPOTHESIS (1) by making more observations of the natural phenomena being investigated, (2) by doing relevant experiments, (3) by doing relevant calculations, and (4) by making predictions and then independently checking the predictions for accuracy.
d. The HYPOTHESIS is either VERIFIED or REJECTED based on the results of the tests. In many cases in "big picture" natural science issues, the verification or rejection can come decades after the hypothesis is proposed. And in some cases, a rejection is reversed to verification after a new test or new technology for testing the idea has been discovered or invented.
e. If VERIFIED, an HYPOTHESIS becomes a THEORY (a theory is simply a well-tested explanation)

Two recent articles on the SCIENTIFIC METHOD are Lipton (2005) and van Loon (2004). These authors emphasize the progression from speculations to hypotheses to models as well as the concept of making predictions that can be independently tested (i.e., without prejudice).

Now let us discuss some of the special characteristics of the subjects of this book: the twin sister planets, Venus and Earth. Figure 1.1 shows a planar view of the geometry of part of the Solar System and this diagram will give the reader some orientation for the discussions in this introductory chapter. Figure 1.2 shows typical images of the terrestrial planets: Earth, Venus, Mars, and Mercury. We will begin our cursory survey of Earth and Venus by listing and briefly discussing some special features of the Earth as a planet and then focus our attention on planet Venus.

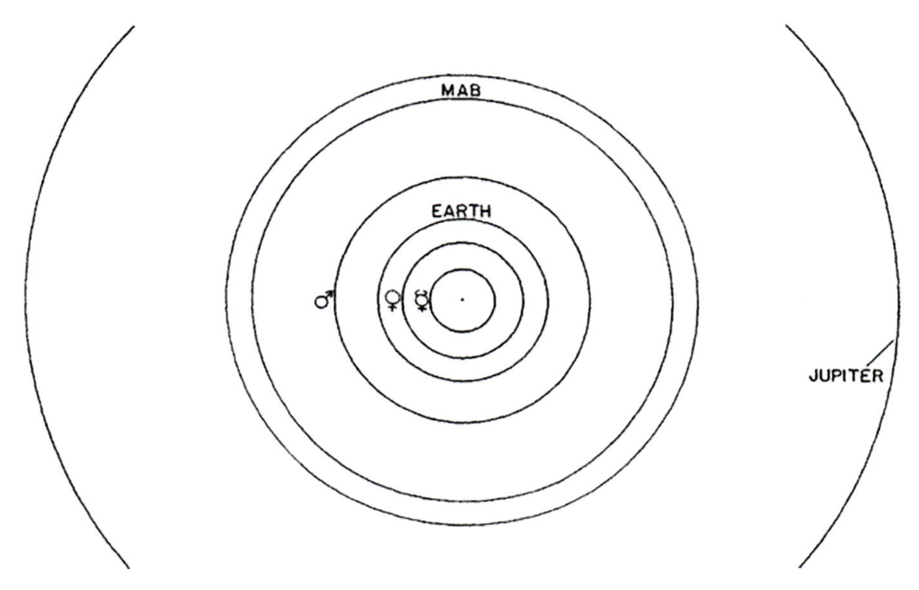

Fig. 1.1 Scale sketch of the orbits of the planets in the inner part of the Solar System. Mars is between Earth and MAB (*Main Asteroid Belt*). Venus and Mercury, respectively, are interior to Earth. For scale, Earth is located at 1 astronomical unit (*AU*)

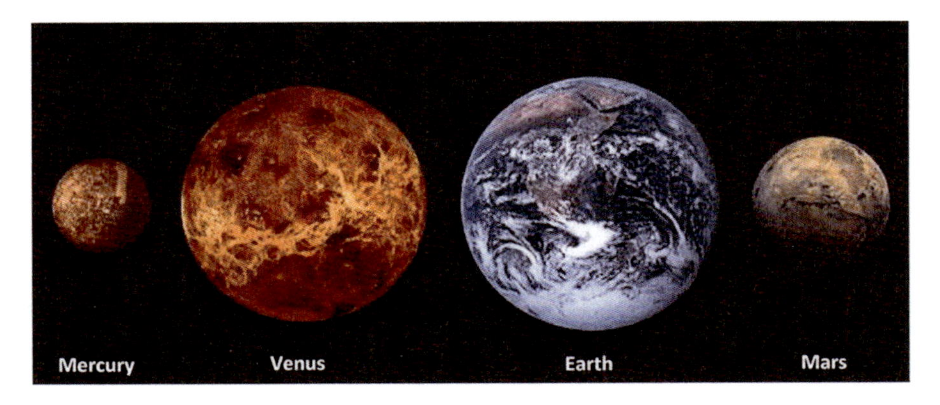

Fig. 1.2 Scale pictograms of the four terrestrial planets. Note that the twin sister planets, Venus and Earth, are very similar in diameter. (From Faure and Mensing (2007, Fig. 11.1), (NASA, LPI)

1.2 Some Special Features of Earth as a Planet

- LIQUID WATER ON THE SURFACE: At present, the Earth is the only planet with standing water on the surface and we have oceans of water as well as lakes, rivers, and streams of various shapes and sizes. By comparison our terrestrial neighbors appear not to have so much as a mud puddle at the present time. Mars does, however, have features which suggest that it did have liquid water on its surface in the past. These features are channel systems with branching tributaries which look very much like stream channels on Earth (Carr 1999). Venus apparently has even less water than Mars. The atmosphere of Venus (over 100 times denser than the Earth's) has less than 1% water vapor (Saunders 1999). The surface rock temperature is about 600 °C, much too hot to permit liquid water on the surface. In the scientific literature there is an ongoing discussion of the "Venus Oceans Problem". The main evidence for water on Venus is the deuterium to hydrogen ratio that suggests that it may have had "oceans" of water in past eons (Young and Young 1975; Dauvillier 1976; Donahue et al. 1982; Donahue 1999).

- FREE OXYGEN IN THE ATMOSPHERE: At present, the Earth is the only planet with free oxygen in the atmosphere. The present composition of the atmosphere is 78% nitrogen (a nearly inert gas), 21% oxygen (a very reactive gas), and 1% inert gases such as neon, crypton, xenon, etc., and carbon dioxide). A small amount of free oxygen can be expected in a planetary atmosphere due to the photo-dissociation of water into hydrogen and oxygen and of carbon dioxide into carbon monoxide and oxygen by solar ultraviolet radiation. The amount of oxygen in the Martian atmosphere is consistent with that expected from photo-dissociation. Oxygen is a very *unlikely* gas to accumulate in a planetary atmosphere because it is very reactive and readily oxidizes the abundant metallic ions (like Fe) at a terrestrial planet surface. There is general agreement in the scientific community that if it were not for the continuous generation of free oxygen by photosynthesis, we would not have our great abundance of free oxygen (Berner et al. 2003). The oxygen content of the Earth's atmosphere has varied considerably over the past 600 million years. It has been as high as 30% in the Carboniferous Period to as low as 12% at the Permian-Triassic boundary (Berner 2006).
- THE ABUNDANCE AND DIVERSITY OF LIFE FORMS: The above two unique features of Earth (the presence of water on the surface and the presence of free oxygen in the atmosphere) have apparently made possible another unique feature—the presence of an abundance and significant diversity of biological forms. We do not yet know how nearly unique our biological system is in the Solar System, but we are very sure (after the data gathered by several lander and rover missions to Mars) that there are no obvious life forms on planet Mars. Nor do we expect much biological activity in the dense Venusian atmosphere. In addition, we have found no evidence for present or former life on our essentially atmosphere-free natural satellite, the Moon.
- THE PRESENCE OF A STRONG MAGNETIC FIELD RELATIVE TO THE SIZE OF THE PLANET: A fourth unusual feature of our terrestrial planet, EARTH, is the presence of an active magnetic field. There are two other bodies in the Solar System that have presently active magnetic fields—the Sun and the giant planet Jupiter. Three terrestrial bodies in the inner part of the Solar System, the Moon, planet Mercury, and planet Mars have magnetized rocks on their surfaces, but these remanent magnetic patterns appear to be the products of long-since decayed magnetic fields. (We have not yet detected strong fossil magnetic patterns or an internally generated field associated with planet Venus).

I note here that although we humans use the terrestrial magnetic field continuously for navigational purposes (as do sea turtles, birds, fish, etc.), we still do not understand the details of the mechanism of its generation. Is it due to (a) the rotation of the Earth, (b) the precession of the Earth, (c) the tidal retardation of the rotation of the mantle-crust complex relative to the inner core via lunar tidal torque, or (d) a combination of these. Concerned scientists know that the polarity of the field changes with an average period of about 1 million years but they are not sure of any definite major biological consequences of these polarity changes. However, studies of oceanic sediments suggest that extinction of some species of micro-organisms may be associated with these reversals of the magnetic field but no large animal extinctions have been associated with magnetic reversals (Strangway 1970; Glassmeier and Vogt 2010).

- THE PRESENCE OF A LARGE SATELLITE RELATIVE TO THE MASS OF THE PLANET: A fifth unique feature of the Earth as a terrestrial planet is the presence of a very large satellite, the Moon. The mass ratio of the Moon to the Earth is about 1–81: this is the highest for any satellite-planet system in the Solar System. A common ratio for satellite-planet systems is 1–5000. The only other terrestrial planet with satellites is Mars which has two very small ones. Thus, we may ask if the presence of this exceptionally large satellite, the Moon, has had anything to do with the development of the previously discussed unique features of planet Earth? My view is that the presence of the Moon in orbit about the Earth *has had a significant effect* on the pathway of evolution of our planet. In some cases the effects are "subtle" and in other cases they are "PROFOUND".

Some additional but important features of the Earth are:

- ONLY PLANET WITH "TRUE" GRANITE: The main rock type separating from the mantle of Earth is basalt. Most of this basalt is being formed at oceanic rises (i.e., ocean-floor spreading centers). In general, granitic rocks are differentiation products of basalt but it takes a good bit of processing of basalt to get much of a yield of granitic product. True GRANITE is even more difficult for nature to generate and it can be generated in significant volume in only certain geotectonic settings. The favored place is in volcanic arcs above subduction zones where basaltic ocean floor is being subducted under either continental crust or under other basaltic ocean floor. Even in this setting, a good bit of subduction must take place to get a small mass of true granite. Indeed, some very small pockets of granitic rock have been found in lunar basalts. These are called granophyres. And I would speculate that small pockets of granitic rocks will be found in the basalts of Mars and Venus also. But to find large masses of granite, like we have on Earth, we apparently need to locate some active or extinct subduction zones on our neighboring planets.
- ONLY PLANET WITH CONTINENTAL CRUST: The Earth is the only planet in the Solar System with continents and these continents are composed of granitic rocks, some of which are true granites. It is still not clear just how all of this continental crust has formed (Rudnick 1995; Polat 2012) but some parts of the continents have been on Earth for over 3 billion years. About one-third of our planet is covered by continental crust and two-thirds is covered by the basalts of

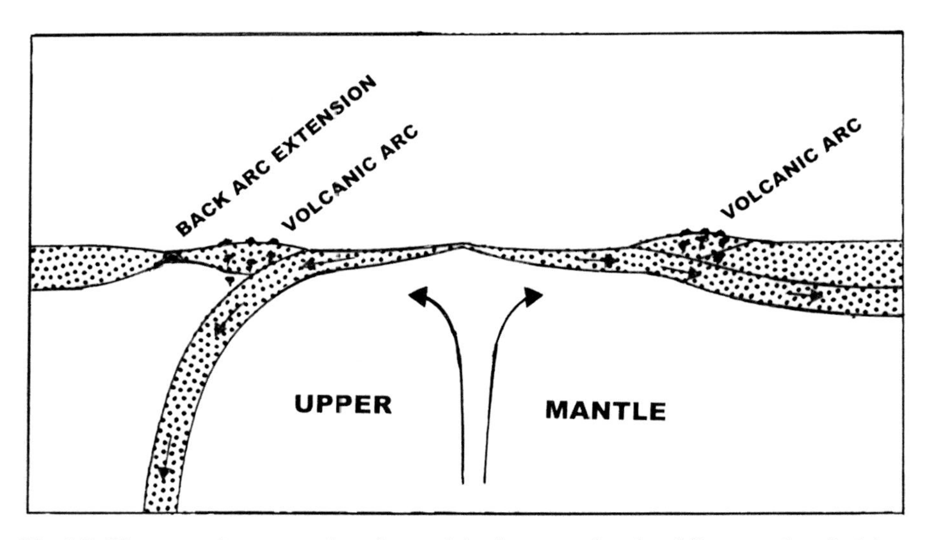

Fig. 1.3 Diagrammatic cross-section of two subduction zones showing different angles of subduction. The main point is that the down-going plate will sink by gravity if the slab is higher density than the surrounding mantle. If the density of the slab is equal to or less than the surrounding mantle, the slab must be forced into the mantle or forced under another surface plate. High angles of subduction are commonly associated with westward dipping subduction zones as well as back arc spreading centers (example: the Japanese arc-trench complex); low angles of subduction are commonly associated with eastward dipping subduction zones (example: the Andes arc-trench complex)

the ocean floors. Although most of the biological system of Earth originated in the shallow waters of the continental shelf areas, it is the continental platforms (i.e., above sea level) that are very important for human civilization. Thus, we not only have oceans of water and a thriving biological system but we also have continental crust above sea level on which to to build our structures.

- ONLY PLANET WITH OPERATIONAL PLATE TECTONICS: We could have granite and granitic rocks and continental crust even without plate tectonics. The granitic rocks and continental crust could have been formed by other processes at an earlier era in Earth history. The surprising thing is that the mechanism that we geologists think generated much of the granite, granitic rocks, and eventual additions to these large masses of continental crust is still operating today. One would think that a warmer Earth would lead to more vigorous plate movements and as the Earth cools, the motion would slow down or cease. But the plate movements are more vigorous today than perhaps 1.5 billion years ago. This is one of the major paradoxes of the plate tectonics paradigm—one would think that plate motion would be slower on a cooler Earth. The explanation probably lies in the mechanism of subduction. There are two major mechanisms for subduction: one is *slab-push* and the other is *slab-pull* (Fig. 1.3). Both of these mechanisms are operated by gravity but the main difference for their operation is the temperature of the associated mantle. For *slab-push* the pressure for subduction is caused by the elevation of an oceanic ridge. The higher the ridge elevation the more effective is this mechanism and the cooler slab is FORCED down the subduction zone; i.e., it does not sink on its own and the slab melts as it is being forced into the mantle. On the other hand, *slab-pull* can only work when the slab can maintain its coolness relative to the hotter mantle. The cooler slab then "sinks" into the warm mantle. If the mantle is too warm, then the sinking slab will melt before it sinks very far. For effective slab-pull the sinking slab should maintain its identity, and excess density, for about 600 km. Since the Earth's mantle is cooler than it was 1 billion years ago, this gravity sinking operation is more effective as geologic time moves forward. Some investigators think that this slab-pull mechanism could only operate during the last 1 billion years of earth history (e. g., Davies 1992; Stern 2005) and any plate tectonics before that time would be dominated by the slab-push mechanism.

1.3 Some Special Features of Venus as a Planet

- *NO* LIQUID WATER ON THE SURFACE: The surface is simply too hot and any water would be in the form of water vapor in the atmosphere.
- *NO* FREE OXYGEN IN THE ATMOSPHERE: The gases of the atmosphere are all reduced gases and any free oxygen formed by photo-dissociation of water and/or carbon dioxide in the top layers of the atmosphere would be consumed quickly by these reduced gases.
- *NO* LIFE DETECTED: The surface of Venus appears to be very hostile for life as we understand it on Earth. There is a possibility that some form of life could live in the outer (cooler) layers of the atmosphere.
- *NO* MEASURABLE MAGNETIC FIELD: There is no measurable internally generated magnetic field associated with planet Venus. If rotation rate is an important factor in generating a planetary magnetic field, the very slow retrograde rotation rate of Venus could be the reason for the lack of a planetary magnetic field.
- *NO* SATELLITE (NOW): If a large satellite is important for the development of habitable condition on planet Earth, perhaps the absence of a large satellite can help explain the lack of habitable conditions for Venus. This simple explanation may not be very effective. Maybe we should consider the possibility that Venus *had* a sizable satellite in the past but that satellite no longer exists. WOULD THAT MINOR SPECULATION MAKE A DIFFERENCE TO THE HISTORY OF OUR SISTER PLANET, VENUS?

Some additional but important features of planet Venus are:

- A TEMPERATURE OF 450 °C AT THE SURFACE: This temperature is hot enough to melt lead (as in angler's gear). And the atmosphere has a significant content of hydrogen sulfide. This hydrogen sulfide along with some water molecules in the atmosphere yields a sulfuric acid mist.
- ESSENTIALLY NO ROTATION RATE: The rotation rate is really very slow in the *retrograde* direction—one day on Venus (one sidereal rotation of the planet

takes 243 earth days of time). Has it always been that way? Probably not in that the primordial rotation rate according to the empirical plot of MacDonald (1963) should be about 13.5 h/day in the *prograde* direction (Fig. 1.4). Perhaps an explanation of how Venus lost its prograde rotation may help to explain the adverse condition of our sister planet.

- YOUNG SURFACE ROCKS: It certainly was a surprise for the Magellan mission investigators to discover that nearly the entire surface of Venus was covered by basaltic volcanic rocks—mainly sequences of massive lava flows on the surface. It was an even greater surprise to find that the impact craters on the surface were all about the same age: i.e., the crater density patterns over the entire surface of the planet were about the same. This led investigators to propose the concept of a "global resurfacing event". Based on crater-count studies on other planets and satellites, the educated guess for the timing of this "global resurfacing event" was that it happened over a short period of time somewhere between 1000 million years ago and 500 million years ago (Herrick 1994). At present there are three models for this condition on Venus. (1) Perhaps Venus undergoes alternating eras of heating up and cooling down and the last heating up period resulted in the global resurfacing we see now (Herrick and Parmentier 1994). (2) Perhaps a "giant-impact" of some sort caused the resurfacing event (Davies 2008). (3) Perhaps the capture of a satellite in a retrograde orbit could cause the damage to the surface of Venus. This retrograde capture scenario was first proposed by Singer (1970) and further work was done by Malcuit and Winters (1995) to demonstrate that retrograde capture was physically possible, and Malcuit (2009) to show that the time scale for post-capture orbit evolution could be up to 3 billion years.
- VERY DENSE CARBON DIOXIDE-RICH ATMOSPHERE: The atmosphere of Venus is about 100 Earth atmospheres of pressure and it is mainly carbon dioxide (Grinspoon 1997). An outstanding question is: WHERE DID ALL THIS CARBON DIOXIDE COME FROM? Well, the main gas coming from the Earth's interior via volcanic eruptions is carbon dioxide. But the Earth's atmosphere at present contains only about 0.6 % carbon dioxide by volume. We know that the volcanic carbon dioxide gets fixed as carbonate rock via a combination of organic and inorganic processes and then gets recycled via the plate tectonic process in which ocean floor gets subducted beneath volcanic arcs. But Venus does not have either ocean water or subduction zones at present, so the carbon dioxide accumulates in the atmosphere. But the question is: WHY IS THERE SO MUCH CARBON DIOXIDE IN THE ATMOSPHERE OF VENUS? Cloud (1972, 1974) concluded that the carbon dioxide content of the atmosphere of Venus contains about 1.6 times the combined carbon dioxide of the carbonate rocks and sediments of Earth as well as that in the atmosphere and in solution in ocean water. His conclusion was that the mantle of Venus is more completely depleted (degassed) than the mantle of Earth. So there is very little doubt as to where this carbon dioxide came from. It came from the mantle of the planet by way of the extreme volcanism that is associated with the mysterious GLOBAL RESURFACING EVENT.

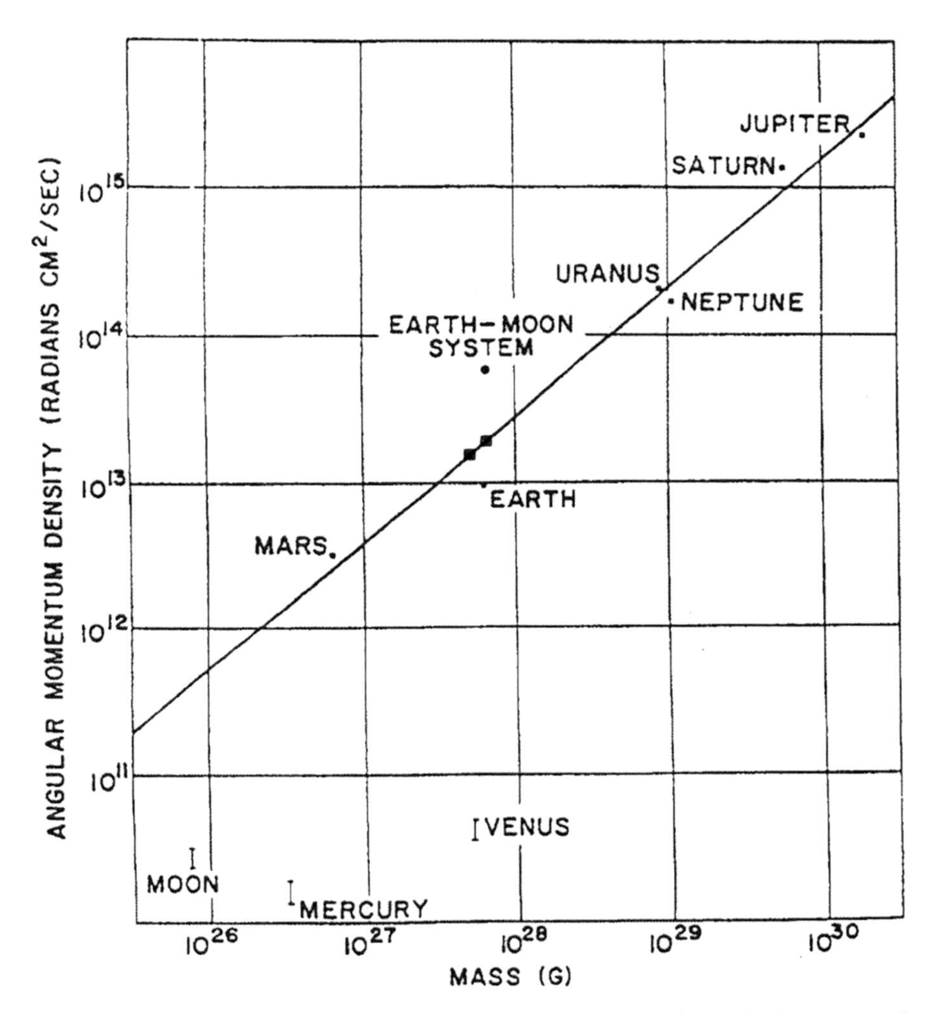

Fig. 1.4 Plot of Angular Momentum Density vs. Mass for the planets of the Solar System as well as for the Earth's Moon and the Earth-Moon system. (Diagram adapted from MacDonald (1963, Fig. 38), with permission from Springer.). The line was placed for the best fit of the information for the rotation rates for the outer planets and Mars. The assumption, which is suggested by the plot, is that Mars has a rotation rate very close to its primordial rotation rate (~25 h/day). If all the angular momentum of the Earth-Moon system is placed in the Earth, the rotation rate would be ~4.5 h/day. If the Earth is rotating ~10 h/day, the Earth would plot on the line between the position of the Earth-Moon System and the Earth (the *red square* symbol on the *right*). This information suggests that the original rotation rate for Earth was ~10 h/day. If the angular momentum of a lunar-mass body in a 30 earth radii circular orbit in the prograde direction is added to the prograde angular momentum of an Earth rotating at 10 h/day, then that combination plots in the position of the Earth-Moon system on the plot. Likewise, if Venus is elevated to the line vertically above its position on the graph (the *red square* symbol on the *left*), then the original rotation rate would be ~13.5 h/day prograde. The primordial rotation rates of the Moon and Mercury can be estimated using the same procedure

The "big question" of this project is "WHY ARE THESE SISTER PLANETS SO DIFFERENT?" What happened to planet Earth to develop it into a "paradise" for bacteria and algae and later for abundant plant and animal life. And what happened to planet Venus to develop it into a "hellish" environment that still persists today?

References

Alfven H, Arrhenius G (1972) Origin and evolution of the Earth-Moon system. The Moon 5:210–230
Berner RA (2003) Phanerozoic atmospheric oxygen. Annu Rev Earth Planet Sci 31:105–134
Berner R A (2006) GEOCARBSULF: a combined model for phanerozoic atmospheric O_2 and CO_2. Geochim Cosmochim Acta 70:5653–5664
Carr MH (1999) Mars. In: Beatty JK, Petersen CC, Chaikin A (eds) The new solar system. Cambridge University Press, Cambridge, pp 141–156
Cloud PE Jr (1972) A working model of the primitive earth. Am J Sci 272:537–548
Cloud PE Jr (1974) Rubey conference on crustal evolution (meeting report). Science 183:878–881
Dauvallier A (1976) The venus oceans problem. J Br Astron Assoc 86:147–148
Davies G F (1992) On the emergence of plate tectonics. Geology 20:963–966
Davies J H (2008) Did a mega-collision dry Venus' interior? Earth Planet Sci Lett 268:376–383
Donahue T M (1999) New analysis of hydrogen and deuterium escape from Venus. Icarus 141:226–235
Donahue TM, Hoffman JH, Hodges RR Jr, Watson AJ (1982) Venus was wet: a measurement of the ratio of deuterium to hydrogen. Science 216:630–633
Faure G, Mensing TM (2007) Introduction to planetary science. The geological perspective. Springer, Berlin, p 526
Glassmeier K-H, Vogt J (2010) Magnetic polarity and biospheric effects. Sp Sci Rev 155:387–410
Grinspoon DH (1997) Venus revealed. Helix Books, Addison-Wesley, Reading (MA) p 355
Herrick RR (1994) Resurfacing history of Venus. Geology 22:703–706
Herrick DL, Parmentier EM (1994) Episodic large-scale overturn of two-layer mantles in terrestrial planets. J Geophys Res 99:148–227. doi 1029/93JE03080.
Lipton P (2005) Testing hypotheses: prediction and prejudice. Science 307:219–221
MacDonald JGF (1963) The internal constitutions of the inner planets and the Moon. Sp Sci Rev 2:473–557
Mackenzie D (2003) The big splat (or how our Moon came to be). Wiley, New York, p 232
Malcuit RJ (2009) A retrograde planetoid capture model for planet Venus: implications for the Venus oceans problem, an era of habitability for Venus, and a global resurfacing event about 1.0–0.5 Ga ago. Geol Soc Am Abstr Progr 41(7):266
Malcuit RJ, Winters RR (1995) Planetoid capture models: implications for the great contrast in planetological features of planets Venus and Earth. Geol Soc Am Abstr Progr 27(6):A–208
Polat A (2012) Growth of Archean continental crust in oceanic island arcs. Geology 40:383–384
Rudnick RL (1995) Making continental crust. Nature 378:571–578
Saunders RS (1999) Venus. In: Beatty JK, Petersen CC, Chaikin A (eds) The new solar system. Cambridge University Press, Cambridge, pp 97–110
Singer SF (1970) How did Venus lose its angular momentum? Science 170:1196–1198
Stern RJ (2005) Evidence from ophiolites, blueschists, and ultrahigh-pressure metamorphic terranes that the modern episode of subduction tectonics began in Neoproterozoic time. Geology 33:556–560
Strangway DW (1970) History of the Earth's magnetic field. McGraw-Hill, New York, p 168
van Loon AJ (2004) From speculation to model: the challenge of launching new ideas in the earth sciences. Earth Sci Rev 65:305–313
Ward PD, Brownlee D (2000) Rare earth (why complex life is uncommon in the Universe). Copernicus (An imprint of Springer), New York, p 333
Young A, Young L (1975) Venus. Sci Am 233(3):70–81

Chapter 2
The Origin of the Sun and the Early Evolution of the Solar System

The direct investigation of such inner regions around protostars and young stars will also provide us with knowledge about the physics and evolution of circumstellar disks. It is within such disks that planetary systems are believed to be formed. We now have reason to believe that, as we progress toward a greater understanding of star formation, we will also begin to unlock the secrets of the origin of planetary bodies.
From Lada and Shu (1990, p. 572).

The origin of the Solar System has intrigued scientists for centuries. As recently as five decades ago the models were still very general (e. g., Cameron 1962) and were concerned mainly with the collapse of a cloud of stellar dust and gas of roughly solar composition and the transformation of that cloud into a rapidly rotating disk-shaped mass around a proto-sun. The next few decades were dominated by calculations of equilibrium chemical condensation models from a cooling nebula of solar composition (e. g., Lewis 1972, 1974; Grossman 1972) based mainly on the temperature and pressure conditions for the solar nebula from Cameron and Pine (1973). Identification of high-temperature condensates [calcium-aluminum inclusions (CAIs)] in the Allende meteorite was a critical event in the development of more sophisticated models for the evolution of the Solar System. After the discovery of CAIs it was important to develop models to explain (a) the origin of chondrules (the main constituent of chondritic meteorites), (b) the origin of CAIs, as well as (c) the origin of the very fine-grained matrix of the chondritic meteorites.

In the decades of the 1980s and 1990s, groups of astrophysicists presented the results of simulations of the dynamics of the early history of the Sun in an attempt to relate the rapidly rotating disk stage of Solar System evolution to observations of T-Tauri stars of roughly solar mass. For example, Lada and Shu (1990) and Shu et al. (1994) published the results of simulations of the dynamic interaction between the material infalling along the nebular midplane and the strong magnetic fields associated with a very hot, rapidly rotating nascent Sun. Major features of the magnetic field action were the formation, breaking, and reconnection of the magnetic flux lines. The breaking (snapping) and reconnection of the flux lines is associated with very high temperature pulses that may relate to the thermally generated features recorded in the various types of CAIs (summaries of the types of CAIs are in Taylor 2001). Shu et al. (1997) attempted to explain the origin of both CAIs and chondrules

Robert J. Malcuit, *The Twin Sister Planets Venus and Earth*,
DOI 10.1007/978-3-319-11388-3_2

via the X-Wind model (the name of the model refers to the 2-D cross-sectional geometry of the intersection of magnetic flux lines and the midplane disk). After critical consideration of the merits of the X-Wind model, Taylor (2001) and Wood (2004) proposed that something like the X-Wind model can be used to explain many of the features of CAIs but that the X-Wind model does *not* relate to the environmental conditions for chondrule formation. Although the X-Wind model has its critics (e. g., Desch et al. 2010), it is generally accepted as a reasonable explanation for the origin of CAIs.

The origin of chondrules (the main features of chondritic meteorites and spherical particles that are much simpler in composition and structure than CAIs) is another story with a long history dating back to Henry Sorby and the petrographic microscope about 1870 (McSween 1999). Although chondrules appear to have less complex features, their origin appears to be more difficult to explain. Most investigators agree that a "flash-melting" process as well as rapid cooling are involved. The rapid heating melts whatever clumps of dust that are in the environment at that time and the resulting features are glass beads (some with crystallites and crystals of identifiable minerals). These chondrules, plus or minus a few CAIs, are the main megascopic components of chrondritic meteorites. These components of chondritic meteorites are bound together by a matrix material. Most investigators agree that the fine-grained material of the matrix is composed of a combination of fine-grained silicate-rich material which contains various quantities of chondrule fragments, CAI fragments, and very fine-grained nebular dust as well as some material that was infalling along the midplane from the molecular cloud (Rubin 2010, 2013). It is interesting to note that the matrix of some enstatite chondrites has a significant quantity of flakes or chips of iron-nickel metal and sulfide minerals embedded in a mainly silicate matrix (Rubin 2010). The chemistry of chondrules varies considerably but there is a trend related to distance from the proto-Sun and the volatile content of the chondrules increases with heliocentric distance. Many investigators think that the Disk-Wind model of Bans and Konigl (2012) and Salmeron and Ireland (2012) looks promising as an explanation for the origin of chondrules. In general, chondrules are a few million years younger than CAIs and were formed by significantly different thermal processes. There is, however, evidence that there may be some overlap in time of formation (Brearley and Jones 1998).

As chondritic meteorites and associated CAI particles are formed, an accretion process begins. There are probably several embryonic planetary nucleation sites early in the accretion process, but in the later stages only a few would remain in the accretion torus (an accretion torus is a heliocentric doughnut-shaped geometric form from which smaller bodies are gravitationally attracted to participate in the planet-building process). The accretion torus, then, constitutes the "feeding" zone for the planet accretion process.

The chemical composition of the resulting planet or planetoid is determined by the composition of the particles in the accretion torus. For example, if the material in the accretion torus is mainly CAI material, then the planet or planetoid will be composed of CAI chemistry. If the accretion torus has particles and agglomerates of particles that are rich in iron, then the resulting planet or planetoid will be rich in iron and have a high specific gravity relative to a body composed mainly of silicates. Thus, the composition and density of an accreted body probably reflects the composition of the particles that were in the accretion torus.

In general, I think that nearly all (and possibly all) features of a Solar System origin model, from the origin of CAIs, chondrules, and the matrix of chondritic meteorites, and their derivative bodies, are involved in the processes that led to the formation of the terrestrial planets (e. g., the twin sister planets, Venus and Earth) and associated Vulcanoid planetoids and Asteroids, as well as the outer (gaseous) planets.

Figure 2.1 is composed of two simplified scale diagrams of the orbits of the planets of the Solar System. Figure 2.1a shows the orbits of the outer (gaseous) planets relative to the orbit of Mars. Figure 2.1b shows the orbits of the inner (terrestrial) planets relative to the orbit of Jupiter. The reader may ponder the following question: Why involve a large slice of the Solar System when discussing the condition of planets Venus and Earth? A reasonable answer is that when dealing with a capture origin for the Moon as well as a capture origin for a satellite for Venus, it is necessary to have a place of origin for the Moon and related planetoids. The best fit, both chemically and physically, seems to be a Vulcanoid Zone (Wiedenschilling 1978; Leake et al. 1987; Evans and Tabachnik 1999, 2002) between the orbit of Mercury and the Sun. Thus it appears necessary to involve at least the zone inside the orbit of Mercury and out to the vicinity of Earth's orbit.

Then we have the problem of a large volume of ocean water on Earth as well as the possibility of water on Venus in an earlier era [i. e., the "Venus Oceans problem" (Donahue 1982, 1999)]. I think that the model of Albarede (2009) explains the origin of ocean water problem fairly well. His suggested source is water-bearing asteroids from the middle to outer Asteroid Zone. Thus, our sphere of influence needs to be extended to include the entire Asteroid Zone.

Then we need a delivery system for the Aquarioid Asteroids (my name for the water-bearing asteroids). The most reasonable delivery mechanism for getting the Aquarioids from the Asteroid Zone to near Earth orbit is a process of gravitational perturbations by a combination of Jupiter and Saturn, a process that has been studied by celestial mechanicians for many decades.

Since there is also an interest in explaining the source of water for planet Mars, we must explain the deuterium to hydrogen (D/H) ratio of the water associated with that planet. [For readers who are not familiar with the importance of the deuterium/hydrogen ratio, some definitions and an explanation are in order. The hydrogen atom (H) (also called protium) has only one proton in the nucleus. Deuterium (D) has both a proton and a neutron in the nucleus and has twice the atomic weight as hydrogen. In many cases thermal processes will cause molecules with the lighter hydrogen to be separated from those with heavier hydrogen. As a result many substances can be characterized by their D/H ratio.] Since the D/H ratio of martian water is much different from that of Earth we must search for a source of martian water. Well, the D/H ratio of martian water is similar to that of comets (Robert 2001). Most investigators think that the effective source of these comets is the Asteroid Zone and that these comets are Jupiter-captured bodies: i. e., captured into heliocentric asteroid-like orbits after a close encounter with Jupiter. The apparent ultimate source of all comets, however, is the Oort Cloud/Kuiper Belt which is

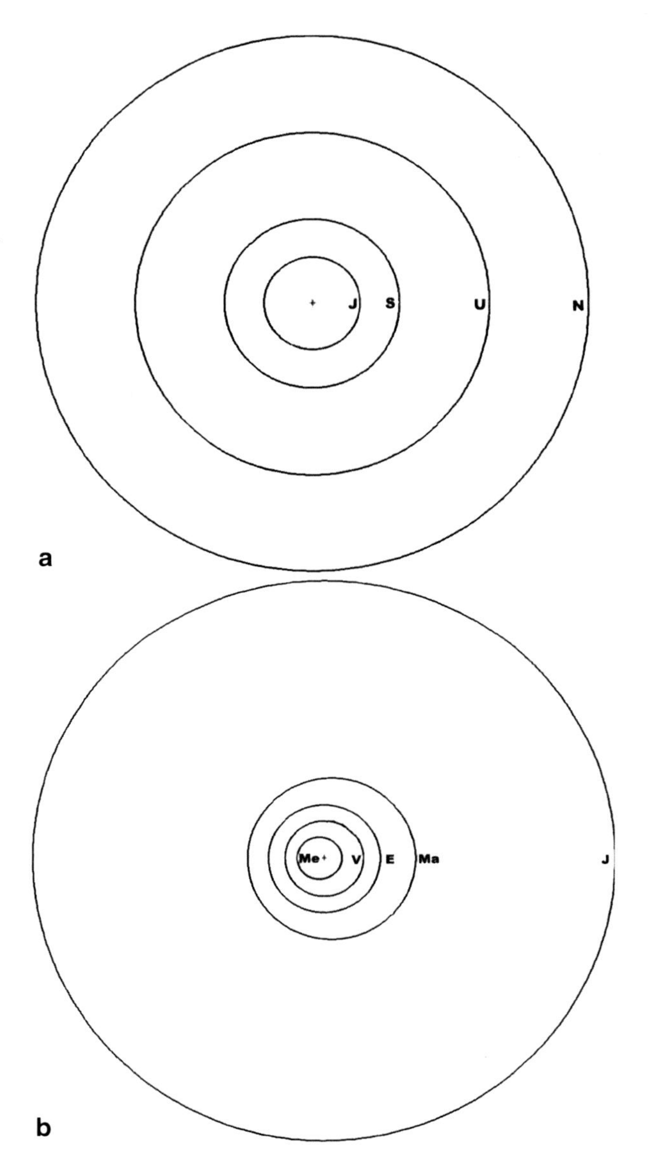

Fig. 2.1 Relative geometry of the orbit of the planets of the solar system as viewed from the North Pole of the Solar System. Note that all orbits are nearly circular except those of Mars and Mercury. **a** Outer Solar System orbits: *N* Neptune, *U* Uranus, *S* Saturn, *J* Jupiter. **b** Inner Solar System orbits: *Ma* Mars, *E* Earth, *V* Venus, *Me* Mercury

beyond the orbit of NEPTUNE. In order to transfer these comets from their place of origin to the vicinity of Jupiter's orbit, it is difficult to exclude gravitational interactions with planets Neptune, Uranus, and Saturn. [Note: The origin of water for Earth and Venus will be discussed further in Chap. 5.]

In addition to aiding in the transfer of the water-bearing bodies for Mars, Earth, and Venus, planet Jupiter, and its accomplice Saturn, apparently have played a significant role in maintaining habitable conditions on Earth for a long stretch of geologic time by powering a major component of the Milankovitch cycles (Berger et al. 1992). The

Earth's periodic variations in heliocentric orbit eccentricity, an oscillation from 0 to 6% in eccentricity with major periods of 100 and 400 Ka, are due to gravitational perturbations by a combination of planets Jupiter and Saturn (Imbrie and Imbrie 1979; Berger 1980). These oscillations of the eccentricity have been operating from a very early era in Earth history (Berger et al. 1989a, b).

Two other components of the Milankovitch Model are the precession cycle (~20,000 years) and the obliquity (or tilt) cycle (~40,000 years). Both of these cycles are dominated by gravitational interactions with the Moon (Imbrie 1982; Berger et al. 1992).

Thus, it seems imperative that we spend some time on the problems associated with THE ORIGIN AND EVOLUTION OF THE SOLAR SYSTEM. For origin and early evolution, I favor a combination of (1) the X-WIND MODEL (Shu et al. 2001), (2) the DISK WIND MODEL (Salmeron and Ireland 2012), (3) the FU Orionis model (Bell et al. 2000), and (4) the T-Tauri model (Calvert et al. 2000). This combination of activity appears to explain most of the "FACTS TO BE EXPLAINED BY A SUCCESSFUL MODEL" for the Solar System (this list of facts is in the next section of this chapter). Although this combination of models may not be perfect (and we must remember that all models in science are amendable to modification, improvement, and replacement), it does explain a good bit. In some cases models can be used to explain much more than for which they were designed. In my opinion this feature is the hallmark of a successful hypothesis (model).

2.1 List of Some Important Facts to be Explained by a Successful Model

1. Chemical composition and body density (compressed and uncompressed) patterns of the Moon, Mercury, Venus, Earth, Mars, the asteroids, Jupiter and the other outer planets and their satellites, and comets in addition to the various groups of chrondritic and achondritic meteorites. (see Fig. 2.2 and Table 2.1 for body density patterns.)
2. Composition and dates of formation of calcium-aluminum inclusions (CAIs) that occur in chondritic meteorites as well as an explanation of the processing that results in the various types of CAIs.
3. Patterns of oxygen isotope ratios for bodies of the Solar System (see Fig. 2.3 for some trends of oxygen isotope ratio patterns).
4. Composition and dates of formation of chondrules and chondritic meteorites as well as the devolatilization patterns associated with volcanic asteroids, planets, and chondritic meteorites. (See Fig. 2.4 for the potassium content relative to uranium for these bodies.)
5. Magma ocean development on the Moon, in particular, but this explanation may include the development of magma oceans on other bodies such as the Vulcanoid planetoids and Mercury.
6. Patterns of magnetization of minerals, rocks, and planetary crusts on various solar system bodies as well as on meteorites and asteroids.

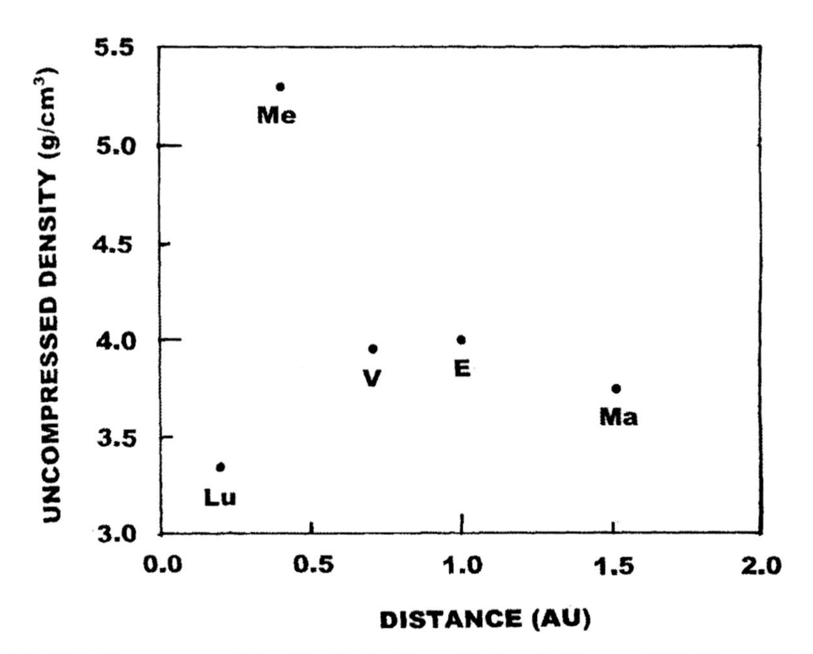

Fig. 2.2 Plot of uncompressed densities for the Moon (*Lu* Luna), Mercury (*Me*), Venus (*V*), Earth (*E*), and Mars (*Ma*). The exceptionally low uncompressed density of Luna as well as the exceptionally high uncompressed density of Mercury need to be explained by a successful model for the origin and evolution of the Solar System. (Numerical values are from Lodders and Fegley (1998, p. 91))

Figure 2.2 is a plot of uncompressed densities for the terrestrial planets and the Moon and Table 2.1 gives numerical values for the bodies in Fig. 2.2 as well as for other Solar System entities. Compressed density of a planetary body is the density of the body as it exists in nature (i. e., in a compressed state). The mineral structures in the lower layers of the planet are denser than they would be at the surface. Small bodies, bodies the size of the Earth's Moon or smaller, have about the same density in the compressed and uncompressed states. The uncompressed density of a planetary body is estimated by theoretically decompressing the mineral structures of successive layers within the planet. The uncompressed density yields a much better estimate of the density of the material of which the body is composed. For terrestrial planets the uncompressed density relates fairly directly to the mineralogical and chemical composition of the meteoritic material that was in the accretion torus before the planetary body was formed.

Before we develop a working model for the evolution of the Solar System we need to discuss some features of the Oxygen Isotope plot in Fig. 2.3 and the Potassium abundance relative to Uranium plot in Fig. 2.4. The hypothesis (model) to be tested is that all volcanic asteroids were formed in heliocentric orbits between ~0.10 AU and the orbit of planet Mercury in the earliest history of the Solar System. The starting material was the composition of calcium-aluminum inclusions (CAIs) of all types (Groups I through VI) (Taylor 2001) that were generated by the thermal and dynamic activity of the X-Wind (Shu et al. 2001). During the early stages of the X-Wind

Table 2.1 Values of compressed density and uncompressed density for various solar system entities. (Sources: Moon through Sun (Lodders and Fegley 1998, pp. 91–95); Asteroids (Kowal 1996, p. 45); Angrite grain density (Britt et al. 2010, 1869.pdf))

Solar system body	Compressed density (g/cm³)	Uncompressed density (g/cm³)
Moon	3.34	3.34
Mercury	5.43	5.30
Venus	5.24	4.00
Earth	5.52	4.05
Mars	3.93	3.74
Jupiter	1.33	0.10
Saturn	0.69	0.10
Sun	1.41	(0.10)
4 Vesta	3.3	3.3
Angra	~3.3	~3.3
Ceres	2.7	2.7
Pallas	2.6	2.6

activities the temperatures may have been too high for the refractory minerals of CAIs to crystallize. As the peak temperatures decreased, the various mineral products of the X-Wind began to collide, fragment, and accrete into larger bodies. According to Taylor (2001) the oxygen isotope ratios in CAIs tend to be fairly uniform and the environment during CAI formation was rich in oxygen-16 relative to that during chondule formation. But the oxygen isotope ratios in the purported Vulcanoid planetoids in Fig. 2.3 are not identical. Thus we need to develop an explanation for the formation of Vulcanoid planetoids that can have somewhat different oxygen isotope ratios.

An Attempt to Explain the Similarities of Oxygen Isotope Ratios of Earth and Moon

In the author's view the oxygen isotope signature of CAIs depends on the mixture of material infalling along the nebular mid-plane. The dust to gas ratio of infalling material is important because the nebular dust is enriched in O-16 and the gas is enriched in O-18 (Clayton 2003). Thus each major pulse of infalling material could yield a somewhat different dust to gas ratio and thus a different oxygen isotope signature. Perhaps the CAIs that are associated with the various types of carbonaceous chondrites (the main depository of nearly all of the extant CAIs) were all from the same phase of the X-Wind activity. Pursuing this concept a bit further, perhaps Luna accreted CAI material that mainly, by happenstance, matches very closely with the oxygen isotope ratios for Earth; perhaps 4 Vesta accreted from a slightly earlier or later batch of CAI material that was even richer in O-16; and perhaps Angra accreted from CAI debris that was intermediate in oxygen isotope ratios. Using this rationale, one can see how two independent bodies, separated by 0.85 AU of space (e. g., Earth and Luna) could end up in about the same position on a $\Delta^{17}O$ vs. $\Delta^{18}O$ plot.

Now we return to a possible explanation for the formation of Vulcanoid planetoids. As soon as a batch of CAI material gets formed, the particles in the mid-plane begin

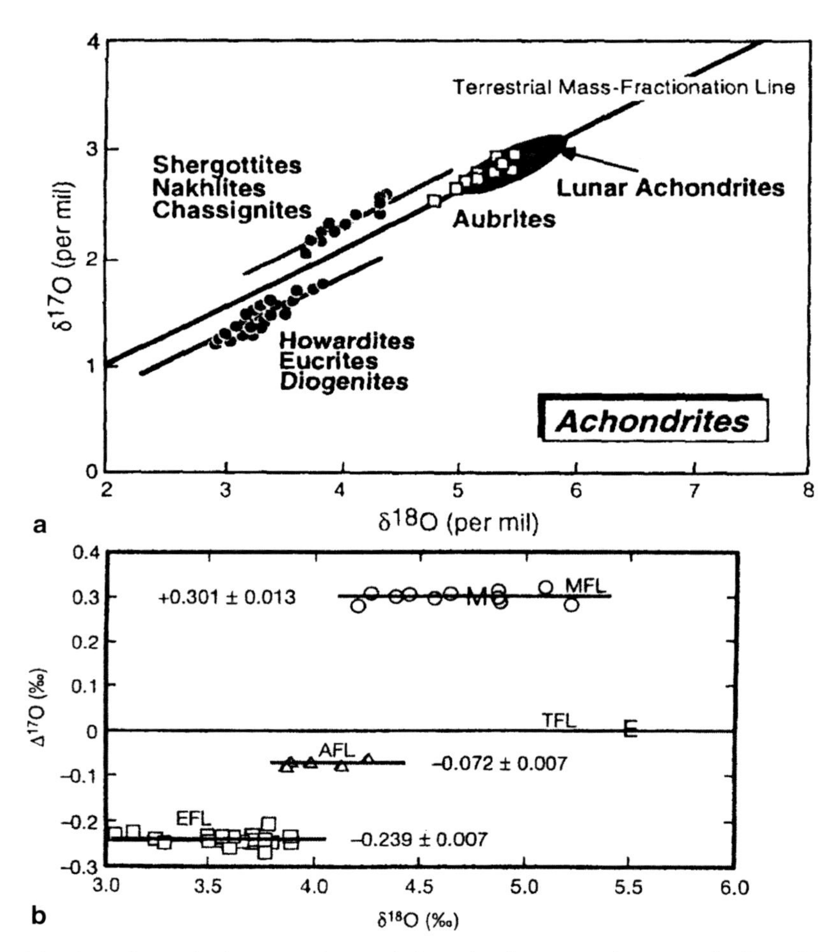

Fig. 2.3 Plots of some basic oxygen isotope information for some solar system bodies. **a** Three-oxygen isotope plot showing the position of the terrestrial fractionation line as well as the position of lunar achondrites (meteorites interpreted to be from the Moon), SNC meteorites (interpreted to be from Mars), and HED meteorites (interpreted to be from 4 Vesta). (Diagram from McSween (1999, Fig. 4.4), with permission from Cambridge University Press). **b** A delta 17 oxygen vs. delta 18 oxygen plot showing the terrestrial fractionation line (*TFL*), the Mars fractionation line (*MFL*), the eucrite parent body fractionation line (*EFL*), and the angrite fractionation line (*AFL*). Note that both Earth (*E*) and Moon (*not shown*) are essentially on the terrestrial fractionation line. Again, all of this information needs to be explained by a successful model for Solar System origin and evolution. (Diagram from Greenwood et al. (2005, Fig. 2), with permission from Nature Publishing Group.)

to collide and accrete into larger planetoid units; the CAIs that are near the edge of the mid-plane can be hurled to various distances in the surrounding solar system. This process of formation, limited ejection, and subsequent accretion continues for many cycles. The yield of the cycles can vary considerably. As more CAI material forms and accretes, some bodies will attain stable orbits as calculated by Evans and

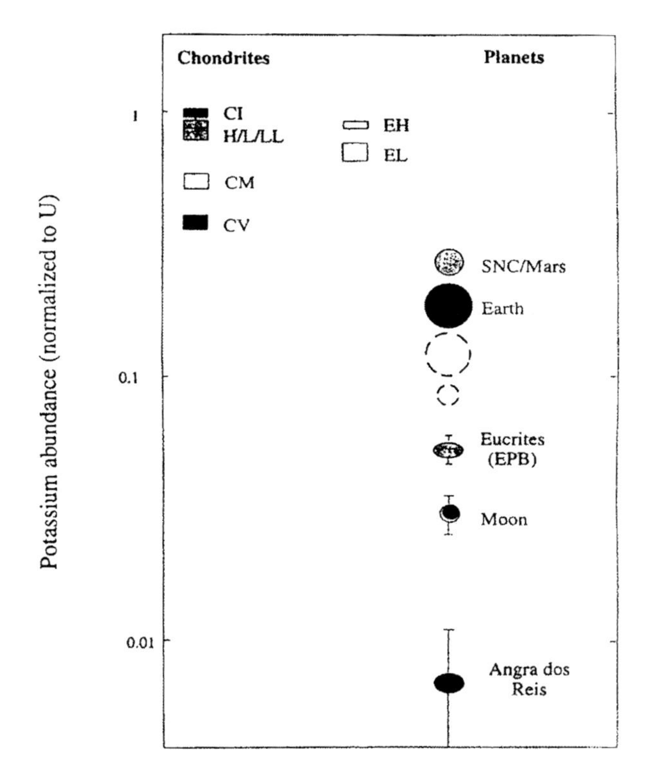

Fig. 2.4 Plot of potassium abundances (normalized to uranium) for various bodies of the Solar System as well as for various classes of meteorites. Angra dos Reis represents the Angrite meteorites; Eucrites are basaltic meteorites that are thought to be from the Eucrite Parent Body which, in turn, is considered to be Asteroid 4 Vesta; *EH* and *EL* are enstatite chondritic meteorites; *H*, *L*, and *LL* are ordinary chondritic meteorites; *CV* and *CM* are types of carbonaceous chrondrites, and the *CI* type is the most primitive carbonaceous chondrite meteorite (see McSween 1999, for more details on classification). The two *open circle* patterns in the Planets and Planetoids Column represent my prediction for the position of the potassium abundance (normalized to U) for planets Venus and Mercury. [Note: There is limited information from Venus via the Venera 8 mission. Vinogradov et al. (1973) report that the K/U ratio for magmatic rocks of Venus is very similar to magmatic rocks on Earth]. (Diagram adapted from Humayun and Clayton (1995, Fig. 1), with permission from Elsevier)

Tabachnik (1999, 2002) and others will collide, fragment, and eventually accrete onto a Vulcanoid planetoid in a more stable orbit.

Concomitant with the generation of the refractory CAI material, thermal pulses associated with the X-Wind systematically devolatilize the material in the proto-solar mid-plane as well as the region of the terrestrial planets and perhaps the Asteroid Zone. Taylor (2001, p. 110) suggests that temperatures as high as 1500 °K at 2.5 AU may have been attained spasmatically during the early stages of infall of gas along the nebular mid-plane. These high temperature pulses can be intimately associated with the solar accretion cycles fueling the X-Wind activities. Perhaps the Potassium abundance normalized to Uranium (the K-Index) in Fig. 2.4 is a product of this devolatilization of the inner part of the Solar System. While solar magnetic field line snapping and reconnection activity is releasing very high energy thermal pulses that systematically devolatilize the material near the nebular mid-plane, some Vulcanoid planetoids have close encounters with sibling planetoids and are perturbed into the proto-sun while others are propelled to more distant regions of the Solar System: i. e., into the realm of the terrestrial planets. As the X-Wind action gradually diminishes due to a decreasing volume of material infalling along the nebular mid-plane, the newly formed CAIs as well as any material left over from previous cycles will accrete into larger bodies that are in somewhat stable orbits.

As the X-Wind phase of solar accretion, radiation, material generation, and devolatilization diminishes, there is a gradual shift to the Disk-Wind (Salmeron and Ireland 2012) activity. The wavelength and intensity of the resulting radiation is lower but the magnetic-field-generated thermal pulses continue. The Disk-Wind proceeds to "flash melt" dusty materials that have sufficient density to be processed into glass beads (chondrules) The dust that results in chondrule formation has already been systematically devolatilized with the volatiles being removed to the outer regions of the terrestrial planet realm, the Asteroid Belt, and on out to the heliocentric distance of the orbit of planet Jupiter.

As the Disk-Wind processes dust into chondrules, flash melting cycle after flash melting cycle, the chondrules along with a low percentage of CAI material, as well as the material still falling in along the nebular midplane accrete to form a multitude of planetoids that eventually accrete into the terrestrial planets. The chemistry of the resulting terrestrial planet is determined by the material in its accretion torus and the state of devolatilization of the material in the torus changes with heliocentric distance from the heat source (the proto-sun).

Following the major phases of terrestrial planet formation, the proto-sun still has material periodically falling in along the mid-plane. The resulting radiation events in the microwave range can be related to the FU Orionis model (Bell et al. 2000; Calvet et al. 2000). The thermal "spikes" (to be discussed in the next section of this chapter) occur with a frequency of a few 10^3 years. I propose that the FU Orionis sequence of thermal-radiation events is responsible for melting, or remelting, all Vulcanoid planetoids between 0.10 AU and the orbit of planet Mercury as well as partially melting the outer portion of planet Mercury. This sequence of heating events could cause thermal (and hydrothermal, if water is available) metamorphism in the region beyond planet Mercury. Future modeling will help to determine the extent of this thermal activity.

A major question raised by this discussion is: Which is more important for determining the original heliocentric distance of formation of a planet or planetoid? Is it the K-Index or is it the ratio of Oxygen Isotopes on a Δ^{17}O vs. Δ^{18}O plot?

Since the X-Wind thermal-radiation events would be the most powerful and the CAIs form only during the waning stages of the X-Wind action, I think that the X-Wind activity, in its earlier stages, is the main cause of the devolatilization. The newly formed and accreting bodies are small at this stage so that the thermally-powered devolatilization process is very efficient. The X-Wind activity would transition into the Disk-Wind heating and "flash-melting" events which are gradually decreasing in intensity throughout the chondrule-forming events and the subsequent small-body accretion events.

In light of the discussion above perhaps the K-Index (Humayun and Clayton 1995) is a record for the progressive devolatilization events. Making this assumption and with reference to Fig. 2.4, Angra dos Reis (representing the Angrite parent body) would form closest to the Sun, then the planetoid Luna, and then the planetoid 4 Vesta. We have no K-Index value for planet Mercury but we can predict that it has a higher volatile content than 4 Vesta. We have K-Index values for Earth and Mars and we can predict with some confidence that Venus will have a value similar or slightly lower than Earth, especially in light of the limited information from the Venera 8 mission (Vanogradov et al. 1973) that suggests that the K/U ratio of the igneous rocks that were tested on Venus are similar to igneous rocks on Earth.

In summary, the K-Index seems like a dependable basis for a starting position for Vulcanoid planetoids as well as for other bodies in the Solar System. In contrast, oxygen isotope values of Solar System bodies tend to vary considerably and yield no systematic pattern. Using this rationale for the interpretation for the K-Index and Oxygen Isotope characterization of planetoids, the similarity of oxygen isotope values of Earth and Moon (the Moon representing the oxygen isotope values of the hypothetical Theia) does not necessarily indicate formation at a similar heliocentric distance from the Sun. Thus a major tenant of the Giant Impact Model for the origin of the Earth's Moon appears to be invalid. In contrast, the K-Index of Luna indicates an origin for Luna as a Vulcanoid planetoid inside the orbit of Mercury.

2.2 A Composite Working Model for Origin and Evolution of the Solar System

The origin and evolution of the Solar System is a complex problem in the natural sciences. Astronomers and astrophysicists have been attempting to develop a model for the early Solar System since the days of Kant and Laplace [see Cousins (1972) for an historical review of the models]. I am presenting here a composite model which is an amalgam of three recently proposed models because I think that this combination of models relates to the *LIST OF FACTS TO BE EXPLAINED* presented above. In general, the X-WIND MODEL of Shu et al. (2001) appears to explain the origin of CAIs fairly well (Taylor 2001; Wood 2004). The DISK-WIND MODEL (Konigl and Pudritz 2000; Bans and Konigl 2012; Salmeron and Ireland 2012) looks promising for explaining the origin of chondrules. A third phase of extraordinary heating in the microwave range of radiation is also needed for my composite explanation of early Solar System history. I will refer to this as a MICROWAVE-WIND MODEL. Such a model was proposed by Sonett et al. (1975) for the development of the lunar magma ocean. The main source of energy for their model of electromagnetic induction heating was the T-Tauri plasma flow from the proto-sun. They thought that both the transverse electric mode and the transverse magnetic mode could be used effectively. The following is a quote of the last sentence of the abstract (p. 231): "Joint excitation of these two modes would modify and intensify lunar heating." Other articles on electromagnetic induction heating are by Sonett and Reynolds (1979), Herbert and Sonett (1980), and Wood and Pellas (1991). In my view, this microwave heating model appears much more attractive as a source of energy for remelting the lunar magma ocean zone if Luna is considered to be formed as a Vulcanoid planetoid. Some enterprising physics student could have fun developing the details of this combination of models. The FU Orionis stage of stellar evolution (Bell et al. 2000) is considered by the author as a good source of electromagnetic radiation to do the melting. (Note: This concept will be pursued further in a later section of the chapter.)

These models represent a progressive decrease in energy as well as a decrease in solar magnetic field activity and each is related to fairly rapid accumulation of debris by way of the proto-solar mid-plane onto the proto-sun in time. The early stage of accumulation is very rapid and is associated with the X-WIND action, and associated bipolar outflow, which in late stages results in the solidification of the first datable solids in the solar system: i. e., the CALCIUM-ALUMINUM INCLUSIONS (CAIs) which appear in chondritic meteorites. Most, if not all of these CAIs formed over a time-span of ~1 million years or less starting at ~4.566 Ga (billion years) ago (MacPherson et al. 2005). This is TIME ZERO for the solidification of solar system solids (Taylor 2001; Liffman et al. 2012). The second stage of accumulation of infalling material is associated with the DISK WIND MODEL (Konigl and Pudritz 2000). The radiation and magnetic activity of the Disk Wind is less intense but more widespread than the X-Wind and is thought by some investigators to explain the origin of CHONDRULES fairly well (Salmeron and Ireland 2012). The chrondrules formed over a period of several million years but their earliest time of formation appears to be about 1–2 million years after the cessation of CAI formation (Kita et al. 2013) [but this gap in time has been questioned by Bizzarro et al. (2004) and more recently by Connelly et al. (2012)]. I would like to use the subsequent FU Orionis stage of solar evolution (Bell et al 2000; Calvet et al. 2000) for the radiant energy for remelting of the magma ocean region of Luna, for remelting all of 4 Vesta, as well as for remelting all or part of the other purported Vulcanoid planetoids. Bell et al. (2000) estimate that this FU Ori stage would occur about 10 million years or more after the end of chondrule formation and was probably finished within 100 Ma after time zero. The subsequent T-Tauri stage of solar evolution is characterized by a longer wavelength radiation and longer duration (Bell et al. 2000; Calvet et al. 2000) and may be a cause for the hydrothermal metamorphism experienced by various groups of asteroids (Herbert et al. 1991). These four radiation episodes which sequentially decrease in intensity but increase in duration appear to be, at the present time, the best candidates for powering the thermal episodes that affected the various extant entities, large and small, in the early Solar System.

The main "events" in my model are the following:

1. Infall of the proto-solar cloud (of solar composition) which results in an increase in rotation rate of the proto-sun [the material is dust and gas from previous stellar explosions (see Fig. 2.5)].
2. Bipolar outflow begins as the proto-sun heats up and the solar magnetic field increases in intensity and eventually the thermonuclear reaction in the proto-sun is ignited.
3. As infall continues at an increasing rate, all processes are intensified and accelerated including bipolar outflow (which features escape of some or much of

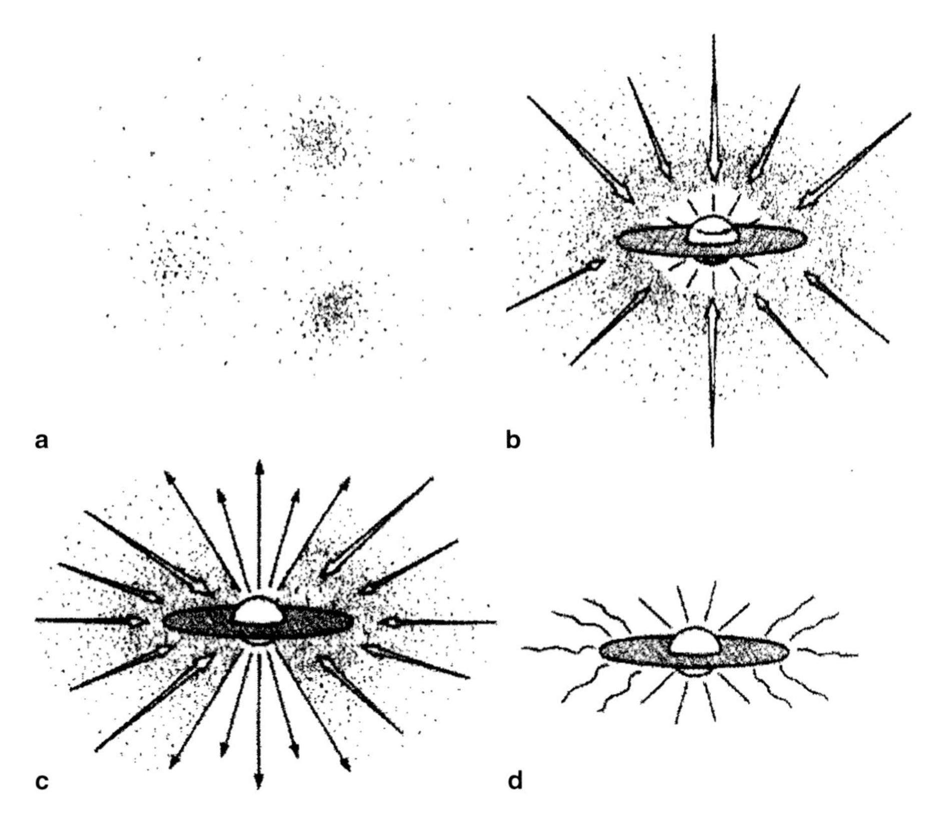

Fig. 2.5 Pictograms of four very general stages leading to the formation of the Solar System: **a** dust and gas from previous stellar explosions, **b** infall of material and increase in rotation rate, **c** infall continues but some material is ejected via bipolar outflow, **d** the star assumes a disk shape, becomes visible, and settles down to the main sequence of burning. (Diagram from Taylor (2001, Fig. 3.1), with permission from Cambridge University Press)

the original infalling material). This is the beginning of the X-Wind stage (see Fig. 2.6).

4. The inner part of the infalling cloud in the mid-plane of the accretion disk is completely vaporized and the infalling material is progressively devolatilized with the water and other volatiles propelled to the Outer Asteroid Zone and probably as far out as Jupiter's orbit where a torus of volatile-rich material begins to accumulate. All materials are in dust and gas phases at this point and this is PEAK HEAT TIME along the proto-solar mid-plane (see Fig. 2.7).
5. At this time volatile-poor silicate dust and iron-nickel vapor is spread out to at least Venus-Earth space. The main concentration of iron-nickel vapor, however, is in the vicinity of 0.4 AU (the location of Mercury's orbit at present).
6. After peak heat time, the dynamic and thermal processes decrease in intensity as the magnetic reconnection activity becomes more moderate in intensity. The gradually decreasing, pulsating thermal activity appears to be recorded in the complex evaporation, condensation, and melting cycles recorded in CAIs. A

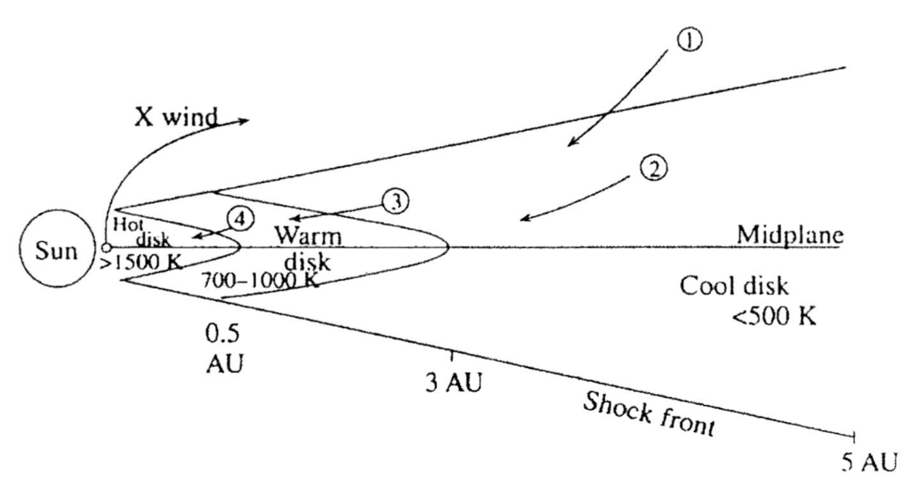

Fig. 2.6 Cross-sectional sketch illustrating some major features of the X-WIND model. Gravitational infall occurs symmetrically along the mid-plane of the nascent Solar System. The temperatures increase as the proto-sun is approached. Inside ~0.5 AU the temperature is over 1500 °K and there are high-temperature heat pulses caused by the breaking and reforming of magnetic field lines powered by the rapid rotation of the proto-sun. Bi-polar outflow is not shown on this diagram. In this model, CAIs can be formed only within a constantly shifting zone centered on 0.06 AU from the center of the Sun [the Reconnection Ring in this model (but not shown in this diagram)]. The X-Wind *arrow* indicates the trajectory by which CAIs can be propelled to outer regions of the Solar system such as the zone of the asteroids and possibly beyond. The trajectory of the CAI debris is guided by a strong magnetic field that is rotating with the proto-sun and the proto-sun is rotating rapidly. All in all, this is a very dynamic scene but it may be a necessary condition for the formation of CAIs. (Figure is from Taylor (2001, Fig. 5.11), with permission from Cambridge University Press)

hurling process at the free-surface X-points (a circular structure in 3-D) (Fig. 2.6) begins and CAIs are propelled above the mid-plane out to at least 3 AU and perhaps to the outer Solar System region or as far as the inner Oort Cloud (Bridges et al. 2012) to mix with infalling material.

7. The diminished heat source associated with the CAI formation process results in a stillstand of devolatilization of infalling materials. Fe-Ni metal vapor is concentrated in the vicinity of Mercury's orbit. Metal-rich planetesimals of various sizes accrete and eventually form a planet with a high specific gravity—Mercury (Fig. 2.7). Figure 2.8 shows my concept of the debris belts from which planet Mercury and some Vulcanoids planetoids accreted.

This May be a Good Time for a Break and Some Remarks by Taylor (1998, p. 176, top, and p. 48 bottom)

"All the evidence from the meteorites is that the primitive mineral components were dry. Water was driven out of the inner solar system and condensed as ice far away from the Sun, at a 'snow line' in the vicinity of Jupiter. Thus,

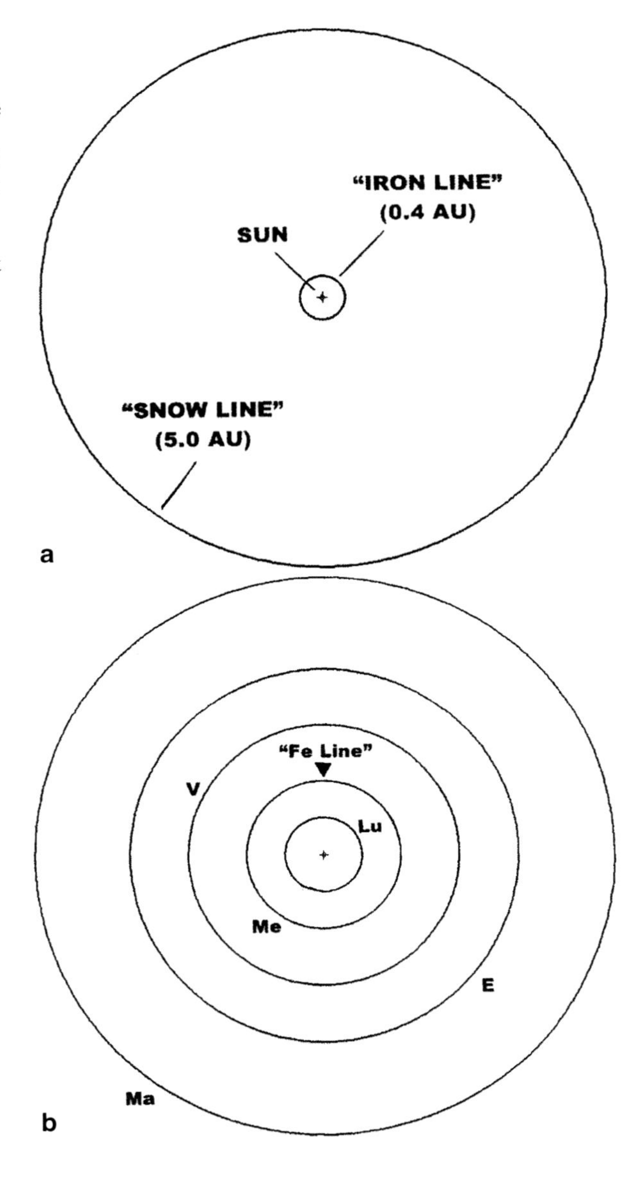

Fig. 2.7 **a** Scale diagrammatic sketch viewed from the north pole of the Solar System showing approximate locations of the "*snow line*" (near the orbit of Jupiter) and the "*iron line*" (near the orbit of Mercury). **b** Location of the "*iron line*" relative to the proposed location of the orbit of Luna (*Lu*) and the current locations of planets Mercury (*Me*), Venus (*V*), Earth (*E*) and Mars (*Ma*)

the terrestrial planets seem to have accumulated from dry rubble. Probably the only way for water to get to the inner planets is from comets coming from the outer reaches of the solar system."

"The inner parts of the nebula were thus cleared of the gas and depleted in volatile elements very early, probably within about one million years of the formation of the Sun. The cause of this clearing of the inner regions of the nebula is twofold. Early on, gas was swept into the Sun. Then, after the

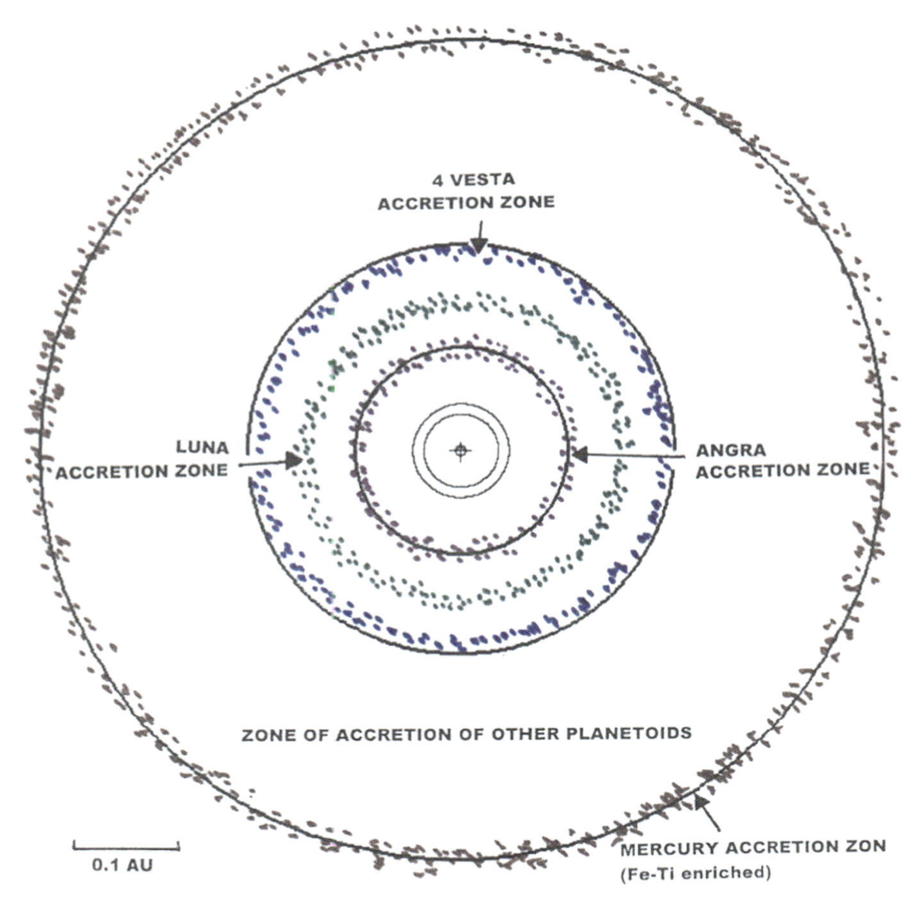

Fig. 2.8 Scale sketch viewed from the north pole of the Solar System showing my concept of the accretion zones for planet Mercury and the Vulcanoids. The debris zones are devolatilized residue from the action of the X-Wind removing Fe and Ni and other more volatile elements from the Vulcanoid Zone. This action causes a buildup of metal in the vicinity of the orbit of Mercury. The inner concentric circles mark the location of CAI material resulting from the X-Wind model (Shu et al. 2001; Wood 2004; Liffman et al. 2012) which subsequently forms Vulcanoid planetoids. (Note that many more Vulcanoid planetoids would be formed than are shown in the diagram)

nuclear furnace ignited, strong winds blew outward from the Sun and swept away any remaining gas. A few bodies, ranging in size from boulders to small mountains, survived even in the strong stellar winds. This was a fortunate circumstance, because we are standing on a planet that formed from this collection of rubble."

8. The CAI material in the vicinity of 0.1–0.2 AU accretes into planetesimals as the X-Wind action diminishes. These early formed planetesimals have the average chemistry of CAIs. It should be noted here that the chemical

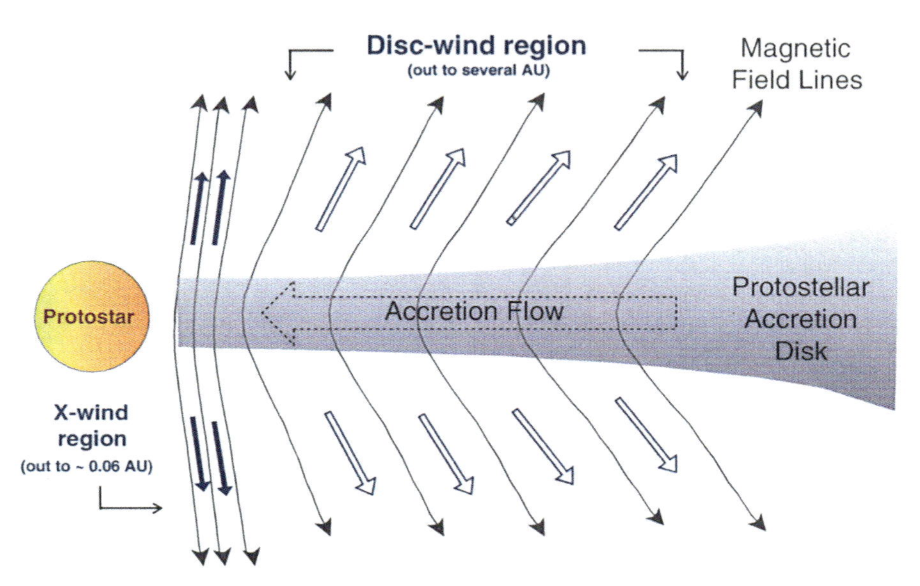

Fig. 2.9 Diagrammatic sketch showing some features of the X-Wind model as well as the Disk-Wind model (Diagram from Salmeron and Ireland (2012, Fig. 1), with permission from Elsevier). The general concept is that X-Winds operate very close to the star (i. e., near the disk truncation region) and that Disk Winds can operate to distances of a few astronomical units from the protosun. The X-Winds are thought to generate CAIs and related very high temperature minerals and mineral assemblages; the Disk Winds are thought to generate chondrules of a variety of compositions depending on the chemicals in the local region

composition of the Moon (Wood 1974; Anderson 1975), Vesta (Ruzicka et al. 1999, 2001; Guterl 2008), and Angra (Taylor 2001) (the parent body of the Angrite meteorites) have much in common and they are all very similar in chemistry to a composite CAI. I note here that this chemical relationship was noticed soon after lunar rocks were analyzed (e. g., Anderson 1973a, b, 1975; Cameron 1972, 1973; Gast 1972; Wood 1974).

9. Comcomitant with the diminishing X-Wind action, the dust, ice, and gas at the "frost line" accretes into various sized bodies in a Jupiter accretion torus. Eventually these icy planetesimals accumulate to form planet Jupiter. Saturn forms in a similar manner from a combination of materials coming in along the proto-solar mid-plane and volatile material going out from the inner solar system.
10. The Disk Wind kicks into action as the X-Wind diminishes (see Fig. 2.9). The Disk Wind sends periodic pulses of magnetically-generated thermal energy into the mid-plane section of the proto-solar cloud. These high-temperature heat pulses flash melt portions of the cloud that have sufficient material density to make glass beads (chondrules). The Disk Wind flash melts whatever material is available in that section of the cloud thus explaining the diversity in composition of chrondrules. The dust in the cloud increases in volatile content systematically from the Venus-Earth region out to the Asteroid Zone because of the previous thermal pulses of the X-Wind and Disk Winds.

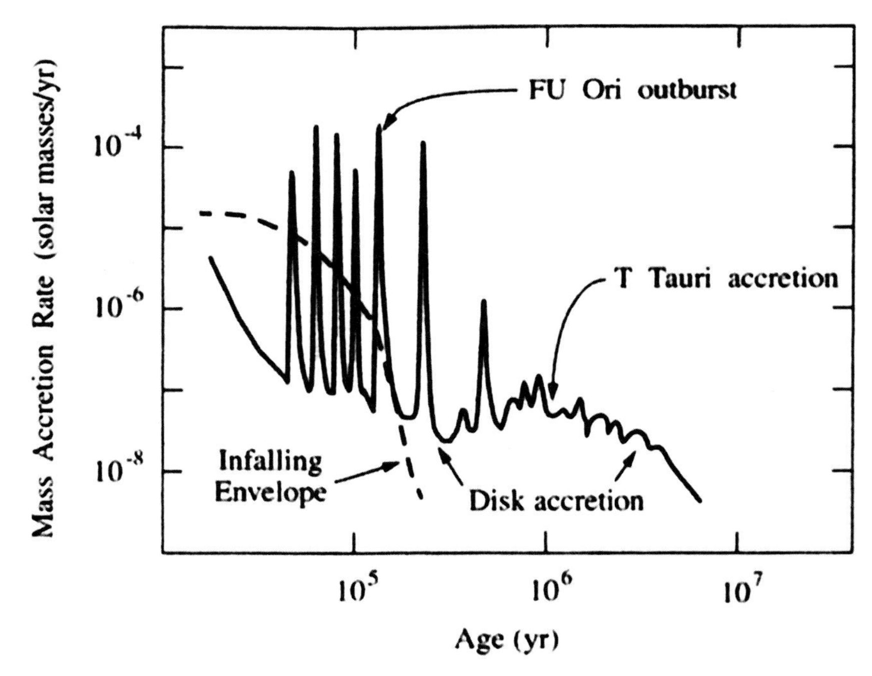

Fig. 2.10 Diagram illustrating the spasmatic nature of the FU Orionis stage of solar evolution. Note that there are intense energy "spikes" associated with the FU Ori stage and that these "spikes" are absent during the T-Tauri stage. The author suggests that the melting of the lunar magma ocean as well as the generation of a lunar magnetic amplifier in the LMO were powered by the thermal energy of the FU Ori stages of solar development. (Diagram adapted from Calvet et al. (2000, Fig. 7), with permission from University of Arizona Press)

The volatile content of the chondrules should reflect this decrease in thermal activity with distance from the proto-sun.

11. The terrestrial planets and asteroids begin to accrete by accumulation of material in their respective accretion tori. The composition of these bodies is essentially inherited from the composition of the chondrules in their accretion space plus perhaps some late infalling material as well as traces of CAI material. Gravitational perturbations by Jupiter and Saturn modifies (or controls) the accretion process for Mars and the Asteroids.
12. At this point in the model, there is very little water in the Venus-Earth space but more is in the Mars space because of a combination of increased volatile element content because of X-Wind action and addition of volatiles from infalling material.
13. THE FU ORIONIS STAGE: At some time after the cessation of the accretion phase of planet formation, the proto-sun goes through a transition which results in an intense microwave radiation which causes melting of the outer regions of any planetoids inside Mercury's orbit and perhaps complete melting of smaller bodies (see Figs. 2.10 and 2.11. In this model the microwave range

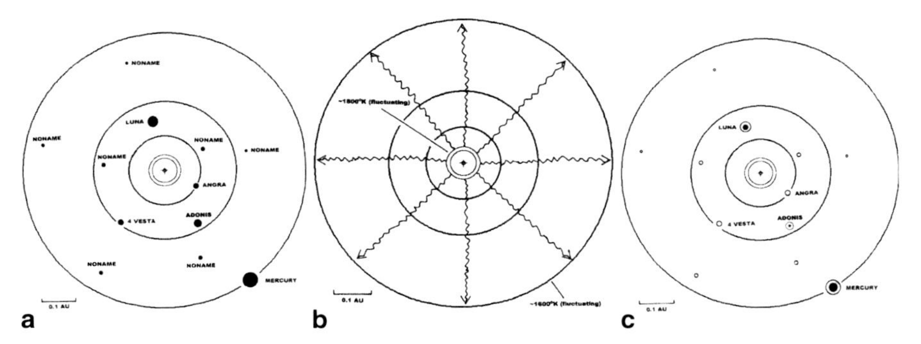

Fig. 2.11 Stages in the "microwaving" process for planet Mercury and the Vulcanoid planetoids. **a** Before the "microwaving" process, **b** During the "microwaving" process, **c** After the "microwaving" process. View is from the north pole of the Solar System

radiation is responsible for melting the outer 600 km of Luna to form the lunar magma ocean. Such microwave radiation could also cause significant melting of the outer parts of Mercury to form a magma ocean from which plagioclase would float to form a planetary crust which would be similar in nature to the ferroan anorthositic lunar crust (Blewett et al. 1997). Some surface melting and local devolatilization could affect materials in the Venus-Earth space as well. In summary, this FU Orionis stage results in magma ocean formation on the Vulcanoid planetoids and perhaps thermal metamorphism of asteroids on the inner zone of the Asteroid Belt. [(NOTE: Melting of andesite by microwave radiation has been reported by Znamenackova et al. (2003). In this case the frequency was 2.45 GHz with a wavelength of 12.2 cm. The microwave oven had an output of 900 watts. The rock specimen, about 150 g, was completely melted in about 30 min when the temperature range was 1260–1280 °C.)]

14. EARLY STAGE OF LUNAR MAGNETISM: As the Microwave Wind intensity decreases and a refractory crust begins to accumulate on the lunar magma ocean (and the near surface of other Vulcanoid planetoids), the rock tidal action of the Sun causes differential rotation between the crustal complex and the solid inner portion (below the magma ocean) of the planetoid. A weak solar magnetic field is then amplified by circulation cells that are tidally induced within the lunar magma ocean (Smoluchowski 1973a,b) and this magnetic field is recorded in the refractory anorthositic crust of Luna as the temperature decreases through the Curie temperature (this model for the lunar magnetic field will be discussed further in Chap. 4). Once the differential rotation between the crustal complex and the lunar mantle ceases, the magnetic amplifier action is over. The Sun then settles down to the Main Sequence of development *BUT THE SHOW IS NOT OVER*!
15. The T-TAURI stage of stellar evolution may be involved in this sequence of events also. The T- Tauri stage is characterized by lower levels of radiation but of longer temporal duration. This radiation stage may be registered in

the extensive hydrothermal metamorphism of the asteroids on the inner edge of the asteroid zone. Herbert et al. (1991) suggest that there is a progressive decrease in intensity of hydrothermal metamorphism in asteroids with distance from the Sun.

16. THE ORIGIN OF WATER ISSUE FOR THE EARTH AND VENUS: The origin of water for planet Earth is a major problem in the natural sciences (Robert 2001). The model that I am outlining here is a slightly modified version of the model of Albarede (2009). The Earth- Venus chondrite-planetoid debris (mainly enstatite chondrites) that accretes to make up the bodies of these planets is depleted in volatiles (including water) during the X-Wind/ Disk Wind Eras. There can be some water in that debris but not much relative to the inventory of water that the Earth has today. Albarede (2009) proposed that the water was brought in by way of water- rich asteroids (e. g., the Themis-type) from the Asteroid Belt within the first 150 million years of solar system history. The main orbital perturber, in this case, is Jupiter with some assistance from Saturn. The main bit of evidence for this scenario is that the deuterium-to- hydrogen (D/H) ratio of the Earth's ocean water is nearly identical to that of this class of water- rich asteroids. The D/H ratio of Mars is significantly different from that of Earth but it is close to that of comets (Robert 2001). So in this model Earth and Venus get their water from Themis- type asteroids and Mars gets its water from one or more Jupiter-captured comets. (Jupiter- captured comets commonly end up in the Asteroid Belt and their subsequent orbital evolution can be treated like that of an asteroid.) There will be a section on THE ORIGIN OF WATER ON EARTH in Chap. 5.
17. THE MIGRATION HISTORY OF THE VULCANOID PLANETOIDS: Fig. 2.12 shows my version of what happened to a few of the Vulcanoid planetoids. This dynamical episode is occurring concurrently with the migration of the water-bearing Aquarioids but it continues over a much longer period of time (about 600 Ma vs. 150 Ma). The main orbital perturber for the Vulcanoid planetoids is planet Mercury with some help from Venus and Earth. The main bit of evidence for the timing of events is that major dynamic and/or thermal events occurred on the Moon and Earth about 3.95 Ga ago. In my model, these dynamo-thermal events are associated with the process of GRAVITATIONAL LUNAR CAPTURE. This event occurred about 600 Ma after the origin of the Vulcanoid planetoids so Luna has a reasonable quantity of time to get perturbed out to the vicinity of Earth's orbit to be in a position to be gravitationally captured by tidal processes. There will be much more on this migration history in Chap. 4.

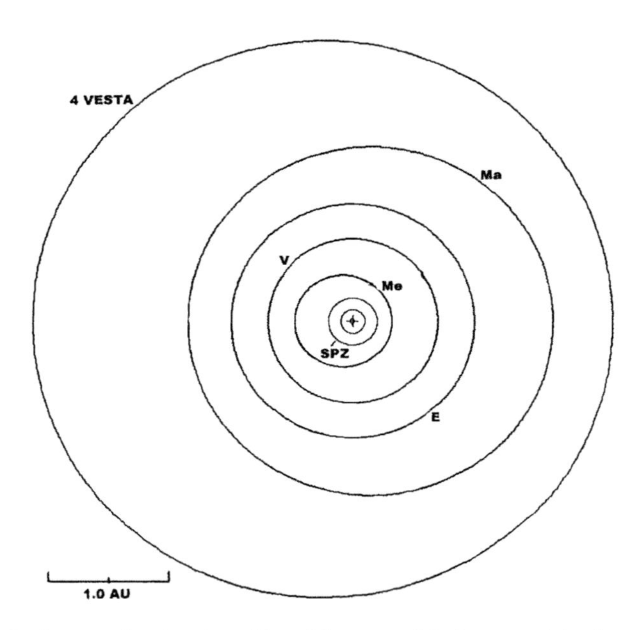

Fig. 2.12 Scale diagram showing the orbits of the terrestrial planets and the Asteroid 4 Vesta as well as the Stable Planetoid Zone (*SPZ*) for the Vulcanoid planetoids in the equatorial plane of the Solar System. In this scenario, 4 Vesta is a Vulcanoid planetoid that is too small to get captured and apparently avoids a "wipe-out" collision. Thus it is a "survivor" that made it to the "safe haven" of the Asteroid Zone. Then sometime later it suffers an impact which gives rise to the Vestoid family of asteroids, some chips of which have impacted on the ice in the Dry Valley region of Antarctica and other places on Earth. These meteorites are referred to as the HED meteorites

Summary

The reader at this early point in this book may get the feeling that these events represent a "long chain of complications" (as suggested by Alfven 1969; Cloud 1978, Taylor 1998, and others). Indeed, I think there is good reason to think this way. Although there may be a number of unusual events affecting the processing of a proto-solar cloud of roughly solar composition into the planetary bodies we have today, I cannot apologize for these complications. One event leads to another to another in a somewhat logical chain of related events. There are very few, if any, *ad hoc* assumptions in this model.

The bottom line is that Earth and Venus appear to have recorded fairly well certain events and that the Earth's Moon (a captured satellite in this scenario) may well turn out to be a "ROSETTA STONE" of the Solar System as Harold Urey suggested at the dawn of the space program (Newell 1973). Now let us continue with this "LONG CHAIN OF COMPLICATIONS".

References

Albarede F (2009) Volatile accretion history of the terrestrial planets and dynamic implications. Nature 461:1227–1233

Alfven H (1969) Atom, man, and the Universe. The long chain of complications. W. H. Freeman, San Francisco, p 110

Anderson DL (1973a) The moon as a high temperature condensate. The Moon 8:33–57

Anderson DL (1973b) The composition and origin of the moon. Earth Planet Sci Lett 18:301–316

Anderson DL (1975) On the composition of the lunar interior. J Geophys Res 80:1555–1557

Bans A, Konigl A (2012) A disk-wind model for the near-infrared excess emissions in protostars. Astrophys J 758(100):17p

Bell KR, Cassen PM, Wasson JT, Woolum DS (2000) The FU Orionis phenomenon and solar nebula material. In: Manning V, Boss AP, Russell SS (eds) Protostars and planets IV. University of Arizona Press, Tucson, pp 897–926

Berger A (1980) Milankovitch astronomical theory of paleoclimates, a modern review. Vistas Astron 24:103–122

Berger A, Loutre MF, Dehant V (1989a) Pre-Quaternary Milankovitch frequencies. Nature 342:133

Berger A, Loutre MF, Dehant V (1989b) Influence of the changing lunar orbit on the astronomical frequencies of pre-Quaternary insolation patterns. Paleoceanography 4:555–564

Berger A, Loutre MF, Laskar J (1992) Stability of the astronomical frequencies over the Earth's history for paleoclimate studies. Science 255:560–566

Bizzarro M, Baker JA, Haach H (2004) Mg isotope evidence for contemporaneous formation of chondrules and refractory inclusions. Nature 431:275–278

Blewett DT, Lucey PG, Hawke BR, Ling GG, Robinson MK (1997) A comparison of mercurian reflectance and spectral quantities with those of the moon. Icarus 129:217–231

Brearley AJ, Jones RH (1998) Chap. 3: chondritic meteorites. In: Papike JJ (ed) Planetary materials. Reviews of mineralogy, vol 36. Mineralogical Society of America, Washington, DC, pp 1–398

Bridges JC, Changela HG, Nayakshin S, Starkey NA, Franchi IA (2012) Chondrule fragments from Comet Wild2. Evidence for high temperature processing in the outer solar system. Earth Planet Sci Lett 341–344:186–194

Britt DT, Macke RJ, Kiefer W, Consolmagno GJ (2010) An overview of achondrite density, porosity and magnetic susceptibility. Abstracts, 41st Lunar and Planetary Science Conference, 1869.pdf

Calvet N, Hartmann L, Strom SE (2000) Evolution of disk accretion. In: Manning V, Boss AP, Russell SS (eds) Protostars and planets IV. University of Arizona Press, Tucson, pp 377–399

Cameron AGW (1962) The formation of the sun and the planets. Icarus 1:13–69

Cameron AGW (1972) Orbital eccentricity of mercury and the origin of the moon. Nature 240:299–300

Cameron AGW (1973) Properties of the solar nebula and the origin of the moon. The Moon 7:377–383

Cameron AGW, Pine MR (1973) Numerical models of the primitive solar nebula. Icarus 18:377–406

Clayton RN (2003) Oxygen isotopes in meteorites. Treatise Geochem 1:129–142 (Holland HD, Turekian KK (eds))

Cloud PE (1978) Cosmos, earth, and man. Yale University Press, New Haven, p 371

Connelly JN, Bizzarro M, Krot AN, Nordlund A, Ivanoya MA (2012) The absolute chronology and thermal processing of solids in the solar protoplanetary disk. Science 338:651–655

Cousins FW (1972) The solar system. PICA Press, New York, p 300

Desch SJ, Morris MA, Connolly HC Jr, Boss AP (2010) A critical examination of the X-wind model for chrondrule and calcium-rich, aluminum-rich inclusion formation and radionuclide production. Astrophys J 725:692–711

Donahue TM (1999) New analysis of hydrogen and deuterium escape from Venus. Icarus 141:226–235

Donahue TM, Hoffman JH, Hodges RR Jr, Watson AJ (1982) Venus was wet. A measurement of the ratio of deuterium to hydrogen. Science 216:630–633
Evans NW, Tabachnik S (1999) Possible long-lived asteroid belts in the inner solar system. Nature 399:41–43
Evans WN, Tabachnik S (2002) Structure of possible long-lived asteroid belts. Mon Not R Astron Soc 333:L1–L5
Gast PW (1972) The chemical composition and structure of the moon. The Moon 5:121–148
Greenwood RC, Franchi IA, Jambon A, Buchanan PC (2005) Widespread magma oceans on asteroidal bodies in the early Solar System. Nature 435:916–918
Grossman L (1972) Condensation in the primitive solar nebula. Geochim Cosmochim Acta 36:597–619
Guterl F (2008) Mission to the forgotten planets. Discover (Feb. issue):48–52
Herbert F, Sonett CP (1980) Electromagnetic inductive heating of the asteroids and moon as evidence bearing on the primordial solar wind. In: Pepin RO, Eddy JA, Merrill RB (eds) Proceedings of a conference on the ancient sun. University of Arizona Press, Tucson, pp 563–576
Herbert F, Sonett CP, Gaffey MJ (1991) Protoplanetary thermal metamorphism: the hypothesis of electromagnetic induction in the protosolar wind. In: Sonett CP, Giampapa MS, Mathews MS (eds) The sun in time. University Arizona Press, Tucson, pp 710–739
Humayun M, Clayton RN (1995) Potassium isotope cosmochemistry: genetic implications of volatile element depletion. Geochem Cosmochim Acta 59:2131–2148
Imbrie J (1982) Astronomical theory of Pleistocene ice ages: a brief historical review. Icarus 50:408–422
Imbrie J, Imbrie KP (1979) Ice ages. Solving the mystery. Enslow, Hillside, p 224
Kita NT, Yin Q-Y, MacPherson GJ, Ushikubo T, Jacobsen B, Nagashima K, Kurahashi E, Krot AN, Jacobsen SB (2013) ^{26}Al-^{26}Mg isotope systematics of the first solids in the early solar system. Meteorit Planet Sci 48:1383–1400
Konigl A, Pudritz RE (2002) Disk Winds and the accretion outflow connection. In: Manning V, Boss AP, Russell SS (eds) Protostars and planets iv. University of Arizona Press, Tuscon, pp 759–787
Kowal CT (1996) Asteroids—their nature and utilization, 2nd edn. Praxis Publishing (Wiley), Chichester, p 153
Lada CJ, Shu FH (1990) The formation of sunlike stars. Science 248:564–572
Leake MA, Chapman CR, Weidenschilling SJ, Davis DR, Greenberg R (1987) The chronology of mercury's geological and geophysical evolution: the vulcanoid hypothesis. Icarus 71:350–375
Lewis JS (1972) Metal/silicate fractionation in the solar system. Earth Planet Sci Lett 15:286–290
Lewis JS (1974) The chemistry of the solar system. Sci Am 230(3):51–65
Liffman K, Pignatale FC, Maddison S, Brooks G (2012) Refactory metal nuggets—formation of the first condensates in the solar nebula. Icarus 221:89–105
Lodders K, Fegley B Jr (1998) The planetary scientist's campanion. Oxford University Press, New York, p 371
MacPherson GJ, Simon SB, Davis AM, Grossman L, Krot AN (2005) Calcium-aluminum-rich inclusions: major unanswered questions. In: Krot AN, Scott ERD, Reipurth B (eds) Chrondrites and the protoplanetary disk. ASP Conference Series, San Francisco, pp 225–250
McSween HY Jr (1999) Meteorites and their parent bodies. Cambridge University Press, Cambridge, p 310
Newell HE (1973) Harold Urey and the moon. The Moon 7:1–5
Robert F (2001) The origin of water on earth. Science 293:1056–1058
Rubin AE (2010) Physical properties of chondrules in different condrite groups: Implications for multiple melting events in dusty environments. Geochim Cosmochim Acta 74:4807–4828
Rubin AE (2013) Secrets of primitive meteorites. Sci Am 308(2):36–41
Ruzicka A, Snyder GA, Taylor LA (1999) Giant impact and fission hypotheses for the origin of the moon: a critical review of some geochemical evidence. In: Snyder GA, Neal CR, Ernst WG (eds) Planetary petrology and geochemistry. Geological society of America, International Book Series 2. Bellwerther, Columbia, pp 121–134

Ruzicka A, Snyder GA, Taylor LA (2001) Comparative geochemistry of basalts for the moon, earth, HED asteroid, and mars: implications for the origin of the moon. Geochim Cosmoshem Acta 65:979–997

Salmeron R, Ireland TR (2012) Formation of chondrules in magnetic winds blowing through the proto-asteroid belt. Earth Planet Sci Lett 327–328:61–67

Smoluchowski R (1973a) Lunar tides and magnetism. Nature 242:516–517

Smoluchowski R (1973b) Magnetism of the moon. Moon 7:127–131

Sonett CP, Reynolds RT (1979) Primordial heating of asteroidal parent bodies. In: Gehrels T (ed) Asteroids. University of Arizona Press, Tucson, pp 822–848

Sonett CP, Colburn DS, Schwartz K (1975) Formation of the lunar crust: an electrical source of heating. Icarus 24:231–255

Shu FH, Shang H, Glassgold AE, Lee T (1997) X-rays and fluctuating X-Winds from protostars. Science 277:1475–1479

Shu FH, Najita J, Ostriker E, Wilkin F, Ruden S, Lizano S (1994) Magnetocentrifugally driven flows from young stars and disks. I. A generalized model. Astrophys J 429:781–796

Shu FH, Shang H, Gounelle M, Glassgold AE, Lee T (2001) The origin of chondrules and refractory inclusions in chondritic meteorites. Astrophys J 548:1029–1050

Taylor SR (1998) Destiny or chance. Our solar system and its place in the cosmos. Cambridge University Press, Cambridge, p 229

Taylor SR (2001) Solar system evolution. A new perspective. Cambridge University Press, Cambridge, p 460

Vinogradov AP, Surkov Yu A, Kirnozov FF (1973) The content of uranium, thorium, and potassium in the rocks of venus as measured by Venera 8. Icarus 20:253–259

Weidenschilling SJ (1978) Iron/silicate fractionation and the origin of mercury. Icarus 35:99–111

Wood JA (1974) Summary of the 5th lunar science conference: constraints on structure and composition of the lunar interior. Geotimes June Issue:16–17

Wood JA (2004) Formation of chondritic refractory inclusions: the astrophysical setting. Geochim Cosmochim Acta 68:4007–4021

Wood JA, Pellas P (1991) What heated the parent meteorite planets. In: Sonett CP, Giampapa MS, Mathews MS (eds) The sun in time. University of Arizona Press, Tucson, pp 740–760

Znamenackova I, Lovas M, Hajek M, Jakabsky S (2003) Melting of andesite in a microwave oven. J Mining Metall 39(3–4)B:549–557

Chapter 3
Models for the Origin and Evolution of the Earth-Moon System

"... in the entire solar system the earth-moon system is completely unique, and the recognized difficulties in all proposed models of origin seem to demand an unusual solution." Baldwin (1966, p. 1936)

"The new interest in the large-impact model is admittedly due in part to the consistent failures of all other models. Three classic theories have been lingering for years, championed by a few individuals, but at the lunar origin conference, ... the classic theories seem to have been finally put to rest." From Richard A. Kerr, 1984, Making the Moon from a big splash: Science, v. 226, p. 1060.

The origin of the Moon remains an unsolved problem. In spite of the claims of the ruling paradigm investigators, the GIANT IMPACT MODEL (GIM) is a disappointment when it comes to explaining many of the major features of the Earth and Moon. It appears to be especially inadequate to explain the facts associated with the COOL EARLY EARTH MODEL.

The main theme of the researchers of the GIM since the Kona Conference in 1984 has been that all of the other models for lunar origin are inadequate, i.e., they all have "fatal flaws" (e.g., see the second quote of this chapter). Furthermore, the "WORST OF THE WORST" model is the GRAVITATIONAL CAPTURE MODEL (e.g., see Kaula and Harris 1973).

The history of science is characterized by unusual twists and turns in the progress of understanding the natural environment. The CONTINENTAL DRIFT controversy is a good example. The STABLE CONTINENT CONCEPT held sway for a long time and really hindered progress in the Earth Sciences. The controversy associated with the MILANKOVITCH MODEL FOR ICE AGES is another good example. But once the technological tools were developed to give reasonable tests for these controversial models, the "ruling paradigms" came crumbling down.

The "tried and true" method for settling these controversies is THE SCIENTIFIC METHOD as outlined in Chap. 1. Once we sort the "facts" from the "interpretations of facts", then we can make progress via THE SCIENTIFIC METHOD. In the next

Robert J. Malcuit, *The Twin Sister Planets Venus and Earth*,
DOI 10.1007/978-3-319-11388-3_3

section of this chapter I will present what I think is a "list of facts to be explained by a successful model". Then we will examine the major hypotheses (models) of lunar origin in light of their explanation of the "facts". In the conclusion of this chapter, we will evaluate two of these models by way of a "REPORT CARD" method used by John Wood (1986) for the Kona Conference on the "ORIGIN OF THE MOON".

3.1 List of Facts to be Explained by a Successful Model

- the mass and density of the Earth and Moon
- the mass ratio of the Earth and Moon
- the similarity of oxygen isotope ratios in rocks of the Moon and Earth
- the enrichment of refractory elements in the Moon relative to those in the Earth
- the internal zonation of the Moon and Earth
- the factual information supporting the concept of magma ocean development on the Moon and Earth
- the large quantity of angular momentum of the Earth-Moon system relative to the rotational angular momentum of the Earth
- the elemental abundance patterns in the mantles of the Earth and Moon
- the factual information supporting the concept of the generation of a lunar magnetic field during two distinctive eras in lunar history:

    1. at the time of formation of the lunar crust (between 4.55 and 4.40 Ga ago)
    2. between 3.9 and 3.6 Ga ago when major impact and volcanic events occurred on the lunar surface

- the factual information associated with the model of the "Cool Early Earth" and the derived interpretation of liquid water on the surface of Earth in that era.

3.2 Fission from the Earth Early in Earth History

The fission model was first proposed by Darwin (1879, 1880). The inspiration for the model was to explain an orbital traceback of the Earth-Moon system while conserving angular momentum. The traceback resulted in an Earth rotating at about 4.5 h/day when the Moon was at the Roche limit (about 2.89 earth radii). The problem with the model is that there is no known mechanism for detaching a lunar-mass body from the primitive planet consisting of the mass of the Earth and Moon.

More recent work has been done on the fission model by Ringwood (1960, 1979) and Wise (1963, 1969). In Ringwood's model a disk of debris is spun off the equatorial zone of Earth early in earth history and the Moon forms by accumulating the debris from circum-terrestrial orbits. It is not clear that a moon-like body can accrete from the debris and it is not clear just how rapidly the proto-planet (containing the mass of the Earth and Moon) must rotate to fission off the debris. This model

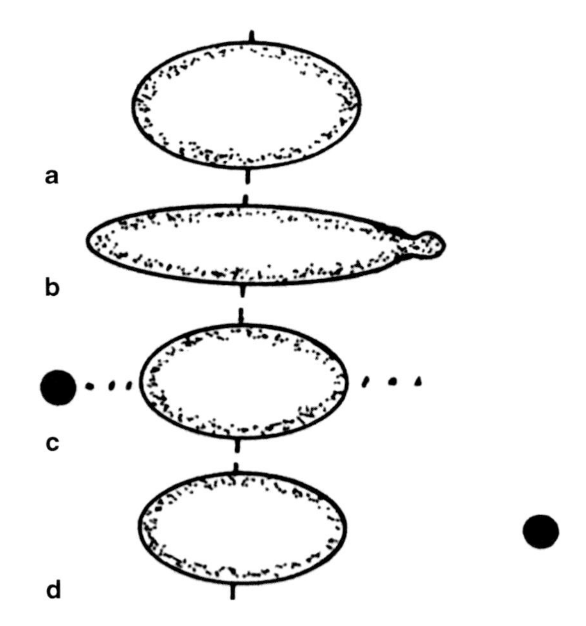

Fig. 3.1 A simplified version of the fission model for the origin of the Moon favored by Wise (1963, 1969). This is a sequence of diagrams illustrating formation of a satellite due to separation from a large body (Earth and Moon mass) due to an increase in rotation rate because of formation of a dense metallic core

does have some favorable geochemical features. This model relates to the similarities of lunar composition with the mantle of Earth as well as the similarities of the oxygen isotope ratios of the Earth and Moon (de Meijer et al. 2013). However, in Ringwood's model as well as those of Darwin (1879, 1880) and Wise (1963, 1969), there is no reasonable explanation for how the Earth would acquire the high rotation rate (~2–4 h/day) to cause the fissioning process to occur regardless of the number of bodies to be fissioned from the main body: i.e., the source of this large quantity of prograde angular momentum for the proto-earth (Earth plus Moon mass) has not been identified. Promoters of fission models, in general, have *not* been able to explain (a) the lunar rock remanent magnetization patterns or (b) the factual information associated with the Cool Early Earth Model.

As a historical note, the fission model of Darwin (1879, 1880) was the favored model until about 1930 when Jeffreys (1929) raised doubts about the physical possibility of the fission process (see Fig. 3.1 for an illustration of the Fission Model).

3.3 Co-formation of the Earth and Moon from the Same Cloud of Dust and Gas

The co-formation model has a long history. A version of the co-formation model was favored by Laplace in the early 1800s. Extensive calculations were done by Harris and Kaula (1975) and Harris (1978) and more recent work was done by Herbert et al. (1986) and Wiedenschilling et al. (1986). The goal of the calculations

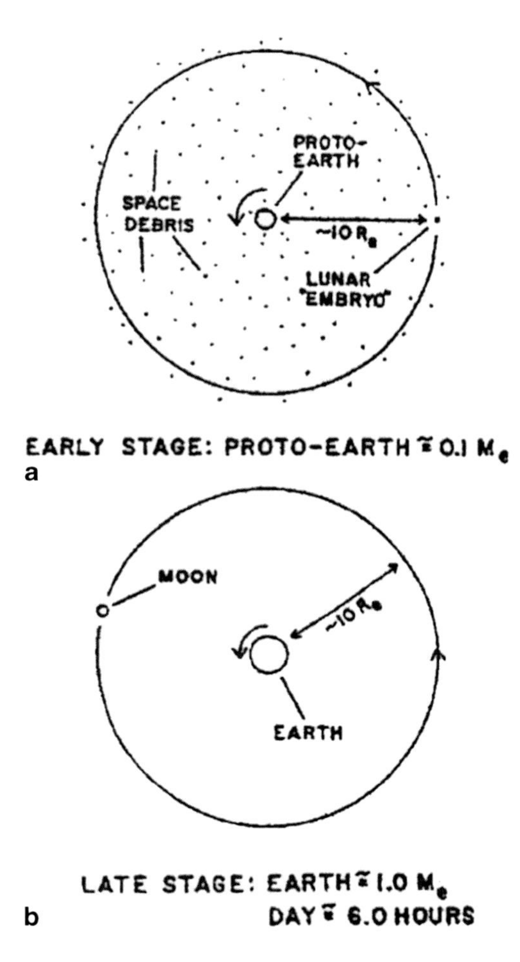

Fig. 3.2 A simplified version of a co-formation model for the origin of the Moon that is consistent with the calculations of Harris and Kaula (1975)

by Harris and Kaula (1975) and Harris (1978) was to show that a lunar-sized body could form from the same dust-cloud as the Earth and simultaneously with the Earth (see Fig. 3.2). They found, after extensive calculations, that they could not find a good combination of lunar embryo size and starting position that would result in the formation of a full-sized Moon at about 20–40 earth radii from the Earth. To account for the angular momentum of the Earth-Moon system, the rotation rate of Earth for a lunar body at 30 earth radii would be about 10 h/day; with the Moon at 40 earth radii the rotation rate for Earth would be ~ 13 h/day. Most simulations resulted in the lunar embryo gaining mass too rapidly and spiraling in to coalesce with the Earth. In other cases, the lunar embryo did not accumulate enough mass to form a lunar mass body. An additional problem with the co-formation model is that the composition of the Moon and Earth are not similar as would be expected if they formed from the same cloud of space debris. A successful co-formation model must somehow explain how the iron-rich debris got concentrated within the earth mass to account for the composition of the Earth's core. Compositional "filters" were employed by Herbert et al. (1986) and Weidenschilling et al. (1986) with very limited success. Proposed co-formation models, in general, cannot explain (a) the lunar rock magnetization patterns or (b) the factual information associated with the Cool Early Earth Model.

As a historical note, a generic co-formation model enjoyed the "favored model" status from the time of the demise of the fission model (about 1930) until the co-formation model was questioned by the calculations of Harris and Kaula (1975) and Harris (1978).

3.4 Intact Capture of the Moon by the Earth (1952–1986)

The intact capture model (Fig. 3.3) was favored by Urey (1952) and Alfven (1954) because of the body density differences between the Moon and Earth. Urey (1952, 1962, 1972) proposed that the moon was a body related to the original population of planetoids that accreted to form the terrestrial planets (i.e., a survivor from the accretion process). He made the statement several times that the Moon could be a "Rosetta Stone" of the Solar System because of its primitive nature (Newell 1973). Cloud (1968, 1972) favored capture for geological reasons. He thought that there were geological signatures on both the Earth and Moon that related to the capture episode. For example, he suggested that both the Earth and Moon underwent mutual thermal events in an era prior to 3.6 billion years ago (later pushed back to ~3.8 billion years ago); the mutual thermal events suggested to him that they were associated with the capture episode. Singer (1968, 1970, 1977, 1986) presented analytical calculations suggesting that a lunar-sized body was capturable. His mechanism for capture was dissipation of energy in the Earth by way of large number of close encounters between the Moon and Earth. In addition to the lack of a focusing mechanism necessary for the repeated encounters, there was a problem explaining an adequate energy dissipation mechanism in Earth to facilitate capture: i.e., the rock tide action in Earth is not a sufficient energy sink for this capture model. In general, these capture models do *not* explain (a) lunar rock magnetization patterns or (b) the factual information associated with the Cool Early Earth Model. In addition, there is the traditional problem for capture models: WHERE DID THE MOON COME FROM?

3.5 Other Recent Attempts at Intact Capture

Other recent attempts to find a suitable capture mechanism (when it was popular to pursue some form of capture) have been presented by Clark et al. (1975), Cline (1979), Szebehely and Evans (1980), and Nakazawa et al. (1983). Some of these capture scenarios involve three or four body interactions in which the effects of

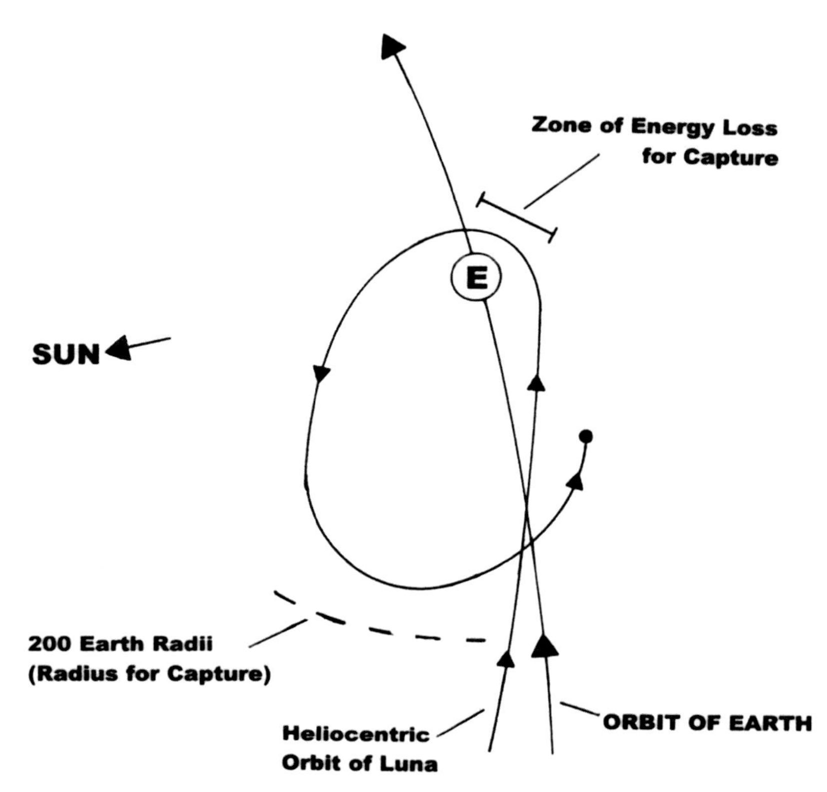

Fig. 3.3 A simplified version of an intact prograde planetoid capture model. View is from the north pole of the Solar System. For "intact" prograde capture a full mass lunar body must be transferred from a heliocentric orbit to a geocentric orbit. The resulting orbit must be small enough to be within the Hill Sphere (i.e., the stability limit) of the Earth-Moon-Sun dynamical system (Roy 1965). If it is not within the Hill Sphere, then the lunar mass body can escape from an unstable geocentric orbit back into a heliocentric orbit. Using a two-body analytical assessment, about 2.3 E28 J of energy must be dissipated in a combination of the planet and satellite to capture a lunar mass body. This quantity of energy is sufficient to melt a zone about 100 km thick in the lunar upper mantle. Tidal energy dissipation is one possibility for the energy sink but collision with a smaller body as the lunar mass body is having a close encounter with the planet is another possibility. The orientation and mass of that colliding body would need to be within narrow limits for successful "intact" capture. Other energy "sinks" are possible and have been proposed in the literature

tidal energy dissipation within the bodies is neglected and hence require very unusual conditions for lunar capture such as solar mass loss (Szebehely and Evans 1980), or satellite-assisted capture (Cline 1979). Others involved simultaneous and instantaneous mass addition to Earth at the time of lunar capture (Clark et al. 1975) or gas drag in a primitive, extensive Earth atmosphere (Nakazawa et al. 1983). A somewhat different approach to lunar capture is presented by Alfven and Arrhenius (1972, 1976). This model involves special spin-orbit resonance states in prograde orbit when the Moon is in a small radius (6–10 earth radii), near-circular prograde geocentric orbit. This model, however, demands a very rapid Earth rotation rate of about 2.5–5.0 h/day and does not suggest a mechanism for dissipating the energy for stable capture. (Note: For a more thorough review of other attempts to develop lunar capture models consult Brush (1986.)) Some of these models *may* relate to (a) patterns of lunar rock magnetization and (b) the factual information associated with the Cool Early Earth Model. But these models also beg the question: "WHERE DID THE BODY OF THE MOON COME FROM?"

3.6 Orbital Traceback Models Suggesting Intact Capture

In general, most of the calculations suggesting that the Moon is a captured body are based on orbital tracebacks, assuming conservation of angular momentum. The calculations are similar to those of Darwin (1879, 1880) but the conclusions were much different. The first such calculation was published by Gerstenkorn (1955) and then more sophisticated calculations were introduced by Kaula (1964), MacDonald (1964), Goldreich (1966) and Gerstenkorn (1967). The main theme that these orbital tracebacks have in common is that the Moon, when it gets near the classical Roche limit of the Earth-Moon system (about 2.89 earth radii) in a near circular orbit, eventually evolves into a highly inclined orbit relative to the equatorial plane of the Earth and eventually becomes retrograde as the Moon orbits away from the Earth. Thus, these orbital tracebacks imply capture from a near earth-like prograde heliocentric orbit into a retrograde geocentric orbit. More recent attempts using the orbital traceback approach by Mignard (1979, 1980, 1981) and Conway (1982) have yielded similar results. In general, all of these orbital traceback models, whether in the prograde direction or in the retrograde direction, suggest evolution into a small radius (about 4 earth radii) near circular orbit before the succeeding orbital expansion. Such models demand an initial rotation rate for Earth of about 4.5 h/day (or faster) as well as a large quantity of thermal energy to be dissipated by rock tidal processes; this tidally generated thermal energy would be sufficient to melt large portions of both the Earth and Moon. It is *not* clear how these traceback models relate to (a) the patterns of lunar rock magnetization or (b) the factual information associated with the Cool Early Earth Model. In addition, these models beg the question associated with all capture models: "WHERE DID THE BODY OF THE MOON COME FROM?"

3.7 More on the Singer (1968) Model of Prograde Capture

Singer (1968, 1970, 1977, 1986) attempted to show, in a schematic manner, that the Moon could stabilize in a prograde orbit after capture from a prograde heliocentric orbit. But Singer's prograde capture model also demands a very high initial rotation rate for Earth because the lunar orbit evolves rapidly into a small circular prograde orbit of about 4 earth radii. This evolution into a smaller orbit involves about 4 times more energy than for the capture itself. Furthermore, he does not suggest a plausible mechanism for dissipating the energy for capturing the Moon into a maximum-sized stable orbit (about 2.0 E28 Joules). Because of its body deformation properties, the Earth is not capable of such energy dissipation. Singer's capture model does *not* suggest a solution for (a) the patterns of lunar rock magnetization or (b) the factual information associated with the Cool Early Earth Model. And again, the question remains (as it does for all capture models): "WHERE DID THE MOON COME FROM TO MAKE IT AVAILABLE FOR CAPTURE?"

3.8 Disintegrative Capture Models

Disintregrative capture models were proposed by Wood and Mitler (1974), Smith (1974), and Mitler (1975). The major concept of disintegrative capture is that bodies encountering the Earth from heliocentric orbit could be tidally disrupted and that a portion of the material that was disrupted at the sub-earth point during perigee passage could be captured by Earth into a geocentric orbit. The circum-terrestrial swarm of captured debris would then coalesce to form the lunar body. *This model explains fairly well the iron depletion of the Moon* (i.e., only the silicate portion of the encountering planetoids was disrupted), but it does not explain many other items in the "list of facts to be explained by a successful model". Furthermore, Mizuno and Boss (1985) and Boss (1986) demonstrated that it is very difficult to tidally disrupt much material from solid planetoids during close encounters with an earth-like planet in heliocentric orbit. There simply is not enough time during the perigee passage through the Roche limit for a solid body for the encountering body to be disrupted (Jeffreys 1947; Aggarwal and Oberbeck 1974; Holsapple and Michel 2006, 2008). These disintegrative capture models, in general, do *not* have an explanation for (a) the patterns of lunar rock magnetization or (b) the factual information associated with the Cool Early Earth Model.

3.9 A Multiple-Small-Moon Model

In order to circumvent to thermal consequences of orbital traceback models for a capture origin for the Earth-Moon system and to explain a ~1.5 Ga age for the Earth-Moon system as suggested by the orbital traceback calculations, MacDonald (1964) proposed that several small planetoids of a variety of sizes could have been satellites that formed in Earth orbit. If there was a larger one that formed fairly close to the planet, then the major axis could be expanded by angular momentum transfer via tidal action with the planet at a much faster rate than for the smaller moons. The larger moon, as its major axis gradually increases, would collect the mass of the smaller ones via collision and mineral isotopic ratios would be reset by the collisions. In this model the Moon would give a much younger date of formation than the Earth. He then suggested that the present Moon might form at 30 to 40 earth radii from Earth and then evolve to the present configuration via normal tidal processes(MacDonald 1963).

Such a model does circumvent the thermal problem of the orbital traceback models but has a set of problems of its own. For example, would these planetoids yield a body of lunar chemistry? Could such satellites accrete to form a lunar mass body or would their mutual collisions result in much of the collisional debris impacting on Earth? An added complication for this model is that the accumulation would have to occur very early in Earth-Moon history to account for the age of the Moon and not at ~ 1.5 Ga as suggested by the traceback models. There is no hint, in this model, for an explanation of (a) the patterns of lunar rock magnetization but (b) it would be compatible with the Cool Early Earth Model because there would be no large satellite until ~ 1.5 Ga. And the chemical question remains: WHERE DID THESE SMALL, VOLATILE-POOR, VERSIONS OF THE MOON COME FROM?

As a historical note, capture models were never very popular because they were considered *AD HOC* and CATASTROPHIC. However, capture models got reasonable attention from the time that the lunar rocks arrived on Earth (~ 1969) until the time of the Kona Conference (1984). After the Kona Conference, the Giant-Impact Model took a position on "center stage" and is at present the ruling paradigm for the origin of the Earth's Moon.

3.10 A New (Post-Kona) View of the Intact Capture Process

More recently Malcuit et al. (1989, 1992) showed conclusively via numerical simulations that a lunar-like planetoid can be captured from a heliocentric orbit into a high eccentricity, large major axis geocentric orbit. Figure 3.4 shows a typical heliocentric orientation and geocentric orbital evolution for a successful intact capture. The key to capture, in their model, is to store and subsequently dissipate the energy for capture (~ 1.0 E28 Joules) *within the lunar body*. Most, if not all, of this energy is dissipated during one encounter (the capture encounter) if the dissipative properties of the Moon are adequate at the time of the encounter. The Moon can not be too hot or too cool. It needs to have an intermediate viscosity. In this model the Earth is mainly a passive planet furnishing a strong gravitational field for deforming the Moon so that the Moon can dissipate the energy for its own capture. Although this model suggested a mechanism for dissipating the energy for capture and a time for capture (about 3.9 Ga), *it did not offer a solution for the place of origin of the lunar body*. This gravitational capture model does,

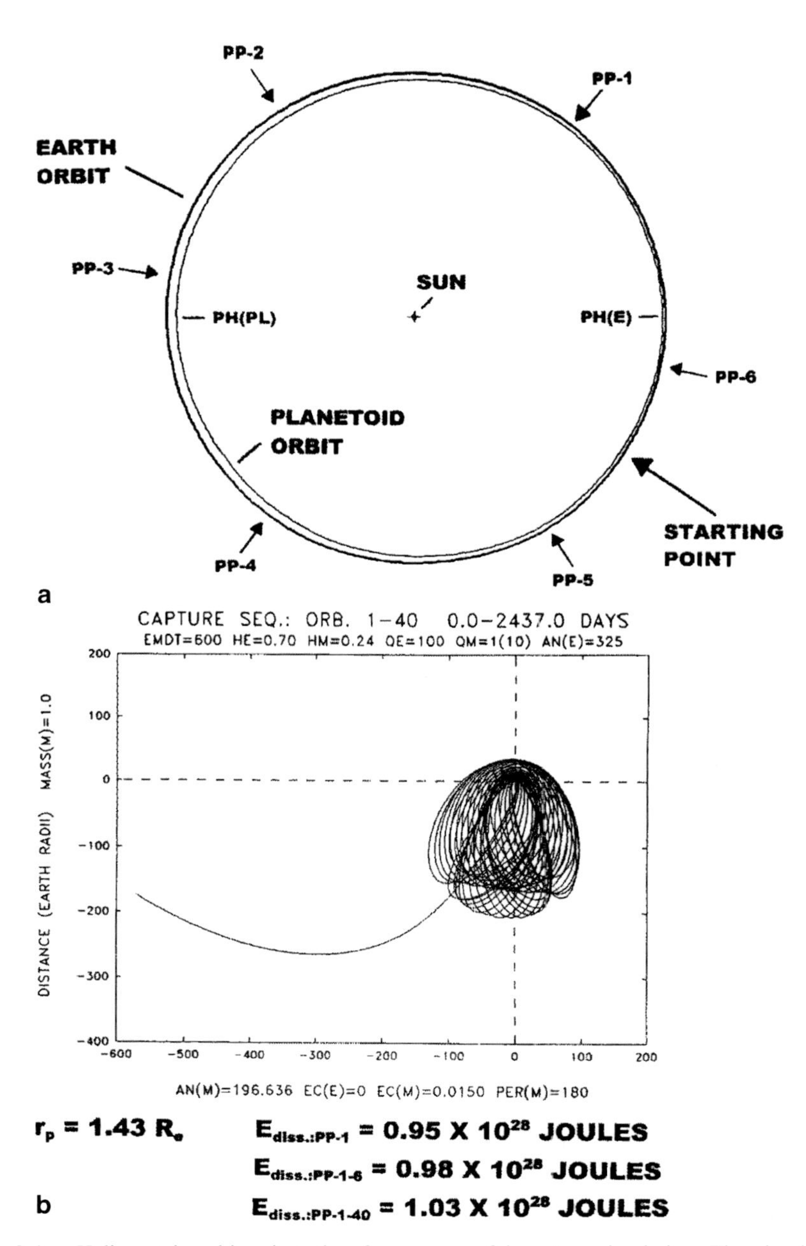

Fig. 3.4 **a** Heliocentric orbit orientation for a successful capture simulation. The simulation begins at the *STARTING POINT*. The lunar body is in an orbit that is inside the orbit or Earth. The lunar body encounters the Earth at PP-1 (perigee passage-1) after about 60 days of the simulated

however, explain the total angular momentum in the Earth-Moon system in a very straightforward way. The angular momentum of an earth-like planet with a 10 h/day prograde rotation rate plus the angular momentum of a maximum size prograde post-capture orbit (equivalent to a circular prograde orbit with a semi-major axis of 30 earth radii) equals the angular momentum of the Earth-Moon system (MacDonald 1963). So where is the angular momentum problem for this version of gravitational capture???? ANSWER: *THERE IS NO ANGULAR MOMENTUM PROBLEM FOR THIS VERSION OF A GRAVITATIONAL CAPTURE MODEL!!!*

This post-Kona version of the Intact Capture Model also offers a qualitative solution for the generation of a lunar magnetic field within the Lunar Magma Ocean (LMO) at two specific times in lunar history by way of a series of magnetic amplifier cells, a mechanism suggested by Smoluchowski (1973a, b). The general concept is that a magnetic amplifier system can be powered by lunar rock tides, caused by solar gravity, which causes differential rotation between the crustal complex and the solid part of the lunar mantle as the anorthositic crustal complex is cooling through the Curie temperature. The "seed" magnetic field for this early version of a lunar magnetic field is the solar magnetic field. The retrograde-rotating circulation cells describe a geometry of Tidal Vorticity Induction cells which are explained in Bostrom (2000).

The second episode of lunar magnetization is powered by the dynamics of gravitational capture and the subsequent orbit circularization process. Since over 90 % of the energy for capture must be stored and subsequently dissipated in the lunar body, the lunar magma ocean zone is partially remelted via the sharply defined energy dissipation of the capture process. Magnetic amplifier cells are once again powered by rock tidal action but for this era of lunar magnetization the rock tides are caused by Earth's gravity and the "seed" field is the Earth's magnetic field. The resulting lunar magnetic field is strongest soon after capture at 3.95 Ga and this magnetic field exponentially decreases in strength as the post-capture orbit circularizes due to energy dissipation in a combination of Earth and Moon. This second episode of lunar magnetic field generation is predicted to operate between 3.95 and 3.6 Ga and would be recorded in any lunar basalts and breccias that are hot enough to be cooling through the Curie temperature during the operation of the magnetic amplifier system. (Note: There will be more on this mechanism for the generation of lunar magnetic fields in Chap. 4.)

I should point out that regardless of where the Moon was formed, this second episode of lunar magnetism could be associated with the remelting of the LMO via the gravitational capture process. The first, and strongest, lunar magnetic

---

encounter. The positions of the first six perigee passages are shown on the diagram. **b** Geocentric orientation for this successful capture simulation. Forty orbits are shown but the orbit was calculated to several 100 orbits at which time the orbit begins to circularize. Note that about 1.0 E28 J of energy is sufficient for stable capture for this set of encounter parameters. Thus, the energy demands for stable capture can be decreased by about 50 % in the three-body simulation via solar gravitational assistance. However, it is possible to choose sets of encounter parameters for which the energy needs for successful capture can be higher than the two-body estimate. This leads to the concept of STABLE CAPTURE ZONES which will be discussed in the next chapter

episode is intimately associated with the remelting of the lunar magma ocean via the FU Orionis radiation events, a strong solar magnetic field as the "seed" field, and a short lived set of magnetic amplifier cells that would operate only for centuries of time. The generation of these two lunar magnetic fields are separated by about 600 Ma.

As we will see in the "report card" and in the last section of this chapter, this post-Kona version of an intact capture model does relate to a large number of the items on the "list of features to be explained by a successful model". Although Malcuit et al. (1992) did not have an explanation for the "PLACE OF ORIGIN OF LUNA" at the time of that publication, a reasonable explanation has now been developed for the formation of LUNA (the pre-capture Moon) as a VULCANOID PLANETOID between the orbit of planet Mercury and the Sun (as indicated in Chap. 2). In addition, a testable model has been developed for explanation of (a) patterns of lunar rock magnetization and (b) the factual information associated with the Cool Early Earth Model.

3.11 Formation of the Moon Resulting from a Giant Impact Early in Earth History

The Giant Impact Model for the origin of the Earth-Moon system (Fig. 3.5) is currently the most popular model for the origin of the Moon and the Earth-Moon system. But when it comes to "the list of facts to be explained by a successful model", very few of these items can be uniquely explained by the Giant Impact Model. Let us examine the three major claims by the promoters of the GIANT IMPACT MODEL in more detail.

3.11.1 The Angular Momentum Problem of the Earth-Moon System

Claim The GIM offers a solution to the traditional angular momentum problem (Boss and Peale 1986). The quantity of angular momentum associated with the Earth-Moon system is too great to be explained by the accretion process alone. Thus, this excess angular momentum must be explained by a lunar origin process and the GIANT IMPACT MODEL offers a unique solution.

Critique The problem is that nearly every rendition of the numerical simulations of the giant impact process results in a significant excess of angular momentum and *it is not clear just how that excess angular momentum can be discarded* (Canup 2008, 2013).

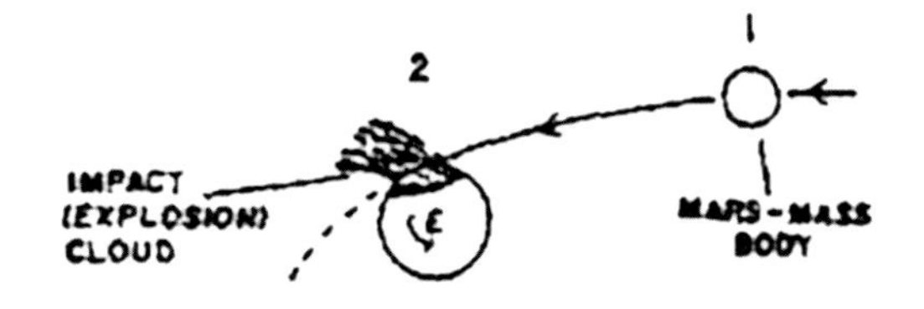

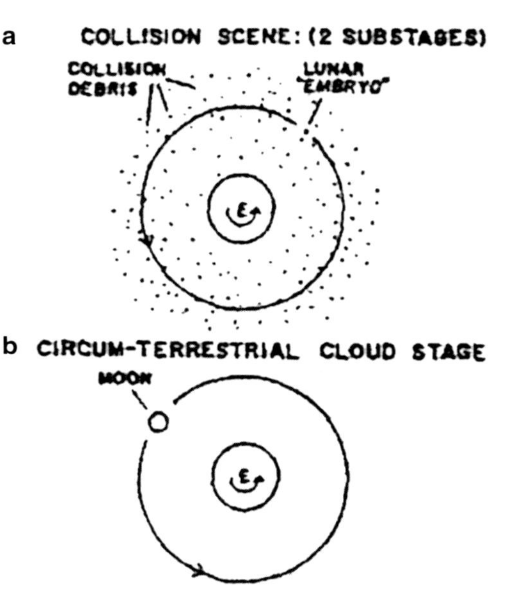

Fig. 3.5 A simplified view of the Giant-Impact model. View is from the north pole of the Solar System. There are two stages to the Collision Scene (*top* pictogram). The first part features a mars-mass body on a tangential collision course with Earth. The second part portrays the collision and the resulting debris cloud. The circum-terrestrial cloud stage (*middle* pictogram) depicts a lunar embryo beginning to accumulate debris that resulted from the collision. The Late Stage (*lower* pictogram) represents the desired result of the giant impact scenario: a lunar-mass body in a circular orbit just beyond 3 earth radii with the Earth rotating at about 5 h/day. Although there have been many attempts, this desired result of accreting a lunar-like body directly from the debris cloud of the collision has never been attained

3.11.2 The Oxygen Isotope Similarities Between Earth and Moon

Claim The oxygen isotope ratios of the Earth and Moon can be explained by the GIANT IMPACT MODEL. The oxygen isotope ratios of the Earth and Moon are very similar. This similarity of oxygen isotope ratios should indicate that the impactor, a mars-mass planetoid called Theia, was accreted in a near Earth-like orbit. But the results of the numerical simulations, both the earlier simulations (Cameron 1985; Benz et al. 1987) and the later ones (Canup 2004a, b; 2008, 2013), suggest that Theia needs a significant velocity at infinity which suggests to the investigators that Theia was formed somewhere between the orbits of Mars and Earth. Pahlaven and Stevenson (2007) and Pahlaven et al. (2011) have suggested that maybe the oxygen isotope ratios of the resulting planetoid (the Moon in this case) can be homogenized within the debris cloud resulting from the giant impact.

Critique This homogenation process within the debris cloud is not all that convincing (Halliday 2007; de Meijer et al. 2013). *Thus, the GIANT IMPACT MODEL appears to have a serious oxygen-isotope ratio problem which translates into a "place of origin of Theia" problem!*

3.11.3 The Mass and Density of the Moon

Claim The mass and density of the Moon can be explained by the GIANT IMPACT MODEL.

Critique In reality, none of the simulations of the giant impact event come anywhere close to making a body of lunar mass directly from the debris cloud of the giant impact (Canup 2008). The only way a lunar mass body results from the simulations is with the use of "special effects", for example, by starting with a debris cloud of special design. The debris for these lunar accretion simulations is not the result of the numerical simulations from a giant collision (e.g., Kokubo et al. 2000).

Thus, the state of affairs is that the promoters of the GIANT IMPACT MODEL do a spectacular job of simulating giant impacts between two bodies in prograde heliocentric orbits. But the results of the simulations *do not look very promising for explaining the origin of the Moon and the Earth-Moon system*.

In addition to the problems listed above, the promoters of the Giant-Impact Model have *not* been successful in developing (a) an explanation for the patterns of lunar rock magnetization or (b) an explanation of the factual information associated with the Cool Early Earth Model.

3.12 A Report Card for Models of Lunar Origin

The Report Card method was used by Wood (1986) to evaluate the various models of lunar origin presented at the Kona Conference on the "ORIGIN OF THE MOON". Since there have been improvements to some models since that conference, especially for the Gravitational Capture Model, another attempt at the Report Card method of evaluation is used here (see Table 3.1). GIM is the evaluation for the Giant-Impact Model. GCM (before X-Wind) is the evaluation for the Gravitational Capture Model before using the X-Wind Model for the early history of the Solar System. GCM (after X-Wind) is the evaluation for the Gravitational Capture Model when using the features of the X-Wind Model in forming the lunar body between the orbit of planet Mercury and the Sun. The reader will notice the critical importance of the X-Wind Model for the Gravitational Capture Model.

Table 3.1 An attempt to evaluate the Giant-Impact Model and the Gravitational Capture Model using the "report card" method of Wood (1986). Note that (before X-Wind) = before the introduction of the concept of the X-Wind Model to the lunar origin problem. X-Wind really stands for a combination of the X-Wind Model, the Disk Wind Model, and the FU Orionis Model because all three are involved in the formation of the lunar body and processing it into a Vulcanoid planetoid. The X-wind Model is very important for explaining the similarities between the chemistry of CAIs and the lunar body. The Disk Wind Model, which appears to be a reasonable mechanism for forming chondrules, appears to be of lesser importance for explaining lunar composition but is very important for explaining the chemistry of the terrestrial planets, especially Venus and Earth. The FU Orionis Model, however, is very important in that it is the main mechanism for melting, or remelting, the Lunar Magma Ocean so that solar gravitational tides can power a magnetic amplifier system for imprinting a magnetic fabric into the lunar anorthositic crust

Feature (or process)	GIM	GCM (before X-Wind)	GCM (after X-Wind)
Lunar mass	C	A	A
Earth-moon system angular momentum	C	A	A
Volatile element depletion in lunar body	C	D	A
Iron depletion in lunar body	C	D	A
Oxygen isotope similarities between earth and moon	C	F	A
Similarities of mantle trace element patterns	?	?	?
Magma ocean formation	C	D	A
Physical plausibility	D	C	C
Remanent magnetization patterns			
Primitive crust	F	F	A
Mare basalts	F	A	A

References

Aggarwal HR, Oberbeck VR (1974) Roche limit for a solid body. Astrophys J 191:577–588

Alfven H (1954) On the origin of the solar system. Oxford University Press, London, p 194

Alfven H, Arrhenius G (1972) Origin and evolution of the Earth-Moon system. The Moon 5:210–230

Alfven H, Arrhenius G (1976) Evolution of the solar system. NASA Special Paper-345, Washington D.C., p 599

Baldwin RB (1966) On the origin of the moon. J Geophys Res 71:1936–1937

Benz W, Slattery WL, Cameron AGW (1987) The origin of the moon and the single-impact hypothesis, II. Icarus 71:30–45

Boss AP (1986) The origin of the moon. Science 231:341–345

Boss AP, Peale SJ (1986) Dynamical constraints on the origin of the moon. In: Hartmann WK, Phillips RJ, Taylor GJ (eds) Origin of the moon. Lunar and Planetary Institute, Houston, pp 59–101

Bostrom RC (2000) Tectonic consequences of earth's rotation. Oxford University Press, London, p 266

Brush SG (1986) Early history of selenology. In: Hartmann WK, Phillips RJ, Taylor GJ (eds) Origin of the moon. Lunar and Planetary Institute, Houston, pp 3–15

Cameron AGW (1985) Formation of the prelunar accretion disk. Icarus 62:319–327

Canup RN (2004a) Dynamics of lunar formation. Annu Rev Astron Astrophys 42:441–475

Canup RN (2004b) Origin of the terrestrial planets and the Earth-Moon system. Physics today 57(4):56–62

Canup RN (2008) Lunar-forming collisions with pre-impact rotation. Icarus 196:518–538

Canup RN (2013) Lunar conspiracies. Nature 504:27–29

Clark SP, Tourcotte DL, Nordmann JC (1975) Accretional capture of the moon. Nature 258:219–220
Cline JK (1979) Satellite-aided capture. Celest Mech 19:405–415
Cloud PE (1968) Atmospheric and hydrospheric evolution on the primitive earth. Science 160:729–736
Cloud PE (1972) A working model of the primitive earth. Am J Sci 272:537–548
Conway BA (1982) On the history of the lunar orbit. Icarus 51:610–622
Darwin GH (1879) On the bodily tides of viscous and semi-elastic spheroids, and on the ocean tides upon a yielding nucleus. Phys Trans R Soc Lond 170:1–35
Darwin GH (1880) On the secular change in the elements of the orbit of a satellite revolving about a tidally distorted planet. Philos Trans R Soc Lond 171:713–891
de Meijer RJ, Anisichkin VF, van Westrenen W (2013) Forming the moon from terrestrial silicate-rich material. Chem Geol 345:40–49
Gerstenkorn H (1955) Uber gezeltenreibung beim zweikorperproblem. Z Astrophys 36:245–271
Gerstenkorn H (1967) The importance of tidal friction for the early history of the moon. R Soc Lond Ser A 296:293–297
Goldreich P (1966) History of the lunar orbit. Rev Geophys 4:411–439
Halliday AN (2007) Isotopic lunacy. Nature 450:356–357
Harris AW (1978) Satellite formation, II. Icarus 34:128–145
Harris AW, Kaula WM (1975) A co-accretion model of satellite formation. Icarus 24:516–524
Herbert F, David DR, Weidenschilling SJ (1986) Formation and evolution of a circumterrestrial disk. Constraints on the origin of the moon in geocentric orbit. In: Hartmann WK, Phillips RJ, Taylor GJ (eds) Origin of the moon. Lunar and Planetary Institute, Houston, pp 701–730
Holsapple KA, Michel P (2006) Tidal disruptions. A continuum theory for solid bodies. Icarus 183:331–348
Holsapple KA, Michel P (2008) Tidal Disruptions: II. a continuum theory for solid bodies with strength, with applications to the solar system. Icarus 193:283–301
Jeffreys H (1929) The earth, its origin, history and physical constitution, 2nd edn. Cambridge University Press, Cambridge, p 346
Jeffreys H (1947) The relation of cohesion to Roche's limit. Mon Not R Astron Soc 3:260–262
Kaula WM (1964) Tidal dissipation by solid friction and the resulting orbital evolution. Rev Geophys 2:661–685
Kaula WM, Harris AW (1973) Dynamically plausible hypotheses of lunar origin. Nature 245:367–369
Kokubo E, Canup RM, Ida S (2000) Lunar accretion from an impact-generated disk. In: Canup RM, Righter K (eds) Origin of the earth and moon. The University of Arizona Press in collaboration with Lunar and Planetary Institute, Houston, pp 145–178
MacDonald GJF (1963) The internal constitutions of the inner planets and the moon. Sp Sci Rev 2:473–557
MacDonald GJF (1964) Tidal friction. Rev of Geophys 2:467–541
Malcuit RJ, Mehringer DM, Winters RR (1989) Numerical simulation of gravitational capture of a lunar-like body by earth. Proceedings of the 19th Lunar and Planetary Science Conference. Lunar and Planetary Institute, Houston, pp 581–591
Malcuit RJ, Mehringer DM, Winters RR (1992) A gravitational capture origin for the Earth-Moon system: implications for the early history of the Earth and Moon. In: GloverJE, Ho SE (eds) Proceedings volume, 3rd international archaean symposium, vol 22. The University of Western Australia, Crawley, pp 223–235
Mignard E (1979) The evolution of the lunar orbit revisited, I. Moon Planets 20:301–315
Mignard E (1980) The evolution of the lunar orbit revisited, II. Moon Planets 23:185–201
Mignard E (1981) The lunar orbit revisited, III. Moon Planets 24:189–207
Mitler HE (1975) Formation of an iron-poor moon by partial capture, or: yet another exotic theory of lunar origin. Icarus 24:256–268
Mizuno H, Boss AP (1985) Tidal disruption of dissipative planetesimals. Icarus 63:109–133
Nakazawa K, Komuro T, Hayashi C (1983) Origin of the moon—capture by gas drag of the earth's primordial atmosphere. Moon Planet 28:311–327

Newell HE (1973) Harold Urey and the moon. The Moon 7:1–5
Pahlaven K, Stevenson DJ (2007) Equilibration in the astermath of the lunar-forming giant impact. Earth Planet Sci Lett 262:438–449
Pahlaven K, Stevenson DJ, Eiler J (2011) Chemical fractionation in the silicate vapor atmosphere of the earth. Earth Planet Sci Lett 301:433–443
Ringwood AE (1960) Some aspects of the thermal evolution of the earth. Geochim Cosmochim Acta 20:241–249
Ringwood AE (1979) Origin of the earth and moon. Springer, New York, p 295
Roy AE (1965) The foundations of astrodynamics. Macmillan, New York, p 385
Singer SF (1968) The origin of the moon and geophysical consequences. Geophys J R Astron Soc 15:205–226
Singer SF (1970) The origin of the moon and its consequences. Trans Am Geophys Union 51:637–641
Singer SF (1977) The early history of the Earth-Moon system. Earth Sci Rev 13:171–189
Singer SF (1986) Origin of the moon by capture. In: Hartmann WK, Phillips RJ, Taylor GJ (eds) Origin of the moon. Lunar and Planetary Institute, Houston, pp 471–485
Smith JV (1974) Origin of the moon by disintegrative capture with chemical differentiation followed by sequential accretion. Abstracts Volume. Lunar Science V. Lunar Science Institute, Houston, pp 718–720
Smoluchowski R (1973a) Lunar tides and magnetism. Nature 242:516–517
Smoluchowski R (1973b) Magnetism of the moon. The Moon 7:127–131
Szebehely V, Evans RT (1980) On the capture of the moon. Celest Mech 21:259–264
Urey HC (1952) The planets. Their origin and development. Yale University Press, New Haven, p 245
Urey HC (1962) Origin and history of the moon. In: Kopal Z (ed) Physics and astronomy of the moon. Academic, New York, pp 481–523
Urey HC (1972) Evidence for objects of lunar mass in the early solar system and for capture as a general process for the origin of satellites. Astrophys Sp Sci 16:311–323
Weidenschilling SJ, Greenberg R, Chapman CR, Herbert F, Davis DR, Drake MJ, Jones J, Hartmann WK (1986) Origin of the moon from a circumterrestrial disk. In: Hartmann WK, Phillips RJ, Taylor GJ (eds) Origin of the moon. Lunar and Planetary Institute, Houston, pp 731–762
Wise DU (1963) An origin of the moon by rotational fission during formation of the earth's core. J Geophys Res 68:1547–1554
Wise DU (1969) Origin of the moon from the earth. Some new mechanisms and comparisons. J Geophys Res 74:6034–6045
Wood JA (1986) Moon over Mauna Loa. A review of hypotheses of formation of earth's moon. In: Hartmann WK, Phillips RJ, Taylor GJ (eds) Origin of the moon. Lunar and Planetary Institute, Houston, pp 17–55
Wood JA, Mitler HE (1974) Origin of the moon by a modified capture mechanism, or: half a loaf is better than a whole one. Abstract Volume. Lunar Science V. Lunar Science Institute, Houston, pp 851–853

Chapter 4
A Prograde Gravitational Capture Model for the Origin and Evolution of the Earth-Moon System

Of the other alternatives, it is perhaps just possible that the moon was originally an independent planet, though it is much less massive than any existing planet.

Jeffreys (1929), p. 37

....capture of the entire moon is an inherently improbable event because of the narrow range of orbital elements for which the relatively slow-working tidal friction could dissipate sufficient energy to prevent escape.

Kaula (1971), p. 224

The basic geochemical model of the structure of the Moon proposed by Anderson, in which the Moon is formed by differentiation of the calcium, aluminum, titanium-rich inclusions in the Allende meteorite, is accepted, and the conditions for formation of this Moon within the solar nebula models of Cameron and Pine are discussed. The basic material condenses while iron remains in the gaseous phase, which places the formation of the Moon slightly inside the orbit of Mercury.

From Cameron 1973, p. 377

The origin of the Moon is one of the outstanding unsolved problems in the natural sciences. Cursory examination of college-level textbooks in the Earth and Planetary Sciences leaves one with the impression that the Moon is simply a "night lantern" and that the Moon effects the Earth in only minor ways, such as controlling the tidal waters on the planet.

In a later chapter there may be some time devoted to the seasonal cycles on Earth and the phases of the Moon. Typically there is a statement that a combination of the Sun and Moon maintain the tilt angle of the planet within a certain range of values. But it is the unusual textbook that has information on the *ROTATION RATE* of the Earth in past eras and that the rotation rate of the planet is *DECREASING* at a rate of about 1 ms/century. This may seem like a small rate of change, but over geologic time it has major implications for the *LENGTH OF THE DAY* and the *HEIGHT OF THE TIDES* over geologic time (see Tables 4.1 and 4.2).

We might wonder if this change in DAY LENGTH and TIDAL RANGE has had any significant effect on the progress of organic evolution on our planet? A shorter

Robert J. Malcuit, *The Twin Sister Planets Venus and Earth*,
DOI 10.1007/978-3-319-11388-3_4

Table 4.1 Orbital history of the Earth-Moon system over the past 4.0 Ga with change of 1 ms/century in rotation rate

Time (B.P)	Hours/Day	Rate of change	E-M Distance
Present	24	1 ms/century	60.00 earth radii
5 Ma (Early hominids)	23.976	"	59.98 earth radii
100 Ma (Dinosaur time)	23.7	"	59.85 earth radii
500 Ma (Metazoan time)	22.6	"	58.61 earth radii
1500 Ma (Eucaryote tme)	19.8	"	55.10 earth radii
3400 Ma (Eobacteria time)	14.3	"	44.60 earth radii

Table 4.2 Tidal parameters for this orbital history of the Earth-Moon system. The "Probable Maximum Tidal Range" is based on the Bay of Fundy tides which have a tidal amplification factor of about 20 relative to the tidal range in the open ocean

Time (B.P.)	Equilibrium spring tidal range (m)	Probable maximum tidal range (m)
Present	0.76	15
5 Ma (Early hominids)	0.76	15
100 Ma (Dinosaur time)	0.77	16
500 Ma (Metazoan time)	0.80	20
1500 Ma (Eucaryote time)	0.92	22
3400 Ma (Eobacteria time)	1.54	30

day would tend to moderate the daily variation of temperatures at a given latitude so that the simple organisms would not need complex internal machinery to cope with temperature extremes. A faster rotation rate would also cause weather system gyres to rotate more rapidly and this would tend to cause more storminess than we have on Earth today.

One can ask whether the tidal regime of Earth would be different for a capture origin than for any of the other origin models such as a Giant Impact origin? The answer is "yes" *but only in the first billion years or so.* For my brand of capture there is no Moon until 3.95 Ga. Therefore there would be no lunar component for rock or ocean tides. Thus, the tidal conditions on Earth would be very placid. Only solar gravitational tides would affect the oceans and the lithosphere, but the rotation rate would be about 10 h/day. A short day, very little organized tidal activity, a significant component of carbon dioxide and nitrogen in the atmosphere, and about 20 % less solar radiation from the Sun (after the Sun settles to the main sequence of development) could be conducive to the development of primitive forms of life in this early era: i.e., the era of the COOL EARLY EARTH.

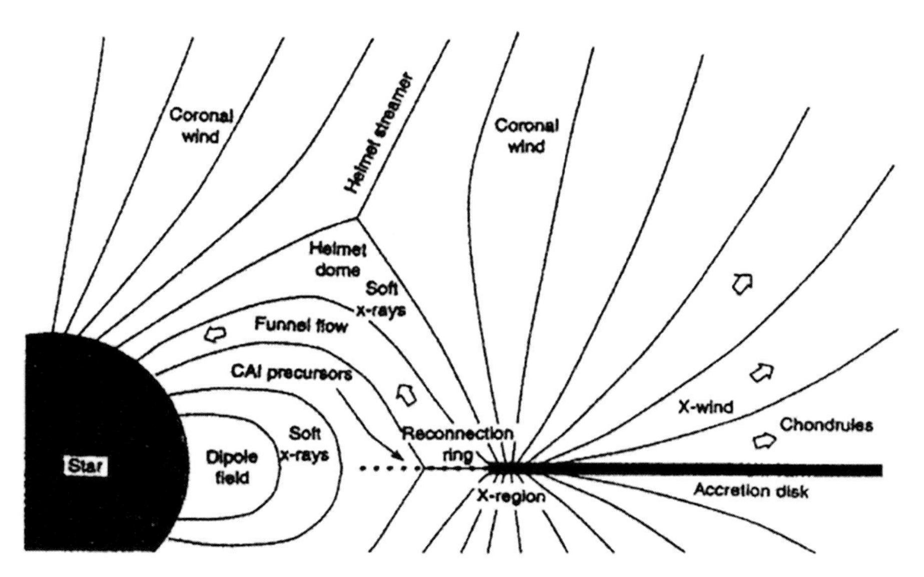

Fig. 4.1 Diagram from Shu et al. (2001, Fig. 1, with permission from Frank Shu) showing place of origin for CAIs. In this model the CAI material can be formed only within the reconnection ring which is oscillating over a radial distance centered on 0.06 AU. Some *CAIs* can be hurled to the outer parts of the solar system (asteroid zone or beyond) via magnetic field activity of the X-Wind. I should note here that the Disk-Wind model of Salmeron and Ireland (2012) appears to be more conducive to the formation of chondrules than is the X-Wind model. The two models have different energy regimes and could overlap a bit in time because there may be a limited time gap between the formation of CAIs and chondrules. (Connelly et al. 2012)

4.1 Place of Origin for Luna and Sibling Planetoids and a Model for Magnetization of the Crust of Luna and Sibling Vulcanoid Planetoids

Since this book is about the similarities and differences of planets Venus and Earth and the general idea is that those differences are somehow associated with planetoid capture, it is imperative to identify a specific place of origin for Luna, Adonis (a name for a pre-capture planetoid for Venus that will be explained in Chap. 6), and possibly other sibling planetoids. I am following the lead of A. G. W. Cameron who ironically was the chief architect of the Giant Impact model. The quote at the beginning of this chapter is from an earlier era and from a time when the study of the processes of star formation was in its infancy (and stellar evolution was one of Cameron's fields of expertise). Nonetheless, I think there is wisdom in his thoughts on where a body of lunar composition might be formed (i.e., INSIDE THE ORBIT OF MERCURY).

The X-Wind model of Shu et al. (1997) and Shu et al. (2001) for the early evolution of the Sun and the processing of the proto-solar cloud is generally accepted for the origin of CAIs (Taylor 2001; Wood 2004; MacPherson et al. 2005; Salmeron and Ireland 2012). Figure 4.1 is a diagram from Shu et al. (2001) which shows the gener-

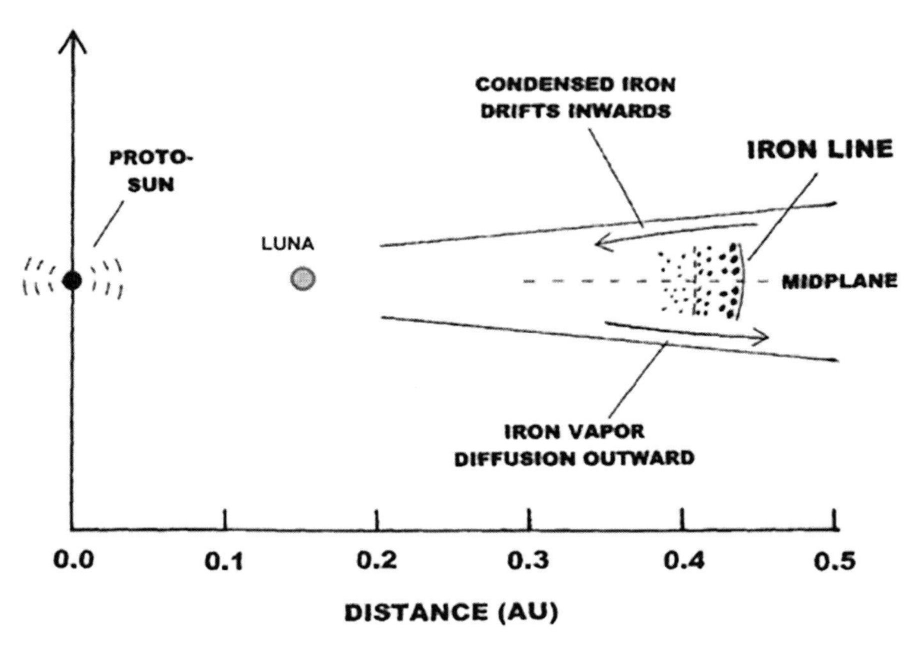

Fig. 4.2 Cross-sectional sketch illustrating how the Iron Line could develop during the devolatization stage of solar system development. Iron and Nickel vapor diffuse outward. As the metal-bearing vapor cools and solidifies, the condensed iron drifts inward. Some of the solidified metallic droplets accrete to form small planetismals and these, in turn, accrete to eventually form planet Mercury with a high specific gravity. The favored place of origin of Luna is indicated on the diagram. This place of origin is consistent with the low Fe content and the high temperature (refractory element) chemistry of the Earth's Moon

al location of formation of CAIs in this very dynamic environment in the proto-solar cloud. Figure 4.2 shows my concept of the development of the IRON LINE which leads to the formation of Mercury (see also Fig. 2.6 for Iron Line and Frost Line geometry). In this model the siderophile elements that are in the Vulcanoid Zone are vaporized and moved outward. Thus there is no need to separate siderophiles from lithophiles to form a planetary core to sequester the siderophiles as suggested by Day et al. (2007). The siderophiles are simply moved to outer (cooler) zones where they crystallize. In this model the metal-poor Vulcanoid planetoids of CAI composition are formed in a zone about mid-way between planet Mercury and the Sun (i.e., well interior to the accretion torus of Mercury). In the Shu et al. (2001) model, CAIs are formed only in the reconnection ring, a toroidal shaped region that shifts within a certain range of values centered on 0.06 AU. Then the CAI material is moved outward by some combination of magnetic, radiational, and mechanical processes. Evans and Tabachnick (1999, 2002) have shown that orbits between 0.1 and 0.2 AU can be stable relative to planetary perturbations for 100s of millions of years.

Soon after formation these Vulcanoid planetoids of CAI composition are affected by the MICROWAVE wind (the FU Orionis sequence of radiation events), Luna is melted to a depth of about 600 km (Taylor 2001) and the other Vulcanoid planetoids are partially or completely melted depending on their specific locations. This is a possible explanation for the formation of the lunar magma ocean. Thus, the magma ocean forms by electromagnetic induction "joule" heating (Sonett et al. 1975) with perhaps some help from the thermal impulse of accretion and the rapid radioactive decay of Al-26 and Fe-60. See Fig. 4.3 for more details.

Does the Moon have a Metallic Core?

In this model for the history of Luna there is no need for a metallic core. There really is no substantial seismic evidence for it, the geochemical evidence is ambivalent, and the coefficient of the moment of inertia is very near 0.40, a value for a homogeneous sphere. Several investigators have stated that a small metallic or metallic sulfide core is permissible. Such a core would imply whole body differentiation but it would be difficult to maintain a coefficient of the moment of inertia near 0.4 with complete body differentiation. Even if Luna had a small metallic or metal-sulfide core, it would be useless for generating a lunar magnetic field (Levy 1972, 1974; Sharp et al. 1973).

> The HED meteorites have O-isotope compositions similar to those of the IIIAB irons, main-group pallasites, mesosiderites, angrites and brachinites and thus may have been formed in the same region of the solar nebula. From Mittlefehldt et al. (1998, pp. 4–104)

> We are presented with some fascinating questions about lunar history: Did the ancient moon have an earthlike field, generated as the earth's is thought to be by an internal dynamo? Was the moon once in an orbit closer to the earth and within the strong terrestrial field? Was the moon magnetized in another part of the solar system and later captured by the earth? From Dyal and Parkin (1973, p. 63)

I think that the last question as well as the preceding one, are appropriate here. I have developed a qualitative model for magnetizing the primitive crust of Luna by way of a shallow shell magnetic amplifier (Smoluchowski 1973a, b) powered by solar gravitational tides operating in a lunar magma ocean on a rotating planetoid (see Figs. 4.4 and 4.5). The systematic tidal action causes formation and rotation of tidal vorticity induction cells (Bostrom 2000) which amplify a solar magnetic "seed" field sufficiently to cause the remanent magnetization of the lunar anorthositic crust (Figs. 4.4, 4.5 and 4.6). The magnetic amplifier cells within the lunar magma ocean cease to operate when the viscosity of the lunar magma ocean reaches a critical value, probably within a few thousand years (more or less) from the time of the peak microwave heating. [Note: The generation of an internal lunar magnetic field at the time of capture relates to the next to the last question in the quote of Dyal and Parkin (1973); this will be discussed later in this chapter].

The sibling Vulcanoid planetoids, in this model, are treated in a similar manner and the prediction is that they could have recorded in their anorthositic crusts either (a) a weak solar magnetic field or (b) a stronger field caused by amplifier action within their respective magma ocean regions with the strong magnetic field of the Sun as the "seed" field.

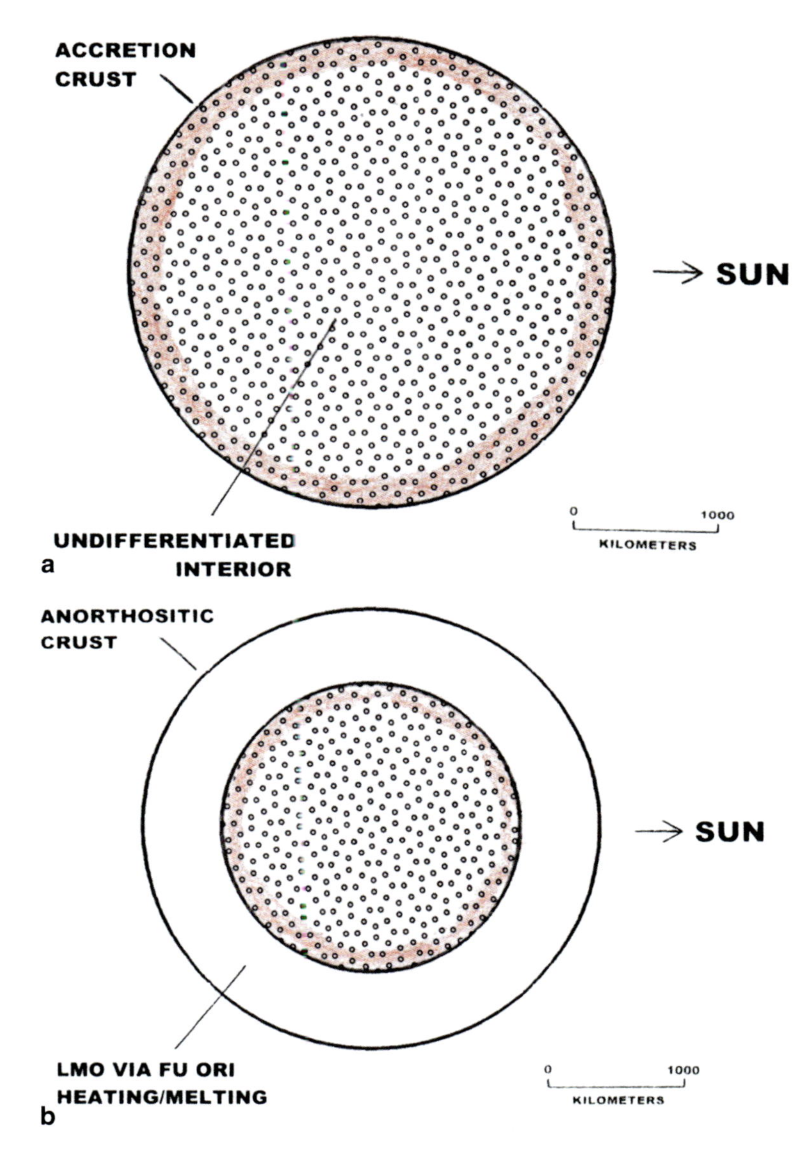

Fig. 4.3 Precursor events for the generation of the primitive lunar magnetic field. **a** Luna before the microwaving event had an original crust resulting from the accretion process. The nature of that crust and the extent of internal differentiation is not known and is not all that important to this model. **b** After the microwaving event Luna has a zone of molten lunar basaltic magma (probably superheated by the FU Ori episode) extending to a depth of ~600 km. A chill crust forms as the lunar magma ocean cools after each microwaving event. As shown in Fig. 2.11, the FU Ori episode consists of several sharply defined joule heating events. The lunar crust accumulates over a period of time even as more radiant energy is added to the magma ocean region. The magnetic

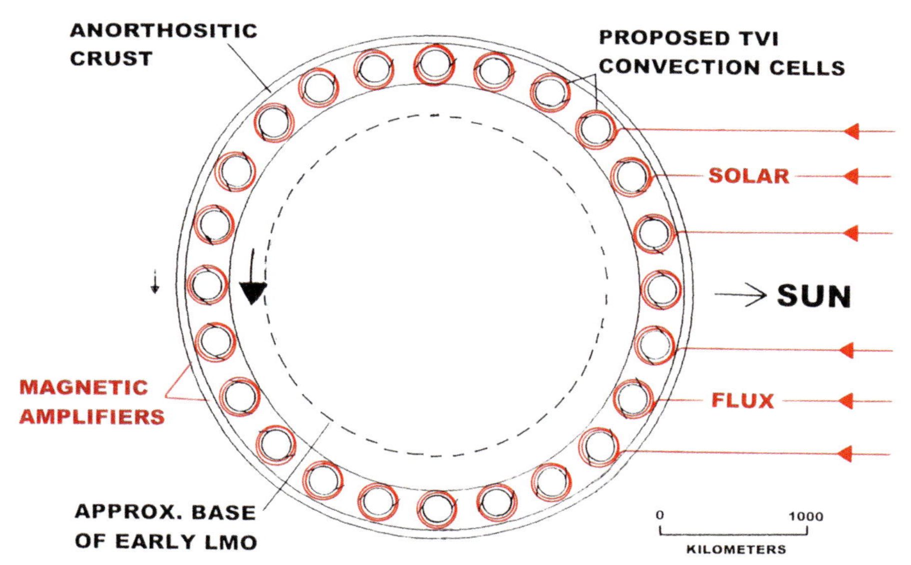

Fig. 4.4 A model for the operation of a shallow-shell magnetic amplifier in the magma ocean region of Luna. The time is soon after remelting of the lunar magma ocean and when the amplifier is operating at maximum strength. View is from the north pole of the solar system. The large *arrow* indicates direction of rotation of the mantle of Luna. The small *arrow* indicates the retardation of rotation of the crustal complex due to solar rock tidal action. Luna is at 0.15 AU and the microwave "rotisserie" has just been turned off. The solar rock tidal amplitude when h of Luna is 0.5 is about 11 m. The rotation rate of Luna is ~20 h/day prograde. If the orbit of Luna has an eccentricity, then the rock tides would be even higher at perihelion; such an extra tidal component would be beneficial in the magnetic amplifier operation. The "seed" field to be magnified is the solar magnetic field which is "strong" at 0.15 AU. The strength of the resulting amplified field of Luna depends on (1) the viscosity of the magma, (2) the angular velocity of the TVI cells, (3) the magnetic susceptibility of the magma, and (4) the strength and reversal characteristics of the "seed" field. The ferroan anorthosite crust records whatever magnetic field is operating at the time that magnetically recording minerals cool through the Curie temperature. The resulting pattern of remanent magnetization is quite complex because the TVI cells are rotating in the retrograde direction but migrating in the prograde direction. The predictable pattern of positive and negative directions of magnetization appears to be registered on the lunar surface magnetization patterns recorded during orbital magnetometer surveys of the Moon (e.g., see Fig. 4.6)

---

amplifier system operates throughout the FU Ori events and the magnetic signature is recorded soon after the last major heating event of the FU Orionis era and as the metal-bearing magnetic recorders cool through the Curie temperature. Note: This model for melting of the outer 600 km of the lunar body is consistent with the geophysical model for the development of the zonation of the Moon proposed by Wood et al. (1970). In their discussion of possible heat sources for causing this melting, they proposed that there was a source in addition to the heat generated by radioactive element decay. This FU Ori sequence of joule heating events may be a solution to their additional heat source

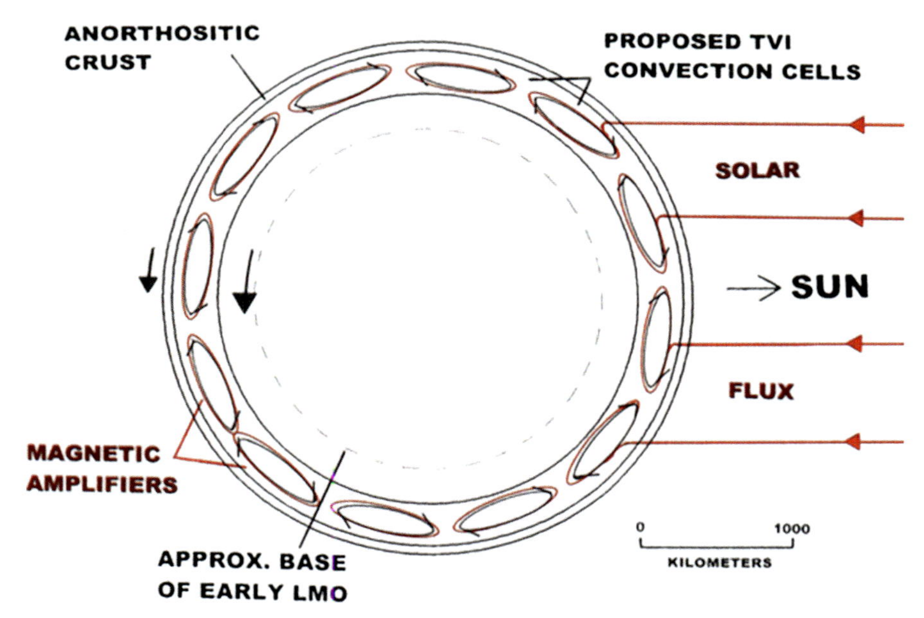

Fig. 4.5 Primitive lunar magnetic amplifier field near the end of the cycle. As Luna cools and the TVI cells diminish in diameter and angular velocity, the brief era of lunar crustal magnetization comes to an end. The main controlling factors near the end of the operation are (1) the monotonically increasing viscosity of the material in the magma ocean as the temperature decreases, (2) the monotonically decreasing thickness of the magma ocean as crystallization occurs on the border zones of the magma ocean, and (3) the monotonically decreasing displacement Love number of the lunar body which controls the vigor of the TVI cells and, in turn, is controlled by the viscosity of the magma. Eventually, the magnetic amplifier ceases to function and the only magnetic field operating is the solar "seed" field

4.2 Migration History of Luna and Sibling Vulcanoid Planetoids

> On the whole, the Moon appears to be more chemically similar to the HED parent body (probably the asteroid Vesta) formed by (1) an impact mechanism and is an escaped satellite, or (2) the Moon is a captured body that formed independently of Earth. Ruzicka et al. (1999, p. 121)

After formation and magma ocean processing and remanent magnetization imprinting, Luna and sibling Vulcanoid planetoids begin their perilous journey away from their zone of formation. The main orbital perturbers are the Sun, planet Mercury, sibling planetoids, and to some degree, planets Venus and Earth. Some of the planetoids are perturbed into highly elliptical orbits and consumed by the Sun. Others are perturbed into orbits that are influenced by Mercury and gain orbital energy so that the major axes of their orbits become larger. Eventually the survivors are perturbed so that they interact gravitationally with Venus, Earth, and Mars. In my

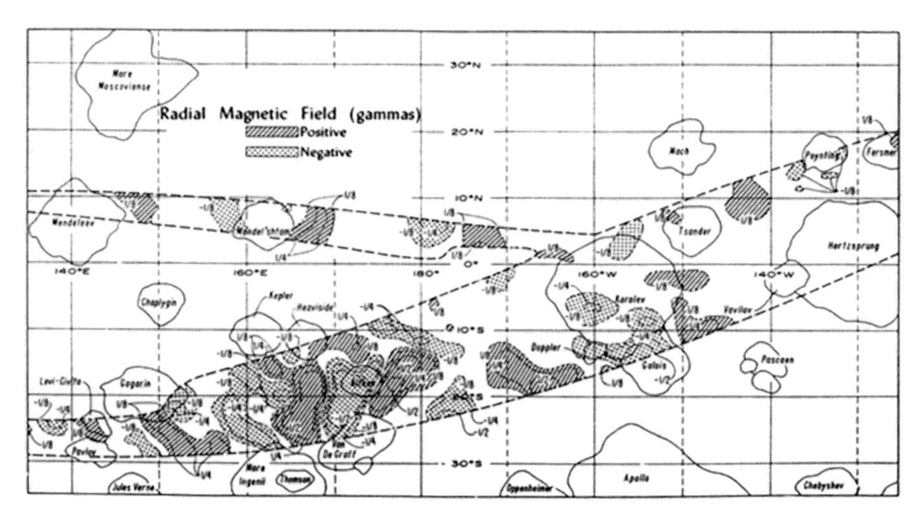

Fig. 4.6 Radial component of the lunar magnetic field mapped by the Apollo 15 and 16 subsatellites from 135° East to 125° West at altitudes between 65 and 100 km. (From Russell 1980, Fig. 25, with permission from the American Geophysical Union). Note the alternating positive and negative remanent magnetic field patterns on the lunar backside crust. This pattern may relate to the TVI cell magnetic amplifiers shown in Figs. 4.4 and 4.5. The "*UP*" direction of the amplifiers is the positive direction and the "*DOWN*" direction of the amplifiers is the negative direction

story Luna is captured into a prograde orbit by Earth (Fig. 4.7a), Adonis is captured into a retrograde orbit by Venus (Fig. 4.7b), Vesta ends up in the Asteroid Belt after non-collisional gravitational interactions with Venus, Earth, and Mars (Fig. 4.7c). Angra (the parent body of the Angrite Family of meteorites which has never been located) follows a similar perturbation history as Vesta and probably ends up in the Asteroid Belt or is so fragmented that the parent body loses its identity (Fig. 4.7d). A prediction is that there may be several other collisional fragments of Vulcanoid planetoids in the Asteroid Belt. These could be identified by way of their CAI-like chemistry. A general prediction of this model is that all "volcanic" asteroids were formed as Vulcanoid planetoids.

Identification of Other Possible Vulcanoid Planetoids or Remnants Thereof

The general premise of this inset is that all V-type (volcanic-type) asteroids originated as Vulcanoid planetoids. A common characteristic of all V-type asteroids is that they were partly or completely melted and probably went through an igneous differentiation process soon after formation (Greenwood et al. 2005). The extent of mineral separation within the body depends on the size of the body and the extent of melting. Another characteristic of V-type

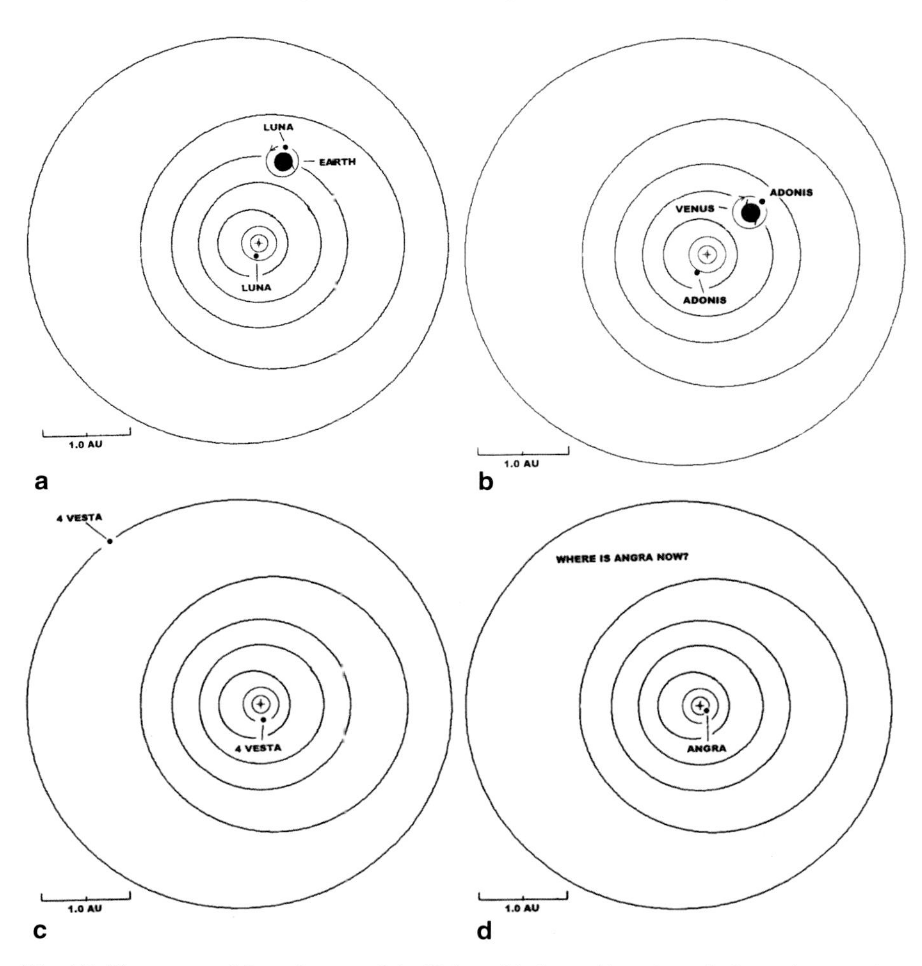

Fig. 4.7 The proposed fate of some of the Vulcanoid planetoids. **a** Luna is formed as a Vulcanoid planetoid and eventually, after about 600 Ma, is gravitationally captured into a *prograde* orbit by Earth; **b** Adonis is formed as a Vulcanoid planetoid and eventually, after about 300 Ma, is gravitationally captured into a *retrograde* orbit by Venus; **c** Vesta is formed as a Vulcanoid planetoid and eventually is gravitationally perturbed by a combination of Mercury, Venus, Earth, and Mars into an asteroid orbit; **d** Angra is formed as a Vulcanoid planetoid and eventually is perturbed by a combination of the terrestrial planets into an asteroid orbit. The angrite meteorites are from the parent body, Angra, but Angra has not yet been located. (View is from the north pole of the Solar System.)

asteroids is that they are much different from typical asteroids like Ceres or Pallas. The typical asteroid does not show much, if any, evidence for whole-body melting or igneous differentiation. If melt is involved, it usually can be associated with an impact process.

Duffard and Roig (2009) identified what they think are two new candidate V-type asteroids. They are (7472) Kumakiri and (10537) 1991 RY 16 which are both located in the outer Main Asteroid Belt with major axes greater than 2.85 AU. Duffard and Roig (2009) claim that these two V-type asteroids plus (1459) Magnya are the only V-type bodies so far identified in the outer Main Belt.

Roig et al. (2008) identified two V-type asteroid candidates from the Middle Asteroid Belt. These are (21238) 1995 WV7 with SMA ~2.54 AU and (40521) 1999 RL95 with SMA ~2.53. Several other possible V-type asteroids have been identified by the Sloan Digital Sky Survey (Roig et al. 2008). These include three positive identifications of V-type asteroids in Mars-crossing orbits (Ribeiro et al. 2014).

Bland et al. (2009) report a recent fall from the Nullarbor Desert (Australia) of a V-type meteorite that has oxygen isotope and orbital properties that suggest a parent body different from 4 Vesta. Their modeling suggests that the meteorite came from a parent body in the innermost Main Belt. The meteorite was assigned the name of Bunburra Rockhole (after a local landscale feature). On a $\Delta^{18}O$ vs. $\Delta^{17}O$ plot this meteorite is located below the Angrite fractionation line and well above the Eucrite fractionation line. Bland et al. (2009) suggested that reconstruction of the pre-impact orbit of the meteorite places the perihelion slightly inside the orbit of Venus and the aphelion slightly outside the orbit of Earth. They suggest an SMA of 0.8514 AU.

Scott et al. (2009) found 5 meteorites out of a group of 18 eucrites and 4 diogenites that have anomalous $\Delta^{17}O$ values. These five outliers are thought by Scott et al. (2009) to be from parent bodies other than 4 Vesta. The mineralogy and chemistry of these volcanic asteroids must also be compared and contrasted with suspected lunar achondrite meteorites. The first achondrite meteorite to be recognized as coming from the Moon was Yamato 791197, a feldspathic regolith breccia that probably came from the lunar highlands because of its similarity to some of the Apollo 16 samples. However, detailed chemical trace element analysis is necessary to differentiate lunar achondites from a mare area from other basaltic achondites (McSween 1999).

The place or origin of metallic meteorites is also a problem in meteoritics. Campbell and Humayun (2005) suggest that many differentiated bodies were involved and that they were later fragmented and subsequently inserted into orbits in the Asteroid belt. Campbell and Humayun (2005) also suggest that the parent melt of IVB meteorites appears to be very similar to the Angrite parent body in that the parent melt was depleted in volatiles and enriched in most refractory elements. Wasson (2013) suggests that there are isotopic similarities between IIIAB iron meteorites and the HED group of meteorites and that the parent body of the IIIAB irons may be the source of some HED meteorites. Goldstein et al. (2009) state that Hf-W systematics suggest that the parent bodies of most iron meteorites formed within about 1 million years after CAI formation and before the accretion of chondrites.

The origin of metallic meteorites can be explained in terms of the Vulcanoid planetoid model. The planetoids that accrete to form the particles crystallizing near the iron line (see Figs. 2.7 and 4.2), on the fringes of the Mercury accretion torus, are enriched in Fe and Ni. Even if the parent bodies are not totally melted by the FU Ori thermal events, sufficient masses of the planetoids are melted to form pockets of molten metal within the silicate material.

These metal-enriched Vulcanoid planetoids then undergo a similar orbital perturbation and collision history as their more refractory sibling Vulcanoid planetoids and some of the metal-enriched bodies end up in the classical asteroid zone beyond the orbit of Mars. After some additional orbital and collision history, some of the bodies move sunward and end up in Mars-crossing orbits and eventually in Earth-crossing orbits.

In summary, I think that all of the volcanic asteroids as well as most, in not all, of the metallic meteorites, as well as Luna and Adonis, had a common origin as Vulcanoid planetoids.

My guess is that we must start out with several Vulcanoid planetoids in order to have at least three survivors. Perhaps the odds are one in six. That is, for every Vulcanoid planetoid that reaches its destination successfully, perhaps five would end up in an eventual collisional scenario with Mercury, Venus, Earth, Mars, or with some other body. This, in my view, would be a hazardous journey and at times a "rough ride".

The implications of this migration are important for Solar System history. First of all, the geological history of the sister planets, Earth and Venus, are impacted severely by the purported capture of Vulcanoid planetoids and the subsequent tidal evolution of the systems (the theme of this book). Second of all, 4 Vesta, the second most massive asteroid and a body that is currently undergoing intense investigation by way of the Dawn Mission, appears to be a major "misfit" for an asteroid. This basaltic asteroid (V-type asteroid) has a similar composition as the Earth's Moon (Ruzicka et al. 1999, 2001), and thus "fits the bill" as a sibling of the Earth's Moon. The mysterious Angrite meteorites, along with their temporarily or permanently lost parent body and all other V-type asteroids need to be explained. These possibly related planetological mysteries associated with the basaltic asteroids, as well as their magnetic signatures, may be the result of their Vulcanoid planetoid origin.

> It has long been recognized that basalts from both the Moon (mare basalts) and the HED parent body (eucrites) have similar abundances of many siderophile and volatile-lithophile elements, implying an overall chemical similarity between the mantles of these two objects.
>
> Ruzicka et al. (2001, p. 980)

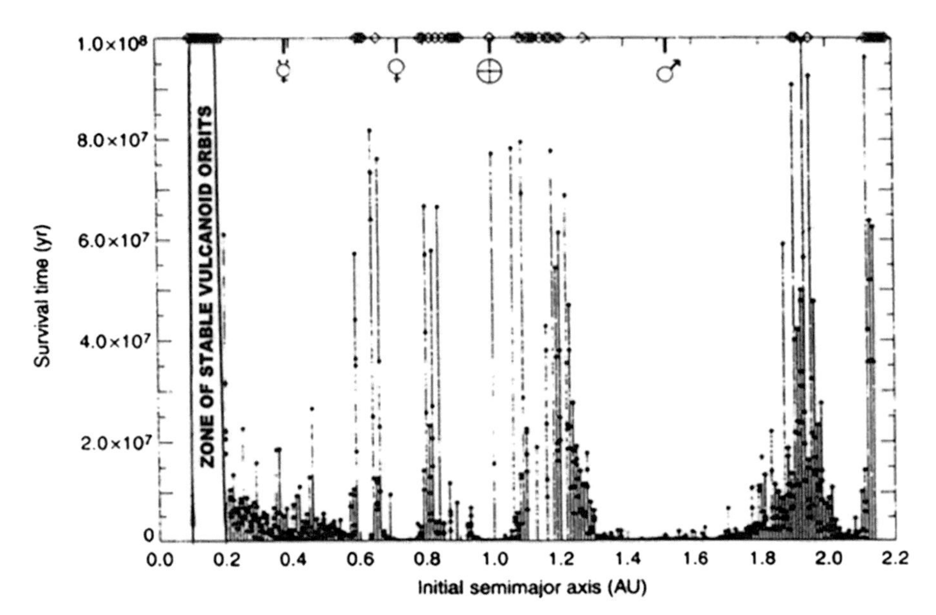

Fig. 4.8 Results of extensive calculations by Evans and Tabachnik (1999) illustrating the stability of orbits of planetoids in the inner Solar System. The positions of the terrestrial planets are shown by symbols at the top of the diagram. *Open diamond symbols* at *top* of diagram denote orbits of particles that are stable for over 100 Ma. The stable orbits between 0.1 and 0.2 were too numerous to be plotted as lines. (Diagram adapted from Evans and Tabachnik 1999, Fig. 1, with permission from Nature Publishing Group)

4.2.1 Stability of Vulcanoid Planetoid Orbits

Figure 4.8 shows the results of numerical simulations by Evans and Tabachnik (1999). They were the first to demonstrate that orbits in the Vulcanoid Zone (the zone between the Sun and planet Mercury) can be stable for 100 Ma or longer. The orbits of Vulcanoids nearest to the Sun (0.1–0.2 AU) tend to be stable for longer periods than those near the orbit of Mercury.

A quote from Evans and Tabachnik (1999, p. 42):

> Even after 100 Myr, some 80% of our Vulcanoid orbits still have eccentricities < 0.2 and inclinations $< 10°$. It is this evidence, together with the low rate of attrition of their numbers, that suggests that they can continue for times of the order of the age of the Solar System.

Since these orbits are so stable, the task is to show how these Vulcanoids can become perturbed (or destabilized) so the Vulcanoid planetoids (Luna and siblings) can gradually be moved from their place of origin to the vicinity of the terrestrial planets and the asteroid zone. The only obvious candidate as a primary planetary perturber is planet Mercury. How can Mercury accomplish this task???

Calculations by Laskar (1994, 1995) suggest that planet Mercury undergoes significant fluctuations in eccentricity (0.1–0.5) and inclination over a timescale of billions of years (Fig. 4.9). Fluctuations in eccentricity in the range of 0.2–0.3 occur over a much shorter timescale.

4.2.2 Transfer of Vulcanoid Planetoids from Orbits of Origin to Venus-Earth Space

My scheme for transfer of Luna from its purported place of origin to a near Earth-like orbit is shown in Figs. 4.10 to 4.16. This is merely a rough outline of the orbital evolution. An enterprising physics/computer science student could do the calculation in segments. I am sure that the person that does such a calculation will find an interesting orbital history. Figure 4.10 shows various orbital eccentricity states for planet Mercury. The present eccentricity of Mercury is 0.206. My speculation is that an eccentricity in the range of 0.3–0.5 is needed to extract Luna from its "birth" orbit and send it on a "bumpy" path to a near Earth-like orbit.

Figure 4.11 shows four diagrams illustrating an orbital scenario that would move Luna from an orbit near it place of origin to the Mercury Eccentricity Belt (MEB) so that Luna can be orbitally perturbed by planet Mercury. (Here again, a numerical simulation of this orbital evolution would shed light on the details.) After numerous encounters with Mercury, the orbit of Luna becomes somewhat similar to that of Mercury.

Figure 4.12 illustrates a possible scheme of transfer of Luna from the Mercury Eccentricity Belt to the orbit of planet Venus. According to the calculations of Laskar (1994), the eccentricity of the orbits of both Venus and Earth have a very small range of variation over geologic time (see Fig. 4.9). Encounters of Luna with Venus would tend to cause the orbit of Luna to become very similar to the orbit of Venus. Figure 4.13 is an illustration of the orbital conditions of Luna which could result in Gravitational Capture of a Vulcanoid planetoid by Venus if the deformation parameters of the encountering body are within the limits for dissipating the energy for capture. In Chap. 7 I will discuss the gravitational capture potential of Venus and the Venus-Adonis "affair".

Figures 4.14 and 4.15 illustrate four possible stages in the transfer of Luna from Venus orbit to Earth orbit. Figure 4.16 depicts an orbit from which gravitational capture is possible for a Luna-sized body by Earth if the deformation properties of the encountering body are within certain limits.

4.2.3 Summary for the Transfer Scheme

Luna forms with sibling Vulcanoid planetoids inside the orbit of Mercury between 0.1 and 0.2 AU. The orbits of the planetoids are destabilized by gravitational interaction with Mercury. Surviving planetoids are then transferred from near the orbit of Mercury to near the orbit of Venus. Then members of the next group of survivors (including Luna) are transferred to the vicinity of Earth's orbit

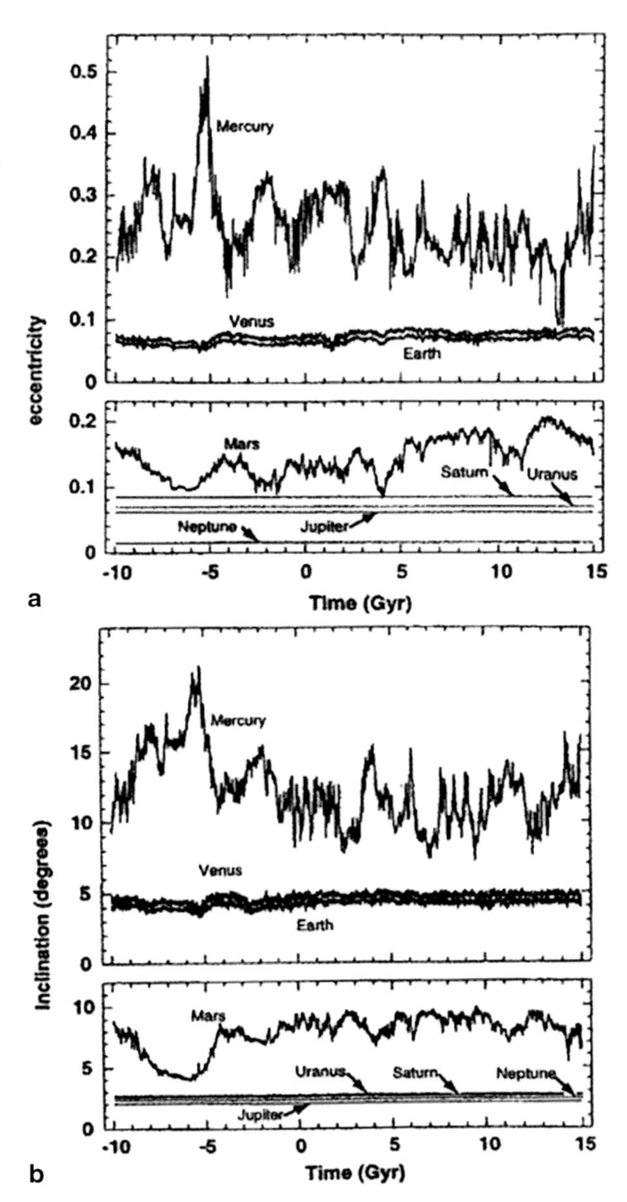

Fig. 4.9 Results of numerical integrations from Laskar (1994, Fig. 1, with permission from Springer) of the averaged equations of motion for various Solar System bodies for 10 Ga backward and 15 Ga forward in geologic time. **a** Plot of orbital eccentricity vs. geologic time; **b** plot of orbital inclination vs. geologic time

to an orbital configuration from which Luna can be gravitationally captured by planet Earth.

Now let us continue with this narrative of a "long chain of complications" that may have been involved in developing the environmental conditions that we enjoy on the third planet from the Sun (Alfven 1969; Alfven and Alfven 1972).

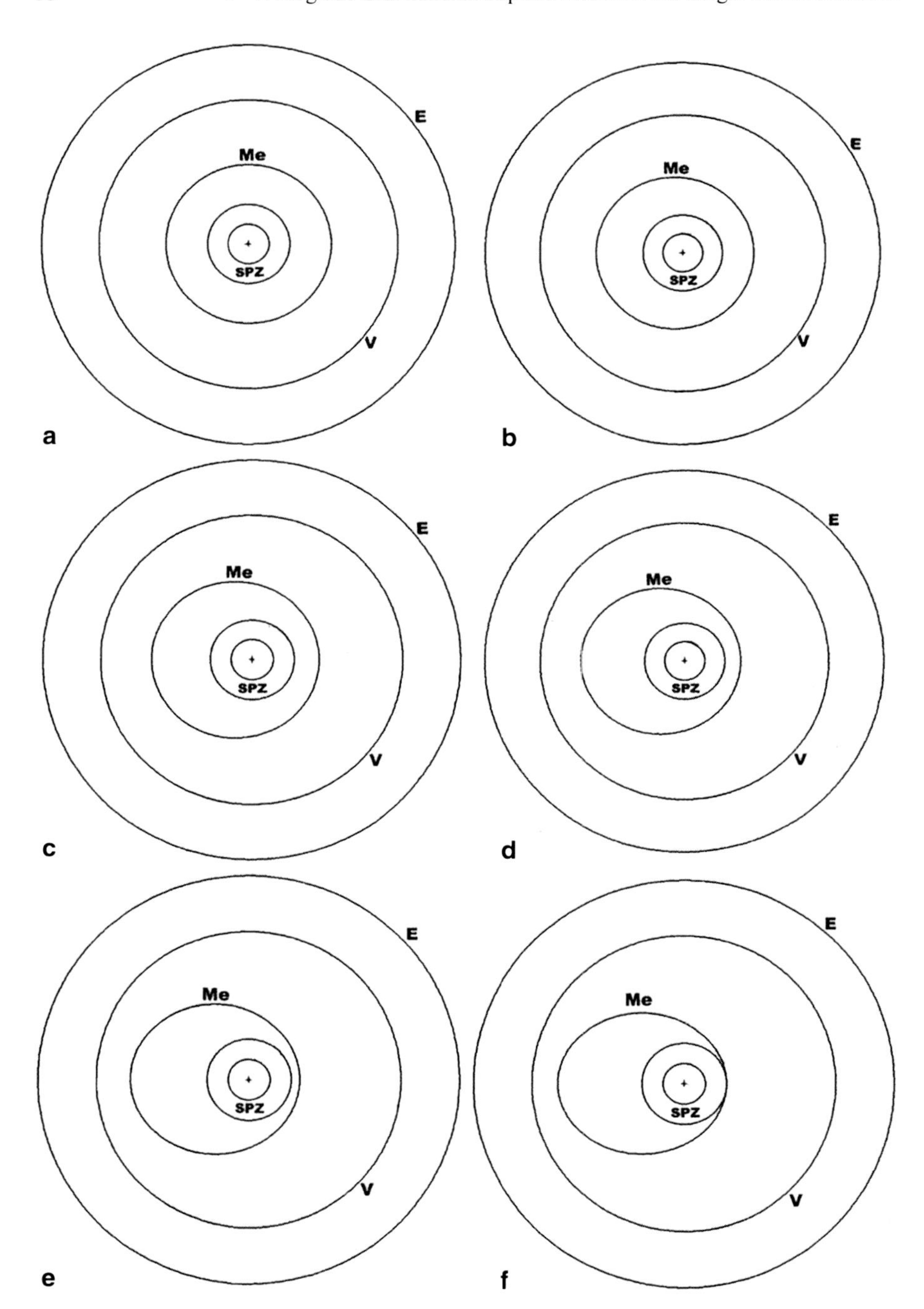

Fig. 4.10 Six eccentricity states for the orbit of planet Mercury. *SPZ* = stable planetoid zone; *Me* = Mercury; *V* = Venus; *E* = Earth. The eccentricity states of **a**, **b**, **c**, **d**, **e**, and **f** are 0.0, 0.1, 0.2, 0.3, 0.4, and 0.5, respectively. Eccentricity values in the range of 0.3–0.5 (**d**, **e**, and **f**) are probably needed to remove Vulcanoids planetoids from the SPZ

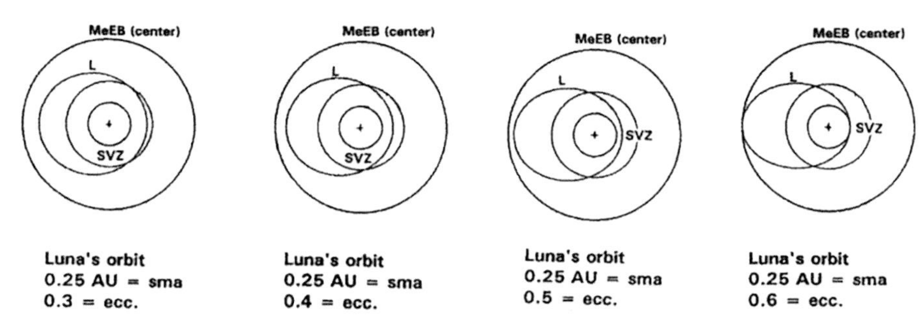

Fig. 4.11 Sequence of orbital diagrams illustrating a possible sequence of events for the transfer of Luna from an elliptical orbit (*ecc.* = 0.3) to an orbit in which the aphelion is in the center of the Mercury Eccentricity Belt (*MEB*) (*ecc.* = 0.6)

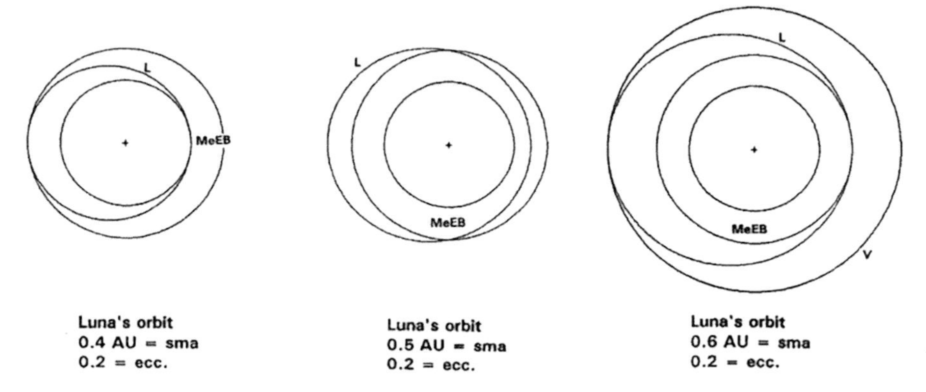

Fig. 4.12 An orbital scheme for transfer of Luna from the Mercury Eccentricity Belt to intersection with the orbit of Venus

4.3 Prograde Gravitational Capture of Luna and the Subsequent Orbit Circularization: A Two-Body Analysis and a Discussion of the Paradoxes Associated with the Capture Process

This is a quote from Pritchard and Stevenson (2000, p. 179) as a prelude to a discussion of PARADOXES.:

> The principal argument favoring a giant impact origin of the moon is dynamical: The angular momentum of the Earth-Moon system can be explained as the outcome of the oblique impact of a body of mars mass or greater on the proto-earth. Alternate explanations of lunar origin such as fission or co-accretion have great difficulty explaining this angular momentum budget, and the capture hypothesis, while not rigorously excluded, requires very special and unreasonable assumptions.

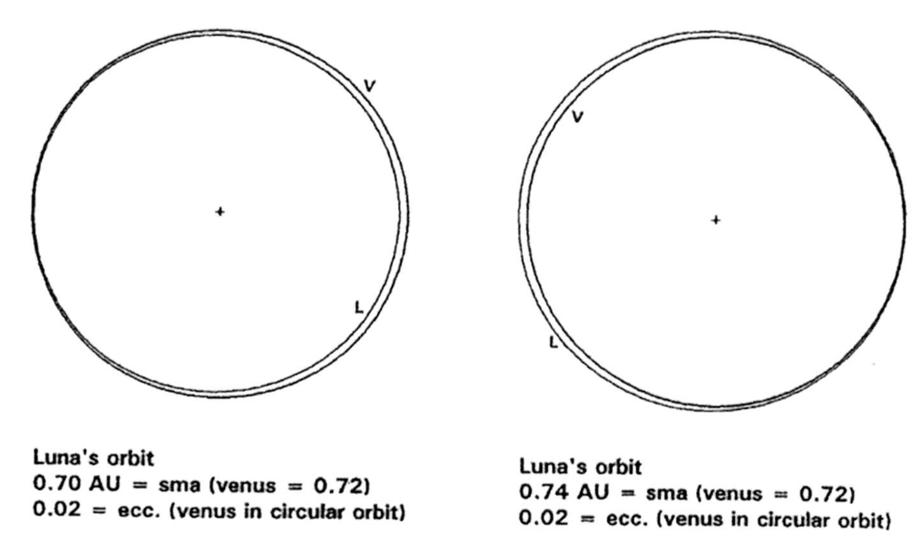

Fig. 4.13 Orbits from which a Luna-like or an Adonis-like Vulcanoid planetoid can be captured by planet Venus into a retrograde orbit. Note: Gravitational capture into a prograde orbit is nearly impossible for planet Venus; this situation will be discussed in Chap. 6

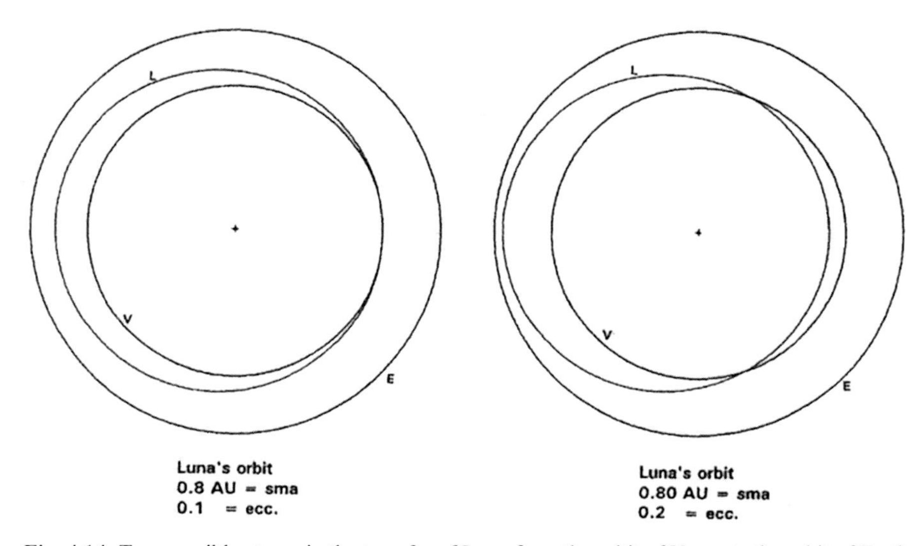

Fig. 4.14 Two possible stages in the transfer of Luna from the orbit of Venus to the orbit of Earth

Gravitational capture of a planetoid by tidal deformation processes is simple in principle but somewhat more complex when we consider the details. There are also FOUR PARADOXES associated with tidal capture of a satellite and three of these are associated with the complications of tidal deformation processes. *It will become apparent that these paradoxes have presented major "stumbling blocks" for all previous attempts for the study of tidal capture processes.*

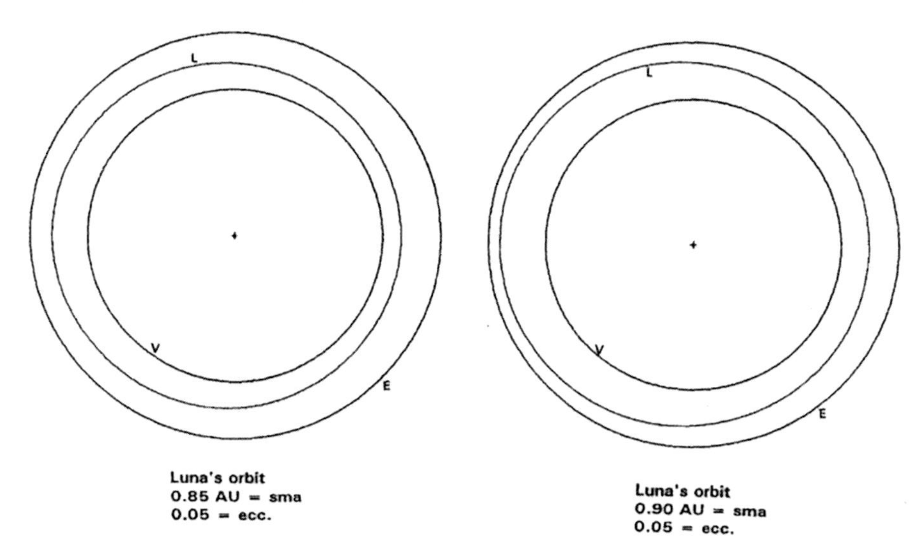

Fig. 4.15 Two orbital configurations that could be considered for "parking" orbits during the orbital transfer from the orbit of Venus to the orbit of Earth. According to the calculations of Evans and Tabachnik (1999), there are several orbital states in this region of the inner solar system that are stable for up to 100 million years

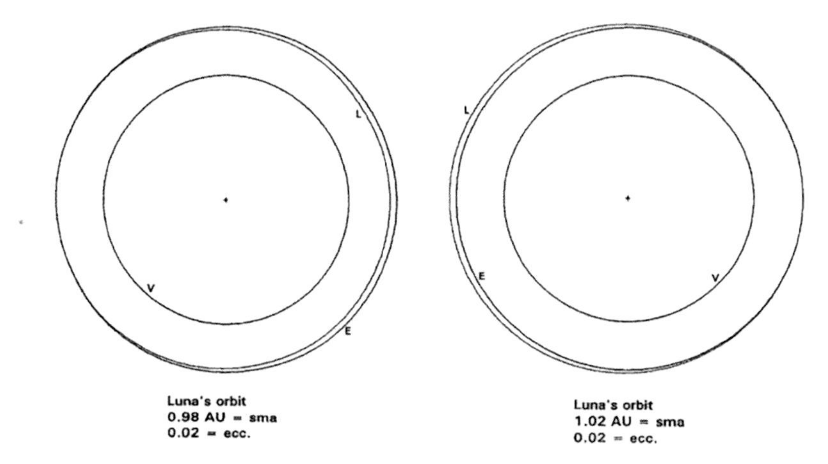

Fig. 4.16 Two orbital configurations for Luna that could permit stable gravitational capture by Earth if the deformation parameters are within certain limits and if the approach from heliocentric orbit is within a Stable Capture Zone. (The Stable Capture Zone concept will be explained in a later section of this chapter.) Note also that capture can occur from an orbit that is slightly smaller than Earth's orbit (diagram on *left*) or from one that is slightly larger than the Earth's orbit (diagram on *right*)

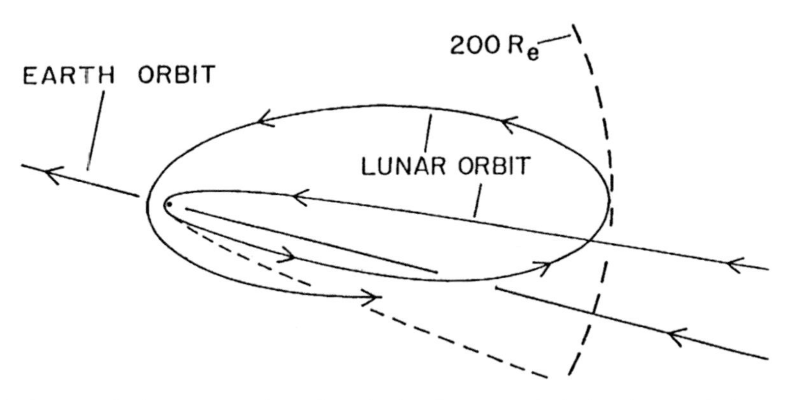

Fig. 4.17 A two-body representation of gravitational capture. Both the planet and planetoid are in very similar, but not identical, prograde heliocentric orbits. The planetoid has a close encounter with the planet. If no energy is dissipated within the two bodies, then the planetoid departs from the planet along the *dashed line* and attains an orbit very similar to its pre-encounter orbit. If sufficient energy is dissipated in a combination of the two bodies, then the planetoid is inserted into a highly elliptical orbit that stays within the stability limit for the Sun-Earth-Moon system [i.e., the Hill sphere for the system (Roy 1965)]. (View is from the north pole of the solar system)

The dictionary definition of a PARADOX, for our purposes, is "a statement that seems contradictory, unbelievable, or absurd but that may be actually true in fact". I will list the first three paradoxes here and then explain them in the following sections of this chapter.

PARADOX 1 The encountering planetoid (the smaller body) must absorb most of the energy for its own capture.

PARADOX 2 Larger planetoids are more capturable than smaller planetoids.

PARADOX 3 "Cool" planetoids are more capturable than "hot" planetoids. (One must distinguish between "cool" and "cold" for this paradox.)

All of these statements seem counterintuitive but, again, they are all true. Such is the nature of a paradox.

Figure 4.17 is a scale sketch of a close gravitational encounter that is a candidate for a successful capture. The essential features for successful gravitational capture by tidal deformation processes are:

1. *AN EARTH-LIKE HELIOCENTRIC ORBIT*: For successful capture a moon-mass planetoid has to be in a heliocentric orbit very similar in geometry to that of the Earth-like planet (i.e., within 3 % in heliocentric orbit eccentricity).
2. *A VERY CLOSE ENCOUNTER*: For successful capture a moon-like planetoid must have a close encounter with the Earth-like planet (i.e., within ~1.5 earth radii, center-to-center distance).
3. *ENERGY DISSIPATION VIA LUNAR ROCK TIDES*: For successful capture a moon-like planetoid must be sufficiently deformable in order to store the energy for capture ($\sim 1.0 \times 10^{28}$ J); the lunar-like body must also be able to dissipate a large fraction of the tidally stored energy during the timescale of the close encounter (about 30 min).

Table 4.3 Energy dissipation requirements for capture into geocentric orbits with specific dimensions. In this case the heliocentric orbit of the planetoid is coincident with that of the orbit of the planet (i.e., the first term in Eq. 4.1 is zero)

Major axis (ER)	Energy necessary for capture
200	Ecapture = 0 – (−2.30 E28 J) = 2.30 E28 J
150	Ecapture = 0 – (−3.06 E28 J) = 3.06 E28 J
100	Ecapture = 0 – (−4.60 E28 J) = 4.60 E28 J
50	Ecapture = 0 – (−9.20 E28 J) = 9.20 E28 J

4. *INSERTION INTO A STABLE GEOCENTRIC ORBIT*: For successful capture the semi-major axis of the post-capture orbit of the lunar-like planetoid must be as large as possible but must be within the stability zone of the Sun-Earth-Moon dynamical system [i.e., the Hill Sphere (within ~200 earth radii)].

Equation 4.1 shows the energy considerations for gravitational capture of a planetoid from an Earth-like heliocentric orbit into a large post-capture orbit. As a first case let us assume for simplicity that the Earth's orbit has zero eccentricity and inclination, is located in the common plane of the planets, and is in an orbit that is coincident with the orbit of Earth. In this case the first term of Eq. 4.1 is zero.

Energy to be dissipated for capture by Earth is:

$$E_{capture} = 0.5\, M_{pl}\, v_{\infty}^{2} - \left(\frac{-G\, M_{e}\, M_{pl}}{2a} \right) \quad (4.1)$$

where M_{pl} is the mass of the planetoid, v_{∞} is the velocity of the planetoid at infinity, G is the gravitational constant, M_e is the mass of Earth, M_{pl} and 2a is the major axis of the planetocentric orbit.

The values in Table 4.3 relate to the second term of this equation. The 200 ER (earth radii) value is for the largest capture orbit possible and this represents the lowest energy dissipation possible (Fig. 4.18, orbit state 1). If we want to capture a lunar-mass body into a smaller orbit, then significantly more energy must be dissipated during the initial encounter for successful capture (Fig. 4.18, orbit states 2, 3, and 4). In this two-body case, capture into this largest orbit is just marginally possible if ~2.30 E28 J of energy is dissipated by some process or combination of processes.

Now let us consider the first term of Eq. 4.1 for the case where the heliocentric orbit of the planetoid is not coincident with the orbit of the planet. In this case there is some value to the velocity at infinity (i.e., the velocity of the planetoid relative to that of the planet when the bodies are well beyond the boundaries of the Hill Sphere). Table 4.4 gives some values for the velocity at infinity (v_{∞}) for the encountering body and the values for the extra energy dissipation needed for successful capture into the largest geocentric orbit possible. Figure 4.19 shows the heliocentric orbits represented in Table 4.4. One can notice that the energy dissipation requirements increase rapidly as the encounter velocity at infinity increases. Even with a value of 0.5 km/s, about 40 % more energy dissipation if necessary for capture.

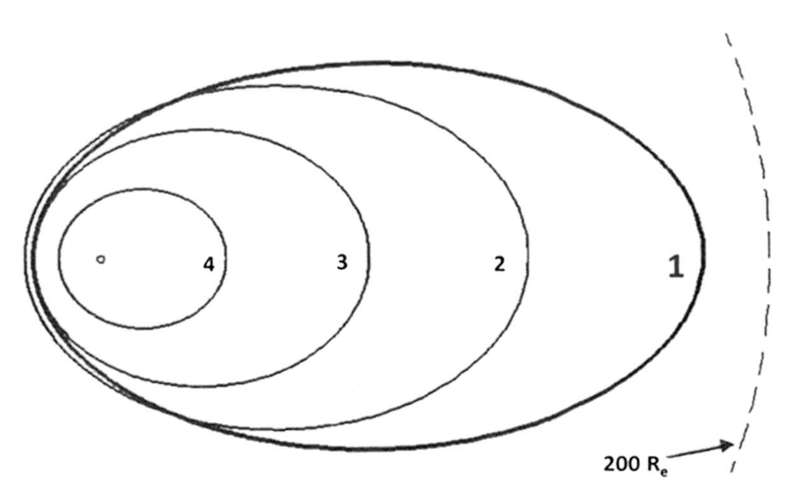

Fig. 4.18 Graphical representation of the geocentric orbits listed in Table 4.3. Orbits 1, 2, 3, and 4 are 200, 150, 100, and 50 ER respectively

Table 4.4 Energy dissipation requirements for capture into a maximum size geocentric orbit when there is a velocity at infinity value for the case of a lunar-mass planetoid encountering an earth-mass planet from an inside orbit

Ecc. (%)	v∞ (km/s)	Energy for capture
~ 6.5	0.5	Ecapture = 0.92 E28 J – (−2.30 E28 Joules) = 3.22 E28 J
~ 13.0	1.0	Ecapture = 3.68 E28 J – (−2.30 E28 Joules) = 5.98 E28 J
~ 19.0	1.5	Ecapture = 8.27 E28 J – (−2.30 E28 Joules) = 10.6 E28 J
~ 25.3	2.0	Ecapture = 14.7 E28 J – (−2.30 E28 Joules) = 17.0 E28 J

Ecc.(%) is the difference in eccentricity between earth's orbit (circular) and the lunar orbit when the orbits intersect at lunar aphelion

The conclusion from this discussion is that we need a lunar-like planetoid in essentially in co-planar Earth-like heliocentric orbit to be inserted into the largest possible stable geocentric orbit (i.e., the planetoid is just barely captured). *These are the necessary conditions for a successful gravitational capture scenario in a two-body context!*

As noted earlier in this section, there are four paradoxes that seem to have clouded a reasonable assessment of the GRAVITATIONAL CAPTURE MODEL (GCM). The first three paradoxes will be discussed in this segment and the fourth will be discussed a bit later. *The first paradox is that the encountering body must absorb most of the energy for its own capture.* To understand this paradox we need some background information on the concept of the displacement number (h) for tidally deformed bodies.

In brief, the displacement Love number (h) of a planet or planetoid relates to the deformability of the body. The Love numbers are dimensionless parameters representing various aspects of elastic tidal deformation (Love 1911, 1927). Peale and Cassen (1978) show that the displacement Love number (h) is the important one for analyzing energy

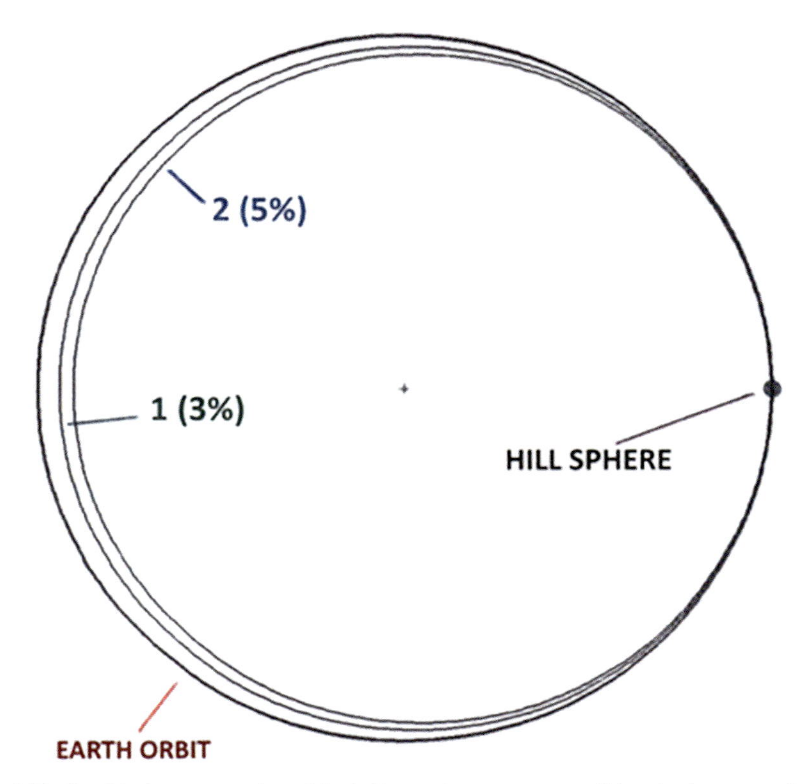

Fig. 4.19 Graphical representation of the heliocentric encounter conditions in the range of values listed in Table 4.4. The energy dissipation necessary for capture increases rapidly as the velocity at infinity increases. That is why successful gravitational capture by tidal energy dissipation can only occur from near co-planar orbits that are within ~3 % difference in eccentricity. (Examples above: 3 % difference in eccentricity results in ~0.23 km/s velocity at infinity; 5% difference in eccentricity results in ~0.39 km/s velocity at infinity.)

storage by radial tidal deformation. The displacement Love number is defined by Stacey (1977) as the ratio of the height of the body tide to the height of the equilibrium (static) marine tide (i.e., the height of the ocean tide if it had time to come to equilibrium with the tidal potential). The displacement Love number of the Moon at present (a cold rigid body) is estimated to be ~0.033 (Kaula and Harris 1973; Ross and Schubert 1986). The displacement Love number for the Earth is estimated to be about 0.54 (Munk and MacDonald 1960; MacDonald 1964). The displacement Love number for the water on a planet covered with a significant thickness of ocean water is 1.

The second paradox is that larger planetoids are more capturable than smaller planetoids. To understand this paradox we need information on the concept of the specific dissipation factor (Q) in addition to the h factor. 1/Q is the fraction of stored energy that is subsequently dissipated during an encounter cycle. For example, if an effective planetary Q equals 10, then about one-tenth of the energy stored during an encounter cycle will be dissipated (mainly in the form of heat within the body). The present effective planetary Q of Earth is estimated to be about 13 (mainly because

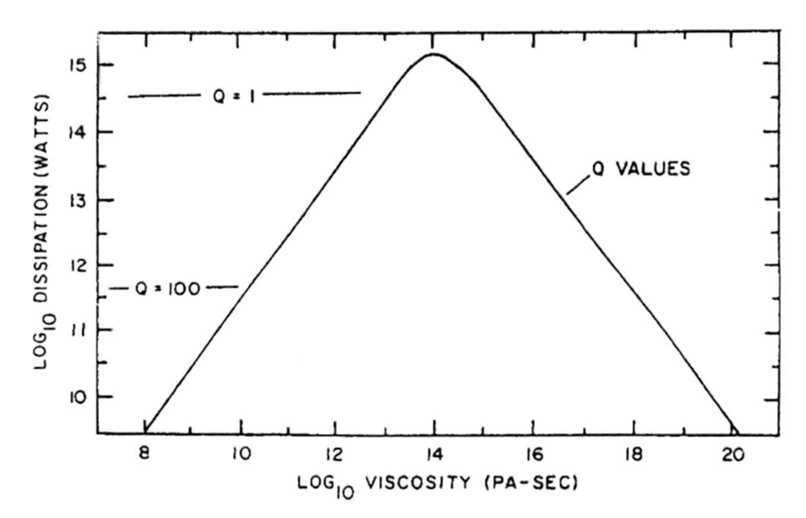

Fig. 4.20 Plot of energy dissipation vs. body viscosity for a lunar-like body. (From Ross and Schubert 1986, Fig. 3, p. D450, with permission from the American Geophysical Union). Note that a Q value near one is associated with an intermediate value of body viscosity

of the energy dissipation by ocean tides) (Munk and Macdonald 1960). The present planetary Q for the solid Earth is estimated to be 100 or higher (Melchior 1978). The present Q value for the Moon is about 25 (Goldreich and Soter 1966). The present Q value of Io (the innermost Galilean satellite of planet Jupiter) is estimated to be about 1 (Ross and Schubert 1986). Figure 4.20 shows a plot of Energy Dissipation vs. Body Viscosity with the associated Q values. In the analysis of Ross and Schubert (1986) very low Q values are associated with intermediate viscosity values (~ 10 E14 Pa-s). Q values of 1, 2, or 3 are necessary for capture of a lunar-like planetoid by Earth.

The displacement Love number (h) is a very important factor for gravitational capture and h is fairly closely related to the thermal content of the body. Small bodies tend to cool much more rapidly than the large bodies. Thus, small bodies become rigid much more rapidly than large bodies and it would be difficult to keep a small planetoid at the viscosity value that would be associated with a low Q value (i.e., it would be difficult to keep them in a physical state to be capturable).

A third paradox is that "cool" planetoids are more capturable than "hot" planetoids. Most of the energy for capture (~90 %) must be tidally stored and subsequently dissipated during the first close encounter of an encounter sequence if the planetoid is to be captured into a stable geocentric orbit. The reason for this is that the viscosity will be considerably lower (and thus the Q factor higher) after significant melting due to energy dissipation associated with the earlier encounter. Distant encounters are not associated with much energy dissipation, relative to that needed for capture, because of the $1/r^6$ dependence. For example, an encounter beyond about 2 earth radii does not lead to much energy dissipation.

Equation 4.2 is for energy storage in the Earth during one close encounter. The energy needed for capture is ~2.3 E28 J. Again, the closeness of the encounter

Table 4.5 Values for energy stored in the Earth during a single encounter ($h_e = 1$)

Pericenter radius		Energy stored (J)
For 1.27 earth radii	(a grazing encounter)	=8.09 E27
For 1.32 earth radii	(a grazing encounter)	=6.42 E27
For 1.35 earth radii	(non-grazing encounter)	=5.61 E27
For 1.43 earth radii	(non-grazing encounter)	=3.97 E27

Table 4.6 Values for energy stored in luna during a single encounter (hm = 1)

Pericenter radius		Energy stored (J)
For 1.27 earth radii	(a grazing encounter)	=8.13 E28
For 1.32 earth radii	(a grazing encounter)	=6.44 E28
For 1.35 earth radii	(non-grazing encounter)	=5.63 E28
For 1.43 earth radii	(non-grazing encounter)	=3.99 E28

is important here because of the $1/r^6$ dependence. With *h* (earth) = 1.0, the energy stored in the earth = the numerator of the equation divided by the denominator. Table 4.5 shows the results for various distances of close approach.

Energy stored per encounter in the Earth is:

$$E_{\text{stored-earth}} = \frac{3\,G\,M_{pl}^2\,R_e^5\,h_e}{5\,r_p^6} \tag{4.2}$$

Where G is the gravitational constant, M_{pl} is the mass of the planetoid (the Moon in this case), R_e is the radius of the Earth, h_e is the displacement Love number of Earth, and r_p is the distance of closest approach during the encounter.

From the above numbers, even using a Q value of 100 (low for the Earth), these numbers should be discouraging for tidal capture enthusiasts (i.e., ~4–6 E25 J per encounter). It would take over 1000 encounters at a very close perigee passage distance to dissipate the energy for stable capture in a two-body context. Singer (1968, 1970) and nearly everyone else that looked at the problem of gravitational capture, including Kaula and Harris (1973), considered the Earth to be the primary energy sink for capture. *Such a capture scenario would be very implausible by orders of magnitude.*

Energy storage in a lunar-like body during a single encounter is:

$$E_{\text{stored-planetoid}} = \frac{3\,G\,M_e^2\,R_{pl}^5\,h_{pl}}{5\,r_p^6} \tag{4.3}$$

Where M_e is the mass of the Earth, R_{pl} is the radius of the planetoid (the Moon in this case), h_{pl} is the displacement Love number of the planetoid, and r_p is the distance of closest approach during the encounter.

The energy needed for capture for this two-body analysis is ~2.3 E28 J. The closeness of the encounter is very important here because of the $1/r^6$ dependence. Energy stored-moon = the numerator of this equation divided by the denominator. h_{pl} of the moon = 1.0. Now let us look at the numbers (Table 4.6) for Luna as we did before for Earth.

From the above numbers, even with a Q value of 1 (the practical lowest value of Q) it is difficult to dissipate the energy for capture during one encounter. Using an h value of 0.5 (upper practical limit) and a Q value of 1, the energy dissipated in the lunar body during one encounter at 1.43 earth radii would be ~2.0 E28 J. This calculation simply makes gravitational capture via energy dissipation in the lunar body marginally attractive in a two-body context.

Equation 4.4 is for energy dissipation in both bodies during a close encounter. The energy needed for capture is ~2.3 E28 J. For a close encounter at 1.43 earth radii and with a Q for the planetoid = 1 and a Q for Earth = 100, capture from an Earth-like orbit (velocity at infinity = 0) would require an h for the Moon of 0.5 or a bit higher) and a displacement Love number for Earth of ~1.0. Using these numbers, about 2 % of the energy of capture is dissipated in the Earth. These results appear to make gravitational capture of the Moon just marginally possible in a two-body context! This fact alone should be enough to discourage most capture enthusiasts!!!

Energy dissipated in both bodies during a close gravitational encounter is:

$$E_{diss.(pl+e)} = \frac{3G}{5r_p^6}\left(\frac{M_e^2 R_{pl}^5 h_{pl}}{Q_{pl}} + \frac{M_{pl}^2 R_e^5 h_e}{Q_e}\right) \tag{4.4}$$

where Q_{pl} is the specific dissipation factor for the planetoid and Q_e is the specific dissipation factor for the Earth.

In summary, even after understanding the paradoxes associated with gravitational capture, there is an unusual history surrounding the development of Eqs. 4.2, 4.3, and 4.4, the energy storage and energy dissipation equations. When Kaula and Harris (1973) derived the equations, they had the numerical coefficient value at 3/10 because they considered only the energy stored, and subsequently dissipated, to be associated with the tidal deformation of the body (and not with the subsequent relaxation of the tidal distortion). Winters and Malcuit (1977) showed that a second quantity, which is equal to the first, is associated with the tidal relaxation process. This changed the numerical coefficient to 3/5, a change which doubled the energy stored, and subsequently dissipated, during a close gravitational encounter. Peale and Cassen (1978) independently arrived at the same conclusion that the numerical coefficient should be 3/5. The bottom line is that these changes to this fundamental equation brought gravitational capture into the realm of the physically possible!!!

4.4 Numerical Simulations of Gravitational Capture of a Lunar-Like Body by an Earth-Like Planet

We had our first successful numerical simulations of the capture process in the September of 1987 and our first report on the simulations was Malcuit et al. (1988) at the LPSC. More capture simulations are in Malcuit et al. (1989, 1992). There were two very favorable surprises associated with the capture simulations.

Surprise # 1: Solar gravitational perturbations during a capture encounter can decrease the required energy dissipation for capture by up to 50% relative to what was expected from a two-body calculation.

Surprise # 2: There are such things as STABLE CAPTURE ZONES (Malcuit and Winters 1996) and the probability of capture for any one encounter can be measured directly from the geometry of the stable capture zone.

4.4.1 Computer Code Information

The computer code for this general co-planar, three-body calculation uses a fourth-order Runge-Kutta numerical integration method. The accuracy is a few parts in 10 E5 for energy and a few parts in 10 E8 for angular momentum. The energy dissipation subroutine is activated for all encounters within 20 earth radii. The density of time-steps (calculation points for orbital elements and energy extraction from the orbit) is inversely proportional to the distance of separation of the planet and planetoid. If the perigee distance is outside 20 earth radii, then the energy is taken from the orbit impulsively at the perigee point.

There are two general orientation schemes for this co-planar calculation. One is a non-rotating coordinate system (Fig. 4.21a) in which the geocentric computer plot theoretically revolves around the Sun in 1 year but the plot does not rotate. There is also a rotating coordinate system (Fig. 4.21b) which is useful for analysis of certain orbital characteristics but it is not as easily understood as the non-rotating plot. The hallmark of a rotating coordinate system is that the computer plot rotates as it revolved around the Sun.

4.4.2 Development of the Computer Code

The N-body numerical simulation code was borrowed from Tom Stoeckley (Michigan State Univ., Astronomy Dept.) by Bob Malcuit in 1976. This code was modified by computer science/physics students David Mehringer (DU, '88) Wentao Chen (DU, '92), and Albert Liau (DU, '93) under the supervision of Ronald Winters (Physics-Astronomy Dept.) and myself from 1987 to 1993. A two-body evolution code was developed by David Mehringer, Ronald Winters, and myself in the 1985–1987 era. Plotting packages for planetocentric orbits, tidal amplitudes, and tidal ranges were developed by David Mehringer, Wentao Chen, and Albert Liau from 1987 to 1993 under the supervision of Ronald Winters and myself.

Figure 4.22 is a plot showing typical heliocentric orbit information for a three-body calculation. The *planet anomaly* is the position of the planet at the beginning of the calculation (starting point) relative to the 360° of the orbit. The *planetoid anomaly* is the starting position of the planetoid relative to the planet (shown in the geocentric orientation at the position marked "start" which is located at ~187° counterclockwise from the 0° position relative to Earth). The *argument of pericenter*

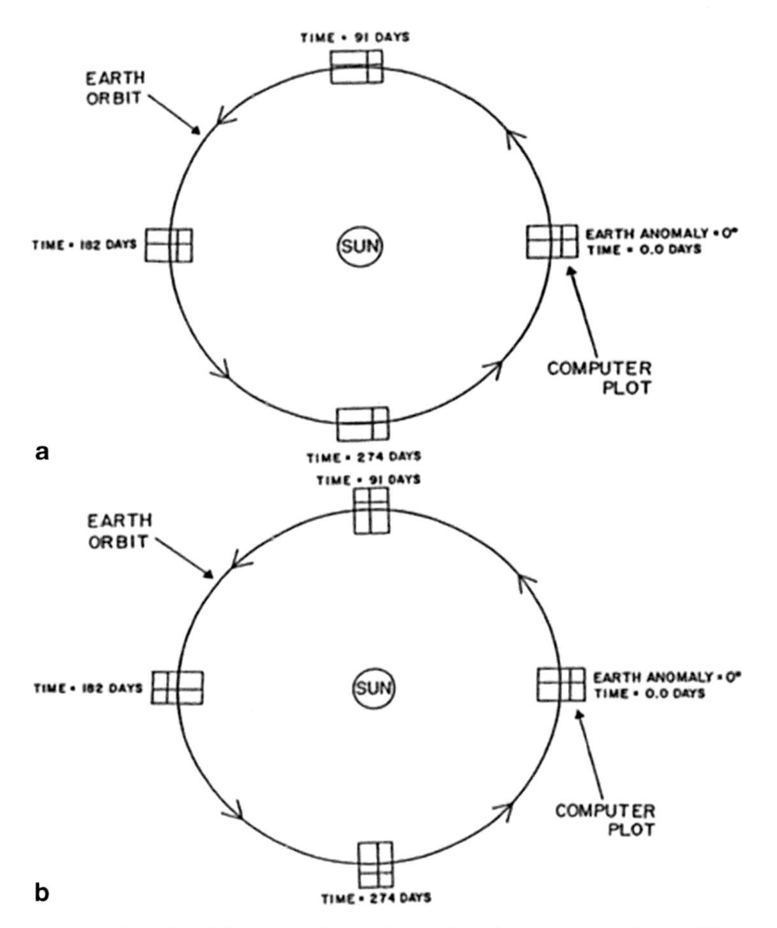

Fig. 4.21 Some orientation information for understanding the computer plots. **a** Diagram illustrating a non-rotating coordinate system for plotting the results of numerical simulations. The computer plot does not rotate as it theoretically revolves around the Sun. The Earth is at the origin of the rotating plot. **b** Diagram illustrating a rotating coordinate system for plotting the results of numerical simulations. In this case the computer plot rotates as it theoretically revolves around the sun. Special symmetry information can be gained from this plot but the plotted information is more difficult to understand. Thus all plots in this book use the non-rotating coordinate system and all orbital plots are viewed from the north pole of the solar system

for the planet's orbit, PH(E), is the position of the pericenter (perihelion) for the planet's orbit (this is 0° in the illustration). The *argument of pericenter for the planetoid's orbit, PH(PL)*, is the position of the pericenter (perihelion) of the planetoid's orbit (this is 180° in the illustration). The eccentricity of the planet's orbit is normally 0.0 % unless otherwise noted. The *eccentricity of the planetoid's orbit* usually has a value from 0.0 to 3.0 % for a successful capture encounter. In this illustration the eccentricity of the planetoid's orbit is 0.015 and the planetoid is encountering

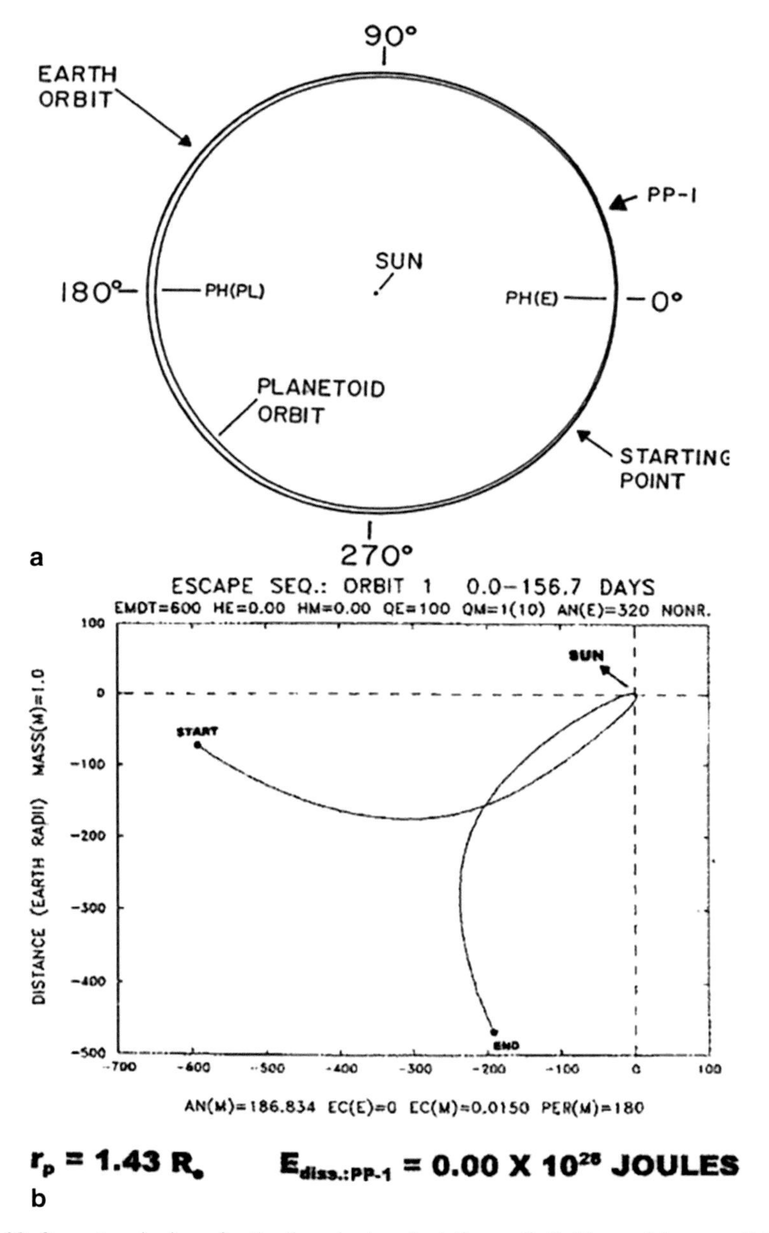

Fig. 4.22 Some terminology for the three-body calculation. **a** Definitions of the essential terms for this diagram are in the text. **b** Organization of the information on the geocentric computer plot is the following: *First title line*: results of simulation (escape, collision, or capture); number of orbits; duration of simulation in earth days. *Second title line*: *EMDT* initial earth-planetoid distance of separation; *HE* h for earth; *HM* h for planetoid; *QE* Q of earth; *QM* Q of planetoid; *AN(E)* earth anomaly at beginning of calculation. *Bottom line*: AN(M) planetoid anomaly; *EC (E)* eccentricity of orbit of earth; *EC(M)* eccentricity of the orbit of planetoid; *PER(M)* pericenter radius position for orbit of planetoid. *Left side line*: Scale for both axes in planet radii; fractional mass of planetoid relative to mass of moon

the planet from an orbit that is inside the planet's orbit (but, in other cases, the planetoid can be encountering the planet from an orbit which is outside of the orbit of the planet). And finally, the *time of the first perigee passage (PP-1)* occurs at some time after the beginning of the simulation (normally between 40 and 60 days depending on the parameters of the simulation). This simulation ends (*END*) in this example ~ 157 days after the starting point. The SUN arrow points to the position of the sun at the initiation of the simulation. There is no energy dissipated during the encounter and the planetoid returns to a heliocentric orbit that is very similar to its pre-encounter orbit.

4.4.3 A Sequence of Typical Orbital Encounter Scenarios Leading to a Stable Capture Scenario

Figure 4.23a shows the results of a simulation in which there is no energy dissipation within either body. The starting positions as well as the ending positions are shown on the geocentric framework. The heliocentric orientation diagram shows the starting position relative to the Earth's orbit and PP-1 is the location of the first (and only) perigee passage in this orbital encounter sequence. The closest encounter distance is 1.43 earth radii; there is no energy dissipated during the encounter. The lunar body returns to the Earth-like heliocentric orbit essentially unchanged. The starting position, as well as the ending position, are shown on the geocentric plot. During this first encounter, the close approach distance is 1.43 earth radii. This brings the lunar surface within about 1000 km of the surface of the Earth. I should note that the first encounter is focused on the value of 1.43 earth radii in order to eliminate a variable in the calculation.

Figure 4.24 shows the results of a simulation in which there is significant energy dissipation within the interacting bodies and the lunar-like body returns for two additional perigee passages after the initial encounter. The encountering body then returns to a slightly changed Earth-like heliocentric orbit. On the heliocentric orientation diagram the starting position is the same as in Fig. 4.23a as well as the position of the first perigee passage (encounter). There is not enough energy dissipated for capture and after three perigee passages the lunar-like body returns to a somewhat changed heliocentric orbit from which it can eventually have another encounter with Earth.

Figure 4.25 shows the results of a simulation in which there is sufficient energy dissipation for stable capture. The first 40 orbits of the stable capture scenario are shown but the calculation was extended to over 100 orbits and it showed no signs of instability. On the heliocentric orientation diagram the starting position is the same and the position of the first six perigee passages are shown. Note that only 0.95 E28 J of energy dissipation was sufficient for stable capture for this particular set of co-planar orbital parameters. This is less than 50 % of what was expected from the two-body analysis. Some orbital resonance motion is obvious in the pattern of orbital geometry. The bottom line is that a favorable orientation of the Sun during

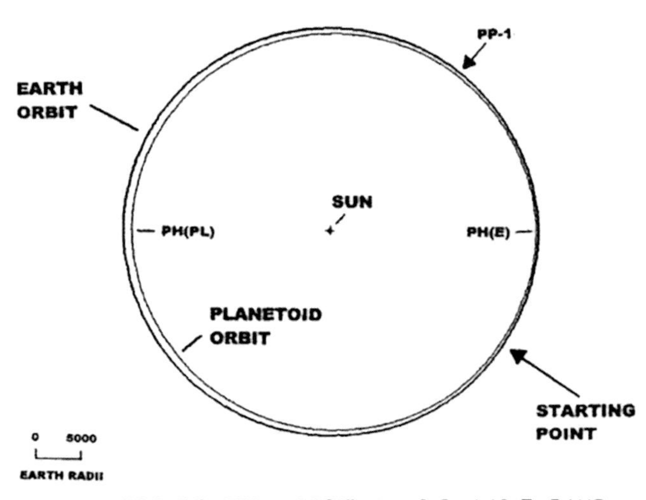

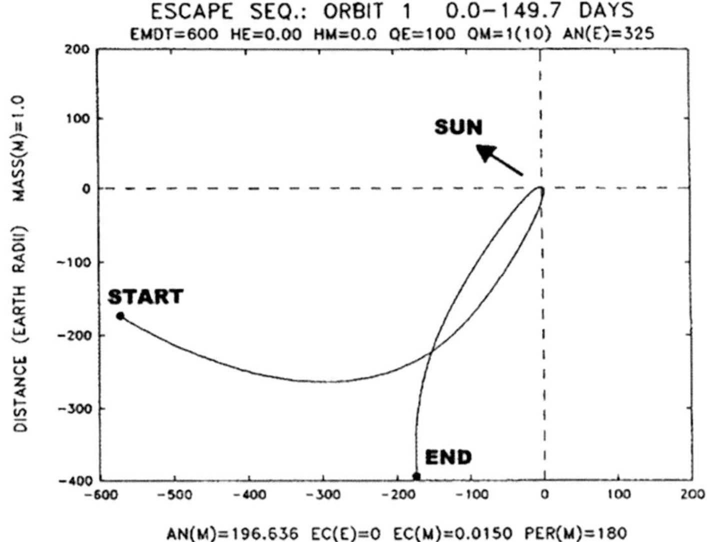

Fig. 4.23 Results of a close encounter in which no energy was dissipated. **a** Heliocentric orientation for the encounter. The simulation begins at the starting point (START). The close encounter occurs about 60 days later at PP-1 (perigee passage number 1). No energy is dissipated by tidal processes and the planetoid escapes to a heliocentric orbit about 60 days after PP-1. **b** Geocentric orientation for the encounter. Earth is at the origin of the plot. The *arrow* on the plot indicated the position of the Sun at the beginning of the simulation. The simulation begins at 600 earth radii, which is well beyond the boundary of the Hill Sphere (which is at about 200 earth radii)

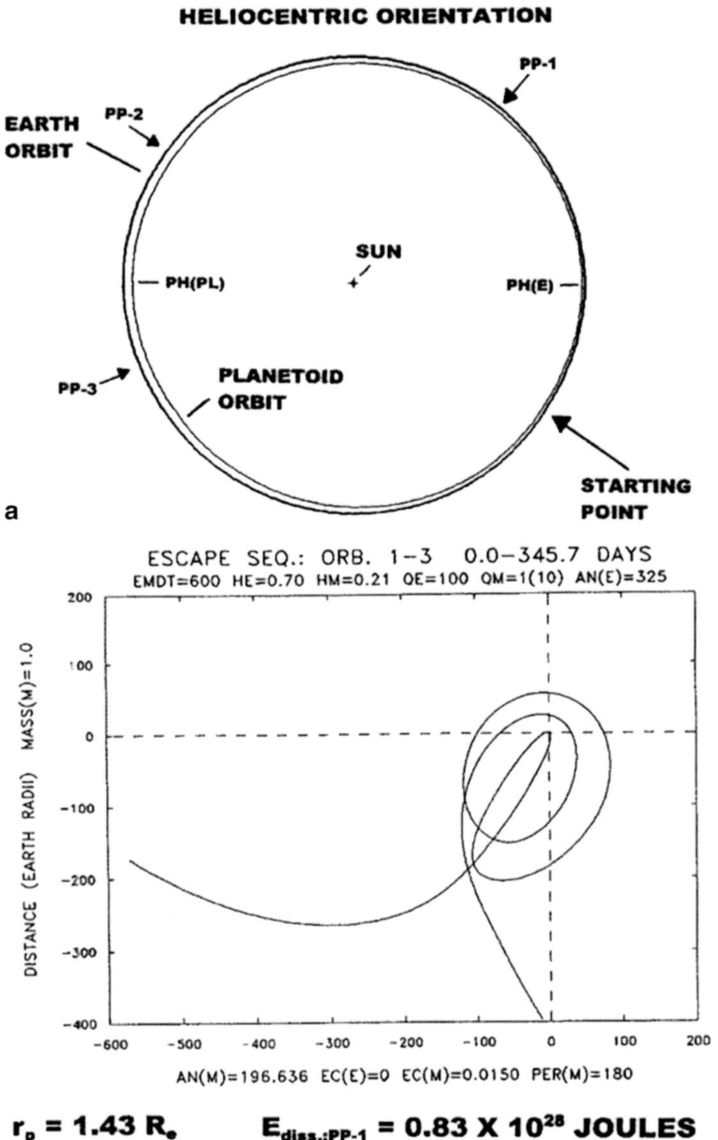

Fig. 4.24 Results of a three orbit escape scenario. **a** The heliocentric orientation shows the starting point and the positions of the three close encounters of this encounter sequence. **b** The geocentric orientation shows the pattern of close encounters.Solar gravitational perturbations cause the variation in the orbital path. A significant quantity of energy is dissipated but not quite enough for stable capture

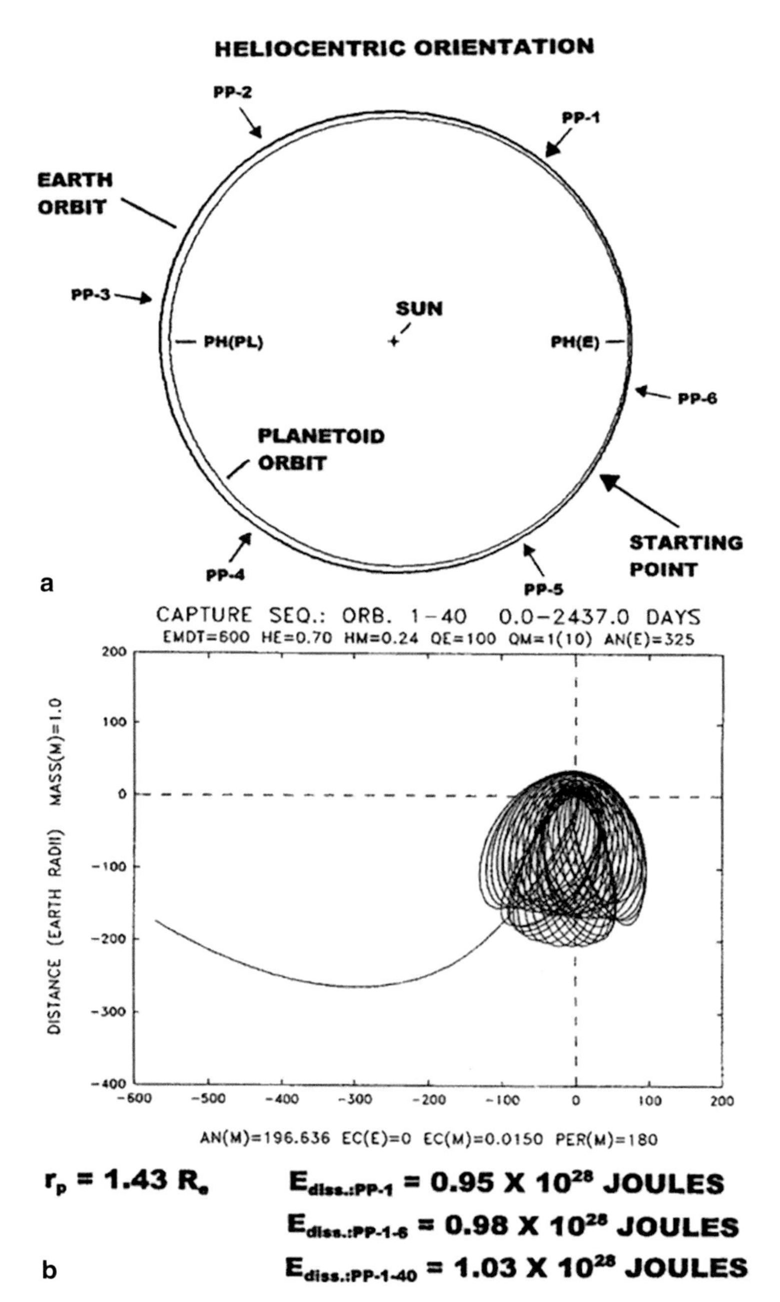

Fig. 4.25 Results of a stable capture scenario. **a** Positions of the first six perigee passages are shown on the heliocentric orientation. **b** Orbital pattern of the first 40 orbits (~6.67 years) are shown on the geocentric plot. After 100 orbits the major axis of the orbit is beginning to decrease. Note that about 1.0 E28 J of energy is sufficient for capture for this set of orbital parameters

the timeframe of the initial encounter reduces the energy dissipation requirements to about one-half.

The pattern of equilibrium tidal amplitudes for Earth and Moon are shown in Fig. 4.26 for the first 8 years of the simulation. The tidal amplitudes for each body are calculated directly from the orbital files. Equilibrium rock tides of up to 18 km occur on the planet during the initial encounter and in this simulation the initial encounter is the closest. But this is not a general case. Many simulations will give similar results but each one is somewhat different. In some cases, just a slight change in input parameters can yield a much different result. In this particular simulation while the Earth is having 18 km rock tidal amplitudes raised on its surface, the rock tides for Luna for the first encounter are nearly 400 km. Within the first 8 years there are 5 perigee passages that raise over 8000 m tidal amplitudes on the Earth and over 150 km tidal amplitudes on the lunar body. Thus, it is clear that in addition to Luna dissipating nearly all of the energy for its own capture, it is undergoing severe tidal distortions during the early stages of a stable capture scenario.

An equation for tidal amplitudes on the Earth is:

$$\text{Earth Tidal Amplitude} = \frac{M_{pl}}{M_e}\left(\frac{R_e}{r_p}\right)^3 R_e\, h_e \tag{4.5}$$

where M_{pl} is the mass of the planetoid (the Moon in this case), M_e is the mass of the Earth, R_e is the radius of the Earth, r_p is the distance of separation between the planet and planetoid, and h_e is the displacement Love number of the Earth.

An equation for tidal amplitudes on the planetoid is:

$$\mathit{Lunar\ Tidal\ Amplitude} = \frac{M_e}{M_m}\left(\frac{R_m}{r_p}\right)^3 R_m\, h_m \tag{4.6}$$

where M_m is the mass of the Moon, R_m is the radius of the Moon, and h_m is the displacement Love number of the Moon.

4.4.4 Geometry of Stable Capture Zones for Planetoids Being Captured by Planets

Each planet-planetoid combination has certain zones in space from which the planetoid can be captured into a STABLE PLANETOCENTRIC ORBIT (Malcuit and Winters 1996). A STABLE CAPTURE ZONE (SCZ) is defined as a region of parameter space (planetoid anomaly vs. planet eccentricity) in which the orientation of the orbit of the encountering planet is favorable for a stable post-capture orbit if sufficient energy is dissipated for its capture into a planetocentric orbit. In this book the SCZ concept is for co-planar encounters: i.e., all encounters are in the common

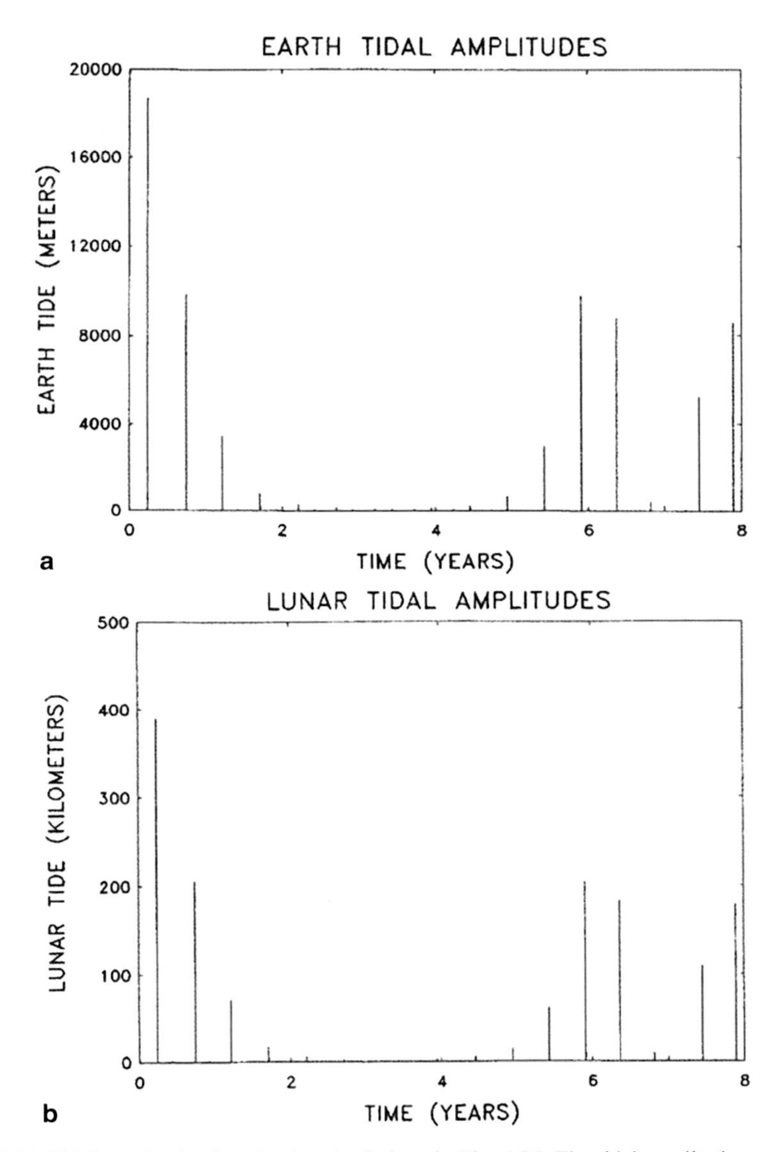

Fig. 4.26 Tidal amplitude plots for the simulations in Fig. 4.25. The tidal amplitudes were calculated directly from the orbital files. **a** Rock tidal amplitudes for earth for the first 8 years of the orbital scenario. The pattern of tidal amplitudes is somewhat irregular because of solar gravitational perturbations of the large major axis orbit with high orbit eccentricity. A 20 km rock tide on Earth causes about 0.3 % radial tidal deformation and very little energy is dissipated within the Earth during this encounter. **b** In contrast, rock tidal amplitudes for Luna are near 400 km for the first encounter and considerably lower for subsequent perigee passages. A 400 km tide on Luna causes about 23 % radial distortion and the energy dissipated during that first encounter is sufficient to melt a zone of magma about 50 km thick in a sub-crustal magma ocean zone of Luna

plane of the planets of the Solar System. Any encounter that is out of this plane would result in a less stable orbital configuration. Eventually, some adventurous character can define these SCZs in three dimensions.

Figure 4.27 shows the geometry of the two sets of Stable Capture Zones. There are SCZs for planetoids encountering the planet from either the prograde (counter-clockwise) and retrograde (clock-wise) direction. To be more specific, there are both prograde and retrograde SCZs for close encounters by planetoids in orbits that are slightly smaller than the orbit of the planet (inside orbits) and there are prograde and retrograde SCZs for close encounters by planetoids that are slightly larger than the orbit of the planet (outside orbits). The zones shaded in red require the lowest h_{pl} for capture (<0.3) and are the most favorable zones for capture. The zones shaded in blue require more energy dissipation for stable capture. The enclosed white zones yield a stable capture orientation but some process in addition to tidal energy dissipation (i.e., a non-tidal energy sink) is required for stable capture. Note that encounters outside the SCZs are possible but the result is an orbital collision or an escape into a somewhat changed heliocentric orbit. Note also that there are no stable capture possibilities from orbits of very low eccentricity that are very similar to Earth's orbit which is zero eccentricity.

There are three important items to consider for determining the geometry of a STABLE CAPTURE ZONE: (1) the ORIENTATION of the orbit of the planetoid relative to the orbit of the planet; (2) ENERGY STORAGE during the timeframe of a close encounter; and (3) ENERGY DISSIPATION during the timeframe of a close encounter. If any of these items is outside acceptable boundaries, then STABLE CAPTURE will not occur. These concepts will become more lucid when reading subsequent sections of this chapter.

An additional advantage of SCZs is that the PROBABILITY OF CAPTURE can be determined directly from the geometry of an SCZ. For any given combination of planetoid orbital eccentricity and planet anomaly, the probability of successful capture can be determined *by the length of the line intercept within the SCZ* divided *by the total length of the line intercept on the plot* (i.e., the 360° of planet anomaly). Cursory examination of the geometry of the two sets of SCZ diagrams strongly suggests that the probability of retrograde capture is favored over prograde capture for values of planetoid eccentricity from 0.3 to 1.5 %. Our planet and our lives would be much different if we had captured a lunar-mass satellite in the retrograde direction. This possibility is the subject of Chap. 7.

A Procedure for Mapping Stable Capture Zones in Two Dimensions

There are three important items for mapping stable capture zones: (1) orientation of the encounter, (2) energy storage during the encounter, and (3) energy dissipation during the encounter.

The importance of ORIENTATION of the planetoid relative to the orbit of the planet can not be overly stressed. Without an encounter orientated within nar-

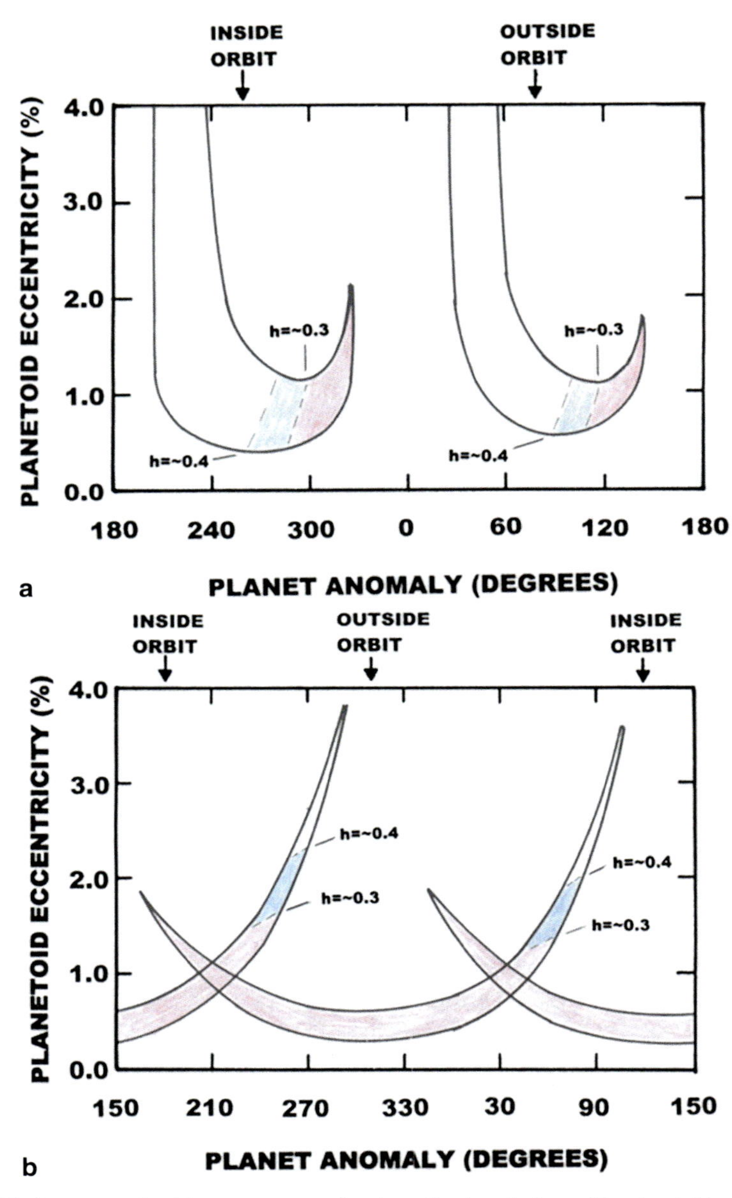

Fig. 4.27 Geometry of stable capture zones for planet Earth and a lunar-like planetoid. **a** Prograde SCZs for lunar-like planetoids encountering Earth in a prograde direction from an orbit that is inside the orbit of the Earth (*left*) and for lunar-like planetoids encountering Earth from an orbit that is outside the orbit of Earth (*right*). **b** Retrograde SCZs for lunar-like planetoids encountering Earth in a retrograde direction from an orbit that is outside the orbit of Earth (*center form*) and for lunar-like planetoids encountering Earth from an orbit that is inside the orbit of Earth (*split form*)

row limits, there is no successful capture. For successful prograde capture, the major axis of the initial geocentric planetoid orbit must be within about + or − 3° to a line that is perpendicular to the tangent of the orbit of the planet. For successful retrograde capture, the major axis of the initial geocentric planetoid orbit must be within about + or −3° of a line parallel to the tangent of the orbit of the planet. These rules are valid for encounters of planetoids from orbits that are either interior or exterior to the orbit of the planet.

For gravitational capture, energy must be stored by tidal deformation processes in a combination of the interacting bodies. Thus the displacement Love numbers of the combination of the encountering bodies must be large enough to temporarily store the energy for capture. If the body of the planetoid is too rigid, then there will be no capture via gravitational (tidal) processes.

And lastly, a very large percentage of stored energy must be dissipated during the time-frame of a gravitational encounter from a heliocentric orbit. Ross and Schubert (1986) demonstrated that bodies of roughly lunar-size can have extremely low Q values when they have values of body viscosity within a certain range of values. In this special case of low Q values, a very high percentage of the tidally stored energy can be dissipated during the time-frame of the encounter.

Now let us consider a procedure for mapping the geometry of an SCZ in two dimensions (i.e., planetoid anomaly along the horizontal axis and planetoid orbit eccentricity along the vertical axis. The process, as I am describing it, is laborious and time-consuming. Defining the boundaries of one SCZ can entail thousands of individual computer runs. First, one must find a tentative stable capture position. Then to be sure that a capture is stable, a simulation of at least 40 orbits (and preferably 100 orbits) is necessary. Once all (or most) of the simulations are stable captures from that position of planetoid anomaly and planetoid eccentricity, the computer operator can move a certain increment of planetoid anomaly or planetoid eccentricity and repeat the process. Eventually all useful locations are to be explored and evaluated and the boundaries of a stable capture zone are defined. Each point entails between 10 and 20 simulations in order to determine if the satellite orbit is stable (see Table 4.7 for an example). One hundred calculation points can give one a sense of the field of stable capture, but it can take about three times more calculation points to define the boundaries in sufficient detail. Thus, 20 simulations per point times 300 points gives over 6000 individual calculations. Six thousand times 4 SCZs yields 24,000 computer runs.

Much of this process can be automated by an industrious physics and/or computer science student. In my view, an automated system would be very important for defining Stable Capture Zones in three-dimensions.

Table 4.7 Displacement Love Number summary for simulations associated with the encounter parameters for the stable capture scenario in Fig. 25 (a somewhat typical scenario within a Prograde Stable Capture Zone). Orbital parameters are: planet anomaly = 325°; planetoid anomaly = 196.636°; planetary orbit eccentricity = 0.0; planetoid orbit eccentricity = 0.015; pericenter radius for earth orbit = 0.0°; pericenter radius for planetoid orbit = 180°; perigee of the initial encounter = 1.43 earth radii prograde; distance of separation at beginning of simulation = 600 earth radii; orbital energy at beginning of simulation = (+)1.945 E28 J; orbital energy at closest approach = (−)1.795 E28 J; displacement Love number for earth = 0.7; Q for earth = 100, Q for planetoid = 1 for first encounter and 10 for all subsequent encounters. Note that all displacement Love numbers for the planetoid lower than 0.17 result in a 1-orbit escape. Two simulations with the planetoid displacement Love numbers of 0.23 and 0.24 resulted in orbital collisions. These types of orbital collisions occur when the orbital parameters for prograde capture are near the border of an SCZ

h (planetoid)	Result of simulation	h (planetoid)	Result of simulation
0.17	1-orbit escape	0.26	100-orbit capture
0.18	1-orbit escape	0.27	100-orbit capture
0.19	2-orbit escape	0.28	100-orbit capture
0.20	3-orbit escape	0.29	100-orbit capture
0.21	3-orbit escape	0.30	100-orbit capture
0.22	*28-orbit collision*	0.31	100-orbit capture
0.23	*13-orbit collision*	0.32	100-orbit capture
0.24	100-orbit capture	0.33	100-orbit capture
0.25	100-orbit capture	0.34	100-orbit capture

Table 4.7 gives a summary of results for the numerical simulations shown in Figs. 23, 24 and 25. In this Table only one variable, the displacement Love number (h) of the planetoid is systematically increased. The results in this Table demonstrate the relatively abrupt change from ESCAPE scenarios to COLLISION scenarios to STABLE CAPTURE scenarios for encounters within a Prograde Stable Capture Zone.

4.4.5 The Post-Capture Orbit Circularization Calculation

A computer program has been developed to calculate a time scale for the post-capture orbit circularization in a two-body context. The computer program averages, then make orbit corrections, then calculates, then averages, makes orbit corrections adiabatically as the orbit progressively circularizes. Depending on the h values and Q values that one uses as input, the resulting time scale is different. Figure 4.28 shows the results of four different sets of parameters for such a calculation and the calculation procedure is explained in more detail in the caption of that figure.

The calculation procedure for the orbit scenarios in Fig. 4.28 yields an unrealistically long time scale for the orbit circularization. The ideal procedure for this calculation in a co-planar context is to start with the three-body capture calculation with all the solar perturbations that it entails and keep calculating until the orbit circularizes. This would require several 100 million years of continuous calculation.

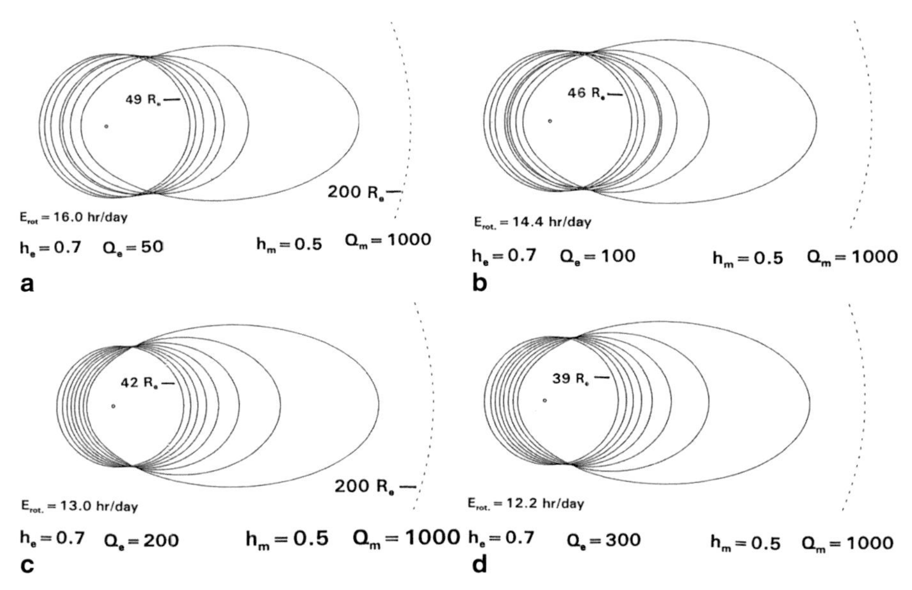

Fig. 4.28 Calculated two-body scenarios in which energy is dissipated within both the satellite (via radial tides) and within the planet (via tangential tides). In all four cases angular momentum of the lunar orbit is increasing as the rotational angular momentum of Earth is decreasing. The h and Q values of the Earth and Luna control the rate of angular momentum transfer. The only variable for this set of four calculations is the Q value of Earth. **a** For $Q_e = 50$, final rotation rate of Earth is 16.0 h/day and the lunar orbital radius is at 10% eccentricity at 49 earth radii (the two-body timescale for circularization to 10% eccentricity is about 1.1 Ga). **b** For $Q_e = 100$, final rotation rate of earth is 14.4 h/day and the lunar orbital radius at 10% eccentricity is at 46 earth radii (the two-body timescale for circularization to 10% eccentricity is about 1.4 Ga). **c** For $Q_e = 200$, final rotation rate for Earth is 13.0 h/day and the lunar orbital radius at 10% eccentricity is 42 earth radii (the two-body timescale for circularization to 10% eccentricity is about 1.4 Ga). **d** For $Q_e = 300$, final rotation rate for Earth is 12.2 h/day and the lunar orbital radius at 10% eccentricity is at 39 earth radii (the two-body timescale for circularization to 10% eccentricity is about 1.4 Ga). In scenario a, more energy is dissipated in the planet than in the satellite; in scenarios **b**, **c**, and **d**, more energy is dissipated in the satellite than in the planet. The quantity of energy dissipated in the satellite during the circularization sequence increases from scenarios **a** to **d**

Such a calculation is possible, but it has not yet been done. However, something more realistic than the two-body calculation can be attained by using the results of the two-body calculation as a base and plugging in the three-body calculation to check the effect of solar perturbations on the orbit which, in turn, effect the energy dissipation per unit of time within the two interacting bodies. This was done at a multitude of positions for the two-body calculation and the results give a more realistic appraisal of the time-scale of orbit circularization. Table 4.8 shows one attempt to develop a condensed time-scale for the orbital scenario in Fig. 4.28b.

Table 4.8 Some orbit and tidal values for the Moon for a quasi-three-body calculation for the post-capture orbit circularization sequence that incorporates some of the effects of solar perturbations. The h and Q values from Fig. 4.28b are used in the three-body "plug-in" calculations, the months are sidereal months. (Note: Some irregularities in minimum perigee and maximum tide in the 2000–4000 a range are due to solar gravitational perturbations of the lunar orbit.)

2-Body T.S (=10E6 yr)	3-Body T.S	Min. perigee (earth radii; 3-B)	Max. tide (Luna; km; 3-B)	Mo/yr
0	0	2.39	400	6.0
1	126 a	2.36	100	
2	281 a	2.53	90	
3	481 a	2.67	80	
4	748 a	2.81	60	
5	1085 a	3.13	50	
6	1515 a	3.20	45	
7	2033 a	3.57	30	
8	2733 a	3.42	40	
9	3467 a	3.44	40	
10	4347 a	3.59	30	6.5
50	347 Ka	7.27	3.0	8.3
100	3857 Ka	10.8	1.2	10.1
200	31.6 Ma	15.6	0.4	13.1
600	306.6 Ma	26.0	0.08	17.5
1000	671.0 M	33.1	0.04	19.0
1400	1030 Ma	42.0	0.0	19.5

4.4.6 A Qualitative Model for Generation of a Mare-Age Lunar Magnetic Field

Using the timescale in Table 4.8b for the post-capture evolution, a general model for generation of a mare-age magnetic field can be developed. First of all, since the vast quantity of energy dissipation for capture must be deposited in Luna, a large portion of the sub-crustal mantle must be melted and perhaps mildly superheated. The capture encounter alone can melt a zone ~ 50 km thick and within a few 10s of years a zone perhaps 200–300 km thick can be remelted. A shallow shell magnetic amplifier, similar to the one proposed for magnetization of the primitive crust, can be reactivated in the remelted magma ocean zone in the lunar upper mantle. An abbreviated version of the tidal and mechanical features of this model is presented in Figs. 4.29, 4.30, 4.31, and 4.32. The main drivers for generating the field are (1) rock tidal action due to close perigee passages associated with the capture encounter and the early stages of the orbit circularization sequence, (2) differential rotation of the lower part of the lunar mantle relative to the crustal complex due to a monotonic decrease in the tidal torque applied to the lunar crustal complex because of the decrease in speed of the lunar body during perigee passages (see Table 4.9 for these numbers).

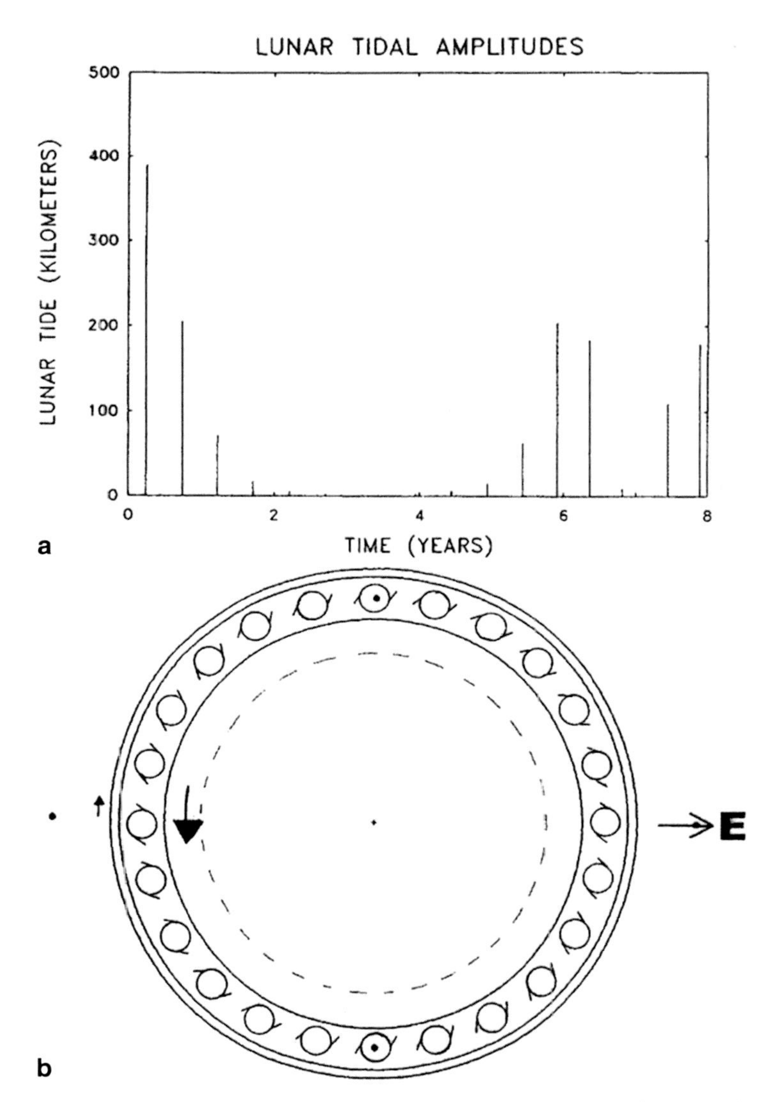

Fig. 4.29 Plot of lunar tidal amplitudes and associated TVI diagram illustrating how a lunar magnetic field may be generated during the orbit circularization sequence associated with the gravitational capture process. **a** Plot of the first 8 years of lunar tidal amplitudes following capture; **b** Diagram illustrating the maximum tidal distortion during a minimum distance perigee passage (400 km). This causes about 23 % radial tidal deformation. There are 6 months per year but many of the perigee passages are not close enough to register on the tidal amplitude plot at the scale shown on the plot. The *black dots* along the axis including the sub-earth point and anti-sub-earth point illustrates the extent of tidal distortion during the closest perigee passage for this orbital state. The *black dots* that are perpendicular to the axis of extension illustrate the extent of rock tidal drawdown during a perigee passage scenario. This tidal action is very episodic in that there is essentially no significant tidal action when the lunar body is beyond 60 earth radii. Note that after about three centuries of perigee passages, the maximum speed of the Moon (at perigee passage) is ~7.0 km/s which would cause the outer shell of the lunar body to rotate at about 4.0 h/day. At this point in time the lunar interior below the magma ocean zone is rotating at about the same rate

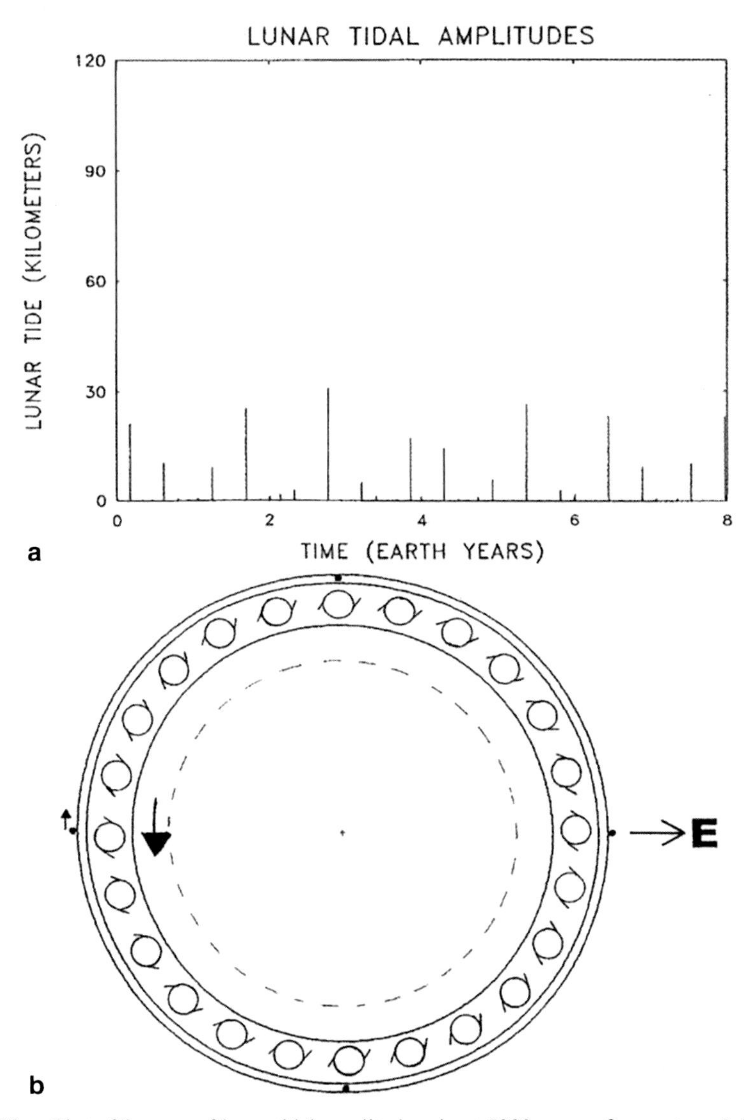

Fig. 4.30 **a** Plot of 8 years of lunar tidal amplitudes about 5000 years after capture; **b** Diagram illustrating the maximum tidal distortion during a minimum distance perigee passage (30 km). This causes about 2 % radial tidal deformation. There are 6.5 months per year but many of the perigee passages are poorly recorded on the tidal amplitude plot (at the scale shown on the plot). The speed of the lunar body during the closest perigee passages during this era is about 6.0 km/s which causes a rotation period for the outer shell of about 6.8 h

as the crustal complex. When moving forward in time, the lunar interior complex begins to rotate slightly faster than the crustal complex because the tidal torque during perigee passages is systematically slowing the crustal complex. This differential rotation powers the TVI circulation cells of the shallow-shell magnetic amplifier

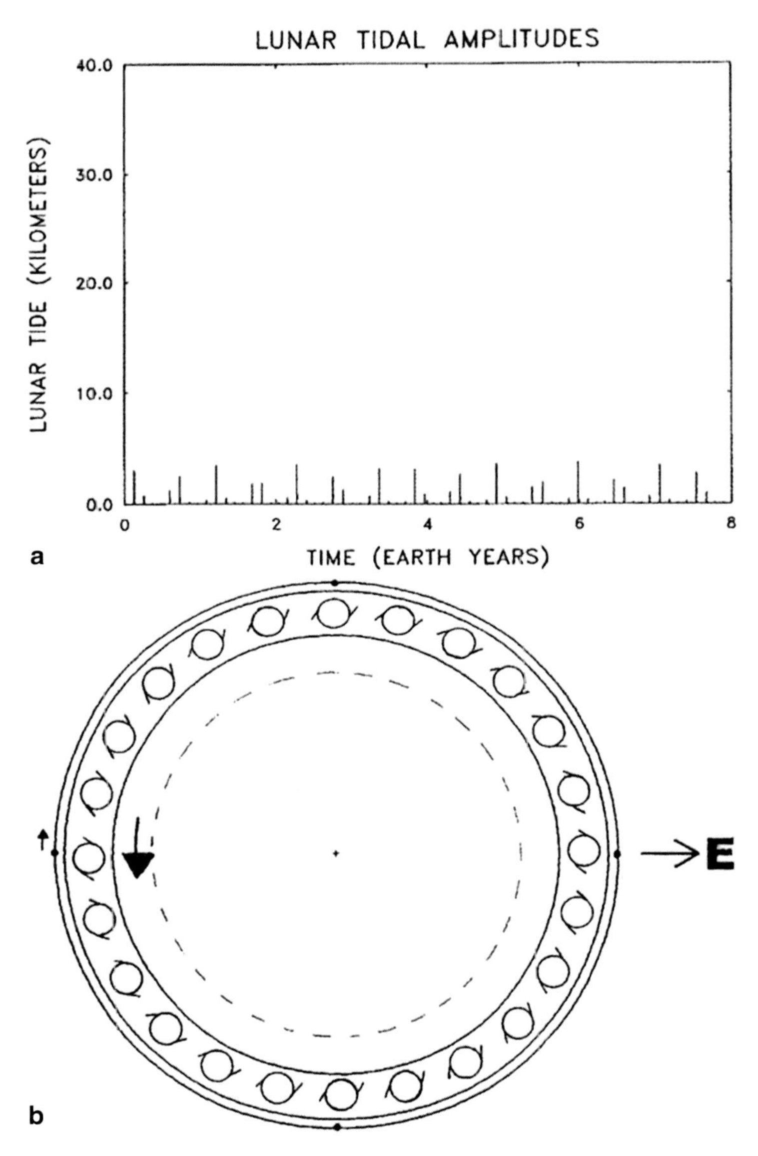

Fig. 4.31 **a** Plot of 8 years of lunar tidal amplitudes about 350 Ka after capture; **b** Diagram illustrating the maximum tidal distortion during a minimum distance perigee passage (3 km). This causes about 0.2 % radial tidal deformation. There are 8.2 months per year but many of the perigee passages are poorly recorded on the tidal amplitude plot at the scale shown on the plot. The speed of the lunar body during the closest perigee passages during this era is about 4 km/s which causes a rotation period for the outer shell of about 20 h

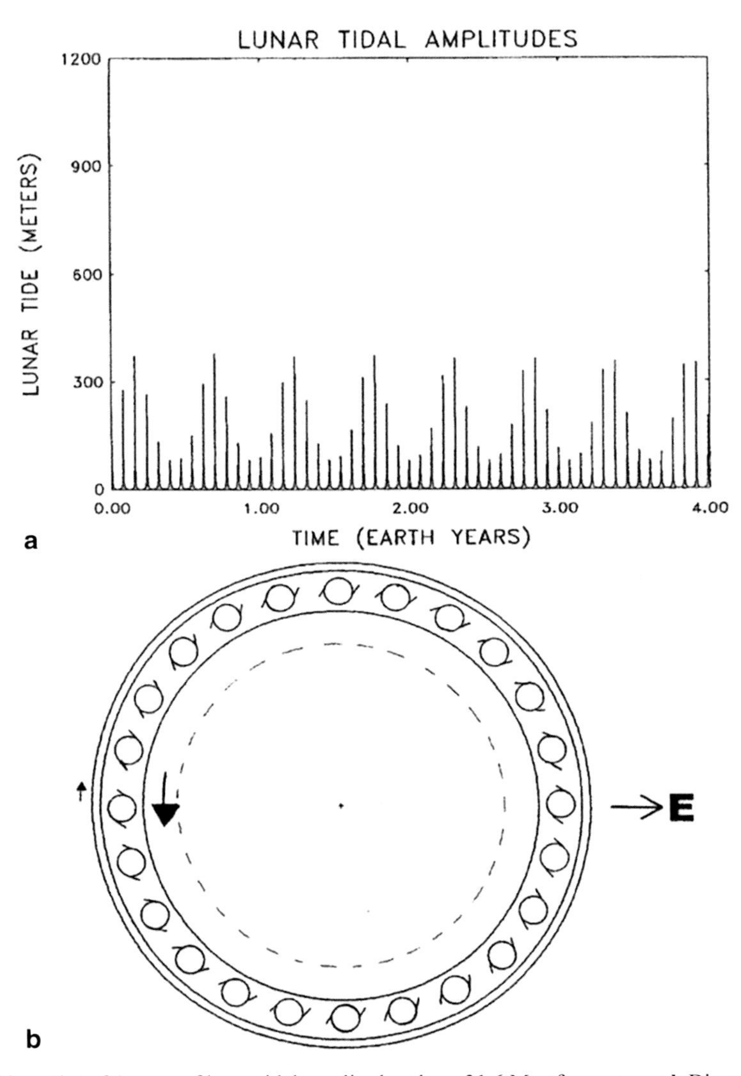

Fig. 4.32 **a** Plot of 4 years of lunar tidal amplitudes about 31.6 Ma after capture; **b** Diagram illustrating the magnetic amplifier operation. Although the tidal distortion is too small to show on the plot there is about 0.02 % radial tidal deformation at minimum perigee distance (400 m). There are 13 months per year and all of the perigee passages show on the tidal amplitude plot. Note that over 30 million years after capture, the pattern of tidal amplitude keeps the magnetic amplifier moving as long as the mantle viscosity is within certain limits. At this stage of orbit evolution, the speed of the lunar body at close perigee passages is about 2.5 km/s and the period of the outer shell is about 78 h. Eventually at about 300 Ma the period of rotation is about 150 h

Table 4.9 Some orbit and tidal values for the Moon for the first 300 Ma of this quasi-three-body calculation that relate to the decrease in rotation rate of the crustal complex during the circularization sequence because of a monotonic increase in the minimum distance at perigee passage. (Note: Some irregularities in v_p (max.) and rotation rate in the 2000–4000 a range are due to solar gravitational perturbations of the lunar orbit.)

3B TS	Input (rp) earth radii	Input ecc	vp (max.) (km/s)	Rot. rate (max.) (h/day)
0	17.07	0.81	7.23	3.7
126 a	17.09		7.27	3.6
281 a	17.11		7.02	4.0
481 a	17.13		6.84	4.3
748 a	17.16		6.66	4.7
1085 a	17.18		6.30	5.5
1515 a	17.20		6.23	5.7
2033 a	17.22		5.89	6.7
2733 a	17.24		6.03	6.3
3467 a	17.27		6.00	6.4
4347 a	17.29	0.80	5.87	6.8
347 Ka	18.19	0.77	4.07	19.9
3857 Ka	19.34	0.72	3.29	36.3
31.6 Ma	21.56	0.64	2.48	78.8
306.6 Ma	28.82	0.42	1.90	151.9

Figure 4.33 illustrates how encounter-driven circulation cells can power a set of shallow-shell magnetic amplifiers and generate a weak lunar magnetic field to get recorded in the mare basalts as they are cooling through the Curie temperature. A weak "seed" field from the magnetic field of the Earth is at maximum strength (relative to the position of Luna) during the closest perigee passages and then decreases exponentially as the perigee distance increases and the speed of the lunar body at perigee passage decreases. (NOTE: The Appendix for this chapter has a full set of diagrams showing (1) multi-year orbital plots and (2) associated lunar tidal amplitude plots for each of the time steps for the orbit circularization sequence in Table 4.8.)

This quote from Dyal and Parkin (1973, p. 63) is appropriate here also:

> We are presented with some fascinating questions about lunar history: Did the ancient moon have an earthlike field, generated as the earth's is thought to be by an internal dynamo? Was the moon once in an orbit closer to the earth and within the strong terrestrial field? Was the moon magnetized in another part of the solar system and later captured by the earth?

4.4.7 Subsequent Orbit Expansion due to Angular Momentum Exchange between the Rotating Earth and the Lunar Orbit

Once the orbit is circularized then the major axis of the lunar orbit is increased as the rotation rate of Earth decreases due to the conservation of angular momentum (Fig. 4.34). There can be "bumps" in this fairly placid orbital pathway that can

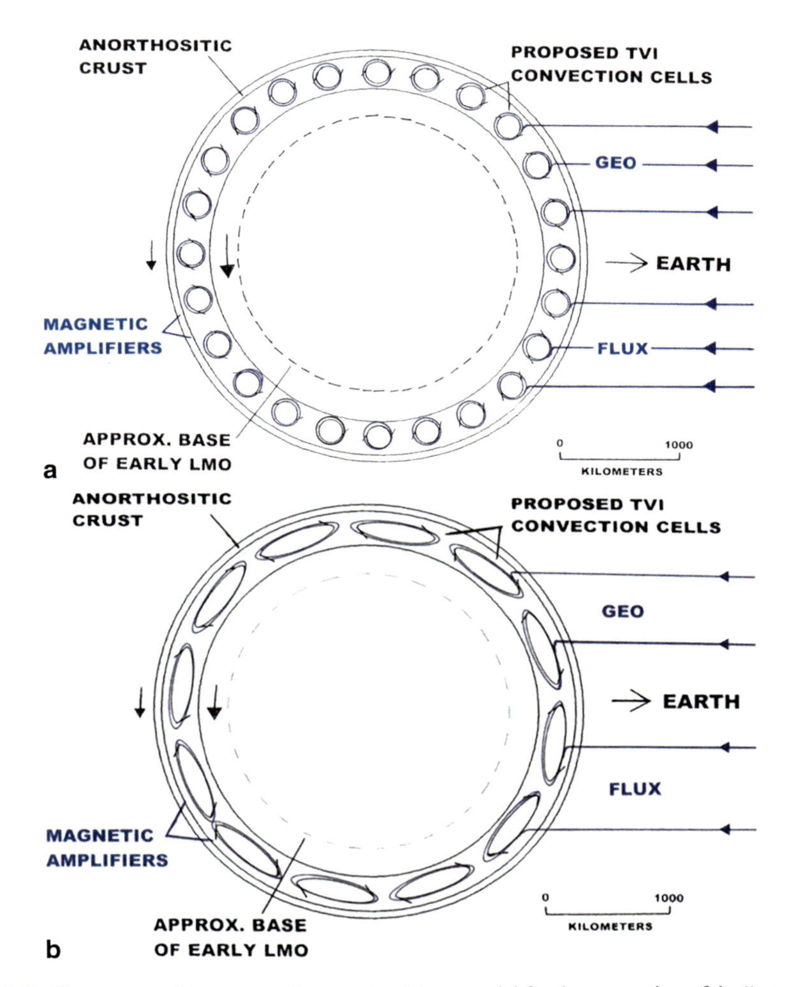

Fig. 4.33 Two stages of the proposed encounter-driven model for the generation of the "mare-era" lunar magnetic field, a magnetic field that is exponentially decaying over about 350 Ma (3.95–3.6 Ga) Cisowski et al. (1983), Fuller et al. (1974) and others. **a** Early Stage soon after capture. **b** Later stage when the magnetic amplifier field is much weaker. Note that for all but the first few perigee passages, the portion of the lunar body beneath the magma ocean is rotating faster than the outer shell. This differential rotation of the outer crustal complex relative to the lunar interior is the driving force for the "tapered roller-bearing" circulation cells for amplification of the terrestrial "seed" magnetic field. Thus the strongest magnetic field is predicted to be in the equatorial zone of the lunar body. In general, the tidal amplitudes decrease exponentially in time as does the differential rotation between the lunar interior and the lunar crustal complex

be caused by PLANET ORBIT—LUNAR ORBIT RESONANCES. These orbital resonances appear to have left an imprint in the rock record of Earth, in particular, and perhaps on the Moon as well. These planet orbit—lunar orbit resonances are the subject of Chap. 8.

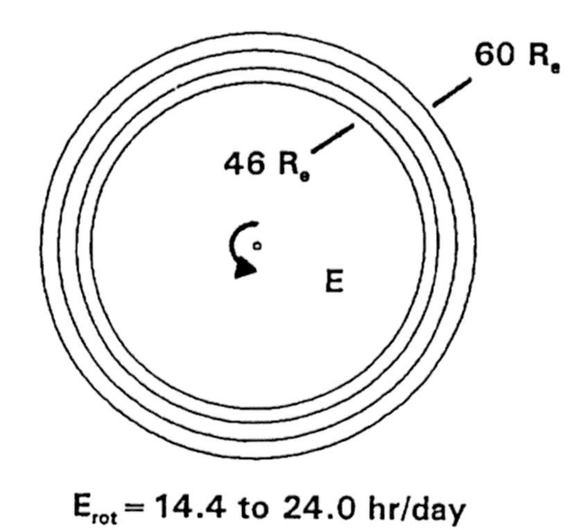

Fig. 4.34 Orbital diagram showing the general sequence of orbital states associated with the expansion of a circular lunar orbit. The first orbital state in this diagram is 46 earth radii (result for the scenario in Fig. 28b) and this is associated with an earth rotation of 14.4 h/day. Then over about 3.0 billion years the lunar orbit gradually expands to the present 60 earth radii as the Earth rotation rate gradually slows to the present, very comfortable, 24 h/day

4.5 Summary and Statement of the Fourth Paradox

Analysis of the process of GRAVITATIONAL CAPTURE in a co-planar, three-body context has added much understanding to the possibility and probability for the ORIGIN OF THE MOON BY GRAVITATIONAL CAPTURE. In the two-body analysis, the capture process was simply on the margin of being physically possible. However, using the three-body simulations, gravitational capture is shown to be physically possible under certain orbital conditions and a probability can be assigned to the capture process by way of the Geometry of Stable Capture Zones.

Future work can be done with the three-body calculation. For example, the Stable Capture Zone geometry can be extended to the third dimension. The early post-capture orbit can be calculated continuously, with all the resulting solar perturbations, in order to get more detailed information for the tidal regime for the Earth and Moon, for the timescale of the orbit circularization era, and for detailed modeling of the operation of an encounter-driven lunar magnetic field.

Now for Paradox # 4: *The Lunar-like body can be captured from an Earth-like orbit but the chemistry of the Moon is not consistent with that place of origin.* The chemistry of the lunar body is much more consistent with an origin near the Sun. This paradox has been a major hurdle for any gravitational capture model (e.g., Cameron 1972, 1973). The essence of this paradox was discussed in an earlier section of this chapter.

4.6 Summary and Conclusions for the Chapter

Progress on a Gravitational Capture Models for the origin of the Earth's Moon has been slow to neutral. The spectacular numerical simulations presented by the investigators of the GIM have captured the imagination of the general public as well as a certain segment of the science community. But the GIM offers very few, if any, unique solutions for the special features of Earth history, Lunar history, or Earth-Moon system history that need to be explained.

Progress on the Gravitational Capture Model has been hindered by four paradoxes. These are: (1) the smaller body must store and subsequently dissipate nearly all of the energy for its own capture, (2) large planetoids are more capturable than small planetoids, (3) "cool" planetoids are more capturable than either "hot" or "cold" planetoids, and (4) the chemistry of the lunar body is not consistent with its formation in the vicinity of the Earth's orbit. All of these statements are contrary to current thinking in the planetary science community but all of the statements, after critical analysis, are TRUE.

Using an equation for the transfer of a lunar-mass body from an Earth-like orbit to the largest stable geocentric orbit, it is shown that ~2.0 E28 J of energy must be dissipated by some physical process(es). Using a standard equation for energy dissipation in the interacting bodies in a two-body interaction context, it is shown that gravitational capture of a lunar-like planetoid by tidal dissipation processes during a single encounter is marginal at best. Numerical simulations of the process of gravitational capture yield some significant surprises. The first surprise is that a solar perturbation during a close encounter can decrease the energy dissipation requirements by up to ~50 % or more. The second surprise is that certain sets of encounter parameters (planet anomaly and planetoid orbital eccentricity) result in stable post-capture orbit configurations: i.e., stable against subsequent solar perturbations.

The main conclusion of this chapter is that gravitational capture of a lunar-mass body by an Earth-like planet IS PHYSICALLY POSSIBLE and that the probability for stable capture for any one orientation can be measured directly from the geometry of the STABLE CAPTURE ZONE.

Appendix

This appendix consists mainly of plots of orbits and tidal amplitudes for Luna during an orbit circularization sequence and for the generation of the second phase of a lunar magnetic field. Another favorable feature of this capture model is that it is compatible with traceback calculations of Hansen (1982) and Webb (1982). They both suggested that the lunar orbit could be as large as 30 to 40 earth radii early in the history of the Solar System. Hansen (1982), in particular, suggested capture of the Moon into a geocentric orbit with the angular momentum equivalent of about 30 earth radii.

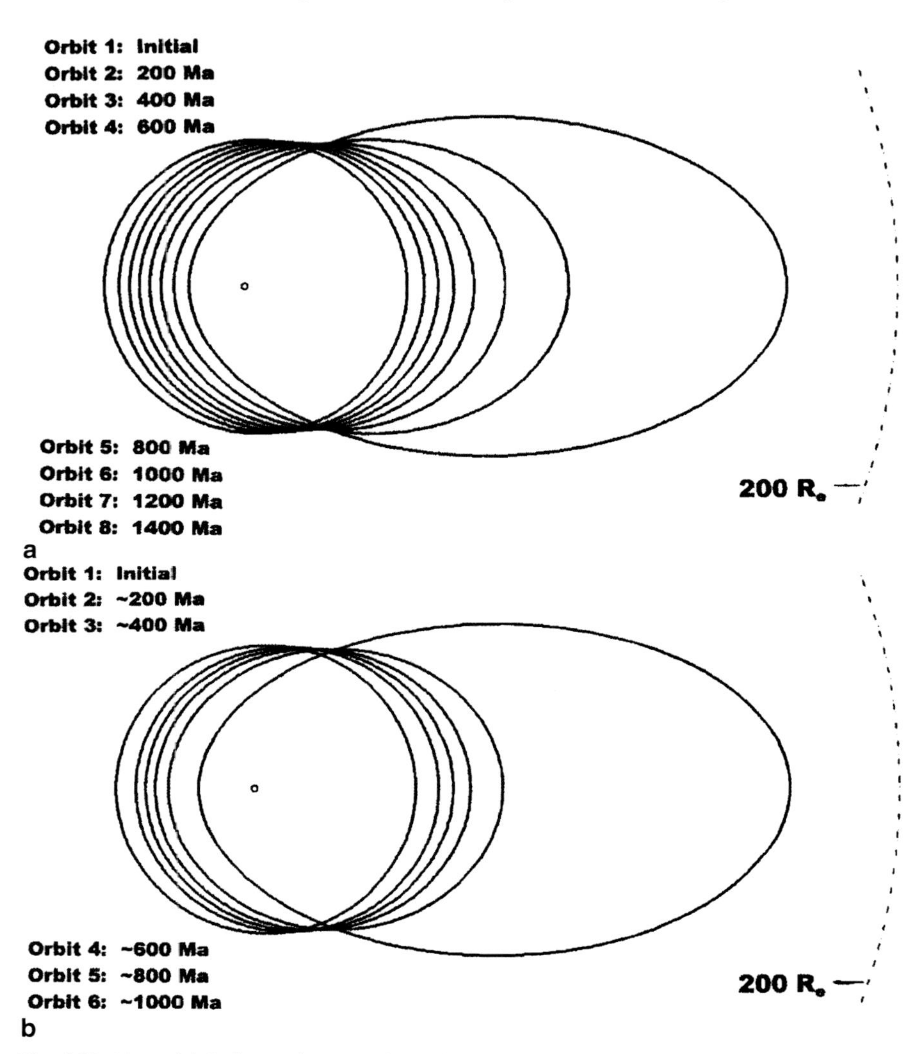

Fig. 4.35 Normal 2-B timescale vs. condensed 3-B timescale. **a** Normal 2-B calculated timescale, 200 Ma intervals. **b** Condensed 3-B timescale, 200 Ma intervals

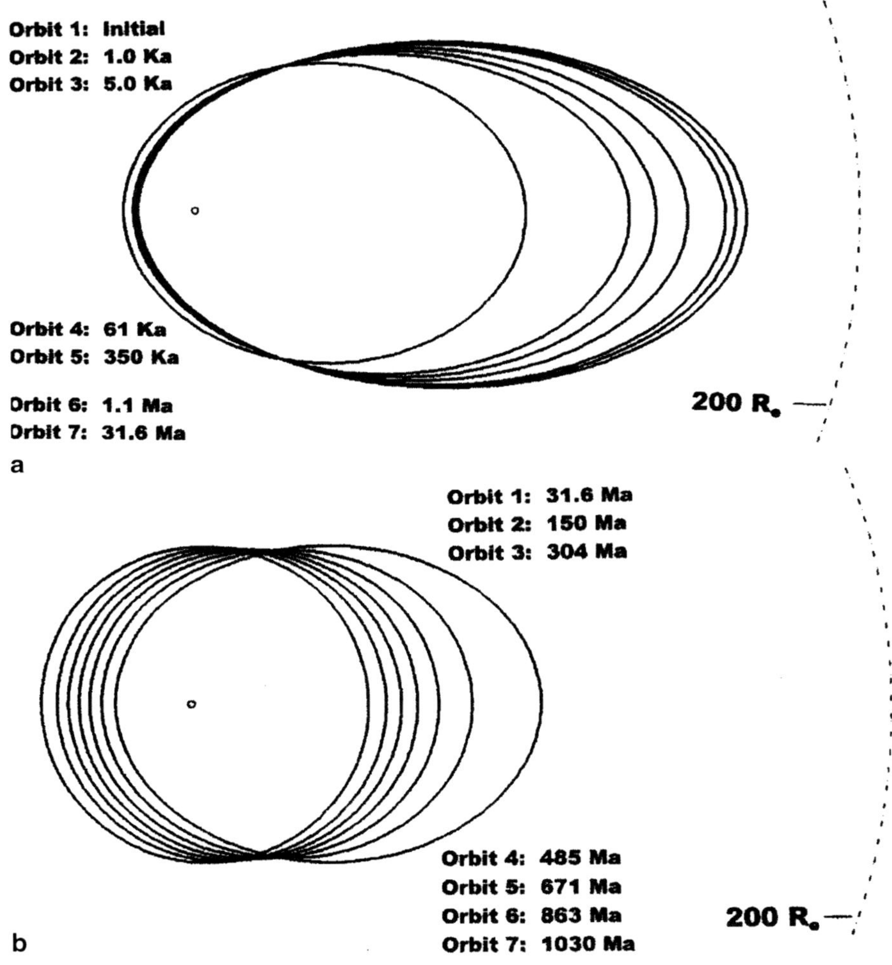

Fig. 4.36 Plots of orbits for condensed timescale. **a** Time of capture to 31.6 Ma. **b** 31.6–1030 Ma (~ 10 % eccentricity)

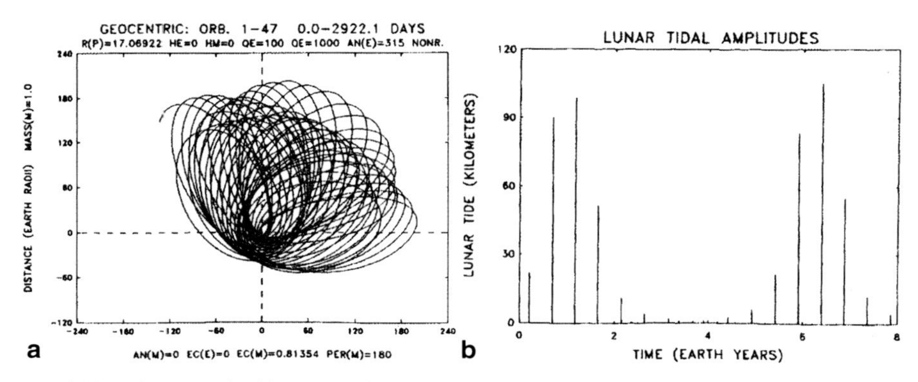

Fig. 4.37 **a** 8 years of orbits soon after capture. **b** 8 years of lunar tidal amplitudes soon after capture

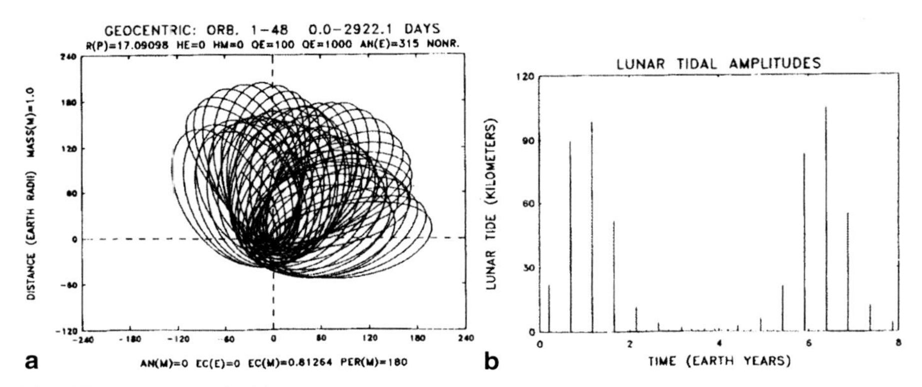

Fig. 4.38 **a** 8 years of orbits ~126 years after capture. **b** 8 years of lunar tidal amplitudes ~126 years after capture

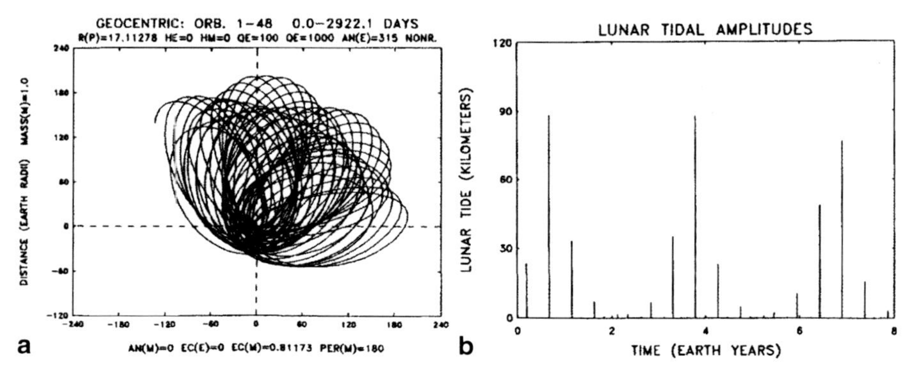

Fig. 4.39 **a** 8 years of orbits ~281 years after capture. **b** 8 years of lunar tidal amplitudes ~281 years after capture

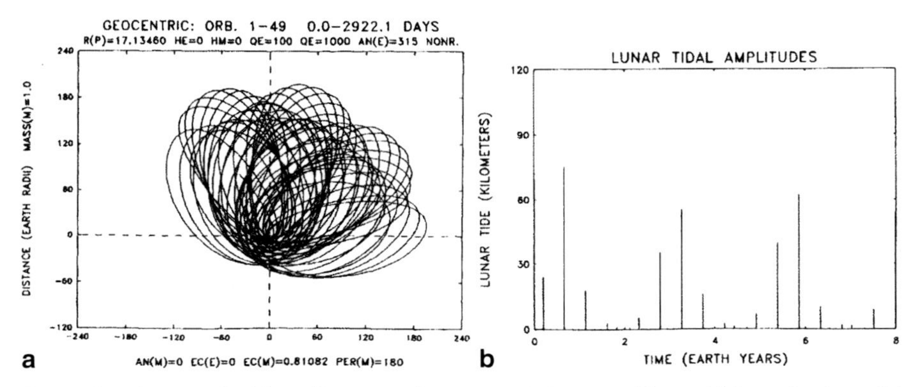

Fig. 4.40 **a** 8 years of orbits ~481 years after capture. **b** 8 years of lunar tidal amplitudes ~481 years after capture

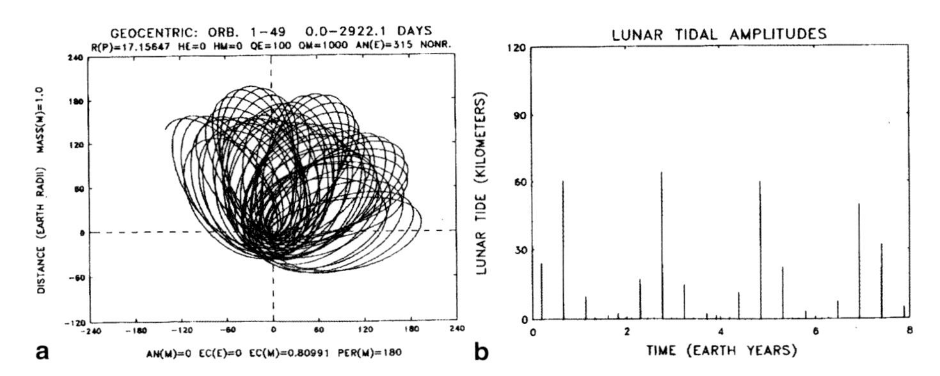

Fig. 4.41 **a** 8 years of orbits ~748 years after capture. **b** 8 years of lunar tidal amplitudes ~748 years after capture

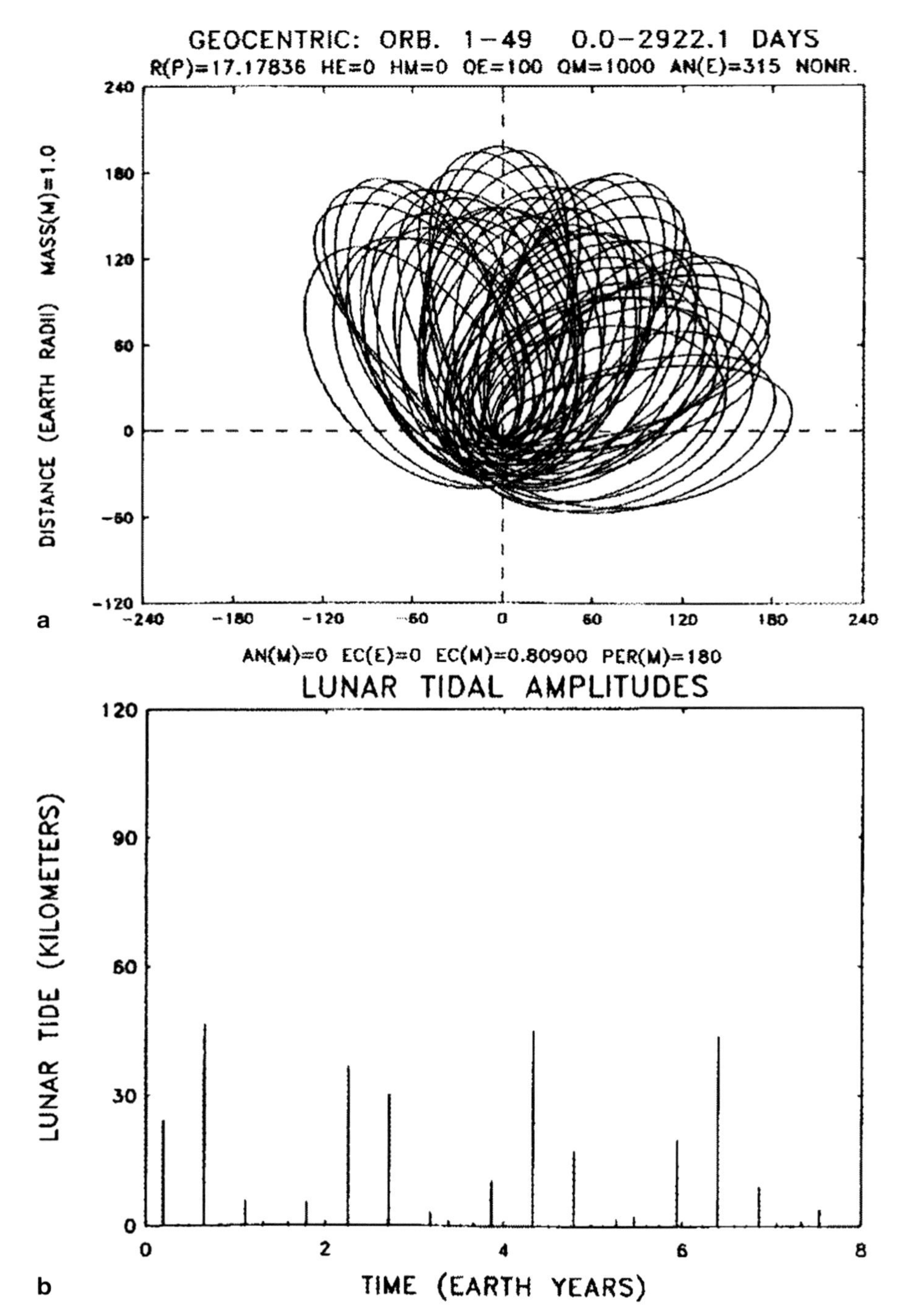

Fig. 4.42 **a** 8 years of orbits ~ 1085 years after capture. **b** 8 years of lunar tidal amplitudes ~ 1085 years after capture

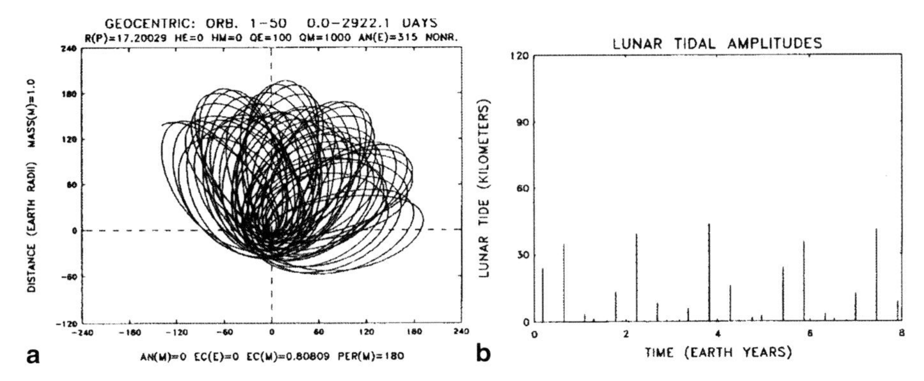

Fig. 4.43 **a** 8 years of orbits ~1515 years after capture. **b** 8 years of lunar tidal amplitudes ~1515 years after capture

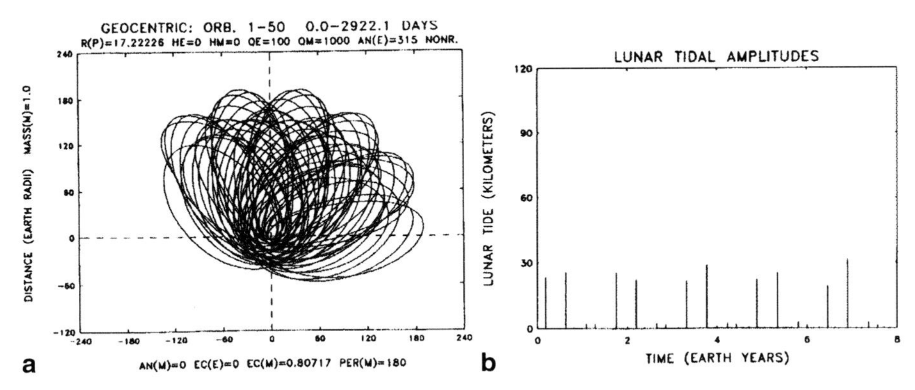

Fig. 4.44 **a** 8 years of orbits ~2033 years after capture. **b** 8 years of lunar tidal amplitudes ~2033 years after capture

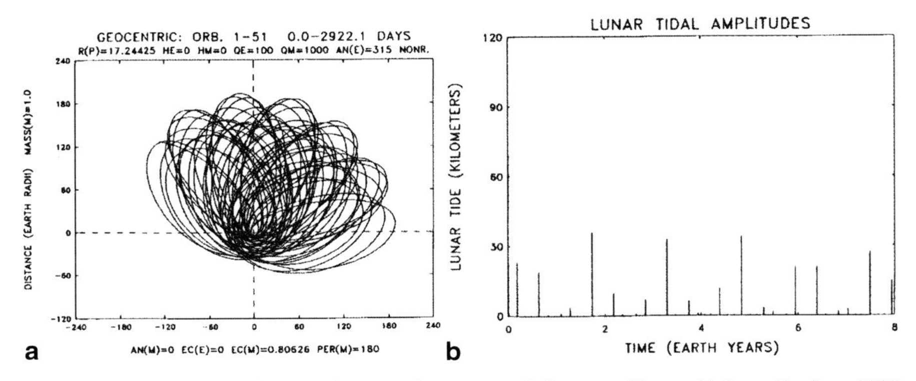

Fig. 4.45 **a** 8 years of orbits ~2733 years after capture. **b** 8 years of lunar tidal amplitudes ~2733 years after capture

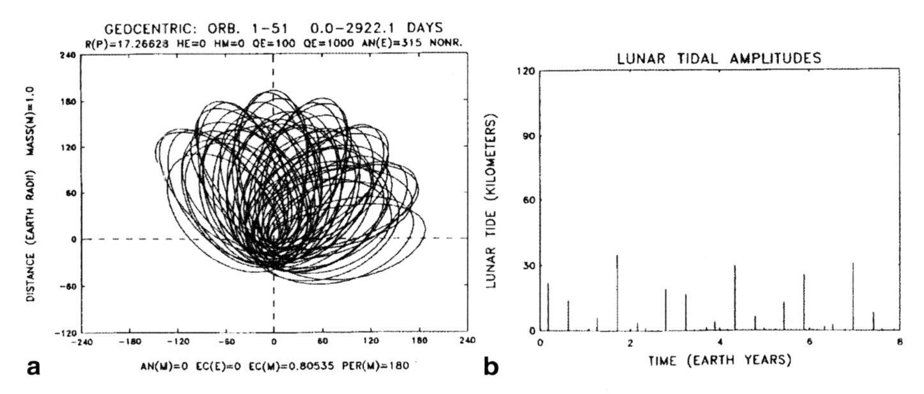

Fig. 4.46 **a** 8 years of orbits ~3467 years after capture. **b** 8 years of lunar tidal amplitude ~3467 years after capture

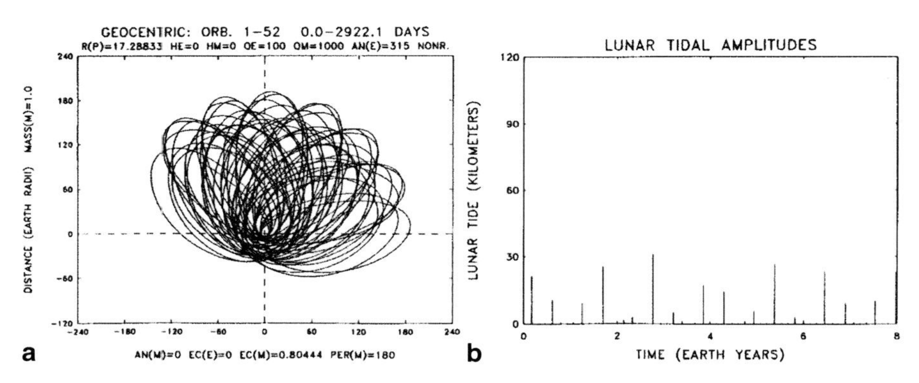

Fig. 4.47 **a** 8 years of orbits ~4347 years after capture. **b** 8 years of lunar tidal amplitudes ~4347 years after capture

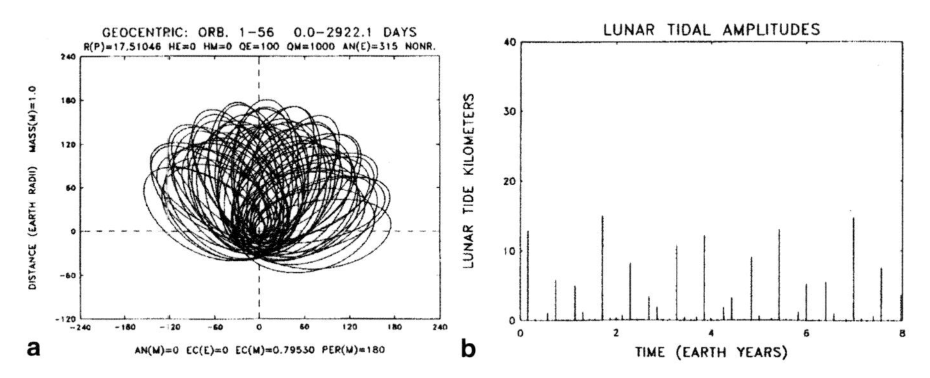

Fig. 4.48 **a** 8 years of orbits ~18.59 Ka after capture. **b** 8 years of lunar tidal amplitudes ~18.59 Ka after capture

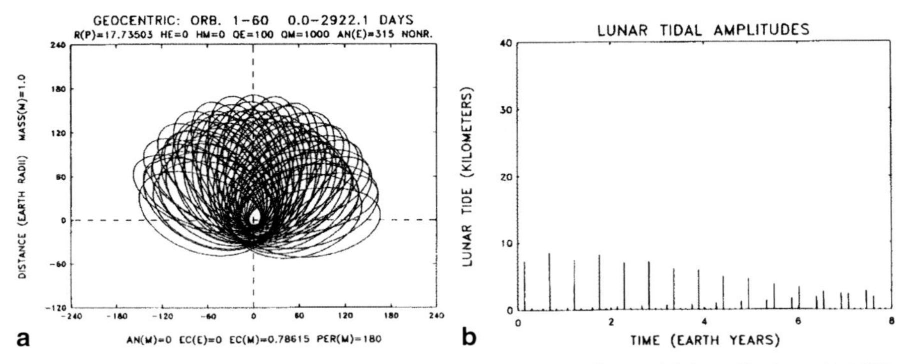

Fig. 4.49 **a** 8 years of orbits ~ 61.04 Ka after capture. **b** 8 years of lunar tidal amplitudes ~ 61.04 Ka after capture

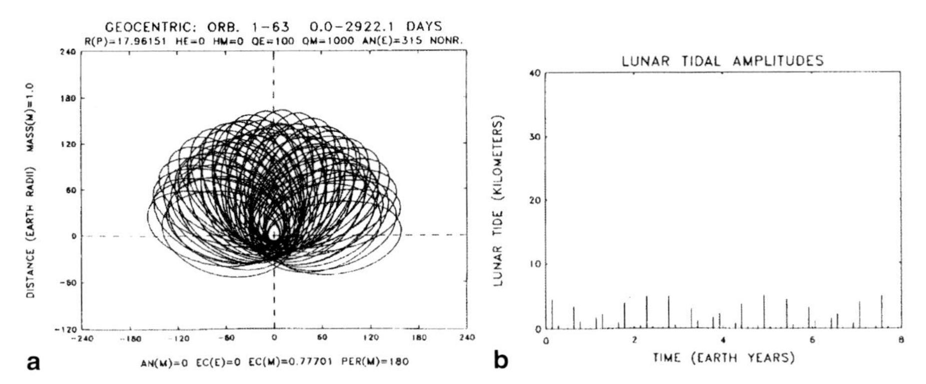

Fig. 4.50 **a** 8 years of orbits ~ 160 Ka after capture. **b** 8 years of lunar tidal amplitudes ~ 160 Ka after capture

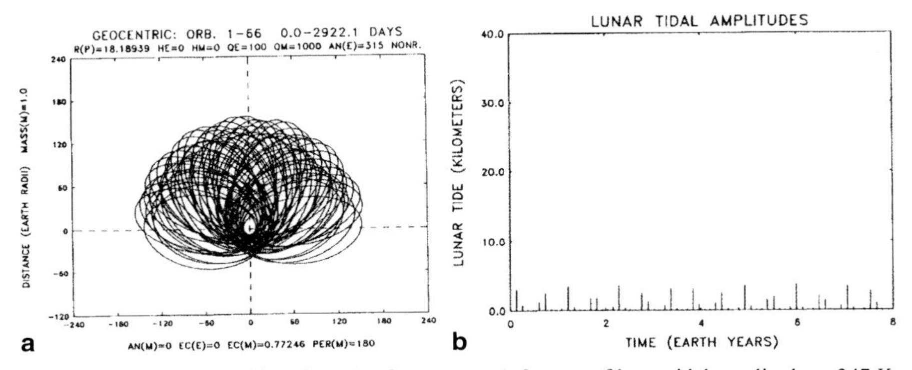

Fig. 4.51 **a** 8 years of orbits ~ 347 Ka after capture. **b** 8 years of lunar tidal amplitudes ~ 347 Ka after capture

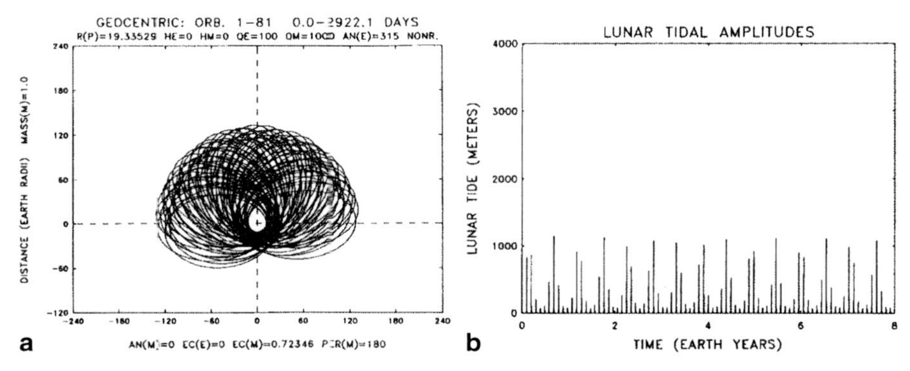

Fig. 4.52 **a** 8 years of orbits ~3.857 Ma after capture. **b** 8 years of lunar tidal amplitudes ~3.857 Ma after capture

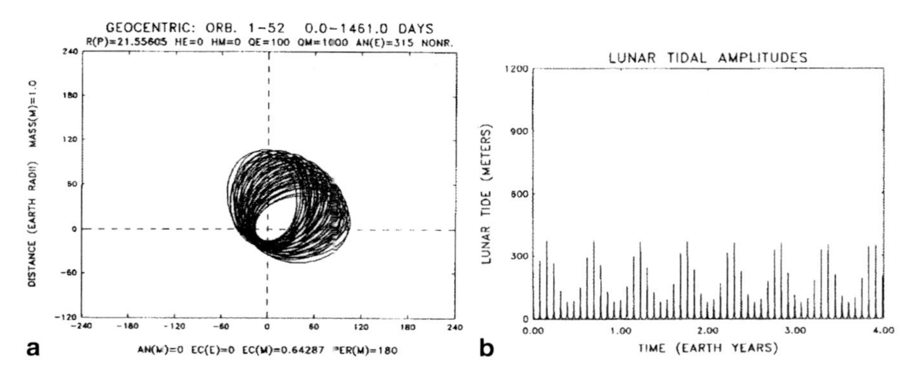

Fig. 4.53 **a** 4 years of orbits ~31.6 Ma after capture. **b** 4 years of lunar tidal amplitudes ~31.6Ma ago

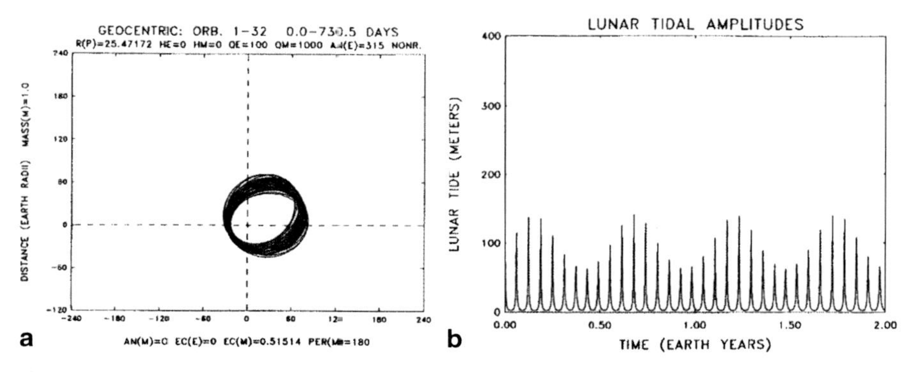

Fig. 4.54 **a** 2 years of orbits ~149 Ma after capture. **b** 2 years of lunar tidal amplitudes ~149 Ma after capture

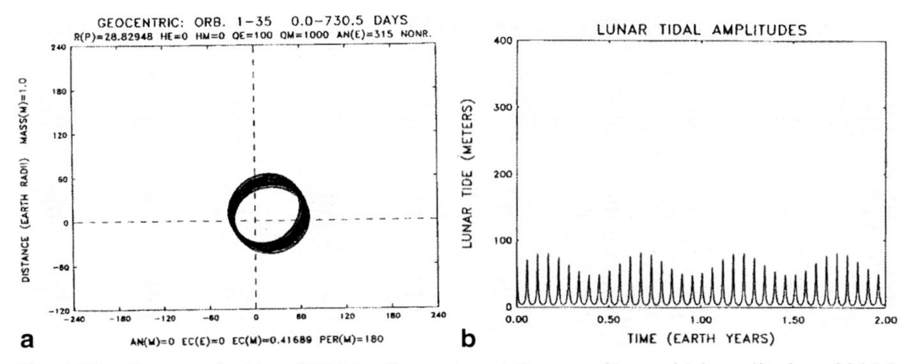

Fig. 4.55 **a** 2 years of orbits ~304 Ma after capture. **b** 2 years of lunar tidal amplitudes ~304 Ma after capture

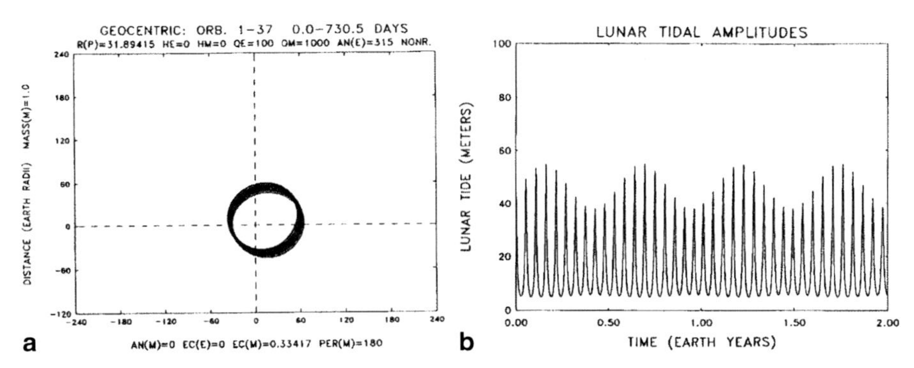

Fig. 4.56 **a** 2 years of orbits ~485 Ma after capture. **b** 2 years of lunar tidal amplitudes ~485 Ma after capture

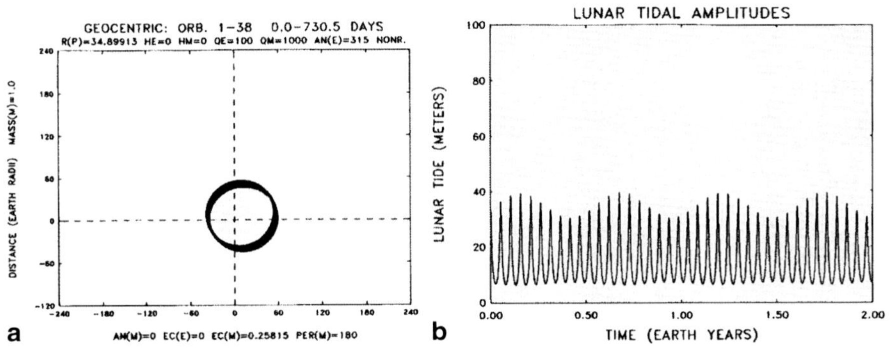

Fig. 4.57 **a** 2 years of orbits ~671 Ma after capture. **b** 2 years of lunar tidal amplitudes ~671 years after capture

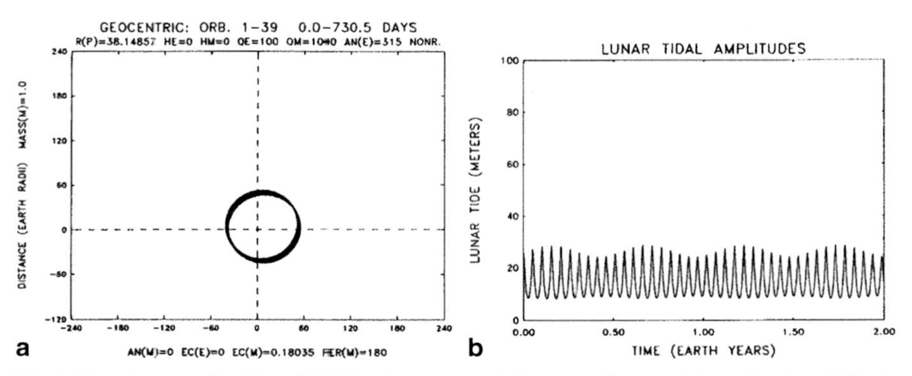

Fig. 4.58 **a** 2 years of orbits ~863 Ma after capture. **b** 2 years of lunar tidal amplitudes ~863 Ma after capture

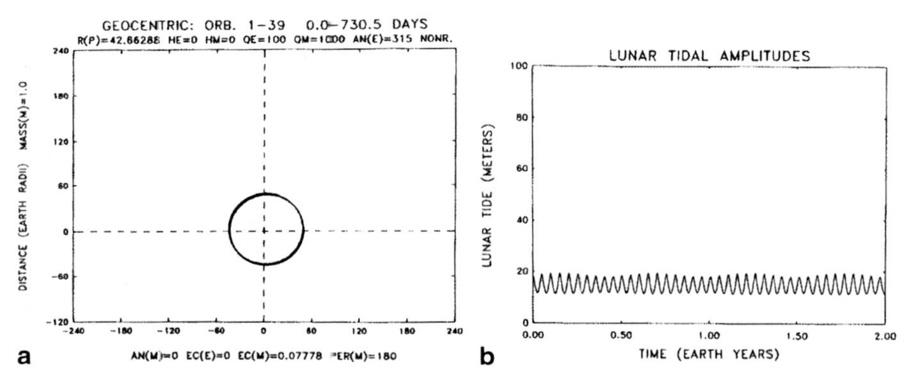

Fig. 4.59 **a** 2 years of orbits ~1030 Ma after capture. **b** 2 years of lunar tidal amplitudes ~1030 Ma after capture

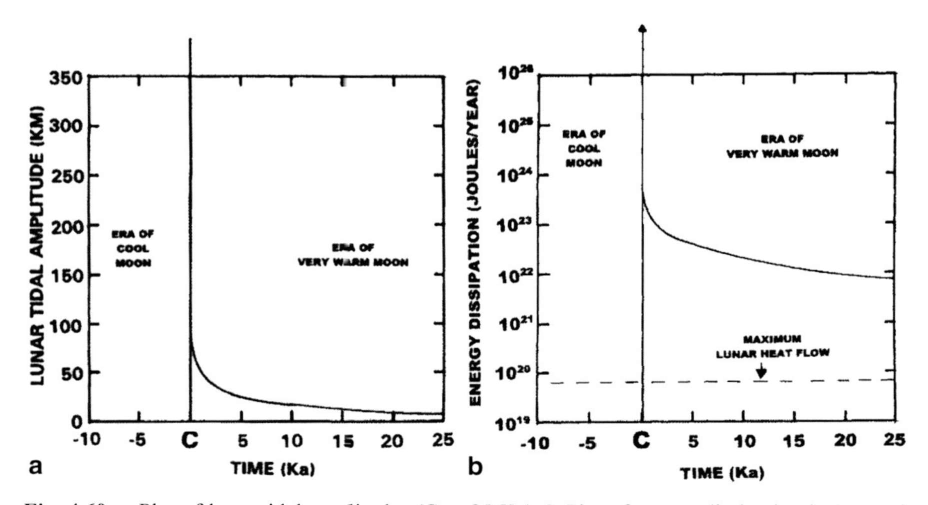

Fig. 4.60 **a** Plot of lunar tidal amplitudes (C to 25 Ka). **b** Plot of energy dissipation in Luna (C to 25 Ka)

References

Alfven H (1969) Atom, man, and the universe: the long chain of complications. W. H. Freeman and Company, San Francisco, 110 p

Alfven H, Alfven K (1972) Living on the third planet. W. H. Freeman and Company, San Francisco, 187 p

Bland PA, Spurny P et al (2009) An anomalous basaltic meteorite from the innermost main belt. Science 325:1525–1527

Bostrom RC (2000) Tectonic consequences of the Earth's rotation. Oxford University Press, London, 266 p

Cameron AGW (1972) Orbital eccentricity of Mercury and the origin of the moon. Nature 240:299–300

Cameron AGW (1973) Properties of the solar nebula and the origin of the moon. Moon 7:377–383

Campbell AJ, Humayun M (2005) Compositions of group IVB iron meteorites and their parent melt. Geochim Cosmochim Acta 69:4733–4744

Cisowski SM, Collinson DW, Runcorn SK, Stephenson A, Fuller M (1983) A review of lunar paleointensity data and implications for the origin of lunar magnetism. J Geophys Res 88:A691–A704

Connelly JN, Bizzarro M, Krot AN, Nordlund A, Wielandt D, Ivanova MA (2012) The absolute chronology and thermal processing of solids in the solar protoplanetary disk. Science 338:651–655

Day JMD, Pearson DG, Taylor LA (2007) Highly siderophile element constraints on accretion and differentiation of the Earth-Moon system. Science 315:217–219

Duffard R, Roig F (2009) Two new V-type asteroids in the outer Main Belt? Planet Space Sci 57:229–234

Dyal P, Parkin CW (1973) The magnetism of the Moon. Sci Am 225(2):63–73

Evans WN, Tabachnik S (1999) Possible long-lived asteroid belts in the inner Solar System. Nature 399:41–43

Evans WN, Tabachnik S (2002) Structure of possible long-lived asteroid belts: Monthly notices. R Astro Soc 333:L1–L5

Fuller M (1974) Lunar magnetism. Rev Geophys Space Phys 12:23–79

Goldreich P, Soter S (1966) Q in the solar system. Icarus 5:375–389

Goldstein JI, Scott ERD, Chabot N L (2009) Iron meteorites. Crystallization, thermal history, parent bodies, and origin. Chem Erde 69:293–325

Greenwood RC, Franchi IA, Jambon A, Buchanan PC (2005) Widespread magma oceans on asteroidal bodies in the early Solar System. Nature 435:916–918

Hansen KS (1982) Secular effects of oceanic tidal dissipation on the moon's orbit and the Earth's rotation. Rev Geophys Space Phys 30:457–480

Jeffreys H (1929) The Earth, its origin, history and physical constitution, 2nd edn. Cambridge University Press, Cambridge, 346 p

Kaula WM (1971) Dynamics aspects of lunar origin. Rev Geophys 9:117–238

Kaula WM, Harris AW (1973) Dynamically plausible hypotheses of lunar origin. Nature 245:367–369

Laskar J (1994) Large-scale chaos in the solar system. Astro Astrophys 287:L9–L12

Laskar J (1995) Large scale chaos and marginal stability in the Solar System: XIth International Congress of Mathematical Physics. International Press, Boston, 120 p

Levy EH (1972) Magnetic dynamo in the moon: A comparison with the Earth. Science 178:52–53

Levy EH (1974) A magnetic dynamo in the Moon? Moon 9:49–56

Love AEH (1911) Some problems in geodynamics. Cambridge University Press, Cambridge, 180 p (reprinted by Dover, 1967)

Love AEH (1927) A treatise on the mathematical theory of elasticity, 4th edn. Cambridge University Press, Cambridge, 643 p

MacDonald GJF (1964) Tidal friction. Rev Geophys 2:467–541

MacPherson GJ, Simon SB, Davis AM, Grossman L, Krot AN (2005) Calcium-aluminum-rich inclusions. Major unanswered questions. In: Krot AN, Scott ERD, Reipurth B (eds) Chrondrites and the protoplanetary disk. Astronomical Society of the Pacific: San Franciso, pp 225–250
Malcuit RJ, Winters RR (1996) Geometry of stable capture zones for planet Earth and implications for estimating the probability of stable gravitational capture of planetoids from heliocentric orbit. Abstracts Volume, XXVII Lunar and Planetary Science Conference. Lunar and Planetary Institute, Houston, pp 799–800
Malcuit RJ, Winters RR, Mickelson ME (1977) Is the Moon a captured body? Abstracts Volume, Eighth Lunar Science Conference, pp 608–610
Malcuit RJ, Mehringer DM, Winters RR (1988) Computer simulation of "intact" gravitational capture of a lunar-like body by an Earth-like body: Abstracts Volume, Lunar and Planetary Science XIX. Lunar and Planetary Institute, Houston, pp 718–719
Malcuit RJ, Mehringer DM, Winters RR (1989) Numerical simulation of gravitational capture of a lunar-like body by Earth. In: Proceedings of the 19th Lunar and Planetary Science Conference. Lunar and Planetary Institute, Houston, pp 581–591
Malcuit RJ, Mehringer DM, Winters RR (1992) A gravitational capture origin for the Earth-Moon system. Implications for the early history of the Earth and Moon. In: Glover JE, Ho SE (eds) Proceedings Volume, 3rd International Archaean Symposium, vol 22. The University of Western Australia, Crawley, pp 223–235
Melchior PJ (1978) The tides of planet Earth. Pergamon Press, New york, 609 p
Mittlefehldt DW, McCoy T J, Goodrich C A, Kracher A (1998) Non-chondritic meteorites from asteroidal bodies. In: Papike JJ (ed) Planetary materials: reviews of mineralogy, vol 36. pp 4–1 to 4–195
Munk WH, MacDonald GJF (1960) The rotation of the Earth. Cambridge University Press, London, 323 p
Peale SJ, Cassen P (1978) Contributions of tidal dissipation to lunar thermal history. Icarus 36:245–269
Prichard ME, Stevenson DJ (2000) Thermal aspects of a lunar origin by giant impact. In: Canup RM, Righter K (eds) Origin of the Earth and moon. University of Arizona Press, Tucson, pp 179–196
Ribeiro AO, Roig F, Canada-Assandri M, Carvano JMF, Jasmin FL, Alvarez-Candal A, Gil-Hutton R (2014) The first confirmation of V-type asteroids among the mars-crosser population. Planet Sp Sci 92:57–64
Roig F, Nesvorny D, Gil-Hutton R, Lazzaro D (2008) V-type asteroids in the middle main belt. Icarus 194:125–136
Ross M, Schubert G (1986) Tidal dissipation in a viscoelastic planet. Proceedings of the 16th Lunar and Planetary Science Conference. J Geophys Res 91:D447–D452
Roy AE (1965) The foundations of astrodynamics. The Macmillan Company, New York, 385 p
Russell CT (1980) Planetary magnetism. Rev Geophys Space Phys 18:77–106
Ruzicka A, Snyder GA, Taylor LA (1999) Giant impact and fission hypotheses for the origin of the Moon. A critical review of some geochemical evidence. In: Snyder GA, Neal CR, Ernst WG (eds) Planetary petrology and geochemistry. Geological Society of America, International Book Series 2:121–134
Ruzicka A, Snyder GA, Taylor LA (2001) Comparative geochemistry of basalts from the Moon, Earth, HED asteroid, and Mars. Implications for the origin of the Moon. Geochim Cosmochem Acta 65:979–997
Salmeron R, Ireland TR (2012) Formation of chondrules in magnetic winds blowing through the proto-asteroid belt. Earth Planet Sci Lett 327–328:61–67
Scott ERD, Greenwood RC, Franchi IA, Sanders IS (2009) Oxygen isotopic constraints on the origin and parent bodies of eucrites, diogenites, and howardites. Geochim Cosmochim Acta 73:5835–5853
Sharp LR, Coleman PJ Jr, Lichtenstein BR, Russell CT, Schubert G (1973) Orbital mapping of the lunar magnetic field. Moon 7:322–341

Shu FH, Shang H, Glassgold AE, Lee T (1997) X-rays and fluctuating X-Winds from protostars. Science 277:1475–1479
Shu FH, Shang H, Gounelle M, Glassgold AE, Lee T (2001) The origin of chondrules and refractory inclusions in chondritic meteorites. Astrophys J 548:1029–1050
Singer SF (1968) The origin of the moon and geophysical consequences. Geophys J R Astron Soc 15:205–226
Singer SF (1970) The origin of the moon and its consequences. Trans Am Geophys Union 51:637–641
Smoluchowski R (1973a) Lunar tides and magnetism. Nature 242:516–517
Smoluchowski R (1973b) Magnetism of the moon. The Moon 7:127–131
Sonett CP, Colburn DS, Schwartz K (1975) Formation of the lunar crust: an electrical source of heating. Icarus 24:231–255
Stacey FD (1977) Physics of the Earth, 2nd edn. Wiley, London, 414 p
Taylor SR (2001) Solar System evolution: a new perspective, 2nd edn. Cambridge University Press, Cambridge, 460 p
Wasson JT (2013) Vesta and extensively melted asteroids. Why HED meteorites are probably not from Vesta. Earth Planet Sci Lett 381:138–146
Webb DJ (1982) Tides and the evolution of the Earth-Moon system. Geophys J R Astron Soc 70:261–271
Winters RR, Malcuit RJ (1977) The lunar capture hypothesis revisited. Moon 17:353–358
Wood JA (2004) Formation of chondritic refractory inclusions. The astrophysical setting. Geochim Cosmochim Acta 68:4007–4021
Wood JA, Dickey JS Jr, Marvin UB, Powell BN (1970) Lunar anorthosites and a geophysical model of the moon. Proceedings of the Apollo 11 Lunar Science Conference, Lunar Science Institite, Houston, vol 1, pp 965–988

Chapter 5
Some Critical Interpretations and Misinterpretations of Lunar Features

...its present face still bears scars and traces of many events which took place in the inner precincts of the solar system not long after its formation. If so, this should make our Moon the most important fossil of the solar system and a correct interpretation of its stony palimpsest should bring rich scientific rewards.

(Zdenek Kopal, 1973, The Solar System: Oxford University Press, p. 82.)

Urey felt that the Moon might well be a primitive body, formed in the early days of the solar system. If this were so, the record of its early history preserved on the surface of the Moon, would provide invaluable clues to the origin and evolution of other bodies of the solar system as well.

(Homer E. Newell, 1973, Harold Urey and the Moon: The Moon, v. 7, p. 1).

5.1 Discussion of Some Speculations of Harold Urey and Zdenek Kopal

This somewhat lengthy chapter is a series of "vignettes" concerning certain concepts or features of solar system science. A "vignette" for our purposes is defined as "a short literary composition characterized by compactness, subtlety, and delicacy". These short topics are:

a. critique of currently accepted interpretations of lunar surface features
b. consideration of the directional properties of lunar surface features
c. review of the controversy over mass concentrations on the lunar surface
d. review of the concept of a "late heavy bombardment" of the lunar surface
e. review of the concept of a "cool early earth" and implications for lunar origin
f. discussion of the "origin of water problem" on Earth and implications

Robert J. Malcuit, *The Twin Sister Planets Venus and Earth*,
DOI 10.1007/978-3-319-11388-3_5

The main questions that Kopal and Urey were asking are still relevant today:

1. Are there surface features that relate to the origin of the Moon and/or the origin of the Earth-Moon system such as the distribution of craters, the distribution of circular mare basins, the distribution of mare basalts, etc.?
2. Are there other physical features that relate to the origin of the Moon and/or the origin of the Earth-Moon system such as (a) patterns of lunar rock magnetization, (b) the coefficient of the moment of inertia of the lunar body, (c) the zonation of the lunar magma ocean as determined by seismic wave information, etc.?
3. Are there chemical features that relate to the origin of the Moon and/or the origin of the Earth/Moon system such as (a) estimates for the chemical composition of all or part of the lunar body (i.e., Gast 1972; Anderson 1972, 1973, 1975; Cameron 1972, 1973; Taylor 1975, 1982, 2001) which strongly suggest that there is a similarity of the composition of the Moon, in whole or in part, to the composition of calcium-aluminum inclusions (CAIs) in chondritic meteorites; (b) the similarities of isotopic patterns of lunar materials to those of other solar system materials such as Earth and various types of meteorites, meteorite components such as chondrules, asteroid parent bodies, etc.?

5.2 Vignette A. Critique of the "Commandments" for Interpretation of Lunar Surface Features

The scientists assigned to study the Moon for the Apollo Missions had to start somewhere. For the most part the rules that they developed for the interpretation of lunar features have worked fairly well. Lunar geologic maps were made based on objective criteria and these maps were used very effectively for selecting landing sites for the Apollo astronauts.

The main paradigm for the project was that most of the features of the lunar surface are the result of *impact of solid bodies at hypervelocities* onto the lunar surface. The concept of overlapping impact ejecta aprons was a key for the interpretation of relative age relationships if such ejecta features could be identified.

But then there were the basalts which are normally interpreted as a product of volcanism. The big question was whether or not these volcanic rocks were (1) the products of internal heat buildup over time or (2) the products of solid body impacts (i.e., impact-generated melts).

Another issue was the use of crater density studies to determine the relative ages of lunar surface features. I think that almost everyone can agree with the principle that the *relative age* of lunar surface features can be determined via crater density studies if all the craters are considered as primary craters.

The main problem with crater density studies is the temptation to relate them to the *absolute age* of lunar surface features. There are many assumptions that go into such an interpretation and the major one is that the craters were formed by a somewhat uniform rain of external bodies of various sizes. Another assumption is that there was an exponential decrease in the rate of impacting bodies over time. A major problem with this approach is that any "spikes" in the rate of impact can cause a major misinterpretation of the absolute age of a portion of the lunar surface.

5.2.1 Purpose

The purpose of this section is to point out some of the inconsistencies in the "commandments" or rules for the interpretation of lunar surface features that had been developed for the geological mapping group of the Astrogeology Unit of the United States Geological Survey in support of the Apollo Program. These rules for interpretation, summarized in Wilhelms (1993), were followed essentially unchanged over the past 50 years or so. Then I want to suggest some amendments to certain of the rules. (Note: Of the 17 rules I have serious disagreements with only *three* and some serious questions *for two others*. Thus I am in agreement with many of the rules.)

We must remember that all of this suggested modification of the rules for interpretation of lunar surface features must be done within the framework of the SCIENTIFIC METHOD. Another major point is that we must distinguish between "fact" and "interpretation".

5.2.2 Dedication of this Section of the Chapter

This section is dedicated to all those keen observers whose observations and interpretations should have been given more time and consideration at the time they were proposed. But because of the "rush" of the Apollo Program, the insight of some scientists was "brushed aside", at least temporarily. However, in many cases, some of these "unused" ideas were recorded (published) in the literature, however briefly, so that a reader decades or centuries later can recover some traces of their intellectual processes.

I have not read or digested all of the literature on lunar features, but I will point out, in appropriate places in this presentation, some of the key observations that could have guided lunar research in a different direction (i.e., what I think were "forks in the road" of interpretation and of "roads" not taken).

5.2.3 The Scientific Method and its Application to this Particular Problem

An outline of the SCIENTIFIC METHOD was presented in Chap. 1. Here I am illustrating how this method may be applied to the problem of the interpretation of lunar surface features.

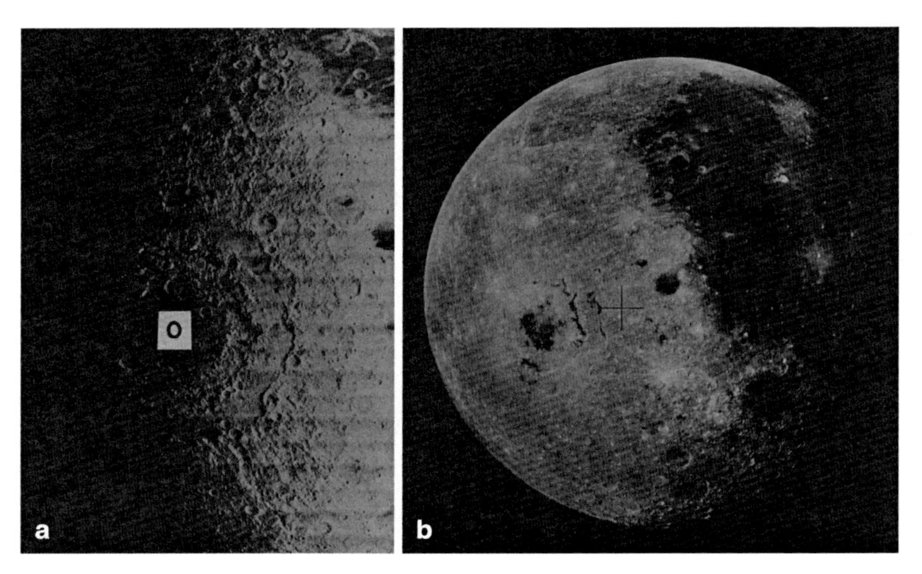

Fig. 5.1 Overview of the western limb of the Moon. *Left photo*: Lunar orbiter photograph IV-187-M. (*O* Mare Orientale); *Right photo*: Zond 8 photograph of the Mare Orientale region

5.2.3.1 Step A: Some Facts to be Explained by a Successful Hypothesis for the Origin of Certain Lunar Features

a. Lunar maria are concentrated mainly along the current equatorial zone of the Moon (Stuart-Alexander and Howard 1970). (See Figs. 5.1, 5.2, and 5.3).
b. Most of the large circular maria are located very nearly on a lunar great circle. These are, from lunar west to east, Mare Orientale, Mare Imbrium, Mare Serenitatis, Mare Crisium, Mare Smythii, and Crater Tsiolkovsky (Franz 1912, 1913; Dietz 1946; Lipskii 1965; Lipskii et al. 1966; Malcuit et al. 1975; Lipskii and Rodionova 1977; Runcorn 1983). (See Figs. 5.1, 5.2, and 5.3).
c. The average diameter of these features decreases systematically from Mare Imbrium eastward to the mare-filled circular Crater Tsiolkovsky (Malcuit et al. 1975). (See Figs. 5.2 and 5.3).
d. The average elevation of these prominent circular features decreases systematically from Mare Imbrium to Mare Smythii (Sjogren and Wollenhaupt 1973). (See Fig. 5.4).
e. The pattern of excess mass (mascons) associated with circular maria shows a decrease from Mare Imbrium to Mare Smythii (Sjogren and Wollenhaupt 1973).
f. Mare Orientale and Mare Marginis have much weaker mascon signatures than expected from the trend registered by the other circular maria (Gottlieb et al. 1969).
g. A number of maria with mascons are located near a second great-circle pattern. These are Mare Humorum, Mare Nubium, Mare Nectaris, and Mare Moscoviense (Runcorn 1983).

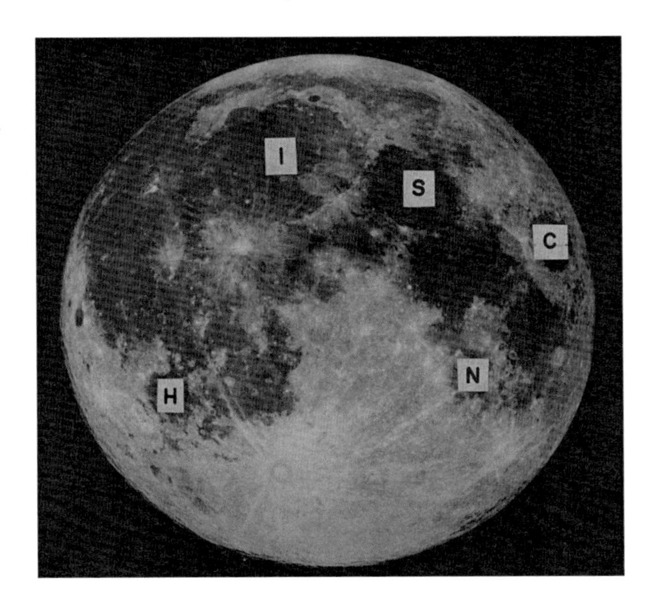

Fig. 5.2 Overview of the front face of the Moon (Hale observatory photo). *I* Mare Imbrium, *S* Mare Serenitatis, *C* Mare Crisium, *H* Mare Humorum, *N* Mare Nectaris

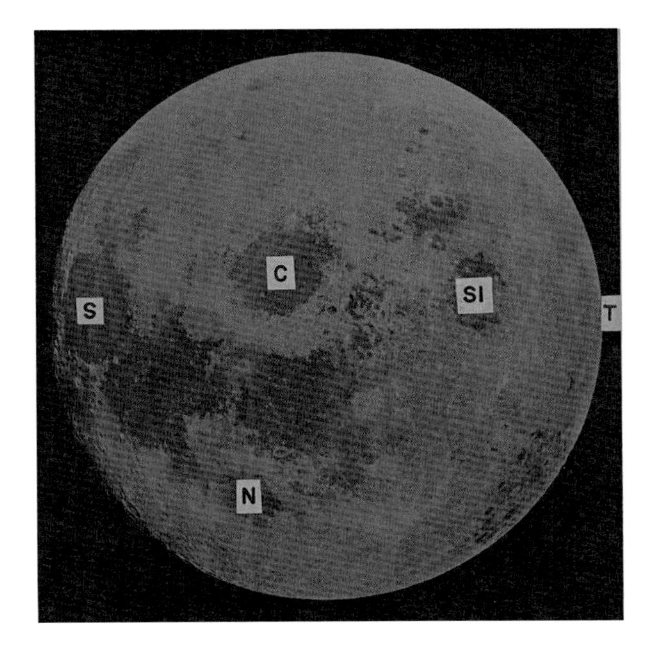

Fig. 5.3 Overview of the eastern limb of the Moon (Apollo photograph 11-44-6667). *S* Mare Serenitatis, *C* Mare Crisium, *SI* Mare Smythii, *T* Crater Tsi-olkovsky, *N* Mare Nectaris

h. Dates of solidification and/or crystallization from many of the older lunar glasses, minerals, and rocks fall within a well-defined time range centered on 3.95 Ga (Ryder 1990; Dalrymple and Ryder 1993, 1996; and Nyquist and Shih 1992) for the era in question.

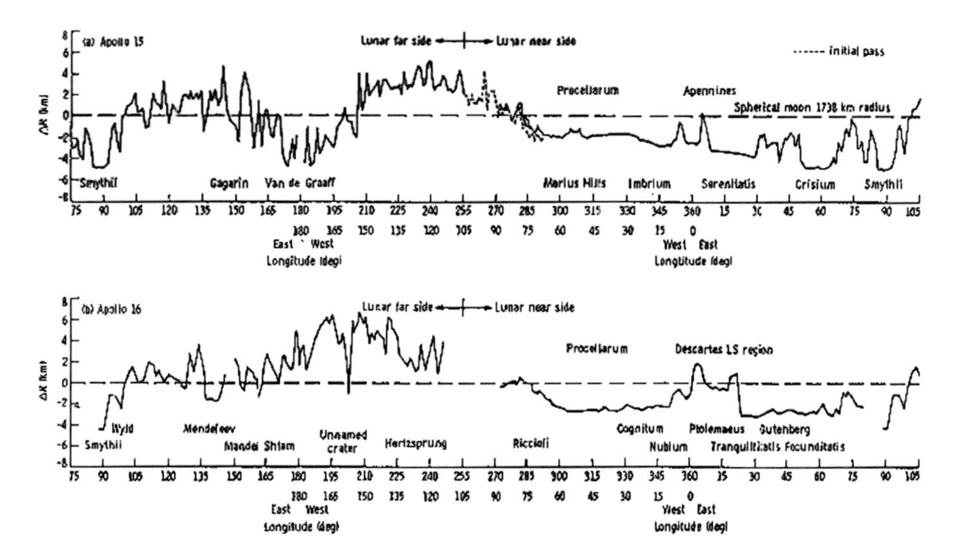

Fig. 5.4 Lunar altitude profiles for several lunar maria on both the lunar front side and the lunar backside from Apollo 15 data (*upper profile*) and Apollo 16 data (*lower profile*). Diagrams show lunar radius deviations from a 1738 km spherical moon. Note decreasing elevations from Maria Imbrium to Serenitatis to Crisium to Smythii. (From Muller et al. 1973, Fig. 2, with permission from AAAS)

i. Surface patterns of remanent magnetization as well as dates of magnetization recorded on the lunar surface and in lunar rocks need to be explained. The primitive lunar anorthositic crust has a pattern of positive and negative directions of magnetization (Russell 1980) and the mare rocks and associated breccias record an exponentially decreasing field of magnetization starting at ~3.9 Ga and ending at ~3.6 Ga (Cisowski et al. 1983).

These are the major facts to be explained but many more facts could be added to this list.

5.2.3.2 Step B: The Hypothesis to be Tested: Tidal Disruption on the 18th Perigee Passage of a Stable Capture Scenario

The hypothesis is that the lunar body was partially tidally disrupted during a very close gravitational encounter with planet Earth, an encounter that was associated with a capture sequence of events somewhere between 4.0 and 3.8 Ga ago (but probably ~3.95 Ga). As stated before, this tidal disruption model is intimately associated with a GRAVITATIONAL CAPTURE MODEL proposed by Malcuit et al. (1988, 1989, 1992). The irony associated with this model is that gravitational capture of a lunar-like body can occur *without* tidal disruption of the encountering body and was not mentioned in the papers cited in the previous sentence. It turns out that it takes a long search of capture scenarios to find one that can result in both a tidal disruption sequence of events and a stable geocentric orbit (i.e., the post-capture orbit does not result in a collision of the two bodies). After many attempts from 1987 onward, it was during the summer of 2006 that I calculated a scenario that fulfilled the requirements for such a scenario. The somewhat ideal requirements are:

1. that the capture encounter is sufficiently close and that the body deformation parameters of the lunar-like body are within limits to dissipate the energy for capture,
2. that during some subsequent encounter, the lunar body gets well within the Weightlessness Limit of the Earth-Moon system (nearly equivalent to the Roche limit for a solid body and defined in the next paragraph) so that the lunar body can be tidally disrupted and magma that was melted via tidal energy dissipation during the capture encounter can be gravitationally extracted from the lunar upper mantle, and
3. that the lunar orbit after the close tidal disruption encounter has long-term orbit stability (i.e., the lunar body will not collide with the Earth during any subsequent encounter).

The Weightlessness Limit (W-limit) is the center-to-center distance between the Earth and Moon at which weightlessness occurs at the sub-earth point on the lunar surface (Malcuit et al. 1975). The W-limit is at about 1.36 R_e (earth radii) for a particle at the sub-earth point on a spherical lunar body. The W-limit is at a greater distance from Earth for a tidally deformed lunar body. A lunar-like body can also have a weightless condition for a particle at the anti-sub-earth point during an encounter with Earth but this condition occurs at a smaller Earth-Moon distance of separation. An equation for the W-limit for the sub-earth point is:

Equation (5.1)

$$\frac{GM_e}{r^2} = \frac{GM_e}{(r - R_m)^2} - \frac{GM_m}{R_m^{\ 2}} \qquad (5.1)$$

where

G	gravitational constant
M_e	mass of Earth
M_m	mass of Moon
r	distance of separation of centers of Earth and Moon
R_m	radius of Moon

The first term in the equation is the force on the body of the Moon toward the Earth; the second term is the force on a particle at the sub-earth point on the moon toward the Earth; the third term is the force on the particle by the gravity of the Moon. If a particle along the earth-moon center line is inside the W-limit, it will be lifted off the surface of the Moon and move toward the Earth. If the particle along the earth-moon center line is beyond the W-limit, then it will remain at its normal position on the lunar surface.

An equation for weightlessness at the anti-sub-earth point is:

$$\frac{GM_e}{r^2} = \frac{GM_e}{\left(r + R_m\right)^2} + \frac{GM_m}{R_m^{\ 2}} \tag{5.2}$$

The first term is as defined above; the second term is for the force on a particle at the anti-sub-earth point on the Moon by the Earth; the third term is the force on the particle by the gravity of the Moon.

As shown in Fig. 5.5, the W-limit for a particle at the sub-earth point on a spherical (non-deformed) Moon is at 1.36 R_e and the distance between the centers of the Earth and Moon is 1.63 R_e. The W-limit for a particle at the anti-sub-earth point (which is not shown on the diagram) occurs when the center-to-center distance of the two bodies is at 1.36 R_e. Thus, with the lunar radius at 0.27 R_e weigthtlessness would occur at the anti-sub-earth point on the lunar surface but the perigee passage would be a near-grazing encounter.

The position of the W-limit, however, is displaced to a greater distance from the Earth as the lunar body is gravitationally deformed during an encounter. For example, with an h value of 0.2, a value commonly necessary for stable capture, the center of the Moon is at 1.85 R_e for weightlessness at the sub-earth point and at 1.54 R_e for weightlessness at the anti-sub-earth point. For an h value of 0.4 for the lunar body, which is a reasonable value for encounters following the initial encounter of a capture sequence, the center-to-center distance of the two bodies for the W-limit is at 2.06 Re and at 1.72 R_e for weightlessness at the sub-earth point and the anti-sub-earth point on the lunar surface, respectively. This shift in the W-limit with increasing deformation of the lunar body during an encounter is very important for analyzing the effects of tidal disruption on the lunar body during a close encounter. For example, during an encounter at ~1.36 R_e and an h value for the Moon of 0.4, a large outward force would occur at the sub-earth area on the lunar surface and somewhat smaller outward force would occur at the anti-sub-earth area of the lunar surface.

This partial tidal disruption scenario may seem a bit catastrophic to some investigators but it really is a minor catastrophe relative to the GIANT IMPACT SCENARIO (e.g., Canup 2004a, b, 2008) that is currently being promoted as the ruling paradigm for lunar origin. In fact, any one of the Giant Impact Scenarios is about 3000 times more catastrophic than the 18th pass scenario described above.

Also note that if I thought that the Giant Impact Model could be useful for explaining the development of lunar features (i.e., the "Facts to be Explained by a Successful Hypothesis"), I would not drag the reader through this presentation. So here it is: THE 18th PASS SCENARIO.

Figure 5.6a shows the heliocentric orbit orientation for a one-orbit non-capture encounter and Fig. 5.6b and c are geocentric depictions of this one-orbit non-capture encounter. Note that the encounter is focused at 1.43 earth radii (Fig. 5.6b); the sub-earth point of the lunar body, in this case of no deformation of the lunar body, is within 1000 km of the surface of earth (Fig. 5.6c). The lunar body, after the encounter, returns to a heliocentric orbit that is very similar to its heliocentric orbit before the encounter.

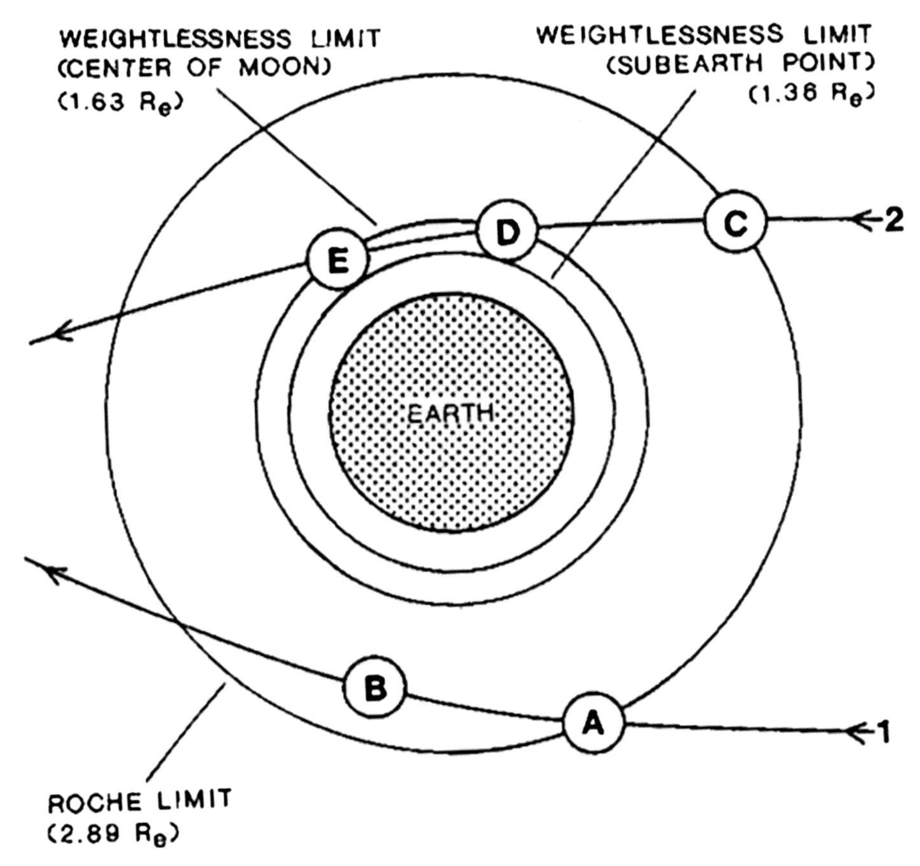

Fig. 5.5 Scale sketch of the weightlessness limit for the case of the Moon encountering the Earth with neither body being deformed. Note that the classical Roche limit is located at a much greater distance from Earth than the weightlessness limit. Scenario 1: The lunar body enters the classical Roche limit at point *A* and the perigee point is at point *B*. A particle at the lunar surface would *not* become weightless under this condition and the lunar body would then exit the Roche limit when it reaches a distance of 2.89 earth radii. Scenario 2: The lunar body passes through the classical Roche limit at 2.89 earth radii (point *C*). At point *D* the lunar body passes through the W-limit for a particle at the sub-earth point on the lunar body. From point *D* to point *E*, any loose bodies or particles on the lunar surface at or near the sub-earth point, would be lifted off the lunar surface and inserted into a sub-orbital trajectories that would either return the particle(s) to the lunar body or the particle(s) would eventually collide with the Earth, the details depending on the position of the lunar body at the time of gravitational liftoff

Figure 5.7a shows the heliocentric orientation for a three-orbit non-capture encounter. Since only about 70 % of the energy necessary for capture was dissipated within the interacting bodies, the lunar-like body returns to a slightly changed heliocentric orbit. Figure 5.7b shows the geocentric orbit configuration for this three-orbit non-capture scenario.

Figure 5.8a shows the heliocentric orientation for the first six perigee passages of a stable capture scenario. Figure 5.8b shows the geocentric orientation for the first 20 orbits.

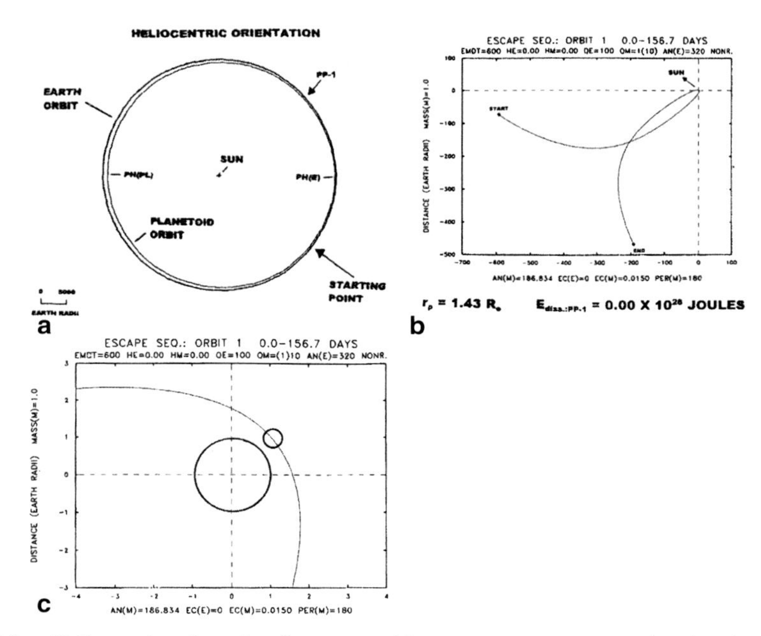

Fig. 5.6 **a** Heliocentric orientation for a one-orbit non-capture encounter showing the starting point of the simulation and the heliocentric position of the first perigee passage. **b** Geocentric orientation for this one-orbit non-capture encounter. No energy is dissipated during the encounter and the lunar-like body returns to an orbit very similar to that before the encounter. **c** Close-up of the perigee passage in **b** Note that the distance of separation of the planet and the lunar-like body, with no deformation of either body, is about 1000 km

Figure 5.9a shows the pattern of rock tidal amplitudes for the Earth and Fig. 5.9b shows the pattern of rock tidal amplitudes on the lunar-like body. Note that the highest tidal amplitudes for both bodies occur about 3.25 years after the initial encounter of this stable capture scenario and that the equilibrium lunar tidal amplitude associated with this encounter is about 30 km higher than that associated with the capture encounter. *This exceptionally close perigee passage is very important for explanation of a tidal disruption scenario for lunar surface features.*

Figure 5.10a shows a geocentric orbit orientation for perigee passages 17, 18, and 19 and Fig. 5.10b shows a close-up of the pattern of perigee passages. Note that perigee passage 18 is much closer than either 17 or 19. The center-to-center distance of closest approach is 1.356 earth radii. With h (moon) at 0.4, a perigee distance of 1.35 earth radii would be a near grazing encounter. Since this 18th perigee passage occurs after the lunar body is thermally processed by previous encounters, especially the capture encounter at 1.43 earth radii, the lunar body would have a somewhat higher displacement Love number, perhaps ~0.4. Thus, a perigee passage at 1.356 earth radii would be a near grazing encounter. At this distance of closest approach a very severe weightless condition would result for any particles in the sub-earth area on the lunar surface as well as a weak, but significant, weightless condition for particles in the anti-sub-earth area of the lunar body.

HELIOCENTRIC ORIENTATION

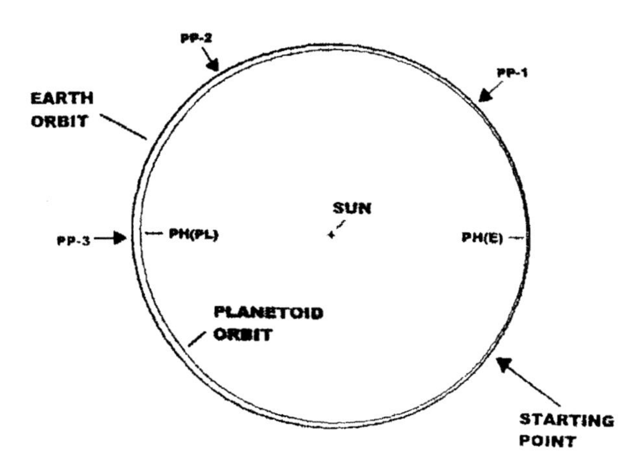

a

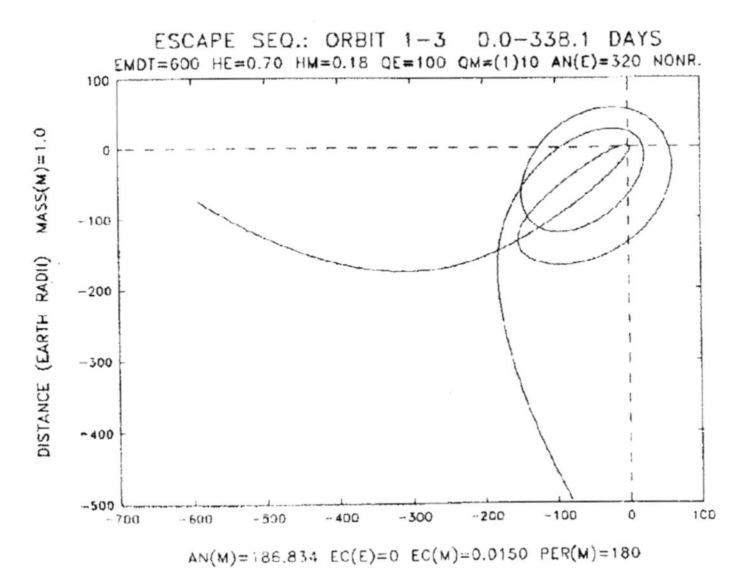

$r_p = 1.43\ R_e$ $\quad E_{diss.:PP\text{-}1} = 0.71 \times 10^{28}$ JOULES

b $\quad E_{diss.:PP\text{-}1\text{-}3} = 0.71 \times 10^{28}$ JOULES

Fig. 5.7 Three-orbit prograde escape scenario. **a** Heliocentric orientation for the encounter; the three perigee passage positions are indicated on the plot. **b** Geocentric orientation for the escape scenario

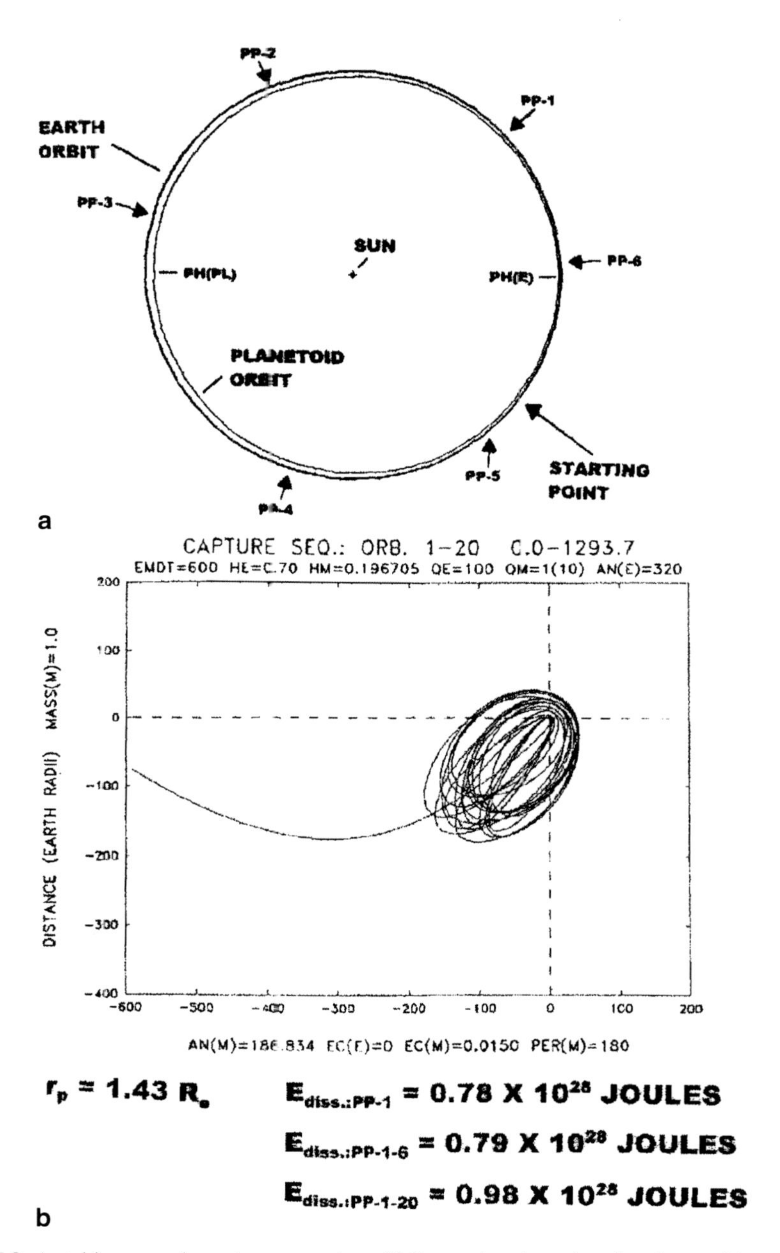

Fig. 5.8 A stable prograde capture scenario. **a** Heliocentric orientation showing positions of the first 6 perigee passages. **b** Geocentric plot showing the first 20 orbits (~3.5 years). Note that energy dissipation during the initial encounter was not sufficient for stable capture but energy dissipation during the 18th pass secured the capture

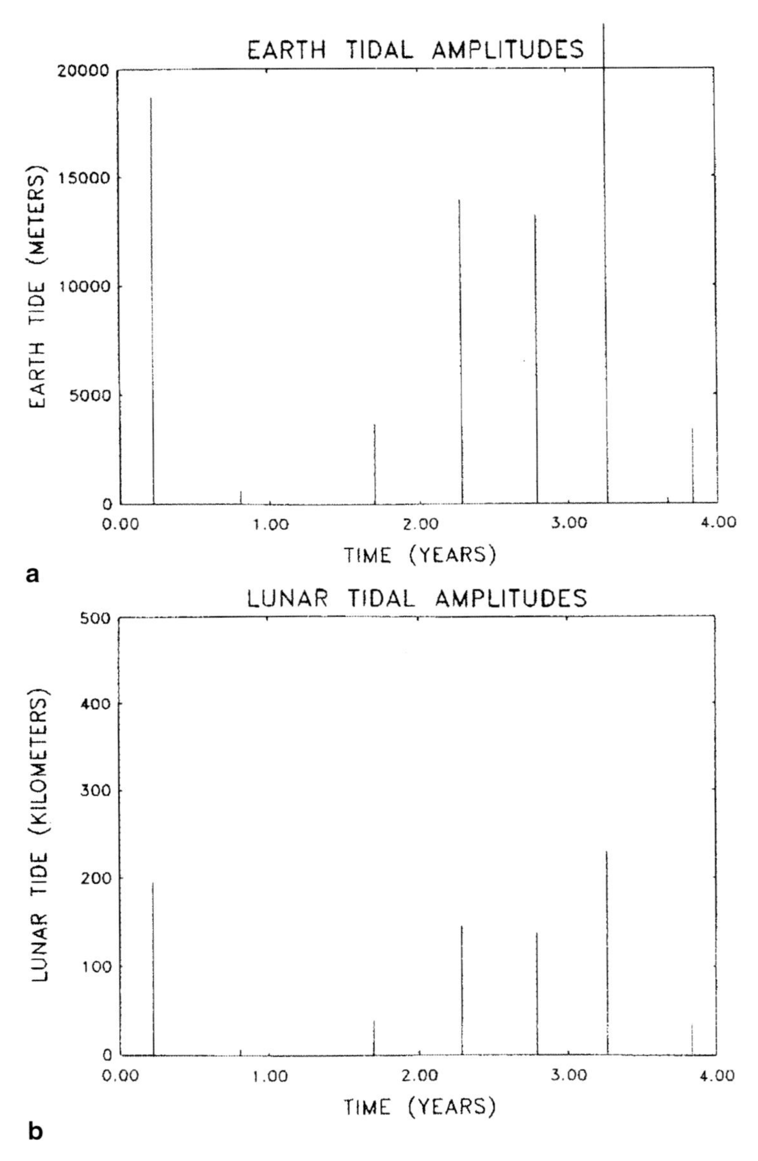

Fig. 5.9 Plots of tidal amplitudes for both bodies for this scenario. **a** Tidal amplitudes for Earth for the first 4 years of this stable capture scenario. Note that an encounter at about 3.25 years is a closer encounter than the capture encounter. **b** Tidal amplitudes on the lunar-like planetoid for the first 4 years of this scenario. The very close encounter at about 3.25 years raises rock tides higher than the capture encounter. This is the encounter that can permit tidal disruption of the lunar-like body if magma is available in the lunar upper mantle. This is perigee passage 18 (PP-18)

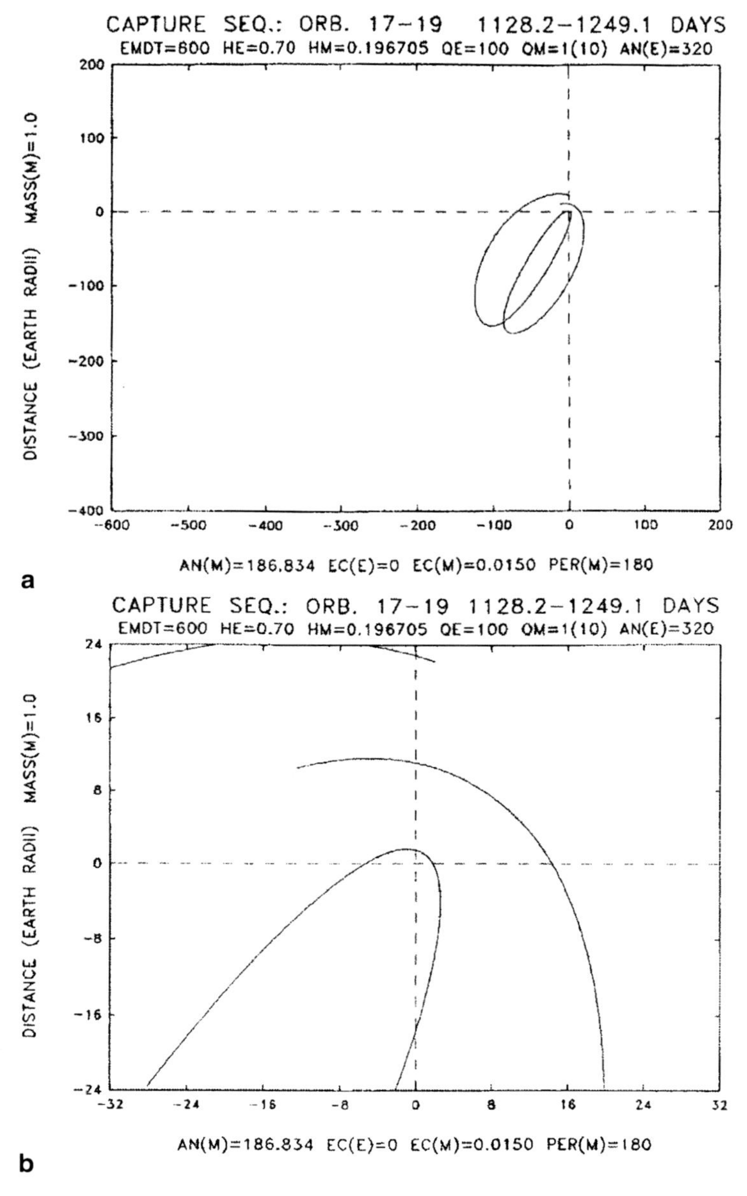

Fig. 5.10 Geocentric orbital setting for the 18th perigee passage. **a** Geocentric plot showing orbits 17, 18, and 19. **b** Close-up view of perigee passages 17, 18, and 19. Note that PP-18 is much closer than the other two

Figure 5.11a shows a geocentric plot of PP-18 and Fig. 5.11b shows the author's concept of what could happen during such a close encounter if magma is available in a sub-crustal magma zone within the lunar-like body.

Figure 5.12 shows a close-up view of a sequence of stages of a tidal disruption and subsequent fall-back sequence of events that could be associated with a close

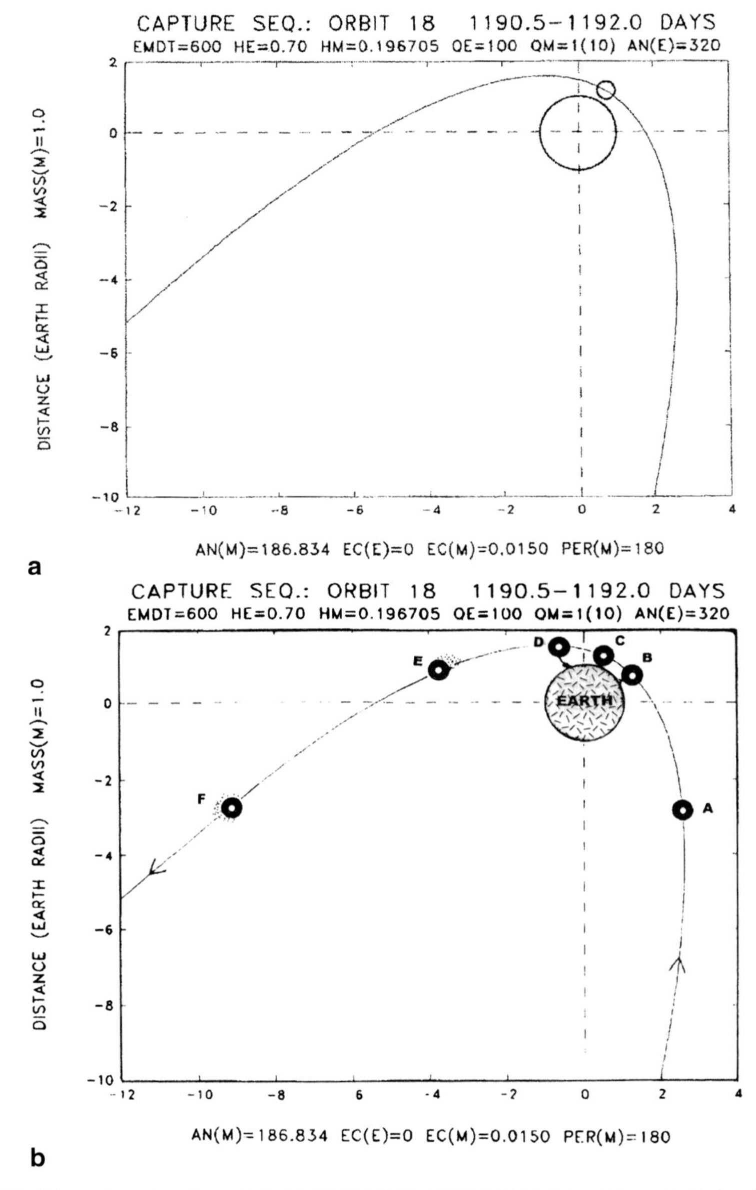

Fig. 5.11 General setting for a TIDAL DISRUPTION SCENARIO. **a** Numerical simulation of PP-18. **b** The author's concept of the broad view of what could happen during a close, tidally disruptive encounter which is well within the W-limit

perigee passage if the near-grazing perigee passage occurs after the capture encounter. Stages A through F (Fig. 5.11b) occur over about 2.0 h of time. At Stage A, the lunar-like body is approaching the Earth. At this point the surface rock patterns are relatively undisturbed and record about 600 million years of normal impact cratering activity. At Stage B (also see Fig. 5.12) the lunar-like body enters the

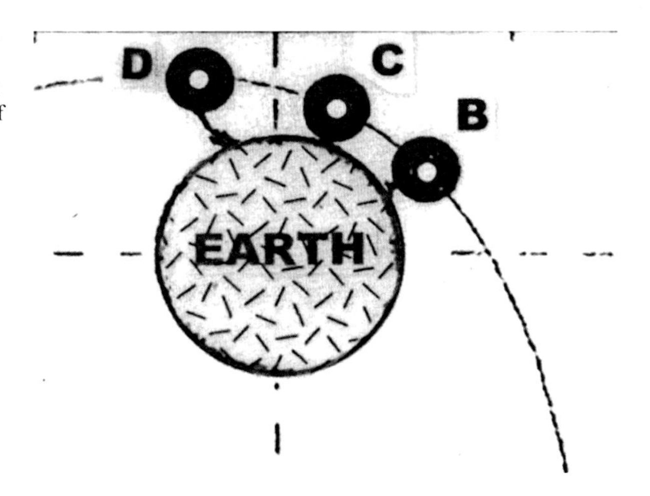

Fig. 5.12 A sequence of pictograms showing a close-up view of stages *B*, *C*, and *D* in Fig. 5.11b. The description of events is in the text

W-limit. At this point, loose particles are lifted off the lunar surface; more importantly, gases such as CO and CO_2 exsolve from the basaltic magma and cause very explosive eruptions to occur. Most of this gas-propelled material is basaltic glass shards that shatter as the gaseous vesicles blow them to "smitherines". Most of this material escapes from the lunar body and is propelled by both gas-jetting and Earth's gravity to the surface of the Earth. This rapid emission of gases and basaltic material continues until Stage C, the closest approach distance. At this stage, a second disruption center initiated with explosive eruptions but after the initial stages settles down to a condition where basaltic material from the lunar upper mantle necks off as basaltic spheroids from a lava column. The first few spheroids are small and probably escape from the lunar-like body and impact on the Earth. As the lunar body moves from Stage C to Stage D, the basaltic spheroids tend to get larger because the Earth's gravitational influence gets progressively weaker and the progressively degassed basalt necks off more placidly under a weaker Earth-influenced gravitational field. After Stage D the DISRUPTION PROCESS is over and the FALL-BACK PROCESS begins. Most, if not all, of the late-stage disruption material returns to the lunar surface. If the lunar body is not rotating or if its rotation is in the plane of the encounter, then the disrupted material tends to fall back along a great circle pattern on the lunar surface. If the lunar body is rotating in a direction that is not in the plane of the encounter or if gas-jetting causes ejected material to be propelled out of the plane of the encounter, then there will be some deviation from a great-circle pattern by the material falling back onto the lunar surface as the lunar body progresses from Stages D to E to F. The time-frame for Stages B to D is about 0.5 h and for Stages D through F is about 1.5 h. Figure 5.13 shows the tidal deformation for both the Earth (Fig. 5.13a) and the Moon (Fig. 5.13b) for this tidal disruption scenario. Figure 5.14 shows the position of this encounter within the prograde SCZ for an encounter from inside the orbit of Earth.

The reader is referred to Figs. 5.1, 5.2, 5.3, and 5.4 for my interpretation of the features that may be associated with a tidal disruption scenario. Mare Orientale (Fig. 5.1) is interpreted as the first disruption zone feature which is associated with

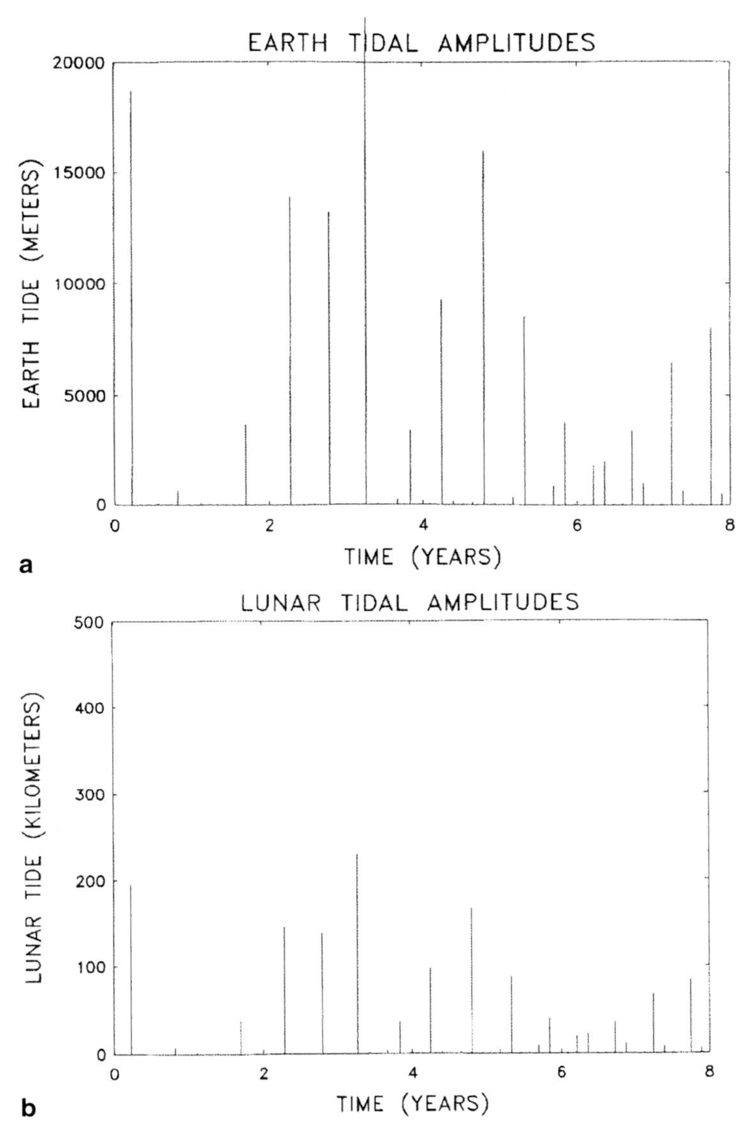

Fig. 5.13 Tidal amplitude plots for the first 8 years of this scenario featuring tidal disruption of the Moon on the 18th perigee passage. **a** Earth tidal amplitudes. **b** Lunar tidal amplitudes

massive gas-propelled ejecta, most of which is propelled toward Earth. Some material is propelled by back-pressure to be smeared onto the lunar surface. The huge disruption-induced volcanic form is gravitationally tugged east-ward and later collapses upon itself as it settles in an eastern direction. This sequence of events is consistent with the features registered in the Zond 8 photograph (Fig. 5.1b). The most prominent circular features that can be interpreted in terms of a tidal disruption scenario are shown in Figs. 5.1, 5.2, and 5.3 and these features define a near

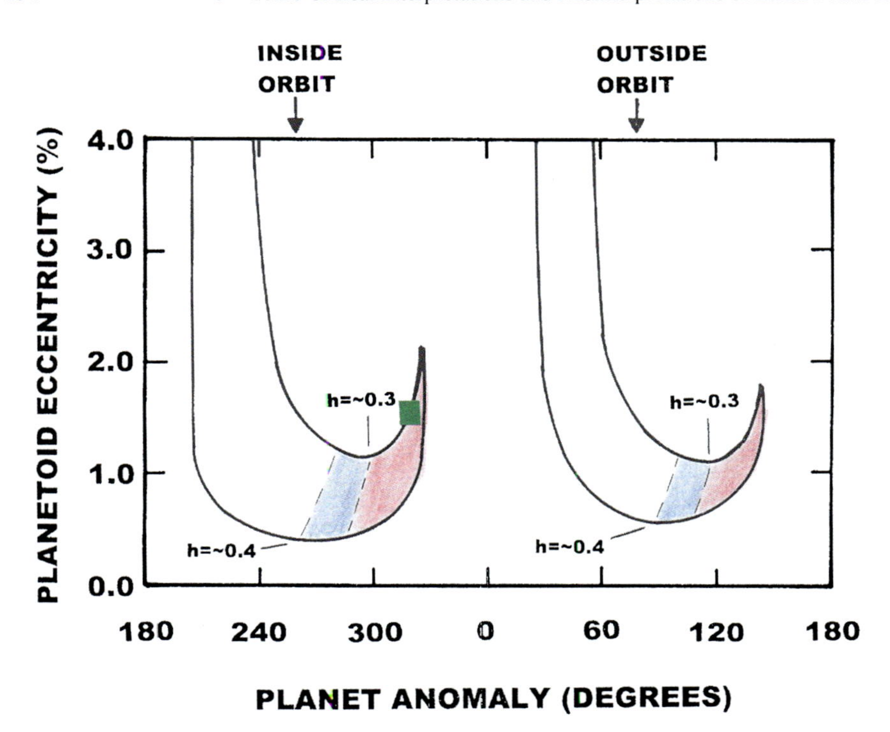

Fig. 5.14 Diagram showing the position of the encounter parameters for this 18th pass scenario within the prograde stable capture zone for a heliocentric orbit that is interior to the orbit of Earth

great-circle pattern of features on the lunar surface. Mare Orientale (Fig. 5.1) is at one end. Then there is, on Fig. 5.2, Mare Imbrium, Mare Serenitatis, Mare Crisium, and on Fig. 5.3, the trend continues to Mare Smythii and the mare-filled crater Tsiolkovsky. These are the features on the most prominent great-circle pattern. Those on a secondary great-circle pattern are Mare Humorum and Mare Nectaris (Fig. 5.2) and Mare Moscoviense (not shown) on the lunar backside.

Figure 5.15 shows the results of numerical simulations of trajectories of bodies released from the Mare Orientale as well as the Oceanus Procellarum regions of a lunar-like body (from Malcuit et al. 1975). These simulated trajectories strongly suggest that a narrow elongated ovoid region in Oceanus Procellarum, located between Mare Orientale and Mare Imbrium, is well-suited as a source region for extracting spheroids of basalt from the lunar mantle. Both of these source regions will be discussed in more detail in the next section of this chapter.

5.2.3.3 STEP C: Critique of the "Commandments" as a Prelude to Testing of the Hypothesis

This step is divided into two sections. The first section is a critique of the "commandments" for the interpretation of lunar surface features and the second is a set

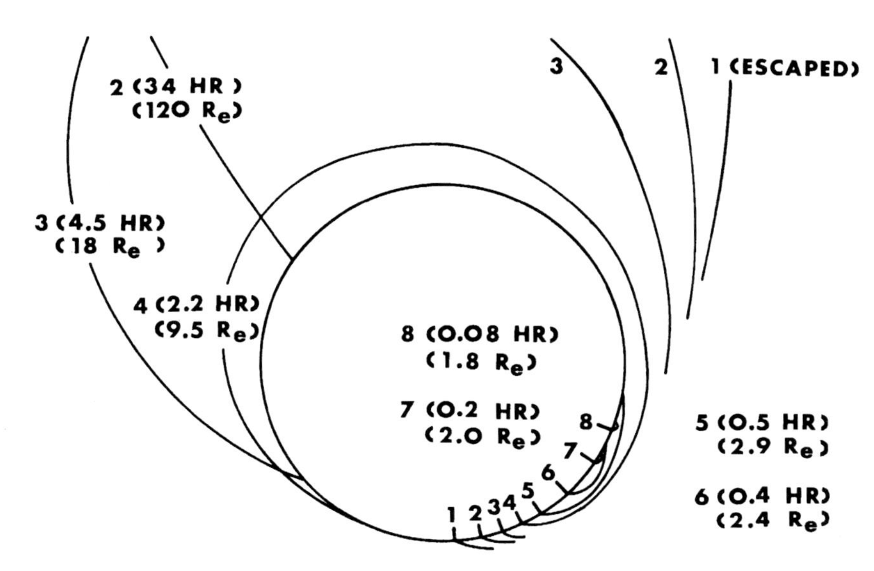

Fig. 5.15 Diagram showing the results of numerical simulations of trajectories of particles launched from specific positions at the sub-earth point on the lunar surface during an encounter that passes within the weightlessness limit of the Earth-Moon system. The launch positions, the flight times, as well as the impact positions are shown. The most favorable location for extracting lunar basalt to neck-off into spheroids appears to be the geometric location of Oceanus Procellarum. The lunar body is not rotating during the timeframe of these simulations. (Diagram adapted from Malcuit et al. 1975, Fig. 5, with permission from Springer)

of PREDICTIONS for testing the hypothesis of "TIDAL DISRUPTION OF THE LUNAR BODY".

This list of guidelines for interpretation of lunar surface features is from a book by Don Wilhelms (1993) titled "TO A ROCKY MOON" (p. 195–196). Wilhelms cited Mutch's book (1972) titled "THE GEOLOGY OF THE MOON: A STRATIGRAPHIC VIEW" as the source of the list. The following statements, then, were the guidelines for interpretation of lunar surface features during the Apollo Program and most, if not all, are followed today.

1. The Moon is heterogeneous and has had an active and diverse history. (I have no problem with this one.)
2. Both impact and volcanism have played important roles in its evolution. (I have no problem with this one.)
3. The regolith formed in all periods in proportion to the impact rate. (I have no significant problem with this one.)
4. The long rays of large Copernican craters are made up of secondary ejecta and demonstrate the great energy required to form them, an energy available from cosmic impacts (not any internal gas releases or the like). (I have no problem with this one.)
5. The maria filled their basins a considerable amount of time after the basins formed: thus the maria and basins *are not genetically related*, as so many early observers thought.

THIS IS A KEY WRONG AND/OR MISLEADING INTERPRETATION!!

CRITIQUE: In most cases the basins *are* associated with the mare fill material. But there can be some fill material added later. The crater density patterns are important in sorting this out.

In the fall-back phase of a tidal disruption model, the mare fill material (a large spheroid of lunar basalt) forms the basin that it occupies! These are *impact-formed basins* but the impact speeds are very low (~0.5–1.5 km/s) relative to hypervelocity impact events (> 7.0 km/s) by solid bodies.

6. The basins, their multiple rings, and their ejecta are the dominant structures of the Moon and control the surface distribution of most other materials and structures. *Their size and range of influence prove their impact origin.*

THERE ARE SOME PROBLEMS WITH THIS ONE.

CRITIQUE: I agree that most basins have rings but, in many cases, the ring structure, in my opinion, has been overly mapped and overly emphasized. A cursory inspection of the terrane around Mare Imbrium, Mare Serenitatis, and Mare Crisium (i.e., the three largest circular maria) suggests that a pre-existing crater pattern has *not* been greatly disturbed. My guess is that many of those craters date from an early era in lunar history (from the time of the formation of the anorthositic crust). Some are flooded with lunar basalt but the craters still show their original morphology.

Solid body impact ejecta is a minor component in most mare basin settings (with the Mare Orientale region being an exception; but my interpretation of the ejecta associated with Mare Orientale is that it is due to tidal disruption rather than solid body impact). Magmatic spheroid impact and subsequent flow of basaltic magma control the distribution of most of the other circular structures. The size and range of influence are very consistent with a *soft-body* impact origin.

7. Basins and craters form a continuous series of impact features. Some physical law of properties of the target causes craters larger than about 20 km to have central peaks, and those more than 250 or 300 km (basins) to have two or more rings.

THIS IS A MISLEADING STATEMENT AND CONCEPT!

CRITIQUE: Craters can form a continuous series of impact features. Some craters have central peaks, others do not. Impact speed may be a critical factor here. We need more modeling on ring structure development for both *solid-body* impacts as well as for *soft-body* impacts.

8. Many craters smaller than 20 km are the secondaries of larger primary craters or of basins.

CRITIQUE: *GENERALLY TRUE but too much emphasis is placed on solid-body impacts*. Many of these small craters may be the result of small bodies of fall-back material from the tidal disruption scenario. Some of these impacts would be low velocity and others more rapid. Some impacts can be oblique impacts that are in no way associated with secondary cratering.

9. An unknown, possibly considerable, number of small craters with irregular shapes or arranged in chains and clusters are endogenic.

CRITIQUE: Generally OK. But some of these chains are irregular features that could result from various types of patterned impacts of small debris from the disruption.

10. Endogenetic origin cannot be excluded for larger craters with smooth ("delta") rims or nondiagnostic features. That is, the absence of sharp features diagnostic of origin may result either from a moonwide process of degradation or from an original lack due to a passive caldera origin.

Note: One region that appears to be of caldera origin is the area around Mare Ingenii. This is the area that is antipodal to the Mare Imbrium—Mare Procellarum region of the front side of the Moon.

11. The depth of the mare-producing layer varies between the mare "province" and the nonmare "province", which includes the southern near side and most of the far side.

THIS IS A KEY WRONG AND MISLEADING INTERPRETATION!

CRITIQUE: For a capture model there is no good reason why the Moon should start out as anything but a generally symmetrical sphere. The front-side and back-side crust should be about the same thickness. (NOTE: Seismic experiments will help to sort this out.)

General Comment: The sharp-eyed reader will have caught a few cases of slightly misplaced emphasis in the above list although it remains mostly valid today. But now consider the following less fortunate elements of the guidelines. Each of these statements contains some "truth", but not in the same sense that was meant in the original statement.

12. Hybrid craters originally formed by impact but then modified by volcanism are common.

GENERALLY TRUE. These hybrid craters may be very important for interpretation. They attest to flooding of craters by mare basalt. Some of these flooded craters attest to *high-mare basalt (flood) levels* (similar to very high tide-water delineations on a beach) due to "tsunami-like flooding" following the impact of large spheroids of very low viscosity lunar basalt. Two good examples of this are on the northern outside edges of Mare Figoris and Sinus Roris and areas surrounding Mare Crisium as well as other areas of the lunar globe. (See Figs. 5.16, 5.17, and 5.18).

In my model, most of the features of the "Imbrium Sculpture" are the result of this huge, tsunami-like flood of very low viscosity lunar basalt over the lunar surface and, in some cases, the reverse flow of the cooler, more viscous lava. (The photos in Figs. 5.19, 5.20, and 5.21 illustrate this concept). Figure 5.22 illustrates some high-lava flow marks along the "shorelines" of Oceanus Procellarum and Mare Orientale.

13. There are two main suites of volcanic rocks, dark and light. Within both suites, morphologic expression, presumably dependent on magma viscosity, ranges from passive (plains and mantles) to positive (domes, cones and plateaus).

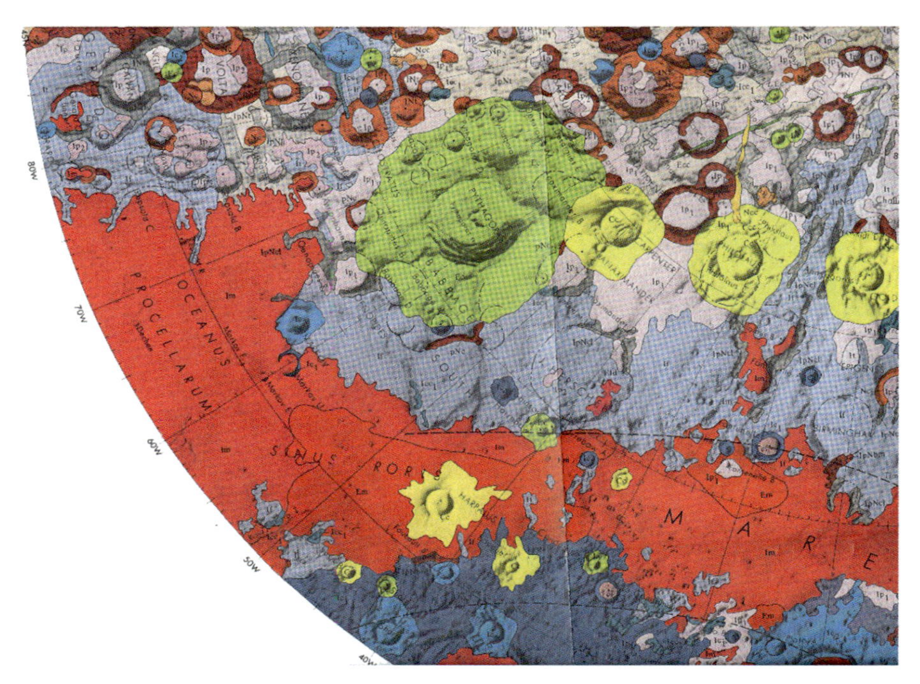

Fig. 5.16 Photo of a geologic map of the western part of Mare Frigoris, all of Sinus Roris, the northernmost part of Oceanus Procellarum, the elevated rim of Mare Imbrium, and a very small portion of Mare Imbrium. Note also the partially flooded craters and lineations oriented in a N-S direction. My interpretation is that basaltic lava from the Imbrium area flowed over the rim area of Imbrium, over the area of Mare Frigoris, and to the northern portion of the map as a "tsunami" wave of basalt. The basalt then settled to the low parts of the terrain and the excess lava helped to form Mare Frigoris, Sinus Roris, and Oceanus Procellarum. (Photo is from "Geologic Map of the North Side of the Moon", by Lucchitta (1978))

GENERALLY TRUE, BUT NEEDS MORE EXPLANATION.

14. The majority of lunar volcanism (mare and terra plains) is of fluid type that seeks depressions, probably the lowest depressions available at the time of extrusion. This type of volcanism seems to have been general and is expressed wherever depressions are available.

IMPORTANT AND GENERALLY TRUE. (LOW VISCOSITY MAGMA IS IMPORTANT FOR THIS SITUATION!)

15. The light plains are mostly Imbrium in age but older than the dark plains (maria): but some, mostly in craters, formed after the maria.

MAJOR QUESTIONS HERE (AND SOME CONFUSION)!

16. Terra (light-colored) volcanics of positive relief are concentrated near mare basins of intermediate (middle to late pre-Imbrium) age. A few of the freshest occurrences are near the Imbrium basin. The positive-relief features predate the nearby terra plains (for example, the Cayley Formation).

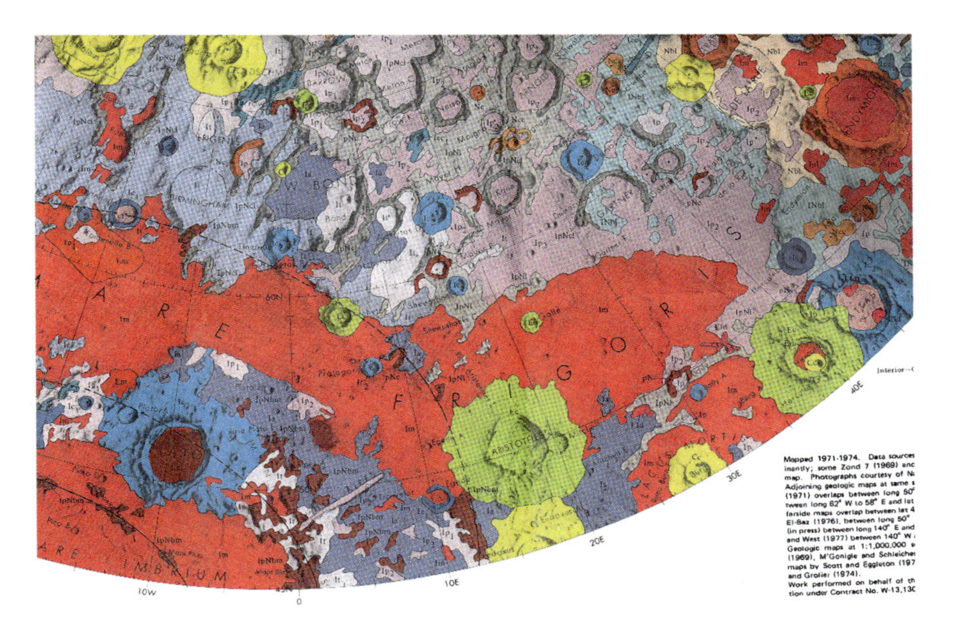

Fig. 5.17 Photo of a geologic map of the eastern part of Mare Figoris showing more of the same features as described in Fig. 5.16

NEED SOME WORK HERE!

17. The only known mare (dark-colored) volcanic of positive relief (for example, the Marius Hills) formed after the mare plains.

Generally true. These basalts and features, in my model, are associated with the Oceanus Procellarum Disruption Zone (see Figs. 5.23 and 5.24). Are they in the Mare Orientale area also??

5.2.3.4 STEP D: Some Testable Predictions for the Model

A MAJOR TEST for any MAJOR HYPOTHESIS is the ability to explain the "FACTS TO BE EXPLAINED" but, even more important, is THE ABILITY TO MAKE ACCURATE PREDICTIONS OF FEATURES THAT WOULD BE UNIQUELY ASSOCIATED WITH THE HYPOTHESIS (which for our purposes = model). Six predictions are listed and discussed here. These predictions are testable on future missions to the Moon.

PREDICTION # 1: There should be no crust associated with the two tidal disruption zones which are located (1) in the central area of Mare Orientale and (2) in a linear, ovoid zone in Oceanus Procellarum between Mare Orientale and Mare Imbrium (see Figs. 5.1 and 5.2).

RATIONALE: The computer simulations of trajectories of spheroids strongly suggest that most of the ejecta from the supra-volcanic form of Mare Orientale would

Fig. 5.18 Geologic map of the Mare Crisium and Mare Smythii regions of the Moon. Note the areas surrounding Mare Crisium appear to be heavily populated with mare basalts. Some of these areas are mapped with the same color symbols as Mare Crisium but other have different color symbols. My interpretation is that both Mare Crisium and Mare Smythii were formed by large spheroids of lunar basalt impacting at a low angle from a westerly direction. The basalt then flowed over the rim area and then flowed over the terrain as a "tsunami" wave of basaltic magma. The magma then settled to the low spots on the terrain. (Photo is from "Geologic Map of the East Side of the Moon" by Wilhelms and El-Baz (1977))

escape from the Moon. On the other hand, any large masses of basalt launched from the Oceanus Procellarum Disruption Zone would return to the Moon in the form of large spheroids within hours of the launch time.

SUGGESTED TESTING: Seismic experiments in the areas of interest could be performed by either robots or humans.

PREDICTION # 2: There should be a full thickness of crust (~60 km thick) under the areas impacted by the large spheroids of lunar basalt. This prediction should hold for Mare Imbrium, Mare Serenitatis, Mare Crisium, Mare Smythii, Crater Tsiolkovsky, as well as for all other regions covered by mare basalt.

RATIONALE: Computer simulations of trajectories of spheroids strongly suggests that most of the impact speeds for these large spheroids would be low, between 0.5 and 1.0 km/s. Since these bodies are not coming in at hypervelocities, there should be very little disruption of the original crust of the Moon. The crust should be displaced downward by tens of km and there should be some normal faulting in the up-range direction.

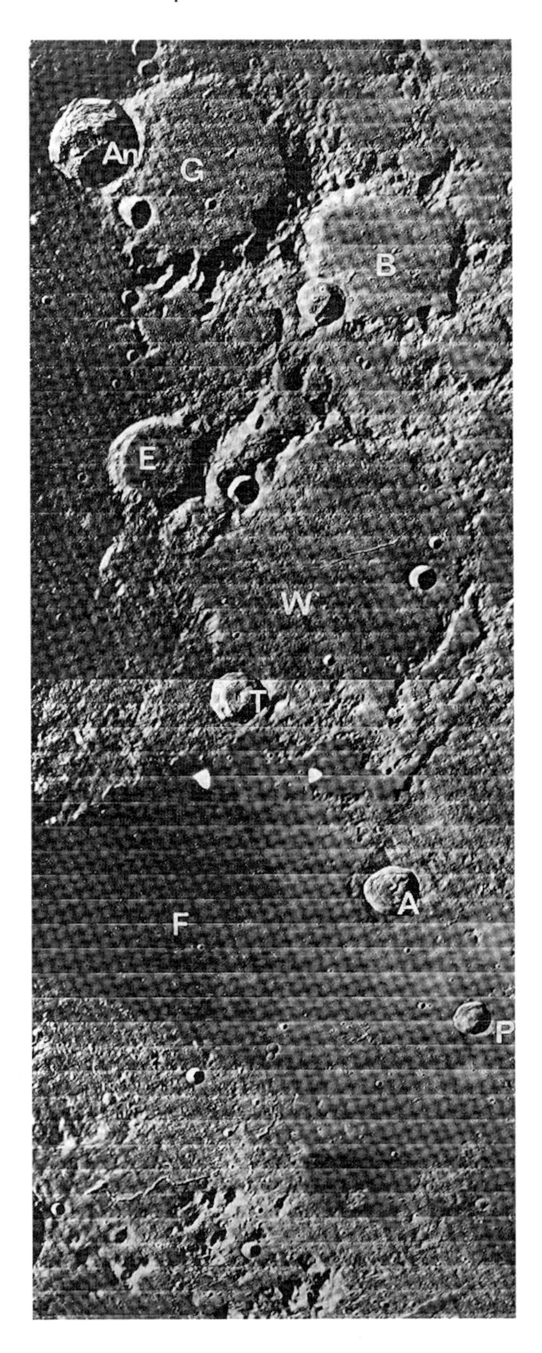

Fig. 5.19 Features of the Imbrium sculpture. *F* is a portion on Mare Figoris. Note the lineations in a NE-SW orientation. My contention is that these lineations could have formed by way of lunar basalt overflowing the rim of Mare Imbrium and moving as a "tsumani" wave of liquid basalt in a northeast direction. Once the wave reached the "high lava" mark, some of the basalt of the wave retreated in a southeast direction but most of the basalt would simply be solidified in place. (Photo is from Wilhelms (1987, p. 204). Lunar Orbiter 4 frame H-116)

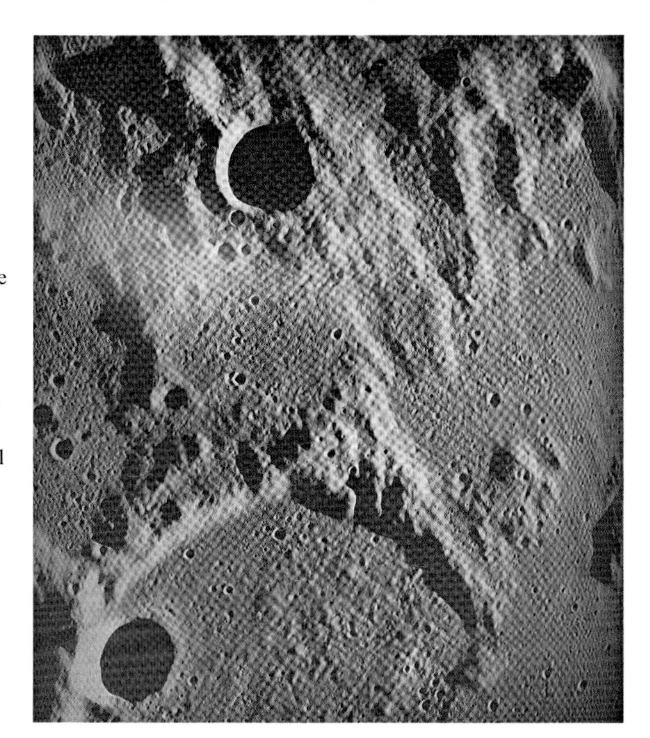

Fig. 5.20 Photo of typical Imbrium sculpture terrain. Note the lineations in the NW-SE direction. My interpretation is that liquid basalt was flowing from the northwest direction as a "tsunami" wave of basalt as it flowed over the southern edge of Mare Imbrium. Some flow indicators are clearly in the SE direction. After the wave passed, the basalt settled to low parts of the terrain and solidified. (Photo is from Wilhelms (1987, Fig. 10.17, p. 206). Apollo 11 frame H-4552)

Fig. 5.21 Photo of typical Imbrium sculpture from a slightly different area. Note the lineations in the NW-SE direction. My interpretation of this photo is the same as for Fig. 5.19. (Photo is from Wilhelms (1987, Fig. 10.27, p. 213). Catalina Observatory photograph # 1907)

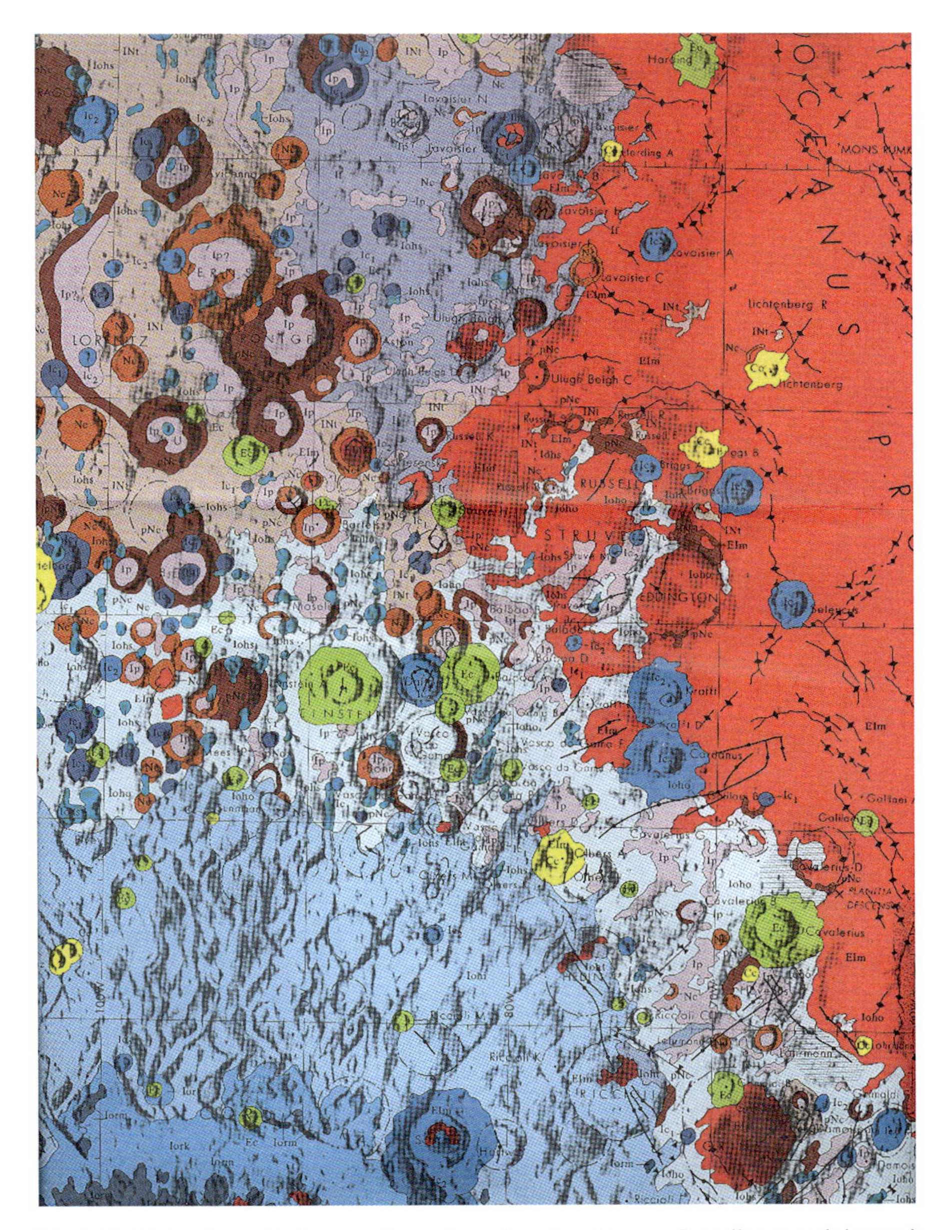

Fig. 5.22 Photo of a geologic map of a portion of western Oceanus Procellarum and the north flank of Mare Orientale. Note that the terrain west of Oceanus Procellarum appears to have been flooded by basalts from Oceanus Procellarum. In addition, some of the debris from the Orientale complex appears to have been flooded by basalts from Oceanus Procellarum. My interpretation is that the basaltic lava from the Oceanus Procellarum area flowed over the “shoreline” areas of Oceanus Procellarum as a “tsumani” wave of mare basalt. The basalt then settled to the low parts of the terrain and the excess flowed back into Oceanus Procellarum. (Photo is from “Geologic Map of the West Side of the Moon”, Scott et al. (1977))

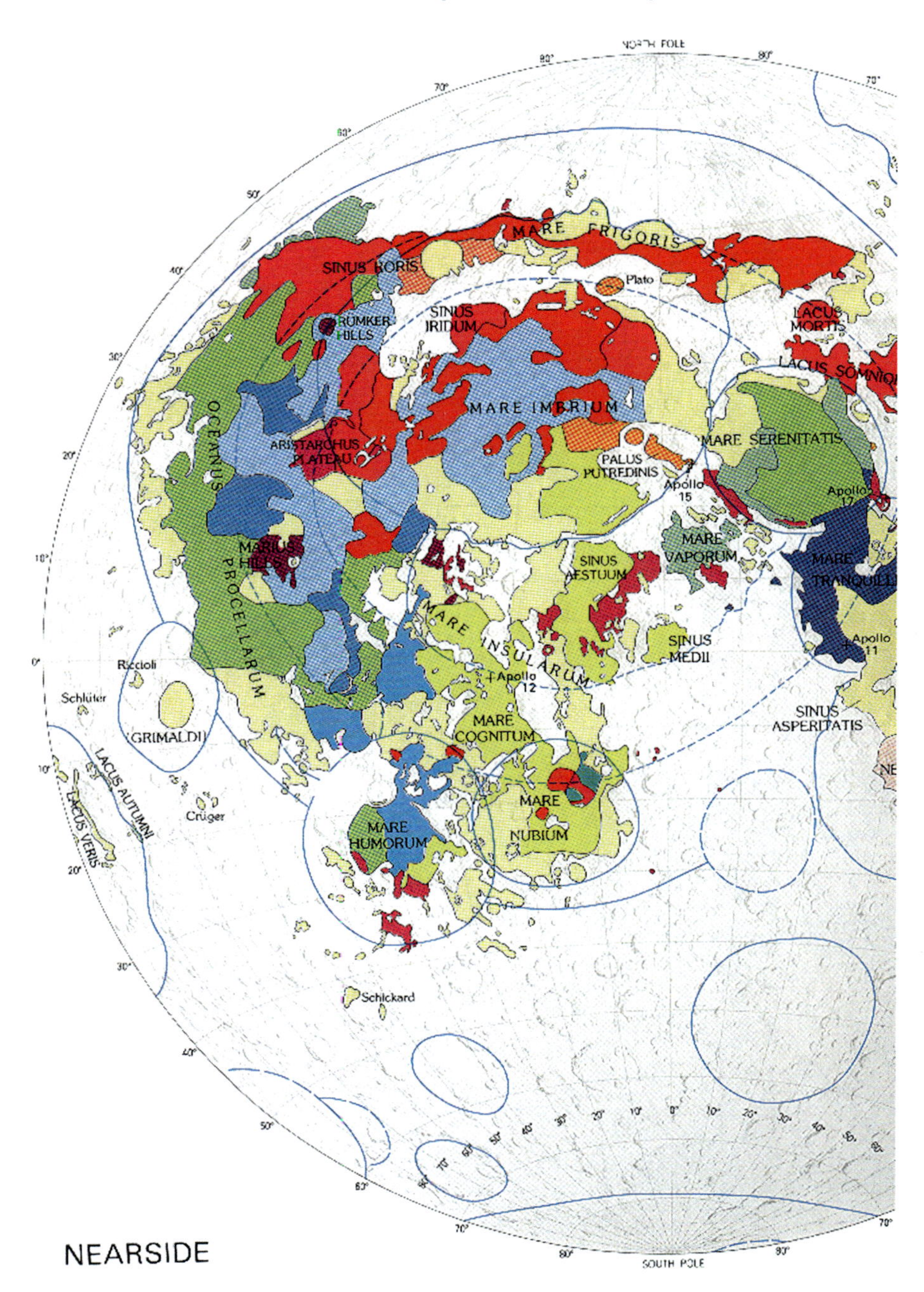

Fig. 5.23 Geologic map showing the western portion of the near side of the Moon in the region of the Marius Hills. The Marius hills are represented by the purple color on the map and were interpreted by the mapper as some of the younger features in the area. It would be useful to attempt to sample these features on future missions to the Moon. (Photo from Wilhelms (1987, Plate 4A))

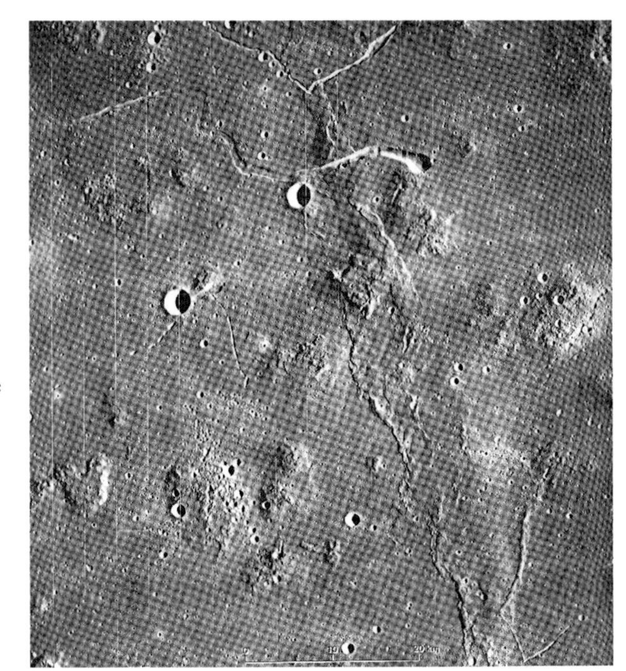

Fig. 5.24 Photograph of the Marius hill area of the Oceanus Procellarum region of the Moon. The region is characterized by diverse, rough-textured domes and cones surrounded by mare materials. My interpretation is that these features were formed over the main tidal disruption zone. My prediction is that there should be no lunar crust (i.e., anorthositic crust) under this zone and lunar basalts could ascent directly from the lunar mantle. Another prediction is that the basalts should be KREEP-rich because of this direct ascent from the lunar mantle. (Photo is from Wilhelms (1987, Fig. 5.7, p. 88). Lunar Orbiter 5 frame M-211)

SUGGESTED TESTING: Seismic experiments, in the areas of interest, could be performed by either robots or humans. I would suggest Mare Crisium as a first location for such experiments.

PREDICTION # 3: There should be raised rims around all circular maria of liquid spheroid impact origin.

RATIONALE: The mare basin depression in the lunar crust is formed by the force of the impact and by the weight of the collapsing spheroid of basalt. Within minutes after the impact the entire area of the basin would rebound and this upward movement would result in a raised rim around the basin. Since several of the maria are asymmetrical because of the purported oblique impact, the raised rim also would have a predicted asymmetry to it. Any lunar mare basalt that is isolated beyond the raised rim would form a "moat-like" structure around the circular maria. Good examples of this are the area surrounding the raised rim of Mare Crisium as well as the areas of Mare Frigoris and Sinus Roris and the Sinus Midii area south of Mare Imbrium.

PREDICTION # 4: The mascons associated with the circular maria of basaltic spheroid impact origin should be proportional to the average diameter of the maria (i.e., there is no need for a dense mantle plug located under the circular maria). The mascon model predicted here is a one-body mass in isostatic equilibrium (Bowin et al. 1975) (i.e., a simple mare model).

RATIONALE: Bowin et al. (1975) devised a two-body mascon model because they needed a basin of lunar basalt about 16 km thick to account for the mascon associated with Mare Serenitatis. Since they could not visualize a 16 km deep basin under Mare Serenitatis, they invented the concept of a dense mantle plug located beneath each typical circular maria. Using this construct, they needed only a 2 km thick mare basalt fill but a 12 km rise of the lunar moho (the dense plug).

SUGGESTED TESTING: The main testing would be via seismic work in the central region of Mare Serenitatis. The basaltic spheroid impact model would be consistent with 16 km of mare fill. The predicted thickness of mare fill for the basaltic spheroid impact model in the central area of Mare Crisium is about 12 km. Any other circular maria, except Mare Orientale, could be tested in this way.

PREDICTION # 5: The Mare Ingenii area is a caldera collapse structure and a number of the circular structures in that area are volcanic vents and the swirls. The prediction is that the swirls are composed of ash-flow tuffs and ignimbrites as suggested by Mackin (1969).

RATIONALE: The Mare Ingenii area is nearly antipodal to the postulated main launch zone for the basaltic spheroids. This antipodal area should have undergone severe tidal expansion but not as much as the lunar front-side bulge. Although the antipodal area would have encountered weightlessness, the gravitational forces would have been significantly less and only gas-rich volcanism is predicted to occur. After the close encounter associated with the tidal-disruption scenario, the tidal bulge area would be restored to near normalcy as the crustal rocks would successively telescope into the mantle via normal faulting. These volcanic features are predicted to be superposed upon any surviving features of the primitive crust in the area.

SUGGESTED TESTING: Perhaps another round of geological mapping would help here. Robots or humans on the lunar surface should be looking for both volcanic and impact features. The volcanic features would be associated with the gas-assisted explosive eruptions associated with partial tidal disruption and some impact features could be associated with the material falling back to the Moon after the tidal disruption scenario.

PREDICTION # 6: The broad Oceanus Procellarum area is predicted to be a drawdown structure in that it is the mantle source region for essentially all the spheroids of basalt for the tidal-disruption model. The entire area of Oceanus Procellarum, except for the narrow region of the second (the main) tidal disruption zone, should be underlain by only a mildly deformed full thickness (~60 km) of lunar crust.

RATIONALE: The general idea is that as the lunar basalt is extracted from the mantle, the crust will simply founder. Then as the tidal deformation decreases, mare basalt floods over the area and covers the area with a thin layer of mare basalt. This scenario can be compared to and contrasted with the proposal by Cadogan (1974). His proposal was that Oceanus Procellarum is the Gargantuan Basin (interpreted as a very large impact structure).

SUGGESTED TESTING: Seismic experiments in the area of interest could be performed by either robots or humans to test for the thicknesses of both the anorthositic crust and the mare fill.

5.2.4 *Summary and Conclusions*

In this presentation I did a "critique" of the "COMMANDMENTS" for interpretation of lunar surface features via an analysis of a version of the Gravitational Capture Model which features partial tidal disruption of the lunar body during a close encounter subsequent to the capture encounter.

There are a few suggestions for revision of parts of the "COMMANDMENTS" but most of them have held up well over the decades and continue to be good guidelines for interpretation. The major issues involve the substitution of "interpretations" for "facts".

Then the Gravitational Capture Model is processed through the SCIENTIFIC METHOD along with a discussion of the "COMMANDMENTS". In the final step of the SCIENTIFIC METHOD *SIX* predictions are made that will help to test a Gravitational Capture Model that features a very close encounter which causes partial tidal disruption of the lunar body.

I think that it would be interesting to process the GIANT IMPACT MODEL for the origin of the Earth-Moon system through the same critical scientific analysis.

5.3 Vignette B. Directional Properties of "Circular" Lunar Maria and Related Structures: Interpretation in the Context of a Testable Gravitational Capture Model for Lunar Origin

Photogeologic study of "circular" maria and mare-filled craters reveals that several of these features are either elliptical and/or asymmetrical. The ellipticity/asymmetry is interpreted as a directional property which may have genetic significance. The features with the most apparent directional properties are Maria Orientale, Crisium, Humorum, Moscoviense, and the mare-filled Crater Tsiolkovsky. Other features with weaker directional signatures are Maria Imbrium, Serenitatis, and Smythii. The asymmetric pattern associated with Mare Orientale is somewhat different from that of the other maria. There are extensional features to the southwest of Orientale and compressional features to the northeast. Thus, movement appears to have been from southwest to northeast. All other maria, except Moscoviense, indicate movement in easterly directions and these vectors are within 30° of the trend of a great-circle pattern of large circular maria described by Malcuit et al. (1975). The vector for Mare Moscoviense is in a north-easterly direction and is oriented nearly perpendicular to the great-circle trend.

Some of the directional properties of these features have been interpreted in terms of at least six models of mare basin formation:

1. that the mare basins were excavated by bodies impacting at high speeds onto the lunar surface (Wilhelms 1987),
2. that the mare basins were excavated by a swarm of fragments from a tidally disrupted planetary body impacting from a westerly direction (Hartmann 1977a),

3. that the mare basins were excavated by a swarm of bodies that were tidally disrupted by an encounter with either Venus or Earth (Wetherill 1981)
4. that the mare basins were excavated by a swarm of bodies from the Asteroid Belt in one-half lunar day (Nash 1963)
5. that the mare basins were excavated by lunar satellites whose orbits decayed sufficiently to cause the satellites to impact on the lunar surface from a westerly direction (Runcorn 1983; Conway 1986),
6. that the basins were formed by large spheroids of molten lunar basalt that were tidally disrupted from the lunar body during a close encounter with Earth and subsequently impacted onto the lunar surface from a westerly direction (Malcuit et al. 1975).

5.3.1 Purpose

The purpose of this "vignette" is to point out some features of the "circular" maria and associated basins that appear to have directional properties. The criteria for interpretation of directionality are those suggested by Moore (1976). These directional properties of the "circular" maria are then discussed in terms of these six different models for mare basin origin.

General Criteria for Interpretation of Directionality of Bodies Impacting at Oblique Angles (from Moore 1976) are:

1. Oblique impacts tend to produce crater morphologies and ejecta patterns that are BILATERALLY SYMMETRICAL with respect of the plane of the impactor's trajectory.
2. Crater rims in down-trajectory and lateral directions tend to be HIGH and IRREGULAR, the details depending on the angle of impact and other factors such as target material and local topography.
3. Crater rims in the up-trajectory direction tend to be LOW and WELL-DEFINED. OPEN CRACKS and DOWNWARD DISPLACEMENT (conditions indicating NORMAL FAULTING) occur only in up-trajectory directions.
4. Craters produced by oblique impacts tend to be ELONGATE PARALLEL to the plane of the trajectory. (However, some are found to be circular or even elongate perpendicular to the plane of the trajectory.)

The following illustrations are the results of impact craters that were formed and documented at the White Sands Missile Range (New Mexico) by impact of high velocity to hypervelocity missiles traveling along oblique trajectories mainly in gypsum sand (Moore 1976 and Figs. 5.25 and 5.26 from his paper). The asymmetrical pattern resulting from these high speed impacts are similar in nature, but not in scale, to the asymmetry associated with several of the large circular maria. By analogy, asymmetry of some circular maria, then, may have resulted from an oblique impact of a projectile.

The standard model for the formation of circular maria is that it is a two-stage process: (1) impact of a bolide at hypervelocity and (2) filling of basalt from the lunar interior at a later time. The model pursued in this book is a one-stage process,

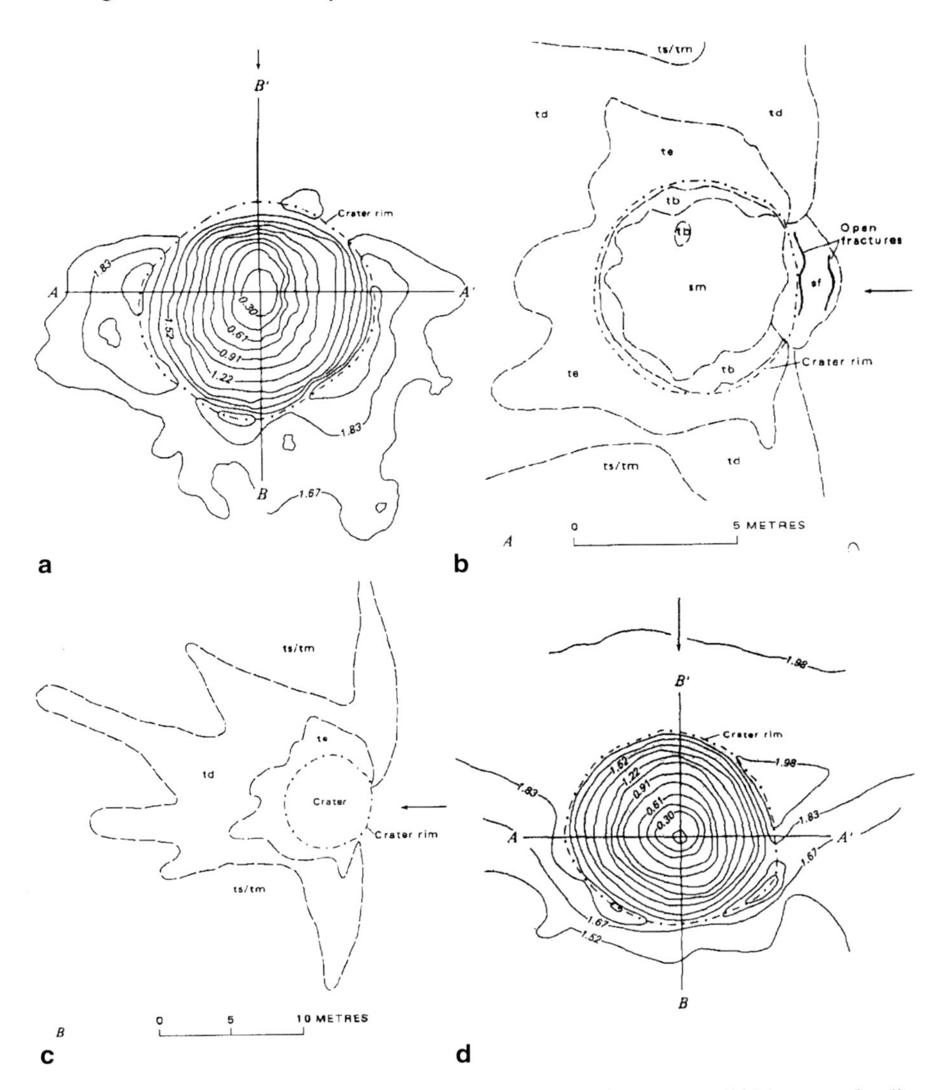

Fig. 5.25 Some examples of bilaterally asymmetrical craters from Moore (1976, composite diagram from Figs. 2, 6 and 7). Note that the bilateral symmetry of the first three craters is striking but that of the fourth one is subtle; also note that fractures in the *top right* figure are in the up-range direction. *Arrows* indicate the direction of impact

the oblique impact of a large basaltic spheroid. Each of these spheroids is extracted from a source region in either western Oceanus Procellarum (the main basaltic source region) or Mare Orientale (a minor basaltic source region). These spheroids then travel in suborbital trajectories over the lunar surface and impact obliquely at low speeds (0.2–2.0 km/s) onto the lunar surface to form a large circular or elliptical lava lake in the depression made by the impact of the magmatic spheroid. Elastic rebound of the region then forms an asymmetrical rim around the circular maria.

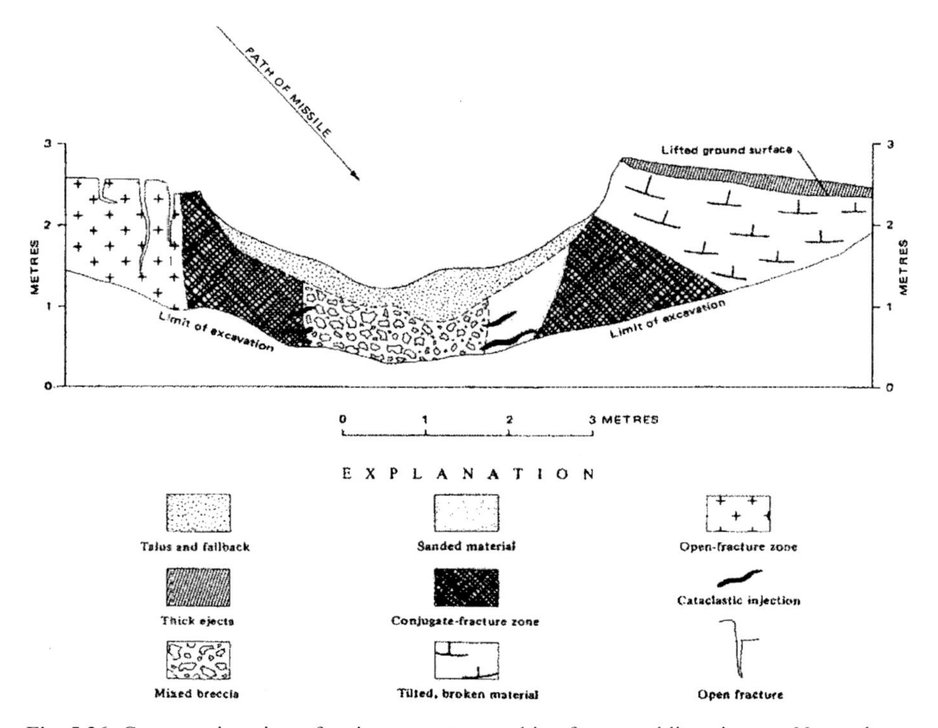

Fig. 5.26 Cross-section view of an impact crater resulting from an oblique impact. Note subsurface deformation zones parallel to the plane of the trajectory of the impactor. Note also zones of mixed breccias, tilted and broken material, sanded material, zone of conjugate fractures, and zone of open fractures (caption information is from Fig. 27 in Moore 1976)

5.3.2 A Cursory Survey of Circular Maria and Some Mare-filled Craters

Figures 5.27, 5.28, 5.29, 5.30, 5.31, 5.32, 5.33, 5.34, 5.35, 5.36, 5.37, 5.38, 5.39, 5.40, 5.41, 5.42, 5.43, and 5.44 consist mainly of photographs of lunar surface features and photographs of lunar charts and maps that represent a list of facts to be explained by a successful model for the origin of lunar maria and related features.

5.3.3 Summary of Observations

A number of the circular maria and associated basins which were briefly described above are located very close to a lunar great circle (Franz 1912, 1913; Dietz 1946; Nash 1963; Lipskii 1965; Lipskii et al. 1966; Lipskii and Rodionova 1977; Stuart-Alexander and Howard 1970; Hartmann and Wood 1971; Malcuit et al. 1975; Schultz 1976; Runcorn 1983). Runcorn (1983) also recognized a more poorly defined great-circle pattern outlined by Maria Humorum, Nectaris, and Moscoviense.

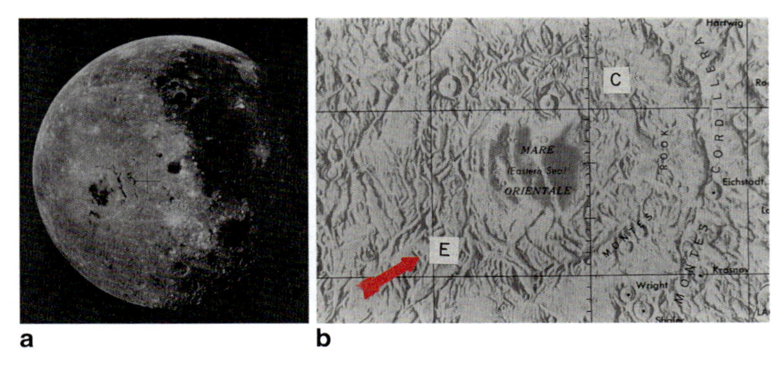

Fig. 5.27 **a** Zond 8 photo of Mare Orientale and surrounding terrain. **b** Photo from "Lunar Chart" (LPC-1, 1st ed. (1970)). Note that asymmetry is expressed by extensional (E) features to the southwest and compressional (C) features on the northeast. Implied directionality is southwest to northeast

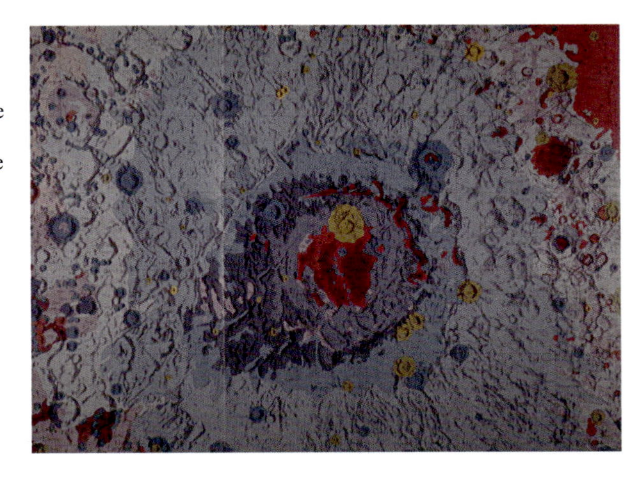

Fig. 5.28 Photo from "Geologic Map of the West Side of the Moon" (Scott et al. 1977). The ellipticity of Mare Orientale is negligible. The asymmetry of Mare Orientale is indicated by a wider belt of units to the southwest and a thinner belt of units to the northeast. Mare basalts are located in the center of Mare Orientale (*red pattern*) as well as in the *upper right* of the photo on the edge of Oceanus Procellarum

Figure 5.45 shows the major features on the primary great-circle pattern as well as two features on the secondary great-circle pattern. The interpreted vector indicators are shown on all of the features except Mare Moscoviense.

5.3.4 *Examination of Models that can be Tested for an Explanation of the Directional Properties of the Circular Maria*

The following is a critical review of some of the models that have been mentioned in the literature of lunar science as possible explanations for linear trends and directional properties of features associated with the lunar maria, craters, and other features.

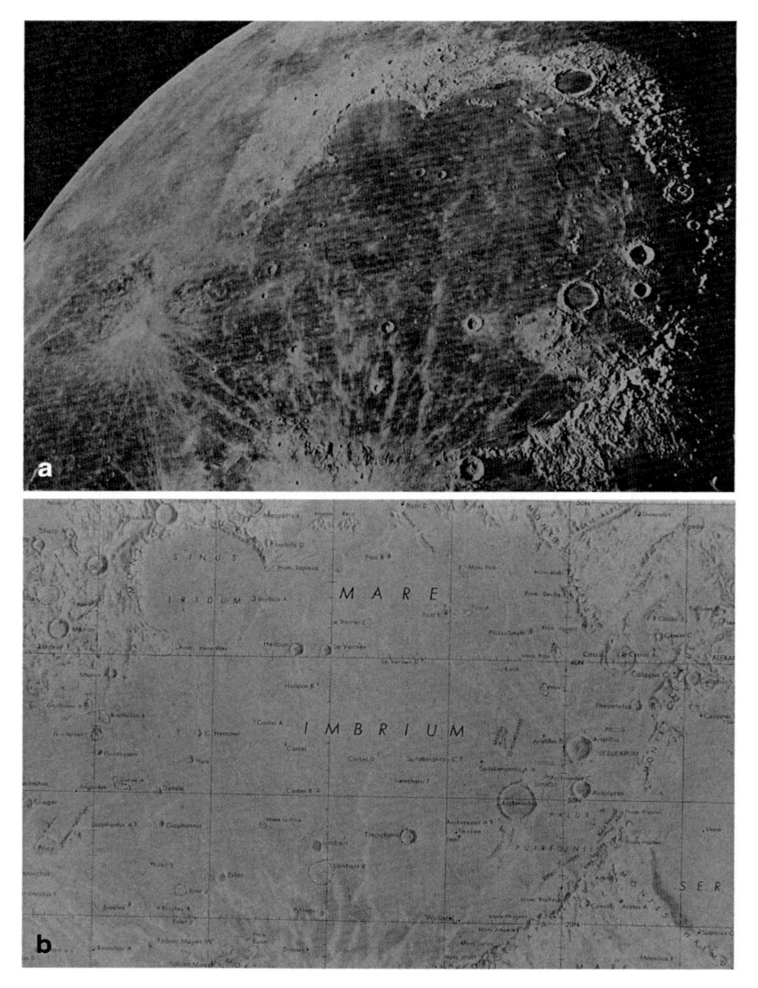

Fig. 5.29 **a** Photo of Mare Imbrium area (Hale Observatory Photo). **b** Photo of portion of "Lunar Earthside Chart" (LMP-1, 2nd ed. (1970)). Note that ellipticity of Mare Imbrium is small with a long axis in the southwest-northwest direction; an asymmetry is expressed by the lack of a rim on the southwest edge of the mare; directional implications are unclear

5.3.4.1 The Random Impact Model (Wilhelms 1987)

The random impact model features the impact of solid bodies of various sizes onto the lunar surface to excavate the lunar basins which eventually fill with mare basalts as the lunar body becomes warmer and the liquid basalt rises to the surface.

QUESTION: What pattern of directional properties should the resulting basins have?

POSSIBLE ANSWER: Something near a random impact pattern with random impact directional properties.

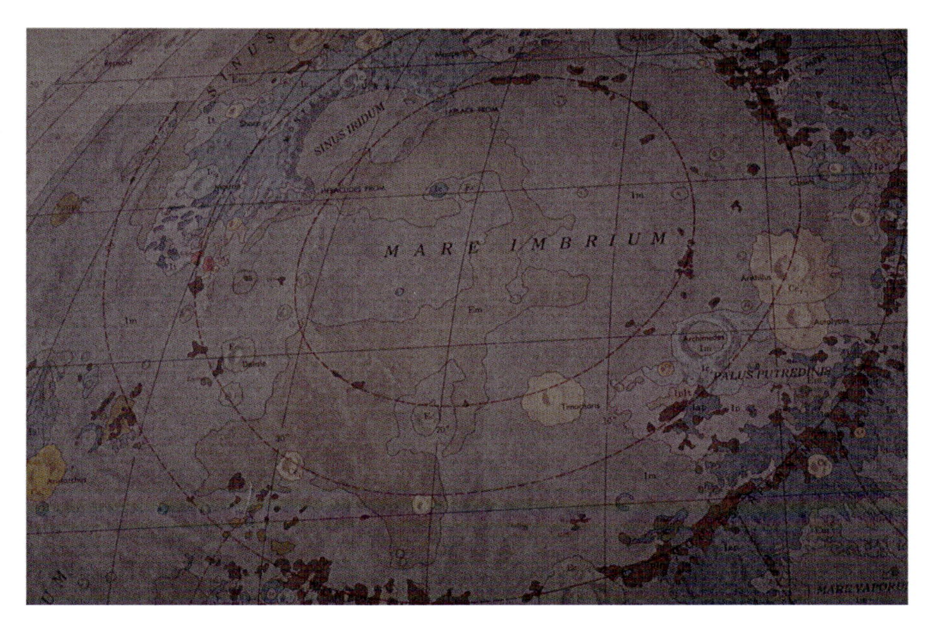

Fig. 5.30 Photo of a portion of the "Geologic Map of the Near Side of the Moon" (Wilhelms and McCauley 1971). The *dashed lines* connect some ridges of a multi-ring structure. There is an apparent elongation in the SOUTHWEST-NORTHEAST direction for Mare Imbrium

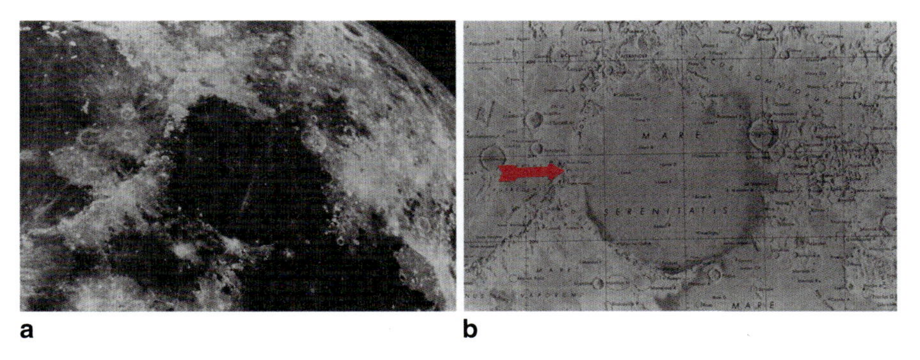

Fig. 5.31 **a** A portion of a Hale observatory photograph of the Mare Serenitatis area. **b** Photo of a portion of "Lunar Earthside Chart" (LMP-1, 2nd ed. (1970)). Ellipticity is negligible; asymmetry is expressed by a sharply defined western rim and an irregular eastern rim. Implied directionality is West to East

QUESTION: What is the probability that the four largest circular basins would be located nearly on a lunar great circle with decreasing diameters from lunar West to East?

POSSIBLE ANSWER: A very low probability.

QUESTION: What is the probability of the basin (original crater) would be anything but symmetrical in map view?

Fig. 5.32 **a** Photo of a portion of the "Geologic map of the near side of the Moon" (Wilhelms and McCauley 1971). The *dashed lines* connect some ridges of a multi-ridge structure associated with Mare Serenitatis. Asymmetry is expressed by a sharply defined western rim and an irregular eastern rim. Implied directionality is WEST to EAST

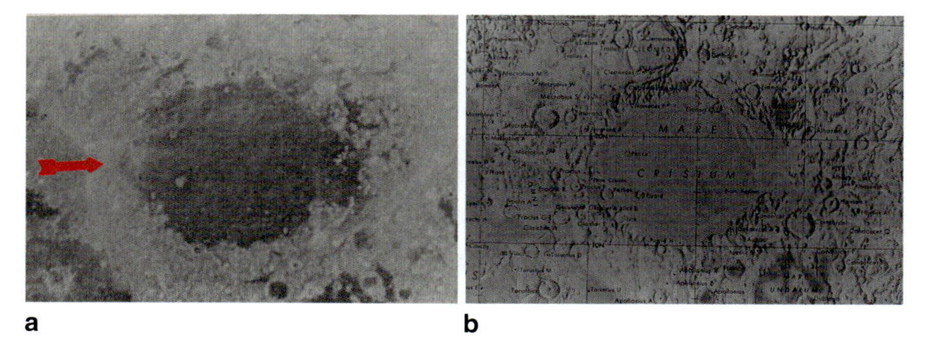

Fig. 5.33 **a** Portion of Apollo photograph 11-44-6667 showing the Mare Crisium area. **b** Photo of portion of "Lunar Earthside Chart" (LMP-1, 2nd ed. (1970)). Ellipticity is about 0.21; asymmetry is expressed by a sharply defined western rim and an irregular eastern rim with many "embayments"; implied directionality is WEST to EAST

POSSIBLE ANSWER: Only low impact speeds will generate a bilaterally symmetrical crater if the body impacts at a sufficiently low angle. If the impactor is moving at hypervelocity, then the rapid melting and vaporization will cause a central explosion that results in a radially symmetrical crater (basin) unless it has a very low angle of impact (Melosh 1989).

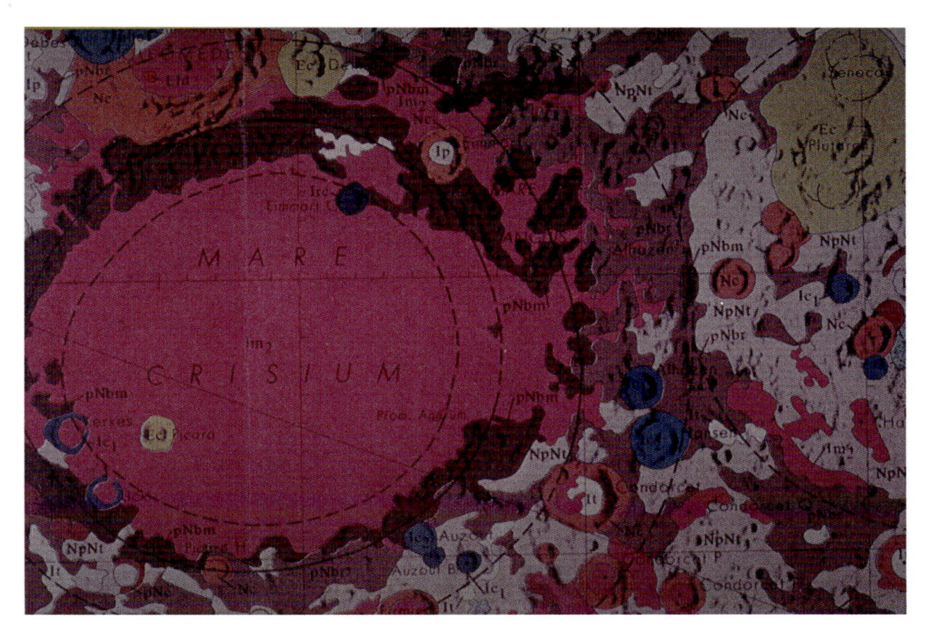

Fig. 5.34 Photo of portion of the "Geologic map of the east side of the Moon" (Wilhelms and El-Baz 1977). Again, ellipticity of Mare Crisium is about 0.21; asymmetry is expressed by a sharply defined western rim and an irregular eastern rim with many "embayments"; implied directionality is WEST to EAST

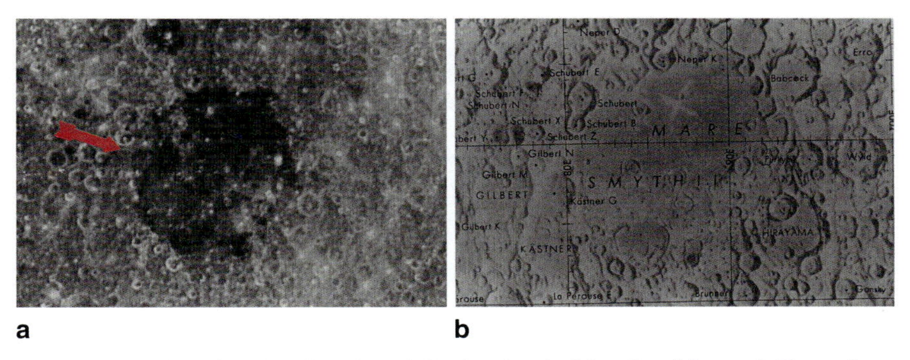

Fig. 5.35 **a** Portion of Apollo photo 8-14-2485 showing the Mare Smythii area. **b** Photo of portion of Lunar Earthside Chart (LMP-1, 2nd ed. (1970)). Ellipticity is negligible; asymmetry is expressed by a sharply defined western rim and a more irregular eastern rim; implied directionality is WEST to EAST. Some "butterfly" pattern is apparent in the Apollo photo

5.3.4.2 Tidal Disruption of a Passing Body Model (Hartmann 1977a)

This type of model features the sequential impact of an organized swarm of bodies of various shapes and sizes onto the lunar surface. The bodies would impact from

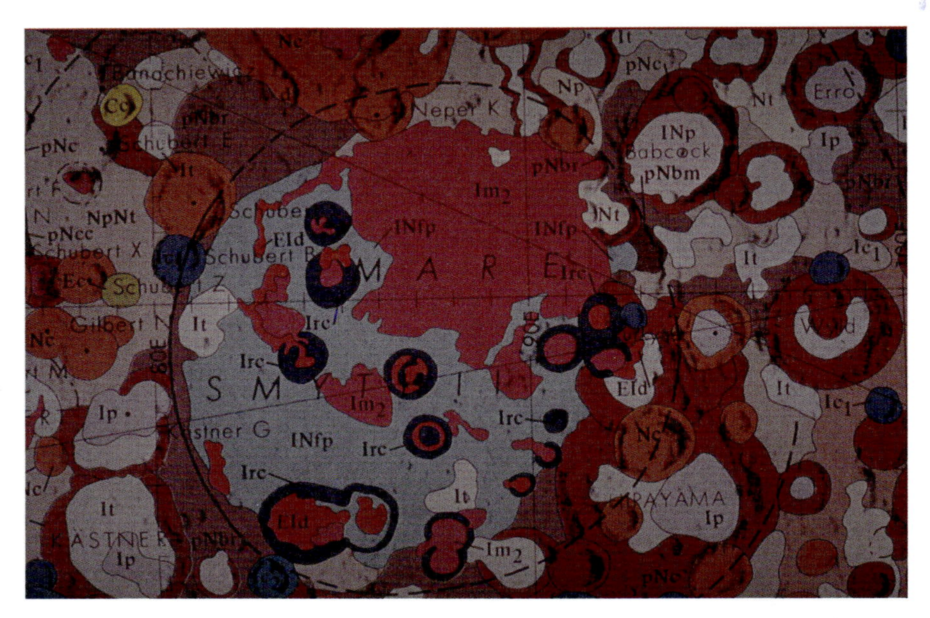

Fig. 5.36 Photo of portion of "Geologic Map of the East Side of the Moon" showing the Mare Smythii area (Wilhelms and El-Baz 1977). Ellipticity is negligible; asymmetry is expressed by a sharply defined western rim and a more irregular eastern rim; implied directionality is WEST to EAST

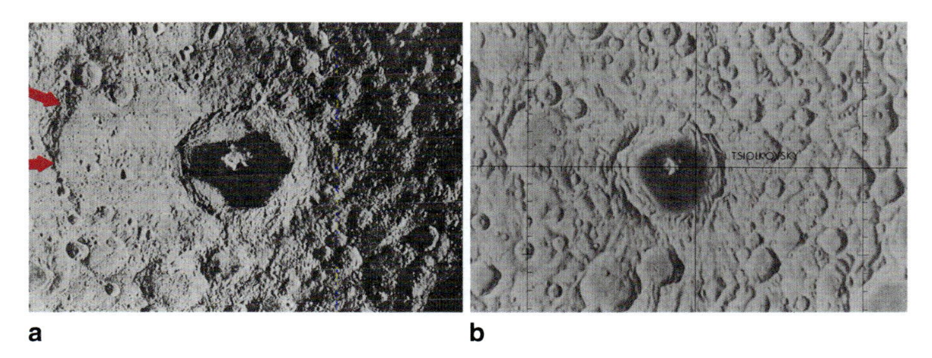

Fig. 5.37 **a** Portion of Lunar orbiter photograph 11-121-M showing the crater Tsiolkovsky area. **b** Photo of portion of "Lunar Farside Chart" (LMP-2, 2nd ed. (1970)). Ellipticity of basin (rim-to-rim) is about 0.11; asymmetry is marked by extensional features on the west (normal fault scarp) and compressional features on the eastern rim; implied directionality is WEST to EAST

a common direction to excavate the basins which would then later fill with mare basalts as the lunar body becomes warmer and the liquid basalt rises to the surface.

QUESTION: Where would these bodies come from?

POSSIBLE ANSWER: They may have come from either inside the orbit of Mercury (the Vulcanoid Zone) or from beyond the orbit of Mars (the Asteroid

Fig. 5.38 Photo of portion of "Geologic Map of the East Side of the Moon" showing the crater Tsiolkovsky area (Wilhelms and El-Baz 1977). Ellipticity of basin (rim-to-rim) is about 0.11; asymmetry is marked by extensional features on the west (normal fault scarp) and compressional features on the eastern rim; implied directionality is WEST to EAST

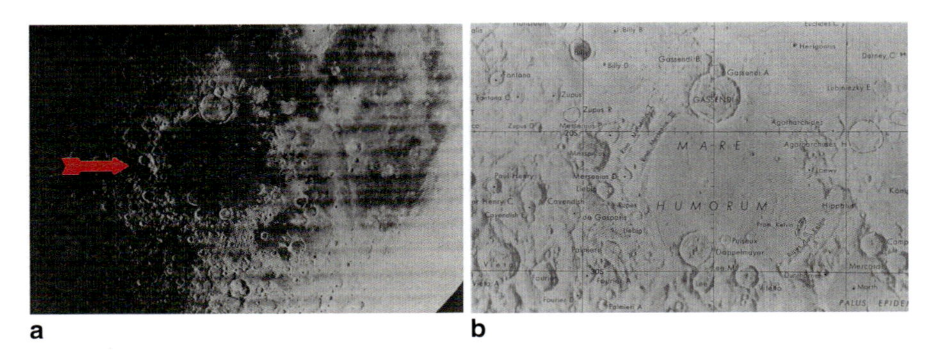

Fig. 5.39 **a** Portion of Lunar Orbiter Photo IV-123-M showing the Mare Humorum area. **b** Photo of portion of "Lunar Earthside Chart" (LMP-1, 2nd ed. (1970)). Ellipticity of mare fill is about 0.17; asymmetry is expressed by a sharply defined western rim and an irregular eastern rim; the implied directionality is WEST to EAST

Zone). In order to be tidally disrupted, the body would have to come very close to a body the mass of the Earth (a near-grazing encounter). It would also help if the material of the body was very loosely bound, a characteristic of some asteroids that are considered to be "rubble piles" (McSween 1999). For this model, the implied

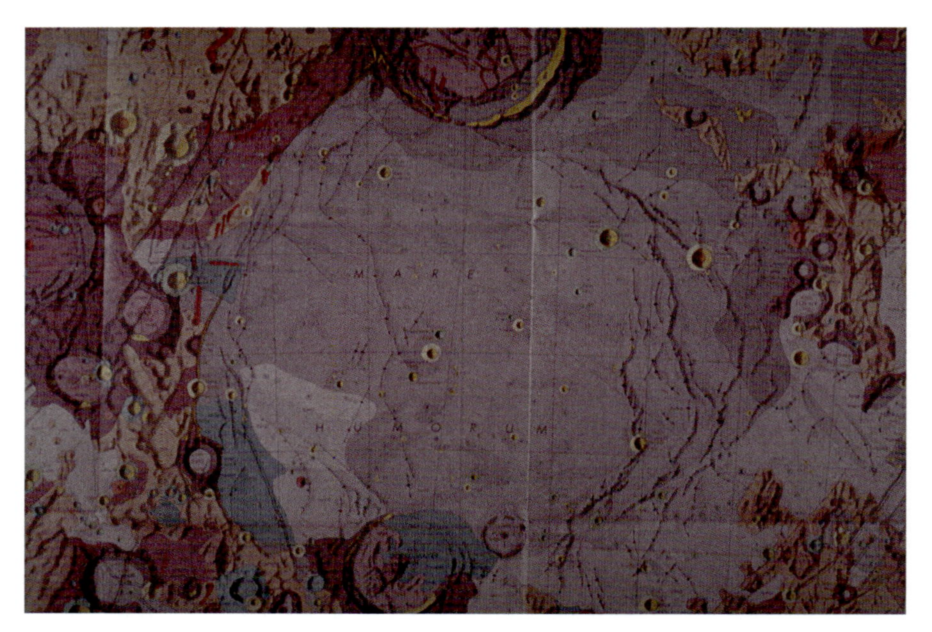

Fig. 5.40 Photo of portion of "Geologic map of the Mare Humorum region of the Moon" (Titley 1967). Ellipticity of mare fill is about 0.17; asymmetry is expressed by a sharply defined western rim and an irregular eastern rim; the implied directionality is WEST to EAST

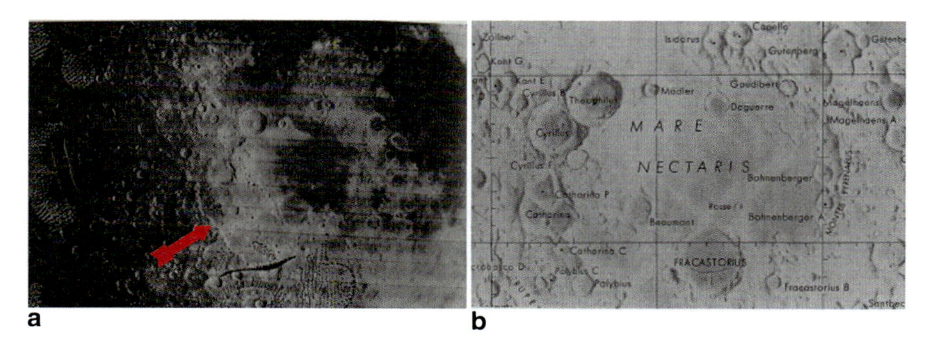

Fig. 5.41 **a** Portion of Lunar orbiter photo IV-84-M showing the Mare Nectaris area. **b** Photo of portion of "Lunar Earthside Chart" (LMP-1, 2nd ed. (1970)). Ellipticity is negligible; asymmetry is indicated by a normal fault scarp southwest of the basin; implied directionality is SOUTHWEST to NORTHEAST

assumption is that the Moon is already in orbit about the Earth before the basins are excavated.

QUESTION: What are the chances of this swarm of impactors remaining together during its subsequent orbital evolution whether in heliocentric or in geocentric orbit?

Fig. 5.42 Photo of portion of "Geologic map of the near side of the Moon" showing the Mare Nectaris area (Wilhelms and McCauley 1971). Ellipticity is negligible; asymmetry is indicated by a normal fault scarp southwest of the basin; implied directionality is SOUTHWEST to NORTHEAST

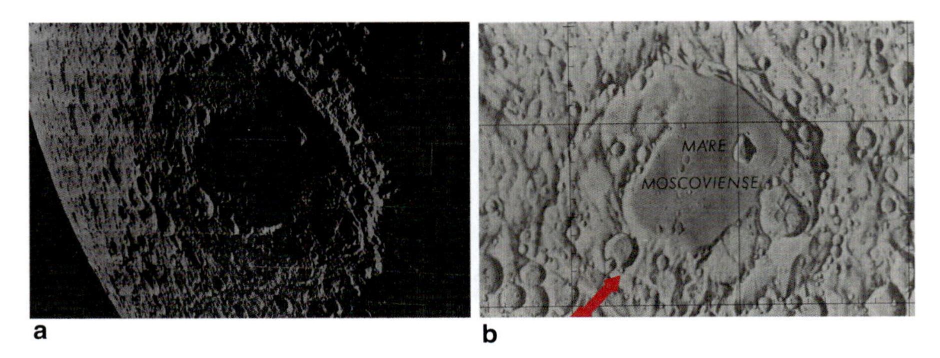

Fig. 5.43 **a** Portion of Lunar orbiter photo V-124-M showing the Mare Moscoviense area. **b** Photo of portion of "Lunar farside Chart" (LMP-2, 2nd ed. (1970)). Ellipticity is difficult to measure but a reasonable estimate is 0.33; asymmetry is espressed by a sharply defined southwestern rim and an irregular northeastern rim; implied directionality is SOUTHWEST to NORTHEAST

POSSIBLE ANSWER: The swarm would have a better chance for staying together if it was generated near the lunar body mainly via a close encounter with Earth. We must keep in mind the series of impacting bodies on planet Jupiter that resulted from a tidal disruption of a small asteroid following a close encounter with Jupiter

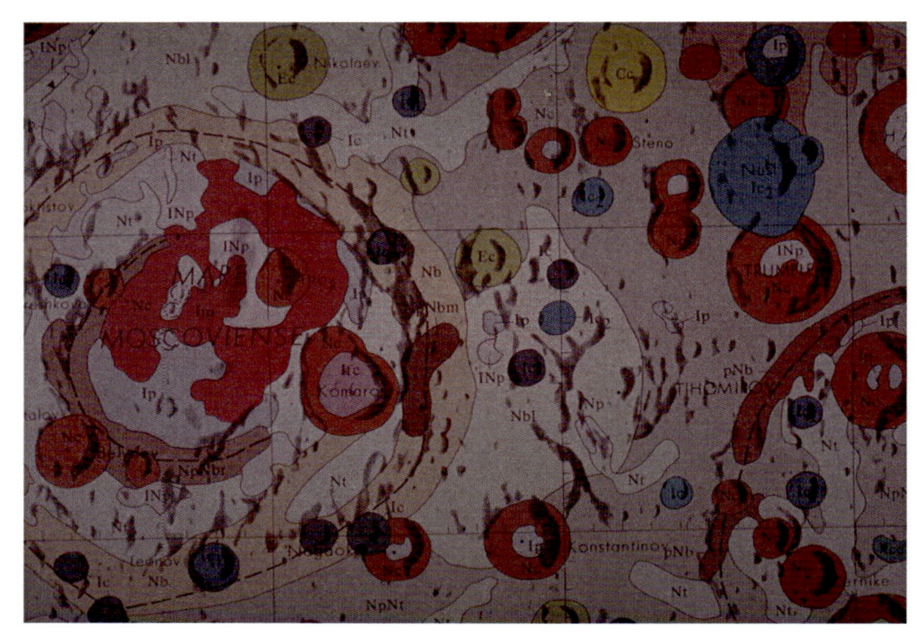

Fig. 5.44 **a** Photo of portion of "Geologic map of the central far side of the Moon" showing the Mare Moscoviense area (Stuart-Alexander 1978). Ellipticity is difficult to measure but a reasonable estimate is 0.33; asymmetry is expressed by a sharply defined southwestern rim and an irregular northeastern rim: implied directionality is SOUTHWEST to NORTHEAST

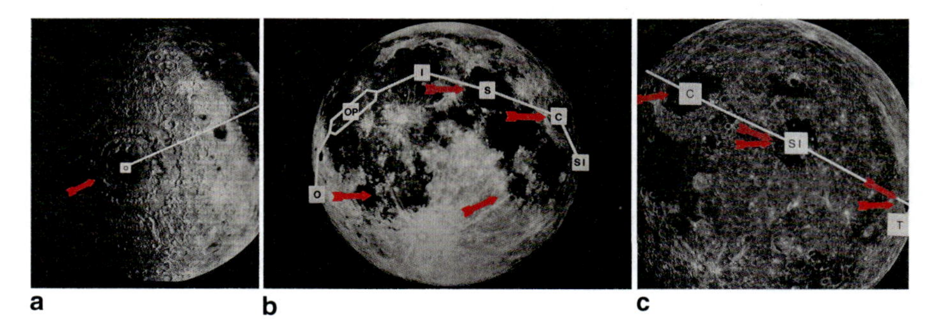

Fig. 5.45 Three lunar photographs that define a primary great-circle pattern of features as well as those that define a secondary great-circle pattern. **a** Western limb of the Moon: *O* Mare Orientale; the white line connects features on the primary great-circle pattern (Lunar orbiter photo IV-187M). **b** Front face of the Moon: *O* Mare Orientale, *OP* Oceanus Procellarum, *I* Imbrium, *S* Mare Serenitatis, *C* Mare Crisium, *SI* Mare Smythii. Features on a secondary Great-Circle pattern are marked only with a *red arrow*: on *left* is Mare Humorum and on the *right* is Mare Nectaris. Mare Moscoviense is on the lunar backside. (Hale observatory photo). **c** Eastern limb of the Moon: *C* Mare Crisium, *SI* Mare Smythii, *T* Crater Tsiolkovsky. (Apollo photo 8-14-2485)

QUESTION: What are the chances that the four largest bodies would impact along a lunar great circle with decreasing crater size?

POSSIBLE ANSWER: A low probability.

QUESTION: What would be the expected time scale for the impacting of a swarm of projectiles?

POSSIBLE ANSWER: Probably a short period to time. All bodies would be impacting within a few hours of time. If the time scale is much longer, then lunar rotation becomes a factor and the impactors would probably not end up on a lunar great circle. Bodies coming from a tidally disruptive encounter with Venus would have even more separation in time (i.e., they would be even more spread out in space).

5.3.4.3 Impact of a Swarm of Bodies Due to a Tidally Disruptive Encounter with Either Venus or Earth (Wetherill 1981)

The same questions pertain to this model as the previous model.

5.3.4.4 Impact of a Swarm of Bodies from the Asteroid Zone (Nash 1963)

This model features the impact of a swarm of solid bodies of various sizes onto the lunar surface in one-half lunar day (the length of the lunar day depends on the lunar orbital radius at the time) to excavate lunar basins which eventually fill with mare basalts as the lunar body becomes warmer and liquid basalt rises to the surface.

QUESTION: What pattern of directional properties should the excavated basins have?

POSSIBLE ANSWER: Since they are coming in at hypervelocity, the resulting craters should be symmetrical with overlapping ejecta aprons.

QUESTION: What is the probability that the four largest circular basins would be located nearly on a lunar great circle and in decreasing order of size?

POSSIBLE ANSWER: There would be a fairly low probability, but not impossible. These impactors would be "focused", theoretically, in both time and space.

5.3.4.5 Impact of Lunar Satellites Model (Runcorn 1983; Conway 1986)

This model features the sequential impact of lunar satellites over a period of time (probably over several million years). The main emphasis for this model is to explain the main GREAT-CIRCLE PATTERN as well as a Secondary Great-Circle Pattern. The explanation of Runcorn (1983) for the Secondary Great-Circle Pattern is that the lunar rotation axis shifted between the first group of satellite impactors and the second group. Runcorn (1983) also speculated that the added force of the impacts could generate a magnetic field within the lunar interior.

QUESTION: Do lunar satellites have stable orbits in Earth-Moon space?

POSSIBLE ANSWER: Conway (1986) did a series of numerical simulations showing under what conditions such a set of satellites would be stable in Earth-Moon space.

QUESTION: Can this model explain the directional properties of the circular mare basins?

POSSIBLE ANSWER: Yes, and this positive answer has two parts. (a) Since the impacting bodies are in lunar orbit, the impact speeds would be relatively low. (b) The bodies would be impacting from a common angle to the lunar surface and from a common direction.

QUESTION: What is the probability that the four largest bodies would impact along a lunar great circle on one side of the moon with the size of the impactors decreasing in a systematic pattern.

POSSIBLE ANSWER: A very low probability.

5.3.4.6 Tidal Disruption of the Lunar Body and Subsequent Fallback Model During a Close Encounter with Earth (Malcuit et al. 1975)

This model features the necking off of liquid basaltic spheroids from a lava column from the lunar mantle during a very close gravitational encounter between Moon and Earth soon after a gravitational capture episode. The magmatic spheroids would then fall back onto the lunar surface to form a series of lava lakes. Under ideal conditions, most of the action of this model would be confined to the plane of the encounter and the predictable pattern would be a great-circle pattern of lava lakes on the lunar surface. In this model the liquid basaltic spheroids would depress the lunar crust to form circular basins and simultaneously deposit a large quantity of basalt within the depressions. Subsequent lava flows, however, could be extruded onto the mare surface through fissures in the chill crust of the lava lakes as well as through crustal fissures from the lunar mantle caused by lunar rock tidal action during subsequent close encounters associated with the orbit circularization sequence.

QUESTION: How close must the Moon get to the Earth for the Moon to be tidally disrupted?

ANSWER: It must come within the Roche limit for a solid body (Aggarwal and Oberbeck 1974; Holsapple and Michel 2006, 2008) or the Weightlessness Limit (Malcuit et al. 1975). For the Earth-Moon system, this limit is located at about 1.63 earth radii (center-to-center distance). If the lunar body is "cold", then only loose debris or solid masses of rock could be lifted off the lunar surface. If the Moon is "warm" enough to have a significant mass of basaltic magma in the lunar mantle, then magma could be extracted from the lunar mantle through a crustal orifice or orifaces by Earth's tidal forces.

QUESTION: What is the time frame of such a lunar encounter with Earth?

ANSWER: The bulk of the action would necessarily take place in a few hours, the launch phase entails about one-half hour and the main part of the fallback phase

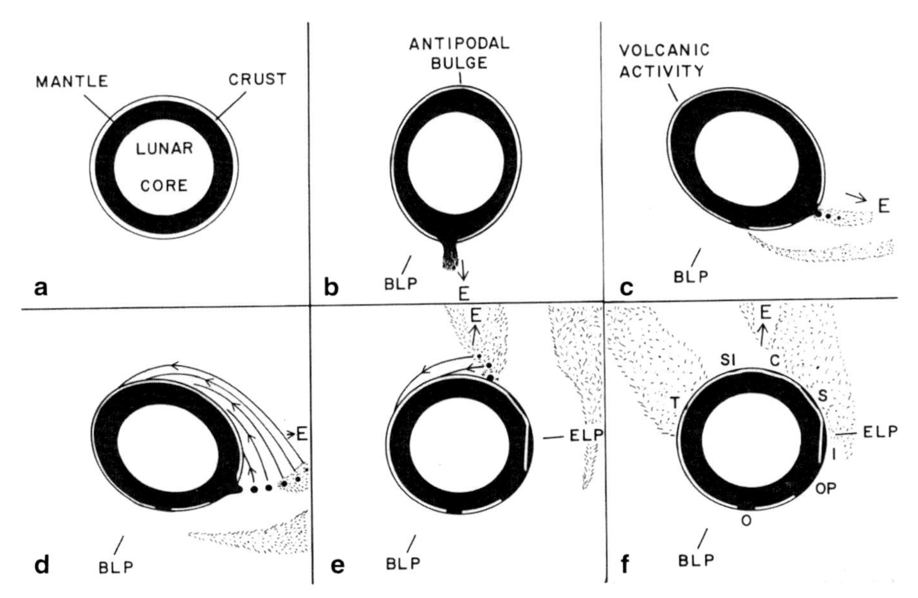

Fig. 5.46 Cross-sectional views of a lunar-like body during a tidal disruption sequence of events. **a** Initial condition before the encounter; the body is essentially undeformed. **b** The lunar-like body enters the W-limit at the *BLP* (= begin launch phase) position; the *E arrow* points toward Earth. Magma in the first disruption zone is loaded with volatiles which cause a very explosive eruption. This is interpreted as the Orientale disruption zone. **c** At this stage the lunar-like body is well within the W-limit and a second disruption zone (in the Oceanus Procellarum region) also erupts very explosively. Gas-rich volcanism also occurs in the antipodal bulge region and this area can also experience a weaker version of weightlessness during this very close encounter. **d** The lunar-like body now exits the W-limit and this is the end of the launch Phase of the very close encounter; *arrows* show the approximate trajectories of orbits of the projectiles. **e** *EPL* end launch phase. At this position the lunar-like body is well beyond the W-limit and the spheroids of basalt and other debris are falling back onto the surface of the body. **f** The lunar-like body is moving away from the Earth at this point but material is still falling back on to the body; the positions of several features in Fig. 5.45 are shown on this diagram as well as the two launch centers; most of the action occurs within about 2 h of time

entails about 1.5 h. Some solid material from the disruption could be falling back to the lunar surface, and to the Earth as well, for millions of years.

5.3.5 Some Testable Features of a Tidal Disruption Scenario That can be Analyzed on Future Mission to the Moon

Figure 5.45 shows the distribution of features on the primary great-circle pattern (along the white line). Only the basic pattern of the proposed model evolution of this feature will be presented in this section. Figure 5.46 shows a sequence of lunar cross-sectional diagrams illustrating some details of how tidal disruption of lunar-like body could occur during a very close encounter. (View is from the lunar north pole.)

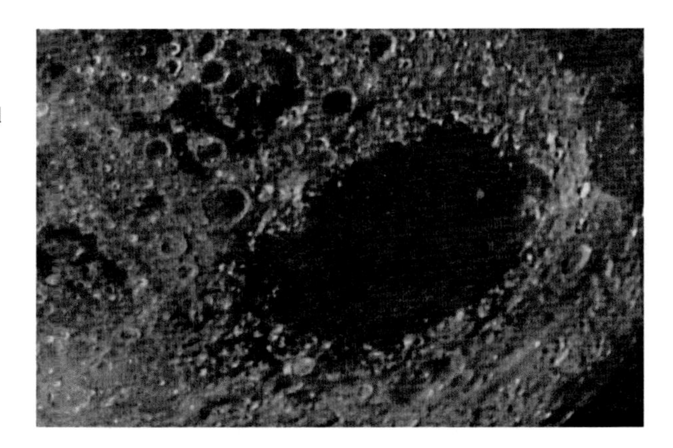

Fig. 5.47 Photograph of Mare Crisium which is oriented to relate to the diagrams in Figs. 5.48, 5.49 and 5.50 (Lunar orbiter photo)

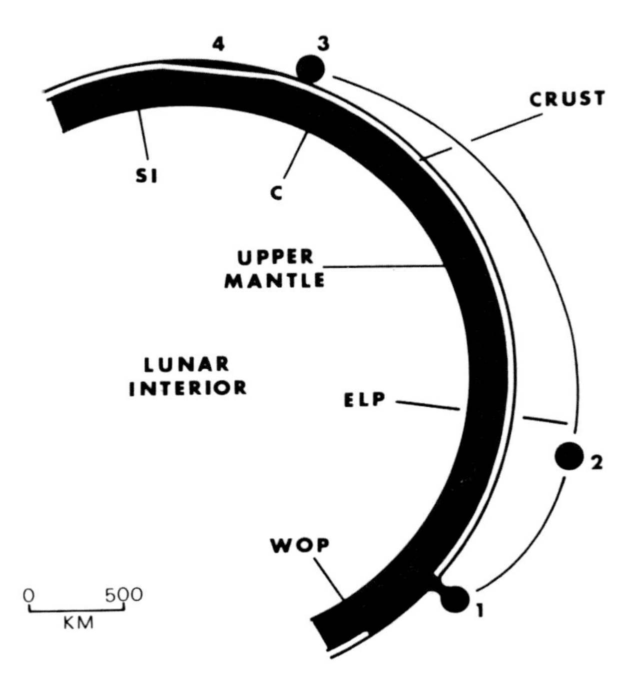

Fig. 5.48 Scale sketch showing four stages in the proposed evolution of Mare Crisium. The trajectory is the result of a numerical simulation. Stage 1: The body is launched from the Oceanus Procellarum disruption zone (*WOP* = western Oceanus Procellarum). Stage 2: This is the highest point of the trajectory. *ELP*=exit launch phase; this is the point at which the surface of the lunar-like body exits the W-limit. Stage 3: The spheroid of basalt makes contact with the surface of the lunar-like body. Stage 4: The spheroid of basalt is fully collapsed onto the surface of the lunar-like body. *C* the center of Mare Crisium, *SI* the center of Mare Smithii

The following series of four diagrams illustrates how a typical mare (Mare Crisium; Fig. 5.47) could be formed by tidal disruption of a lunar-like body during a close encounter with Earth. In this case we are following the launch of a single spheroid of lunar basalt about 150 km in diameter from its launch center in Oceanus Procellarum (Fig. 5.48, position 1) to its impact center near the position of Mare Crisium (Fig. 5.48, position 4). The spheroid of lunar basalt travels in a suborbital path above the lunar surface and impacts at a low angle and at a low speed (about 1 km/s) onto the lunar surface (Fig. 5.49). The lunar basalt then spreads over the surface to form an asymmetrical lava lake on the surface of the lunar-like body (Fig. 5.50). As the surface cools, it begins to record impacts of small bodies of debris falling back onto the surface.

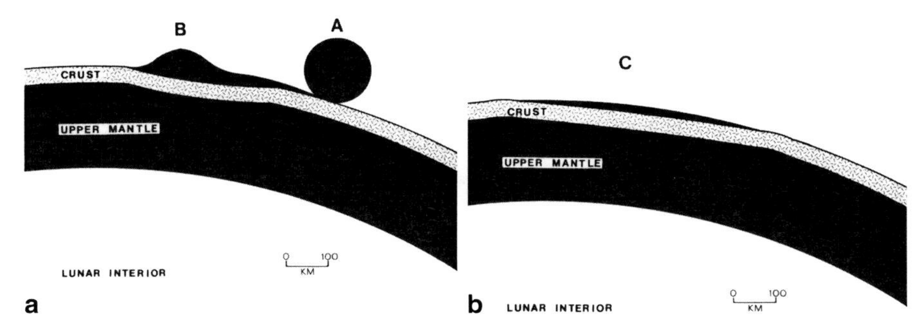

Fig. 5.49 A close-up of a cross-sectional view of the origin of Mare Crisium by way of a tidal disruption Model. **a** Position *A* shows the spheroid of basalt making initial contact with the crust of the lunar-like body. Position *B* shows the spheroid in a partially collapsed condition. Note the depression of the crust under the weight of the mass of basalt. **b** Position *C* shows a *cross-sectional* view of the spheroid of basalt completely collapsed, then the crust rebounds from the dynamics of the oblique impact. Note the raised rim on the edge of this cross-sectional view

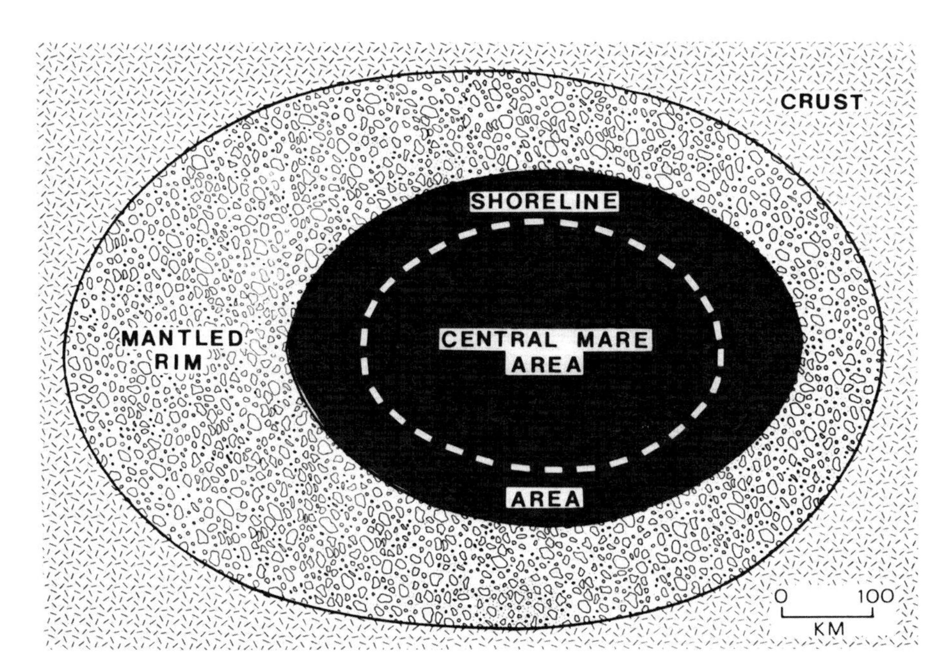

Fig. 5.50 Plan view of the impact site of the large basaltic spheroid. The mantled rim is the area overwhelmed by the "tsunami"-like wave of basalt. There would tend to be much more overflow of basalt in the down-range direction. The mantled rim would become a crater recorder immediately after the overflow activity. The shoreline region would become a crater recorder as the basalts solidify. The central mare area would be the last region to be become a crater-counter

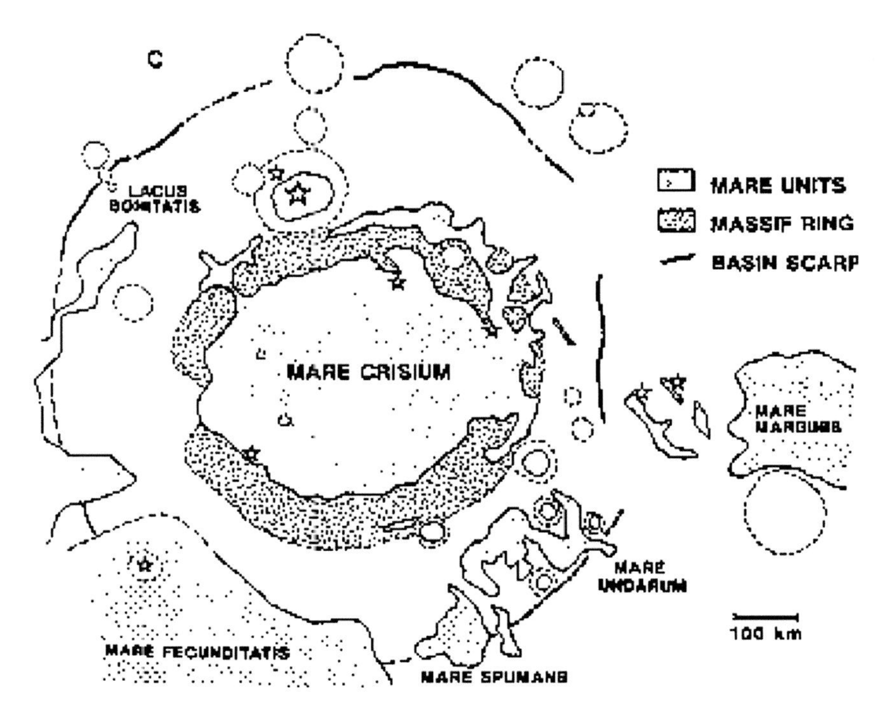

Fig. 5.51 Diagram from Wichman and Schultz (1994, Fig. 3c and their original caption, with permission from the Geological Society of America). "Geologic sketch map of the Crisium basin showing location of mare units, platform massifs and inferred outer scarp ring. Stars denote the location of floor-fractured craters." NOTE: The only change that I would add here is that the impactor is a spheroid of lunar basalt and not a solid external body. In my view, the decapitation of the magmatic spheroid would occur in a similar manner

On an Oblique Impact for the Origin of the Crisium Basin

Wichman and Schultz (1994) suggested an oblique impact model for the origin of Mare Crisium. They were thinking in terms of an oblique solid-body impact scenario for their model but I think that a low-speed impact of a soft body, as presented in Figs. 5.48, 5.49 and 5.50 may be an even better explanation for the features that they were trying to explain.

Here I quote two paragraphs from Wichman and Schultz (1994), p. 64 as well as their Fig. 3c (reproduced here as Fig. 5.51):

> The accepted signatures of an oblique impact are well illustrated by the Crisium Basin. In appearance, both the elongated basin outline and the distribution of ejecta at Crisium resemble the much smaller lunar crater Messier (Wilhelms 1987, p. 171), which is a type example of planetary oblique impact structures. (Gault and Widekind 1978; Schultz and Lutz-Garahan 1982; Wilhelms 1987, p. 32; Schultz and Gault 1990)

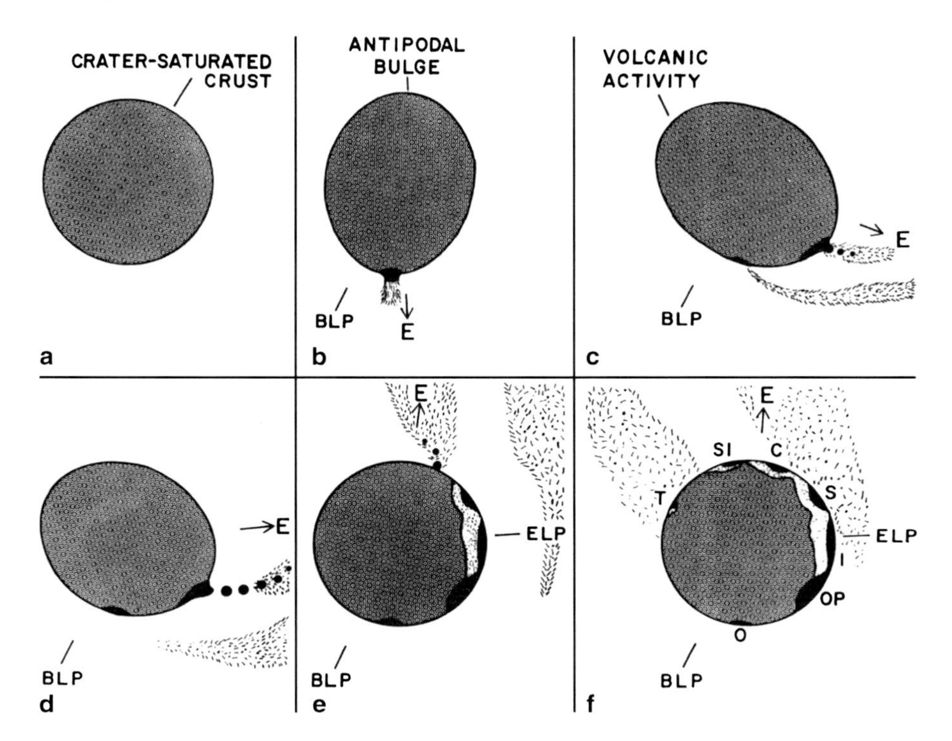

Fig. 5.52 **a** Crater-saturated anorthositic crust. **b** Lunar body enters the W-limit at *BLP* (begin launch phase) position. *Arrow* with E points toward the Earth. First disruption center is an interpretation of the Orientale supra-volcanic form. **c** Initial eruptions of the Oceanus Procellarum disruption zone. A few spheroids of basalt are necking off a lava column from this second disruption zone and there is volcanic activity in the antipodal region (the Mare Ingenii region). **d** This is the apex of the tidal disruption episode and the lunar body is departing from the W-limit. **e** The fall-back phase is well underway and some of the spheroids of lunar basalt have impacted obliquely onto the lunar surface. *ELP* end launch phase. **f** At this stage all the large spheroids have impacted on the lunar surface but much lunar-borne material is in the lunar environment. Most of this multi-sized material will eventually impact on the lunar surface

> Crisium also exhibits a broad, shelf-like breach through the eastern massifs (Fig. 3c), consistent with the failure of a projectile at low impact angles, and Mare Marginis may record the downrange impact of decapitated debris from Crisium (Schultz and Gault 1991). Based on the three-pronged ejecta pattern at Crisium (Wilhelms 1987, p. 171) and the observed variations in ejecta distribution as a function of laboratory impact angle (Gault and Wide-kind 1978), the Crisium basin apparently records an oblique impact from west to east with an impact angle of less than 15°.

Figure 5.52 shows a sequence of six scenes of the surface effects of tidal disruption during a close gravitational encounter with Earth (scenes are the same as in the cross-sectional diagrams of Fig. 5.46. View is from the lunar north pole).

Figure 5.53 is a sequence of twelve scenes of the surface effects of tidal disruption during a close gravitational encounter. View is from the *lunar north pole*. The additional scenes are used in an attempt to show the effects of gas-propelled eruptions that can move material off the encounter plane. Many details of the scenario are in the figure caption.

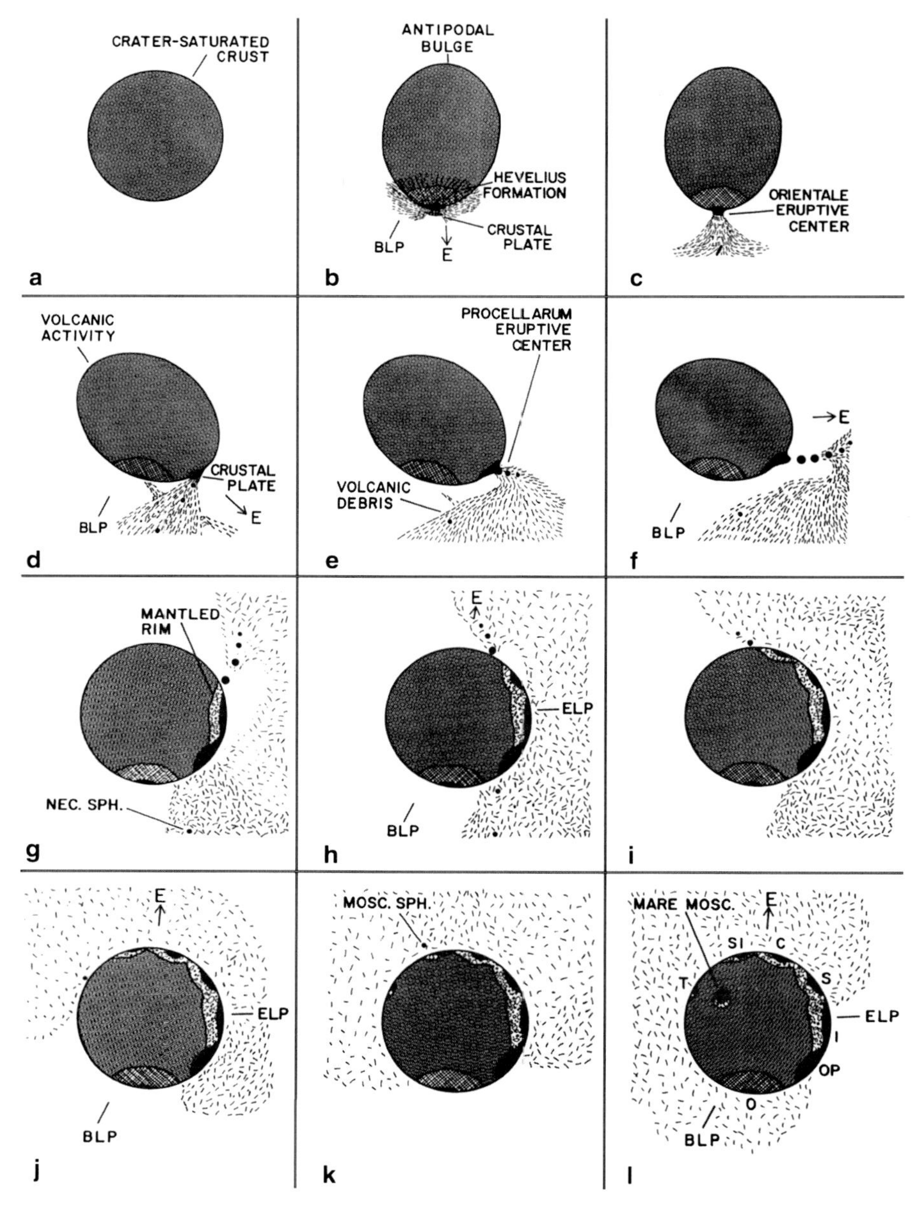

Fig. 5.53 View of activity from the lunar north pole. **a** Crater-saturated anorthositic crust. **b** The gas-propelled eruption of the Orientale Supra-Eruption center which lead to the development of the Hevelius formation. The arrow with *E* points toward Earth. A crustal plate temporarily deflects some material back to the lunar surface. BLP = begin launch phase. **c** Apex of activity associated with the Orientale Supra-Volcanic Center. Most of this material will go to the Earth. The crustal plate is clearly visible and would go to the Earth. **d** The initial eruption phase of the Oceanus Procellarum Supra-Volcanic Center. A crustal plate deflects material both backward from the normal gravitational flow as well as off the encounter plane. In this scene some spheroids are deflected to the south of the encounter plane and some will fall back in the lunar southern hemisphere and some in the lunar northern hemisphere. **e** Normal gravitational flow is occurring from

Figure 5.54 is a sequence of twelve scenes of the surface effects of tidal disruption during a close gravitational encounter. View is from the *lunar south pole*. The additional scenes are used in an attempt to show the effects of gas-propelled eruptions that can move material off the encounter plane as well as in a backward direction relative to the flow of material from the gravitational encounter.

5.3.6 Major Predictions from the Tidal Disruption Model for the Formation of Some Lunar Features

- THERE SHOULD BE NO CRUST UNDER THE TWO TIDAL DISRUPTION ZONES (Mare Orientale and Oceanus Procellarum)
- THERE SHOULD BE A FULL THICKNESS OF CRUST UNDER THE MAJOR SOFT-BODY IMPACT SITES
- ALL OF THE MAJOR ACTIVITY SHOULD OCCUR OVER ABOUT 2 H OF TIME
- THERE SHOULD BE A MASCON (MASS CONCENTRATION) ASSOCIATED WITH EACH OF THE SOFT-BODY IMPACT SITES (and there should be an asymmetry to the mascon that matches the asymmetry of the associated mare)

5.3.7 Discussion of the Predictions

1. There is *one* necessary source region. This is an ovoid shaped region in Oceanus Procellarum. There should be no anorthositic crust under this source region. Tidal distortions of the lunar body subsequent to the tidal disruption would cause periodic upwellings of mare basalt from the magma ocean zone of the lunar body. The Marius Hills volcanic field (Lawrence et al. 2013; Spudis et al. 2013) may be a manifestation of this continued volcanism.
2. According to the numerical simulations in Malcuit et al. (1975), a secondary source region is possible and this area would be located in the Mare Orientale region of the lunar surface. Under normal encounter conditions, all of this ejecta would escape from the lunar body and impact on Earth. Under abnormal

the Procellarum Supra-Volcanic Center. A spheroid is visible in the volcanic debris. **f** The termination of the Procellarum Eruptive Center. Note the stack of spheroids from this center as well as the spheroid with the other debris. **g** A few of the basaltic spheroids have impacted obliquely onto the lunar surface and are causing "tsunami" wave action, mantled rims, etc., as a result of their impact. Note position of the Nectaris spheroid that will impact in the lunar southern hemisphere. **h** Spheroids are falling back onto the lunar surface along a great-circle pattern. *ELP* = end launch phase position. **i** Fall-back phase continues. **j** All major spheroids have impacted on the surface. Note the position of the Tsiolkovsky spheroid (not yet impacted). **k** All major spheroids have impacted onto the Great-Circle Pattern. The Moscoviense Spheroid is moving from the lunar Southern Hemisphere into the lunar Northern Hemisphere. **l** The Moscovience Spheroid has impacted to form an elliptical lava lake on the lunar surface with a vector about 90° to the trend of the Great Circle of Large Circular Maria

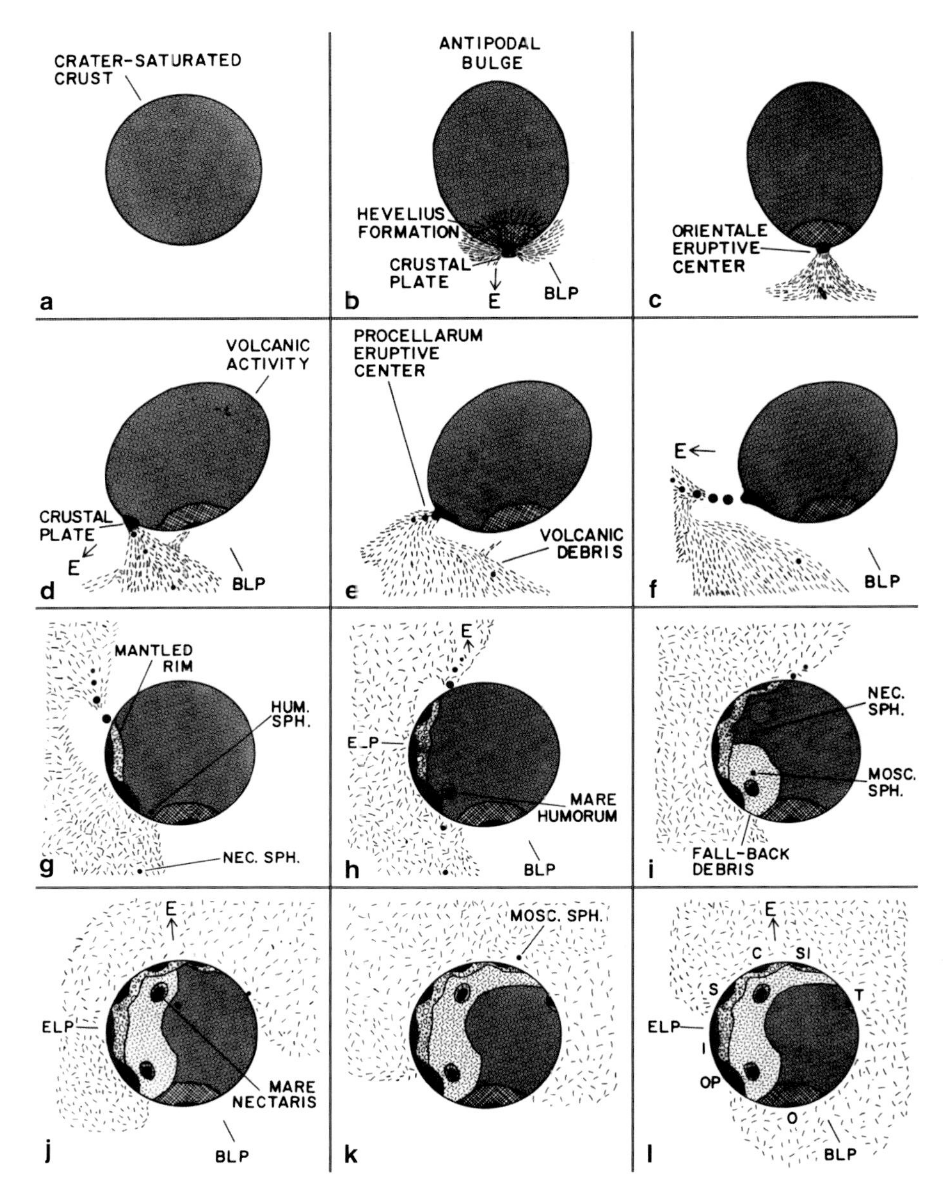

Fig. 5.54 View of activity from the Lunar south pole. Note that the action is in the opposite direction to that in Fig. 5.53. **a** Crater-saturated anorthositic crust. **b** Lunar body enters the W-limit and a gas-charged eruption occurs. A crustal plate deflects material back onto the lunar surface to form the Hevelius formation. *BLP* = begin launch phase. *Arrow* with *E* points toward Earth. **c** Orientale supra-eruption center is in full bloom. Most of this material goes to the Earth. The crustal plate is visible and would go to the Earth. **d** The Procellarum supra-eruption center opens with a very explosive eruption that propels material, including spheroids, both backward relative to the gravitational influence and into the Lunar Southern Hemisphere. Note the volcanic activity in the antipodal bulge area. Note that the anti-sub-earth region is only slightly within the W-limit but this

conditions some of the material may find its way back to the lunar surface. The most reasonable way for the Moon to sample this ejecta is via "a blow-back" (gas-propelled) of ejecta during an explosive episode early in the launch phase of the tidal disruption as illustrated in Figs. 5.53b and 5.54b. The explosive volcanic ejecta would be "smeared" onto the area surrounding this secondary source region and immediately underneath the plastered-on material would be something like "original" cratered highland crust. This would not be an apron of impact ejecta but it would be explosive volcanic ejecta. The best description that I can think of, and this is an anachronistic idea, is that it is lunar basaltic NEUEE ARDENTE material as envisioned by Mackin (1968). Furthermore, nearly all of the Hevelius Formation (the material surrounding the Orientale feature) is composed of this material. And the material erupted in the first phase of the Procellarum eruptive center is of this composition as well. This second batch of material, in my model, eventually is "plastered" onto the "southern highlands" of the Moon as it returns to the lunar body along with some lunar basaltic spheroids that constitute the Secondary Great-Circle Pattern. Returning to the discussion of the Mare Orientale region, there should be no anorthositic crust under the mare basalts. Because of the lack of anorthositic crust, tidal pumping could cause volcanic constructs like the Marius Hills in the central area of Mare Orientale since the lava can be transported directly from the magma ocean of the lunar body to that area. On the "Geologic Map of the Near Side of the Moon" (Wilhelms and McCauley 1971), some units in the southern highlands of the Moon were mapped as mainly volcanic and pyroclastic deposits mixed in locally with mare-type basalts. The description and interpretation of these units by Wilhelms and McCauley (1971) (mainly Ip, *plains material*; and Ihp, *hilly and pitted material*) fits well with my concept of the type of hot airborne pyroclastic debris that would be falling back to the lunar surface to form a thin blanket over the lunar terrain. The best exposures of this pyroclastic material, in my view, would be located in the southern highlands between Mare Nubium and Mare Nectaris (and some of this may have been sampled by the Apollo 16 astronauts).

---

leads to a gas-charged eruption in the Mare Ingenii region. **e** The Procellarum supra-eruption center is settling down to a normal gravitational necking process to produce basaltic spheroids. **f** Lunar body passes out through the W-limit. This is the end of the Launch Phase. Note that a spheroid is mixed in with the small debris. **g** The fall-back phase is well underway and some spheroids of lunar basalt have impacted to form "tsunami" waves of basalt to flow over the surface around the impact sites to form mantled rims. The Humorum spheroid is about to impact and the Nectaris spheroid is visible in the lunar debris field. **h** Regular fall-back activity continues. The Humorum spheroid has impacted and the Nectaris spheroid is visible. *ELP* = end launch phase. **i** The Smythii spheroid as well as the Tsiolkovsky spheroid are about to impact along the lunar great-circle. The Nectaris spheroid is visible above the lunar surface. The Moscoviense spheroid is visible above the Mare Humorum area. **j** The Nectaris spheroid has impacted and the Tsiolkovsky spheroid is about to impact but the Moscoviense spheroid is visible above the lunar surface in the Mare Nectaris area. **k** All of the major spheroids are down but the Moscoviense spheroid. It will impact on the lunar northern hemisphere (as shown in Fig. 5.53). l. The bulk of the Fall-Back Phase is over but much small lunar debris is returning to the lunar surface. Note that Mare Humorum and Mare Nectaris are located off the great-circle pattern and are in the lunar Southern Hemisphere. These two features along with Mare Moscoviense are on a secondary great-circle in this model

Detailed studies of these types of units using a variety of spectral data from lunar orbital missions are consistent with a pyroclastic deposit interpretation and, indeed, with the interpretation of special and massive NUEE ARDENTE activity as envisioned by Mackin (1969). Examples of both earlier and more recent analyses of such deposits are Gaddis et al. (2003), Gustafson et al. (2012), Besse et al. (2014), and Lena and Fitzgerald (2014). These types of remote sensing analyses also can be used to delimit the purported "tsunami"-like advances and retreats of mare basalts surrounding most large circular maria.

3. A volatile-rich explosive eruption at the primary tidal disruption zone (the initial eruption of the Oceanus Procellarum eruption center) can explain many of the features of a secondary great-circle pattern as described by Runcorn (1983). The details depend on how the crust responds to tidal forces. The volatiles could shoot straight out toward Earth or they could be deflected in some direction by a crustal plate. Since tension is on the lunar west and compression on the lunar east, the most likely direction of the volatile-rich basaltic material flow is to the lunar west. Such an eruption could lead to abundant particulate material spewing out followed by some columns or ribbons of lunar basalt. These columns or sheets of basaltic lava could be transformed into basaltic spheroids of various sizes. These spheroids, on impact, could account for Mare Humorum, Mare Nectaris, Mare Moscoviense, and other smaller circular structures. The associated hot, particulate material [the neuee ardente material described by Mackin (1969)] could account for the apparent "coating" of basalt, basaltic glass, and breccia on the lunar highlands south of the Humorum-Nubium-Nectaris region (a region depicted in Fig. 5.54 i–l and sampled by the Apollo 16 mission). This material is described in detail by Ulrich et al. (1981).
4. The normal primary disruption zone (launch center) for this tidal disruption model features a necking-off process from a lava column that is sourced in the lunar magma ocean. The basalt launched early in the process, and perhaps that launched later as well, would be rich in the KREEP component. Basalt in the first spheroids is also more volatile-rich and lower viscosity than that launched later in the sequence. In general, the major factors which determine the size of the spheroids being launched from the primary launching zone are: (a) the gas content of the lava, (b) the viscosity of the magma [which may be related to (a)], and (c) and the ever decreasing gravitational tidal force operating on the lava column. Numerical simulations, in the future, by students of fluid mechanics will help to elucidate the details

   Once space-borne, then the lunar basaltic spheroids are guided to their impact zone by only gravitational forces. Under normal conditions if there is no lunar rotation during the encounter (i.e., the Moon is not rotating relative to the Cartesian coordinates of the encounter), the spheroids would impact along a great-circle pattern soon after launch and they would impact at low angles in an easterly direction. Thus, the impact zone should be characterized by asymmetry and the asymmetry should reflect the directionally.
5. Under ideal conditions the launch zone for tidal disruption would be the locus of points defining an ovoid region centered on the trace of the line connecting the

centers of the Earth and Moon. If the Moon does not enter the W-limit, then there is no trace. If the Moon is rigid, then only loose particles could be lifted from the surface and then transported above the lunar surface to impact at an angle to the lunar surface. If the Moon can yield magma, then the conditions can be suitable for the generation of a great-circle pattern of large circular maria.

5.3.8 An Epilogue to This Section on "Directional Properties of Lunar Maria"

I have presented this paper, or something like it, several times as a poster paper (including a poster paper at the Kona Conference) and I got mixed responses. The general comment is that it is "too complex". So why am I still pursuing this model on "Directional Properties". WHAT IS NEW AS FAR AS ANCILLARY INFORMATION OR CONCEPTS????

Well, there is still no reasonable solution for the cause of the LATE HEAVY BOMBARDMENT (LHB). Most interested investigators agree that something BIG happened to the Moon about 3.85–3.95 Ga ago. What was it? Were the impacting bodies from a distant source or sources or were these impacting bodies from a local source or sources? Did all of this impacting occur over about *200 million years* of time or over about *2 h* of time??? And what was happening on Earth at this time? Was it bombarded by external bodies and is there evidence of impact debris from these external bodies in the oldest rock sequences on Earth??

Three recent papers suggest a short time scale for these event(s) and they are backed up by many DATA papers. The three that I am thinking about are Harrison (2009), Koeberl (2006), and Trail et al. (2007). The interpretation of age data from both lunar glasses and terrestrial zircons, by the authors of these three papers, suggest that these events happened over a very short period of time centered on 3.9 Ga ago. When considering the error bars for the lunar glasses in particular (e.g., Dalrymple and Ryder 1993, 1996), many events could have happened essentially simultaneously!

My interpretation of the directional properties of the large circular maria (in this section of this chapter) suggests that they are the result of the impact of lunar-borne materials falling back onto the Moon to form a GREAT-CIRCLE PATTERN OF LARGE CIRCULAR MARIA. Most of this action could have happened in a matter of 2 h of time about 3.95 Ga ago. The time-frame of a TIDAL DISRUPTION EPISODE cannot be altered much, but the absolute geologic timing is open for discussion.

In my view, a tidal disruption scenario appears to be uniquely associated with a gravitational capture model. It is ironic, however, that (1) gravitational capture *can* occur without tidal disruption of the lunar-like body and (2) tidal disruption *cannot* occur during the capture encounter (i.e., the initial close encounter that dissipates the large quantity of energy necessary for capture). The bottom line is that if TIDAL DISRUPTION does occur, it is uniquely associated with a sequence of events associated with a LUNAR CAPTURE MODEL!

5.4 Vignette C. On the Origin of Lunar Maria and Mascons: The Case for a One-body, Isostatic Equilibrium Model Revisited

The association of mascons with circular lunar maria, as well as the subsurface structure of these features, remains one of the unsolved problems of lunar science. The lunar crust must have been either (1) very rigid or (2) near isostatic equilibrium to maintain the observed mass concentrations over billions of years. The argument for an early warm Moon is supported by the recurrence of lava flows until at least 3.1 Ga ago. A two-body model for lunar mascons (lower body, a mantle plug; upper body, surface lava flows) was chosen by many investigators because they could not visualize how deep basins (about 10–15 km) could be formed on the lunar surface and then subsequently filled with lunar basalt (e.g., Bowin et al. 1975).

A one-body, isostatic equilibrium model, however, can be explained in the context of a very close, tidally disruptive encounter between Earth and Moon as part of a gravitational capture sequence of events (Malcuit et al. 1975). During such an encounter, large magmatic spheroids (about 10–150 km radius) of lunar basalt could be necked off a lava column from a disruption zone. The spheroids are then transported above the lunar surface and subsequently impact along a great-circle pattern on the lunar surface. The soft magmatic bodies impact obliquely at low velocities (0.5–2.0 km/s) onto the surface from the direction of the disruption zone (the source region for the basalt). The crust in the impact zone is depressed simultaneously with the deposition of a large proportion of the mass of the spheroid within the crustal depression. Crustal rebound of the affected region results in a raised rim.

5.4.1 *Purpose*

The purpose of this section of the chapter is to demonstrate that a one-body isostatic equilibrium model is compatible with the major features of lunar mass concentrations (mascons) that are associated with large circular maria. The relatively simple one-body case has been mentioned several times in the literature but no one has pursued it because it is difficult to explain in terms of a solid-body impact model for the formation of lunar basins.

The depth-of-excavation problem associated with the solid-body impact model is by-passed if the circular maria are considered to result from the impact of large soft (magmatic) bodies onto the lunar surface. In the case of soft-body impact, *there is no open excavation* (crater) because a large percentage of the basaltic mass in the circular mare is deposited at the time of impact of a large basaltic spheroid into the lunar surface. As described in the previous section of this chapter, such large spheroidal masses of lunar basalt can result from tidal disruption of a "warm" lunar body during a near-grazing gravitational encounter with Earth. According to Mizuno and Boss (1985) a lunar-sized body is about the optimum radius for such a tidal disruption scenario if the thermal state of the body is favorable.

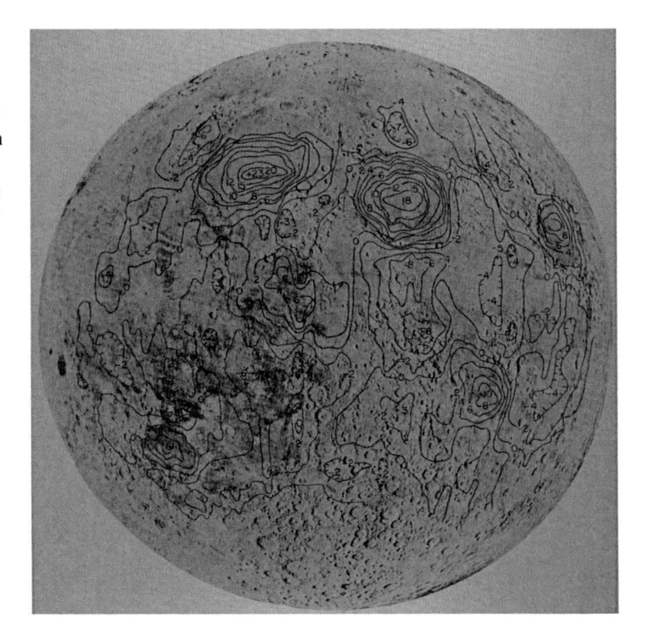

Fig. 5.55 Mascon pattern on the front side of the Moon. Gravimetric and acceleration contour map of the lunar nearside showing the location of mascons. Note that all large mass concentrations are associated with large circular maria. Contour numbers are in milligals. (From Mutch 1972, Figure VIII-34, with permission of Princeton University Press)

5.4.2 Some Special Features of Large Circular Maria and Associated Mascons

1. The four largest circular maria and associated mascons (Imbrium, Serenitatis, Crisium, Smythii) are located near a lunar great circle (Malcuit et al. 1975; Runcorn 1983).
2. The mascon pattern appears to match very closely with the basaltic fill of the maria.
3. Most of the circular maria are either asymmetrical and/or elliptical in map view (Kaula 1969; Malcuit et al. 1975).
4. The gravity profiles of Mare Serenitatis and Mare Crisium are notably asymmetrical with the greatest mass concentration associated with the western side of the features (Janle 1981; Wichman and Schultz 1994).
5. Mare Orientale has a much different mascon pattern (e.g., the pluses and minuses cancel out) and Mare Marginis has no significant mascon (Gottlieb et al. 1969).

Figure 5.55 shows the distribution of mascons on the front side of the Moon.

5.4.3 Some Previously Proposed Models for Mascons

A major question concerning the origin of mascons is whether the lunar body at the time of mascon formation was "warm" or "cold"? A "warm" lunar body could not deviate much from isostatic equilibrium for even short periods of geologic time

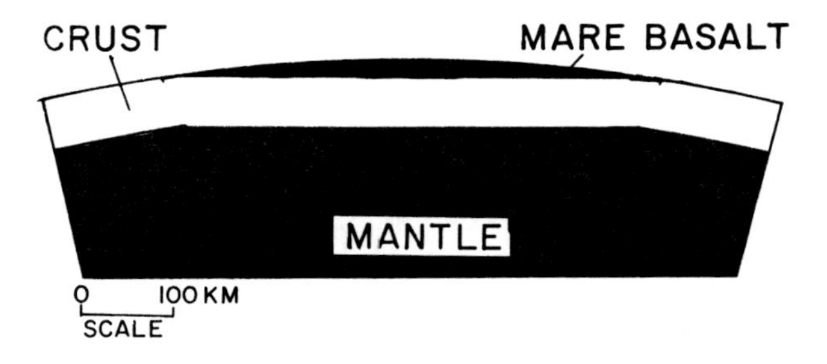

Fig. 5.56 Scale Diagrammatic sketch illustrating the major features of a one-body isostatic equilibrium model described by Bowin et al. (1975). See text for explanation

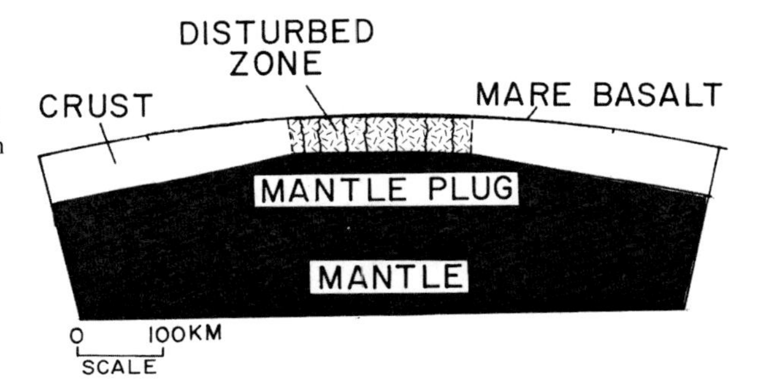

Fig. 5.57 Scale diagrammatic sketch of the currently accepted two-body isostatic equilibrium model of mascon formation described in Bowin et al. (1975)

(Kunze 1974). Even a "cold" Moon would not preserve non-isostatic loads for a very long period of geologic time (Kunze 1974). The presence of a differentiated crust-mantle system as well as the presence of substantial amounts of lunar basalt on the lunar front-side strengthen the case for an early "warm" Moon. In addition, Kunze (1976) suggests that the negative gravity anomalies associated with and surrounding the positive gravity anomalies of mascons are about equal in magnitude. This implies regional isostatic equilibrium for the mascon areas.

Figure 5.56 is an illustration of a one-body isostatic equilibrium model. This is a fairly simple model. Mare basalt (density = 3.3 g/cm^3) is perched upon a full thickness of lunar anorthositic crust (density = 2.9 g/cm^3). The lunar mantle is slightly denser than lunar basalt at ~3.34 g/cm3. The main mystery here is what mechanism can be used to get the mare basalt deposited on a full thickness of lunar crust without an impact crater being there initially?

Figure 5.57 is an illustration of the currently accepted Two-Body Isostatic Equilibrium Model. This model is much more complex than the one-body model and the anorthositic crust is either eliminated or highly deformed (or brecciated) by the energy of the impacting body. The energy of the impact causes melting in the lunar mantle as well. Then a partial melt from this mantle zone oozes up through the brecciated crust to form a thin layer of mare basalt. The elevated mantle plug resulting mainly from rebound of the material in the target zone compensates for the thin mare basalt in the central uplift area of the circular structure.

The clearest statement of the "Conventional Wisdom" Model for the origin of maria and associated mascons is in Wilhelms (1993, p. 171):

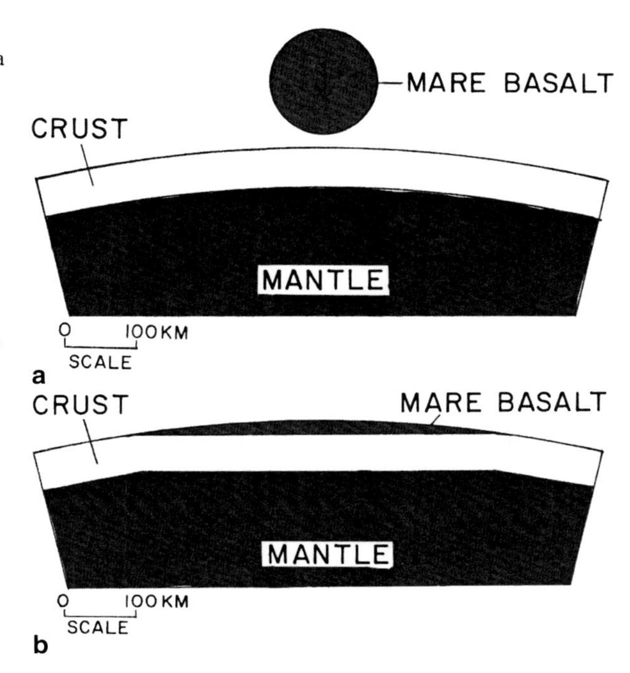

Fig. 5.58 Cross-sectional view (somewhat to scale) of a very simple case of a molten basaltic spheroid impacting onto the lunar surface at perhaps ~ 0.5 km/s. **a** The scene just before impact. **b** The scene after impact of the spheroid and after the basalt settled into the crustal indentation caused by the impacting spheroid. This is the simplest version of a one-body isostatic equilibrium model

> Let Ralph Baldwin have the last word on mascons for the moment: In 1968 he put forth a model for their origin, based on observations beginning in the 1940s, close to the one accepted today: Mare Imbrium (i.e., the Imbrium basin) was formed by a giant impact, remained "dry" for a while, began to adjust isostatically, and before it could flatten out completely was filled in its low spots by the mare basalt. The basalts did not flood in all at once but in many flows over a long period of time. Being denser than the rest of the Moon, the mare rocks sank a little, cracking their peripheries (forming arcuate rilles) and compressing their interiors (forming wrinkle ridges). They could not sink completely, hence the mascons. Add the presence of a mantle uplift beneath the mare suggested later by Don Wise, geophysicist Bill Kaula, and others, and you have the current model of lunar and planetary mascons.

There you have it, the CONVENTIONAL WISDOM MODEL FOR MASCON FORMATION!!!

5.4.4 A Soft-Body Impact Model for Mascons

Figure 5.58 is a diagram showing a very simple version of the essentials of the formation of a mascon by way of a Soft-Body Impact Model. In Fig. 5.58a, a basaltic spheroid is about to make contact with the lunar surface and in Fig. 58b the basaltic spheroid has impacted at a low impact speed on to the lunar surface.

Figure 5.59 shows a few more complications than the simple model of Fig. 5.58. In Fig. 5.59a a molten basaltic spheroid is about to impact obliquely at a low impact

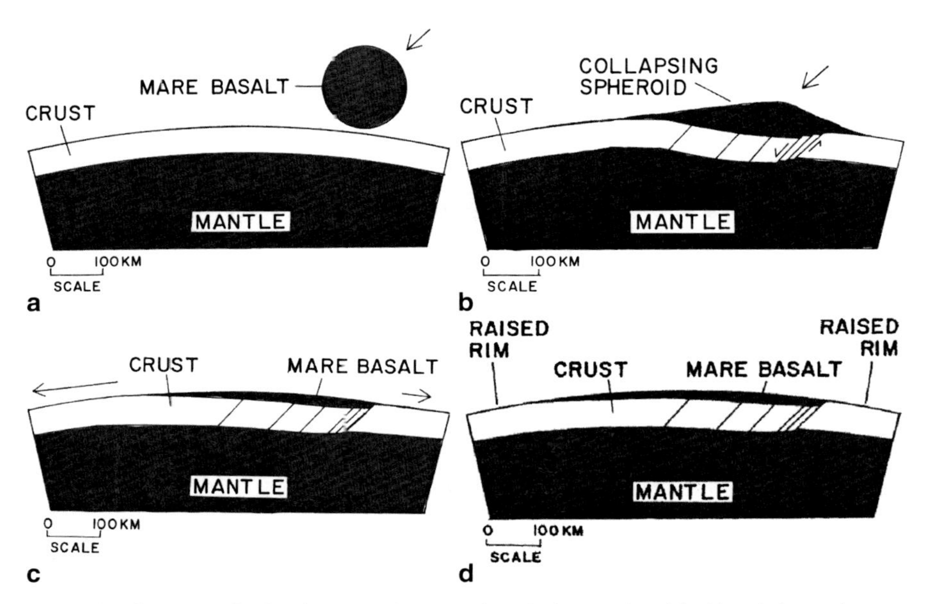

Fig. 5.59 Diagrammatic sketch (somewhat to scale) of a large spheroid of basalt impacting at a 45° angle to the lunar surface. **a** The scene just before impact. **b** Spheroidal collapse and overload stage. **c** Over-rim flooding stage. **d**. Isostatic equilibrium adjustment stage. There is additional description of the process in the text

velocity (0.5–2.0 km/s) onto the lunar surface. In Fig. 5.59b the spheroid is partially collapsed. In Fig. 5.59c the basaltic spheroid has impacted and has caused faulting in the up-range direction and significant rim overflow of basalt in the down-range direction. Figure 5.59d shows the scene after isostatic rebound with a raised rim surrounding the large lava lake of lunar basalt. Note that the mascon is in isostatic equilibrium and is caused by a mass of higher density basalt perched on the lower density anorthositic crust as suggested by Conel and Holstrom (1968).

5.4.5 Some Predictions from the Soft-Body Model for the Formation of Circular Maria and Mascons

1. The circular maria and mascons, with the exception of Mare Orientale, should have lava “lakes” located on a normal thickness of lunar anorthositic crust.

This prediction can be checked by way of seismic experiments on future mission to the Moon.

2. The three-dimensional asymmetry of mare basalt fill in mare basins should relate directly to the gravity anomaly pattern of each of the circular maria (again, with the exception of Mare Orientale).

This prediction can be tested by way of detailed lunar seismic and gravity surveys.

3. The central depth of Mare Serenitatis should be about 16 km for a one-body isostatic equilibrium model (Bowin et al. 1975). The central depth of Mare Imbrium should be about 30 km for a one-body isostatic equilibrium model (Kunze 1974). A reasonable guess for the central depth of Mare Crisium is 12 km.

Note: These numbers contrast greatly with the published estimates of mare fill (for example, Head 1982) which suggests a central depth of about 3–5 km.

5.4.6 Summary

A one-body isostatic equilibrium model is presented for the explanation of the association of circular lunar maria and mascons. This model also relates to the asymmetry of lunar mass concentrations.

This model features emplacement of the bulk of the mare basalt by way of impact of "soft" bodies (magmatic spheroids) onto the lunar surface at low impact speeds. This model circumvents the "depth-of-excavation" problem that caused other investigators to discard one-body models for more complex two-body explanations. The soft-body model has no open cavity during emplacement of the mare basalts and associated mascon.

Brief Epilogue

Although the possible effects of soft-body impacts are the main feature of this treatment of the origin of circular maria and mascons, I should stress here that the vast majority of the surface features of the lunar surface have been shaped by solid-body impact processes. If, indeed, soft-body impact processes occurred, they could have occurred during only a brief episode (a few hours of time) in the history of the lunar body. However, soft-body impact processes, like those described in this section could leave an indelible signature on the lunar surface—a great-circle pattern of large circular maria and associated mass concentrations.

5.5 Vignette D. The Late Heavy Bombardment of Earth, Moon, and Other Bodies: Fact or Fiction?

A quote from Cisowski et al. (1983, p. A700) and some "food for thought" on the topic:

> The apparent coincidence of three major events in lunar history, the end of the late heavy bombardment, the onset of mare volcanism and the period of intense magnetization of lunar samples invite speculation on a common cause for these events.

The concept of a "Terminal Lunar Cataclysm" (Tera et al. 1974) and the later terminology, a "Late Heavy Bombardment", is a current paradigm in the lunar and planetary sciences. Some of the promoters of this paradigm are attempting to extend it to all parts of the inner part of the Solar System (e.g., Kring and Cohen 2002) as well as to the outer Solar System (Charnoz et al. 2009; Matter et al. 2009).

5.5.1 Purpose

The purpose of this "vignette" is to evaluate the evidence associated with this concept via the Scientific Method. The first step in the evaluation is to list the facts to be explained by a successful hypothesis. If the hypothesis under evaluation does not explain a basic set of facts, then another hypothesis (model) can be substituted and the process of the Scientific Method can be repeated. This is the pathway of progress in the endeavors of natural science.

5.5.2 Some Facts to be Explained and Questions to be Answered by a Successful Model

1. Patterns of lunar rock magnetization associated with lunar mare rocks and lunar breccias strongly suggest that a magnetic field was operating probably within the Moon and in an exponentially decaying mode from ~3.95 to 3.6 Ga (Cisowski et al. 1983).
2. Many of the dates for either mare basin formation and/or mare basalt fill occurred over a time range of 3.8–4.0 Ga (Wilhelms 1987).
3. Dates of solidification of many of the older breccias and glasses in breccias occurred over a short period of time in the range of 3.85–3.95 Ga (Dalrymple and Ryder 1993, 1996).
4. Oxygen isotope information from zircon crystals from Western Australia and other geochemical evidence relating to the recycling of an ancient "enriched" crust into the Earth's mantle indicate that the Earth was cool enough to have ocean water on the surface as early as ~4.4 Ga ago (Valley et al. 2002) and that this ancient crust commenced to be recycled into the mantle of Earth ~3.95 Ga ago (Harrison et al. 2009).

And some questions are:

1. Was the Moon the only body involved?
2. Were the Moon and Earth the only two bodies involved?
3. Were most, or all, of the terrestrial planets and the Moon involved?
4. Was the entire Solar System involved?

5.5.3 A Series of Quotes, Mainly in Chronological Order, Concerning Unusual Events on the Moon (and Earth) Between 4.0 and 3.5 Billion Years Ago

The following quotes are from a 17 May 1968 article (page 732 with permission from AAAS) by Preston Cloud published in the journal *SCIENCE*. This was over a year before the lunar landings in July 1969. In my view, this is a fairly accurate PREDICTION of thermal events on the Earth and Moon based on the rock record on Earth.

> Thus the limited chronology of the oldest rocks, and the hints from lead isotope data of the concomitant beginning or peaking of internal differentiation of the earth, can be seen as consistent with a near approach of the moon to the earth about 3.5–3.6 aeons ago. What little is known about lunar evolution also seems to be consistent with such an interpretation (although by no means proof of it). Just as the oldest terrestrial rocks, and conceivably the initiation or climaxing of core and mantle formation, may be related to processes incident to lunar capture, so also may the surfaces of the lunar maria, indicating as they do, from observations by Shoemaker (38), a welling-up from beneath to a "nearly common level." What would account for such a widespread welling-up to fill the presumably collisional maria and some smaller preexisting impact craters (for example, Archimedes) to "an approximately hydrostatic level" (38)? This would seem to require both a major episode, or a sequence of related episodes, of partial melting and a widely interconnected series of conduits along which molten or fluidized particulate matter could travel to such a nearly general level. In the absence of a core or other evidence of major internal differentiation of the moon, tidal friction provides a source of heat for fluidization otherwise difficult to visualize. In addition, stresses incident to capture might well account for an extensive interconnected lunar fracture system, along which lava, ash, or both, could ascent [as well as the impacting masses that produced the maria (27, 28)]. Difficulties arise in trying to explain the preservation of large-scale relief in the lunar highlands, the absence of surface offsets due to faulting, and the lesser relief within the maria themselves. If a major melting episode did occur as postulated, it could not have been a general liquefaction of the interior of the moon nor could it have been narrowly synchronous at all places. Something more like pervasive partial melting is called for. Interconnected fractures along which effusive matter would have travelled outward cannot have had large offsets. And results of statistical treatments suggesting high relief within maria (39) would require some other explanation. These difficulties circumscribe the model envisaged. Like it, they must undergo further critical examination, hopefully with more abundant and better evidence.
> The rocks we now know cannot of themselves tell us how the moon originated. Among hypotheses available to choose from, however, the testimony that can be wrung from the rocks is consistent with Singer's suggested prograde lunar capture and not consistent with a Moon in relatively late pre-Paleozoic time. This testimony can also be interpreted as suggesting that, however the moon originated, a near approach of moon to earth may have taken place closer to 3.5–3.6 aeons ago than to the 4–4.5 aeons suggested by Singer from tenuous geochronological inferences. And, after all, a considerable pre-mare (or pre-Imbrian) lunar history would be in order to account for the extensive lunar highlands whose infrapositional relations to debris seemingly splashed from the maria imply a significantly greater age for these highlands (38).
> A special problem that arises in connection with a date as late as 3.5–3.6 aeons ago for lunar capture is that of where the moon could have been stored for a whole aeon if the earth is about 4.6 aeons old.
>
> If these conjectures should happen to be correct (and they certainly have their shaky aspects), then, when samples of the mare fillings are obtained from beneath any superficial covering debris of cosmic or local origin, such samples should give radiometric ages of about 3.5–3.6 aeons.

My comment on this conjecture: The scientific conjectures by Preston Cloud have all but been "swept under the rug" in the lunar origin debate. Cloud's dates were somewhat fluid and he knew that they would probably be pushed back in time by the dating of terrestrial rocks. When I last talked to Preston about this in 1989, he said that a date of 3.90–3.95 Ga looked like a reasonable number from what we know of the rock records of both the Earth and Moon. The main point is that I think that he was right for good reasons. He was thinking that only the Earth and Moon (and Sun) were involved in these mainly contemporaneous major thermal episodes on the Earth and Moon. We will return to this discussion after we look at several other quotes by concerned scientists on these events on the Moon, the Earth, and possibly other bodies in the Solar System.

5.5.4 Some Quotes on the Concept of the "LATE HEAVY BOMBARDMENT" from 1974 up to 2007

This is a quote from the paper on the initial concept of "A TERMINAL LUNAR CATACLYSM" (Tera et al. 1974, p. 18):

> A terminal cataclysm occurred on the moon and very possibly on earth at ~3.9 AE which caused widespread lunar (and terrestrial) metamorphism and element redistribution. We associate this with the formation of several major basins by late impacts which took place over a relatively short time interval (less than 0.2 AE).

This is a quote from an article resulting from a symposium a few years later with a firmer date for the "TERMINAL LUNAR CATACLYSM" (Wasserburg et al. 1977, p. 21):

> An epoch of massive bombardment occurred at about 3.95 Ga. This was the terminal lunar cataclysm which produced sufficient debris of metamorphosed materials as to dominate the age pattern over wide areas of the Moon. After 3.8 Ga the bombardment rate suddenly became quite low, thereby preserving the features of younger lava flows.

This is a series of quotes resulting from a symposium on Multi-ring Basins in the early 1980s (Wetherill 1981, p. 3). Wetherill suggests that planet Venus may be involved in this scenario.

> It may be expected that many ideas regarding solar system formation that are fashionable at present will have to be abandoned in the future. Nevertheless we must start from where we are, and a disciplined approach combining serious observational, experimental and theoretical work, however imperfect, is the only way to ultimately produce a satisfactory solution of this fundamental scientific problem.

This is another quote from the same article (Wetherill 1981, p. 9):

> This clustering of basin formation in the vicinity of 3.85 b. y. could be explained as a result of the tidal breakup of a large ~ 10^{23} g body following a close encounter to Earth or Venus, followed by a sweep up of all fragments by the terrestrial planets and the moon on the relatively short (~50 m. y.) time scale associated with the dynamics of interplanetary debris in the inner solar system. This would imply that the younger multi-ring basins on other terrestrial planets represent a "marker horizon" dated at 3.85 b. y.)

This is another quote from the same article (Wetherill 1981, p. 12):

> Nevertheless, within the framework of explaining basin-formation in terms of the declining Earth-Venus swarm, the hypothesis of Roche limit breakup still appears to be required to explain the size distribution of the basin-forming projectiles. Thus even with this new

interpretation, it is likely that the Orientale, Imbrium, and possibly Serenitatis impacts were associated with the Roche limit breakup of a single large planetesimal. In this case a near-contemporaneous 3.8–3.86 b. y.. basin forming epoch should also have occurred on all of the terrestrial planets.

The following is a summary of a meeting, by a science reporter, Richard Kerr, concerning the subject of the LATE HEAVY BOMBARDMENT by the year 2000 (Kerr 2000, p. 1677 with permission from AAAS) (since this is a very good and concise review of the problem, I include all of it here):

The inner solar system quieted down considerably within a few hundred million years after that, and relative peace prevailed. But early analyses of lunar rocks returned by the Apollo astronauts hinted at a sudden violent episode 600 million years after Earth's birth. Seemingly out of nowhere, a hail of objects pummeled Earth, the moon, and perhaps the entire inner solar system. Now this "late heavy bombardment" is getting strong support from analysis of rocks the astronauts never saw: meteorites that fell to Earth from the moon's backside.

Lunar meteorite analyses reported on page 1754 of this issue reveal a burst of impacts on the moon 3.9 billion years ago and nothing before that. Cosmochemists Barbara Cohen, Timothy Swindle, and David Kring of the University of Arizona, Tucson, concluded that the moon and Earth endured a storm of impacts 100 times heavier than anything immediately before or after. Such a lunar cataclysm would have scarred the moon with the great basins that now shape the man in the moon. On Earth, the same bombardment would have intervened in the evolution of life, perhaps forcing it to start all over again. 'To me it seems highly likely there was a lunar catastrophe,' says cosmochemist Laurence Nyquist of NASA's Johnson Space Center in Houston. But skeptics wonder how the solar system could have held off delivering such a devastating blow for more than half a billion years.

Then in the early 1990's, geochronologist Brent Dalrymple of Oregon State University in Corvallis and planetary scientist Graham Ryder of the Lunar and Planetary Institute in Houston determined precise ages of 12 bits of Apollo rock apparently melted in 12 different impacts. They found a flurry of impacts 3.9 billion years ago but none older. If the impacts had simply tailed off from the formation of Earth and the moon about 4.5 billion years ago, as dynamical astronomers insist it must have, Dalrymple and Graham (Ryder) should have found impact melts as much as 4.2 billion or 4.3 billion years old. Failing that, they concluded that a burst of impacts 3.9 billion years ago had overwhelmed the few impacts that preceded it.

The lunar cataclysm wasn't immediately accepted, however. Critics such as planetary scientist William Hartmann of the Planetary Science Institute in Tucson, Arizona, pointed out that the apparent surge might just mean that the cratering was obliterating all traces of earlier impacts until it gradually slowed to a point where some could survive. All of the dated moon rocks had come from the equatorial region of the moon's nearside, Hartmann noted, where one or two of the huge, basin forming impacts could dominate the record.

The latest results from the moon are pushing even the doubters toward a lunar cataclysm. 'When we started this study,' says Swindle, 'I thought this would be a way to disprove it. We haven't proved there was a cataclysm at 3.9 billion years ago, but it passes the test.' Hartmann agrees, and he now concedes that obliteration of an earlier impact record may be harder than he had thought. 'The way out may be a compromise scenario.' he says. 'Maybe there was a fairly high spike [superimposed on the tail] 3.9 billion years ago, and we're just arguing over how big that spike was. But you would still have a serious problem of where you store this stuff for 600 million years before dropping it on the moon.

> Astronomers still don't have any good ideas of the cataclysm's source. Simulations show that the gravity of Earth and the other terrestrial planets would have cleared the inner solar system of threatening debris within a few hundred million years. Collisions in the asteroid belt can shower Earth with debris, notes Brett Gladman of the Observatory of Nice, but a cataclysm would require the breakup of a body larger than 945-kilometer Ceres, the largest asteroid. The chance of that happening any time in the past 4.5 billion years is nearly nil, he notes.

> As a last resort, researchers look to the outer reaches of the solar system. Dynamical astronomer Harold Levison of the Boulder, Colorado, office of the Southwest Research Institute and colleagues show in a paper to appear in Icarus how the newly formed Neptune and Uranus could have tossed ice debris, along with some asteroids, inward in sufficient quantities to resurface the moon, give Mars a warm and wet early atmosphere, and sterilize Earth's surface with the heat of the bombardment (Science, 25 June 1999, p. 2111). The only catch, says Levison, is that the two large outer planets would have had to have formed more than half a billion years later than currently thought. Levison is toying with the idea that Uranus and Neptune started out between Jupiter and Saturn, where his simulations suggest they could have orbited for hundreds of millions of years before flying out into the lingering debris beyond Saturn and triggering a late heavy bombardment. 'That's my fairy tale.' He says. Maybe that's just what young planetary bodies need.

A comment by the author: This LHB started with information from the Moon and then the heavy impact feature was extended to the Earth. Now for a few quotes from an article whose authors want to involve the inner part of the Solar System (Kring and Cohen 2002, paragraphs 23–25 with permission from the American Geophysical Union):

> The cataclysmic bombardment that occurred throughout the inner solar system ~3.9 Ga dwarfs the Chicxulub impact event at the K/T boundary and its environmental consequences. Its coincidence with the oldest surviving rocks on Earth is due to the destruction or metamorphism of preesisting crust. Because of the destruction of the rock record, it is not yet clear whether the coincidence of the bombardment with the earliest isotopic evidence of life on Earth implies the event was an environmental bottleneck through which life survived or whether the bombardment was intimately associated with the origin of life.

> In this note, we argue the projectiles that dramatically altered surface conditions in the Earth-Moon system were asteroids. Previously, Ryder (1990) suggested the cataclysm was restricted to the Earth-Moon system, envisioning additional moons in geocentric orbit which collided to form a swarm of projectiles that then hit the Earth and Moon, leaving Venus, Mercury, and Mars unaffected. However, a large number of ~3.9 Ga impact degassing ages among meteorites from the asteroid belt and an impact age in our single sample of the ancient crust of Mars suggests the cataclysm affected the entire inner solar system. Thus the dynamical source of the material was likely in heliocentric orbit in the asteroid belt. Potential sources of debris in this type of orbit include rocky material remaining from the accretion process, the breakup of small bodies in the asteroid belt, chaotic diffusion of asteroids into Earth-crossing orbits, and a scattering of asteroids as Jupiter migrated inward to cause the v6 resonance to sweep through the belt.

> The impacts affected the environment and likely affected the early evolution of life on Earth, both by causing devastation at the surface of the planet and, in contrast, producing habitable conditions in subsurface environments. The impactors likely delivered biogenic materials, although it is not clear if these were essential for life's origins. Comets were not important during this time.

Comment by the author: Perhaps there is a "focusing" problem with the above scenario when bringing impactors from the Asteroid Zone to impact over a brief period of time to excavate large lunar basins.

5.5.5 View of the Late Heavy Bombardment in 2006

This is a quote from Koeberl (2006, p. 214, with permission from the Mineralogical Society of America) on how the LHB relates to the terrestrial rock record:

> The question arises as to whether or not any evidence of this bombardment is preserved on Earth. In any given time span, Earth is subjected to a higher impact flux than the Moon because Earth has a larger diameter and much larger gravitational cross-section. The older rocks on Earth, from Acasta in Canada and Isua/Akilia area in Greenland, are the obvious places to conduct such a search. Unfortunately, searches for an extraterrestrial signature based on platinum-group elements and for shocked zircons in the earliest rocks from Greenland have so far been unsuccessful (Koeberl 2006). There could be several reasons. First, the number of samples studied was probably too small. Second, it is possible that very large impacts form more melt than shocked rock, leading to a preservation problem. Third, it is not certain that the rocks at Isua really have an age > 3.8 Ga (e.g., Lepland et al. 2005).
>
> Tungsten isotope anomalies found in ca. 3.85 Ga metasedimentary rocks from Greenland were interpreted to indicate an extraterrestrial component as a result of the LHB (Schoenberg et al. 2002). However, it is difficult to understand why a similar meteoritic signal would not show up in the platinum-group elements abundances, as high W abundances are not common in meteorites. In fact, Frie and Rosing (2005) used the chromium isotope method on Isua rocks and found no trace of an extraterrestrial component. Thus, little or no evidence of an LHB phase has been found so far on Earth.

5.5.6 Review of the Situation of the Late Heavy Bombardment in 2007

This is a quote from Chapman et al. (2007, p. 235):

> It appears that 10–12 basin-forming impacts happened during the nectarian (Wilhelms 1987) from just ~3.90–3.85 Ga but that the bombardment ended sharply thereafter, with only Orientale occurring a little bit later (perhaps ~3.82 Ga).
>
> Impacts were so abundant during the LHB that little earlier geology remains for stratigraphic studies, crater counts, or link-ages with rock ages.
>
> The most persuasive argument (Ryder 1990, 2002) has been that the absence of old impact melts, which should have been abundantly produced by early basin impacts, means that such impacts were rare.

Comment by the author: The following article is a potentially important source of information on the timing and the THERMAL NATURE of the LHB; the authors place limits on the process of thermal metamorphism that was involved in the LHB. The quote is from Pidgeon et al. (2007, p. 1370):

The zircon U-Pb system has not been affected by the ca. 3.95 Ga thermal pulse that accompanied formation of the host breccia although this event has largely reset the K-Ar systems.

5.5.7 Summary

There is no doubt that something major happened to both the Earth and Moon about 3.95 billion years ago and the processes involved decreased in intensity from 3.95 to 3.60 Ga. There is no substantial evidence that any other bodies were involved and it is not clear that bodies or material outside the Earth-Moon system were involved in the episode. It certainly could be that the concept of a "late heavy bombardment" caused by impacting bodies *external* to the Earth-Moon system is *FICTION*.

Thus, the Late Heavy Bombardment debate for the Moon appears to be a discussion on the timing of the 3.8–4.0 Ga events:

- Was the LHB an episode featuring the impact of solid bodies onto the surface of the Moon over a period of about 200 million years of time (Chapman et al. 2007)?
- Was the LHB associated with gravitational capture of the Moon and partial tidal disruption of the lunar body in which the major action takes place in about 2 h of time (Malcuit et al. 1992)?

A short quote is appropriate here: **"An episode of massive bombardment occurred at about 3.95 Ga."** (Wasserburg et al. 1977, p. 21).

The following is a brief summary of an Earth-Moon interaction model, first suggested by Cloud (1968, 1972), that is a main theme of this chapter and of this book, especially Chap. 4.

Now let us consider a model that involves two main interacting characters—the Moon and the Earth and, of course, the Sun to maintain control over the interactions. In this model the Moon undergoes a major episode of tidally induced volcanism and significant impact activity as a result of the gravitational capture process. Tidal disruption *cannot* occur during the capture encounter because the lunar body is too cool to yield magma at this time. In a favored numerical simulation scenario, the 18th perigee passage is a near grazing encounter and at that time magma would be available in the lunar upper mantle. Most, but maybe not all, of the material extracted from tidally induced disruption centers on the lunar body would eventually impact on the Earth or Moon. The bulk of the lunar-borne material that is extracted from the main disruption center at the surface of the moon as it penetrates the WEIGHTLESSNESS LIMIT of the Earth-Moon system will fall back onto the lunar surface along a great-circle pattern. A great-circle pattern is simply the intersection of the encounter plane and a spherical lunar body. The magmatic spheroids, after launch, are transported above the lunar surface to an impact site. Upon oblique impact, each of the lunar basaltic spheroids, depresses the crust and deposits most of the volume of the spheroid in the crustal depression caused by the low speed impact. As the lunar crust rebounds from the impact, some of the lunar basalt flows over the raised rim to form an asymmetrically mantled rim around each of the newly formed mare basins. Concomitantly, a lunar magnetic amplifier is activated in the lunar upper mantle by the dynamics of the capture process. After several perigee passages which occur before the exceedingly close encounter associated with tidal disruption, the iron-rich basaltic magma in the lunar mantle is "threaded" by a terrestrial magnetic field. The weak terrestrial field is amplified by a "shallow shell amplifier" (Smoluchowski 1973a, 1973b) to yield a strong magnetic field for a short era in lunar history. This strong magnetic field imprints a remanent magnetic signature in the surface and near surface lunar basalts and breccias as they cool through the Curie point. Thus there is a magnetic field operating at the time of magmatism on the lunar surface. The magnetic field decays exponentially as the lunar orbit circularizes due to tidal energy dissipation within a combination of the Moon and Earth.

This scenario then combines the three major events in lunar history, as a successful explanation should do, as suggested by Cisowski et al. (1983). The Late Heavy Bombardment, the onset of major mare volcanism, and a period of intense lunar magnetism are all associated, in this explanation, with a GRAVITATIONAL CAPTURE SCENARIO for the origin of the Earth-Moon system as Cloud (1968) suggested many decades ago. Note that at the time of Cloud's conjecture, there was no knowledge of remanent magnetism in lunar rocks.

5.6 Vignette E. A Cool Early Earth, Recycled Enriched Crust at ~3.95 Ga, and the Subduction Mechanisms Associated with a Tidal Capture Model for the Origin of the Earth-Moon System

Our views of the PRIMITIVE EARTH have been changing in recent years. When Preston Cloud (1968, 1972) proposed that this "dark era" in earth history be called the "HADEAN EON", the common belief was that there was no interpretable rock and mineral record for that era.

This all changed when John Valley and colleagues (2002) interpreted the oxygen isotope information from the Jack Hills zircon crystals as signifying that the Earth was cool enough to have ocean water on the surface during most of that era: i.e., that we probably had "A COOL EARLY EARTH" rather than a "Hadean Inferno" from 4.4 to 3.6 Ga.

5.6.1 Purpose

The purpose of this section is to demonstrate that a gravitational lunar capture scenario is very compatible with (1) the concept of a "COOL EARLY EARTH", (2) the concept of the accumulation of an enriched "basaltic" crust with a granitic component (containing zircons) on essentially a one-plate planet (a stagnant-lid planet) developed over a terrestrial magma ocean, and (3) the concept that this enriched crust was partially (or mostly) destroyed by a forced subduction mechanism associated with the dynamics of the capture process. The suggested commencement time for the crustal recycling is ~3.95 Ga.

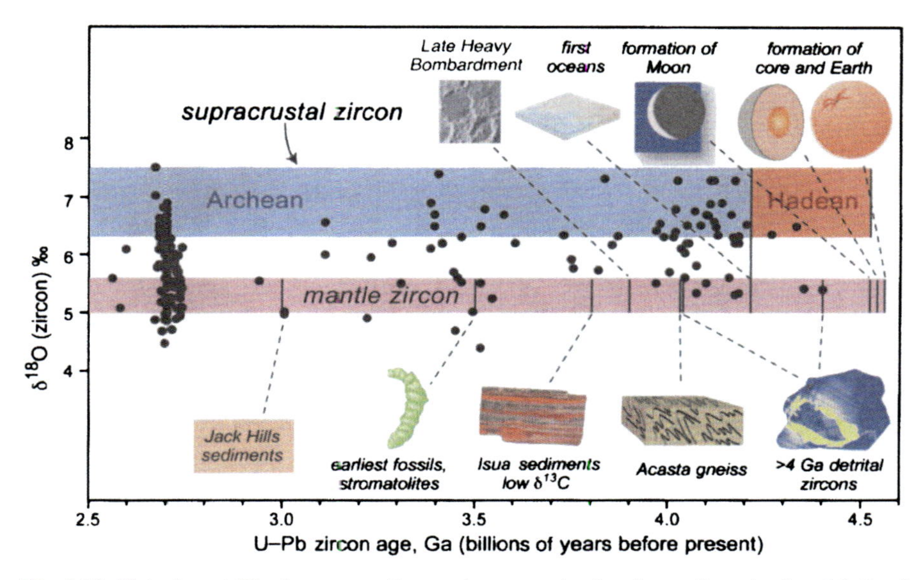

Fig. 5.60 Plot of crystallization age vs. Oxygen isotope ratios for zircons from the first 2.1 Ga of earth history. Note the decrease in data points in the supra-crustal zircon zone between 3.95 and 3.6 Ga. (From Valley 2006, Fig. 2, with permission from Mineralogical Society of America). Note that the decrease in density of data points between 3.95 and 3.6 is also obvious in the more recent analysis of Griffin et al. (2014)

5.6.2 Evidence for a Cool Early Earth

Figure 5.60 is a plot of the oxygen isotope information from the zircon crystals from the Jack Hills Formation of Western Australia. The $\Delta^{18}O$ values of zircon in the range of supracrustal zircon (6.5–7.5 parts per thousand) strongly suggests that the Earth was cool enough to have ocean water on the surface from about 4.4 Ga.

5.6.3 The Bedard (2006) Model for Processing a Basaltic Crust on a Stagnant-Lid Planet

In general, most, if not all, of the zircons that plot in the supra-crustal zone on Fig. 5.60 can only be formed in a granitic magma. The paradox is that it is difficult to form granitic magma directly from the partial melting of a peridotitic mantle. Furthermore, most investigators agree that under ideal conditions, any early crust on Earth should be of basaltic composition and that it takes much processing to get any significant quantity of "granite" from basalt. In the modern planetary rock processing operation, it is the volcanic arc setting above a subduction zone that produces large quantities of granitic magma.

Bedard (2006) developed a petrologic model for processing a basaltic volcanic crust to yield some granitic rocks (and zircons). He refers to his model as a "catalytic

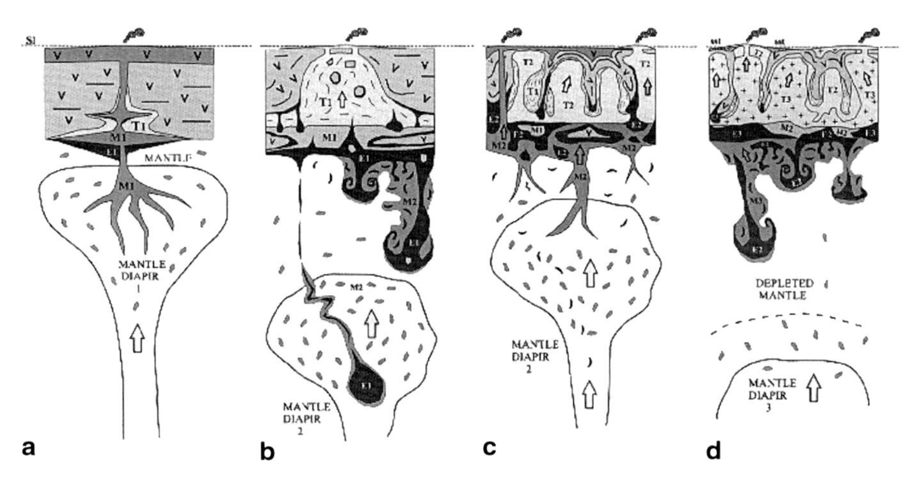

Fig. 5.61 The Bedard model for developing some granitic rocks (with zircons) from a mafic volcanic precursor. This is essentially vertical tectonics. On the *left* (*A*) is a crust of basaltic volcanics with a mantle diaper as a subjacent heat source. After several cycles of processing, the end result (*D*) is a felsic crust above and mafic material sinking into the mantle. Felsic rock units that protrude above sea level (Sl in the diagram) erode to form sandstones and conglomerates. These erosional products contain the supra-crustal zircons. (From Bedard 2006, Fig. 10, with permission from Elsevier)

delamination-driven tectonomagmatic model" for making granitic rocks from a mafic precursor (Fig. 5.61).

A Few Quotes Concerning the Nature of the Missing Crust from the "Cool Early Earth Era" are:

> A final and perhaps most intriguing question is the mechanism whereby the signatures of extreme differentiation in the mantle-crust system during the first 500 million years were almost completely erased. Although there is some evidence for early differentiation in the oldest rocks, the average $^{176}Hf/^{177}Hf$ evolution curve defined by most mantle-derived rocks over the rest of Earth's history diverges from the chrondritic curve at about 4 billion years, which is when the rock record begins (10). From Amelin (2005, p. 1915)

> The existence of a depleted mantle reservoir in the Limpopo's hinderland is reflected by the ~3.6 Ga zircon population, which shows epsilon $Hf_{3.6\ Ga}$ between −4.6 and +3.2. In a global context, our data suggest that a long-lived, mafic Hadean protocrust with some tonalite-trondhjemite-granodiorite constituents was destroyed and partly recycled at the Hadean/Archean transition, perhaps due to the onset of modern-style plate tectonics. From Zeh et al. (2008, p. 5304)

5.6.4 A Unidirectional Earth-Tide Recycling Mechanism Commencing with the Capture Encounter at ~3.95 Ga

First we need to consider energy dissipation in the Earth and Moon during the process of Gravitational Capture of the Moon from a heliocentric orbit into a geocentric orbit. One of the paradoxes of the capture process is that the smaller body, in this case the Moon, must dissipate nearly all (over 90 %) of the energy for its own capture [see Malcuit et al. (1989, 1992) and Chap. 4 for the details of numerical simulations of gravitational capture]. Thus, the Earth stays cool during the capture process as well as during the subsequent orbit circularization process.

Let us begin with the stable capture scenario of Figs. 4.25 and 4.26 of Chap. 4 (shown here again as Figs. 5.62 and 5.63).

Since the emphasis for this "vignette" is on the Earth and for recycling the primitive crust of the Earth I will show tidal diagrams for the Earth only. Figs. 5.64, 5.65, 5.66 and 5.67 show 8-year-long snapshots of orbits and associated tidal amplitudes on Earth at estimated times of 100, 1000, 4000, and 350,000 years after capture (the timescale is from Fig. 4.28b and Table 4.8). Note that the Earth tidal amplitudes decrease from about 4000 m to about 100 m during that time.

To give the reader a more realistic concept of the rock and ocean tidal regime in the equatorial zone of the planet, I present a set of tidal range diagrams for a typical month about 350 Ka after capture. Figure 5.68 shows a typical year of lunar orbits and the resulting year of tidal amplitudes about 350 Ka years after capture. The year is the same as the present year (8766 h; 365.2422 anomalistic 24-h days); (era day at this time is 10.2 h; era year = 859.4 era days; era month = 103.5 era days; era year ~8.3 sidereal months).

Figure 5.69 shows a series of tidal range diagrams for the equatorial zone of the planet the data for which is calculated directly from the orbital files for the scenario shown in Fig. 5.68a. For tidal range diagrams the Earth is viewed as rotating under the lunar body as the lunar body is orbiting around the planet. Significantly high tides are raised on the planet only during perigee passages. These tidal episodes last only about 5 era days which is about 10 semi-diurnal tidal cycles. In Fig. 5.68b the maximum rock tidal amplitude is about 120 m which makes the maximum rock tidal range about 180 m (Fig. 5.69a). But this unusual burst of tidal activity has a duration of only about 5 era days (10 tidal cycles) and then there is no significant tidal activity for about 100 era days. Then for month 2 (Fig. 5.69b) the maximum rock tidal amplitude is about 20 m (rock tidal range about 30 m). During months 3 and 4 (Figs. 5.69c and d), the tidal amplitudes (and ranges) are negligible. For months 5 and 6 (Figs. 5.69e, f), the tidal activity is significant, but for months 7 and 8 (Fig. 5.69f and h), the tidal activity is again negligible. And this goes on, year after year and millennia after millennia.

Now we continue with the Circularization Sequence. Figs. 5.70, 5.71 and 5.72 show the tidal amplitude regimes at ~4 Ma, ~30 Ma, and ~300 Ma after capture, respectively.

5.6.5 *A Proposed Mechanism for Recycling an Enriched Primitive Crust in the Broadly Defined Equatorial Zone of the Planet Beginning ~3.95 Ga*

Destruction of a thick primitive crust on the primitive Earth would be difficult to impossible without external assistance. For this Herculean project the newly captured lunar-mass planetoid is the assistant. For the recycling process I am using the concept

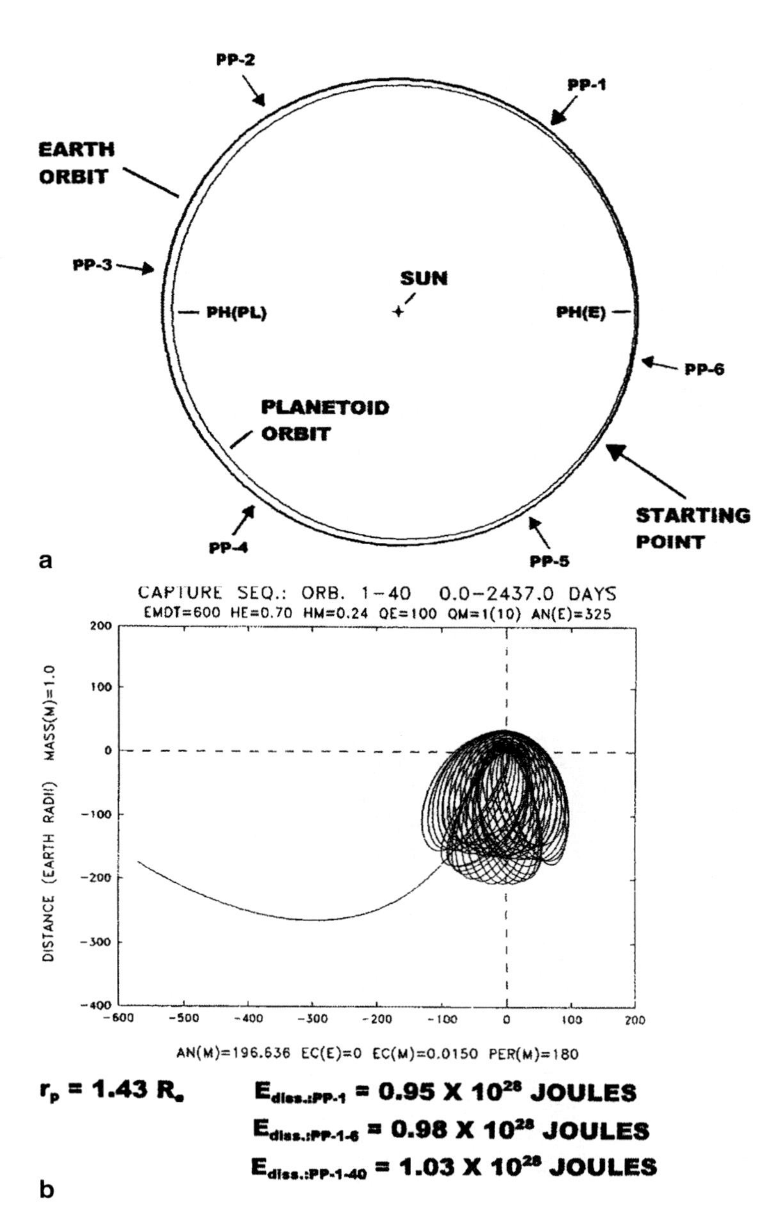

Fig. 5.62 **a** Heliocentric orientation for a numerical simulation that results in a stable capture situation. **b** Geocentric orientation for the first 40 orbits of a stable capture scenario

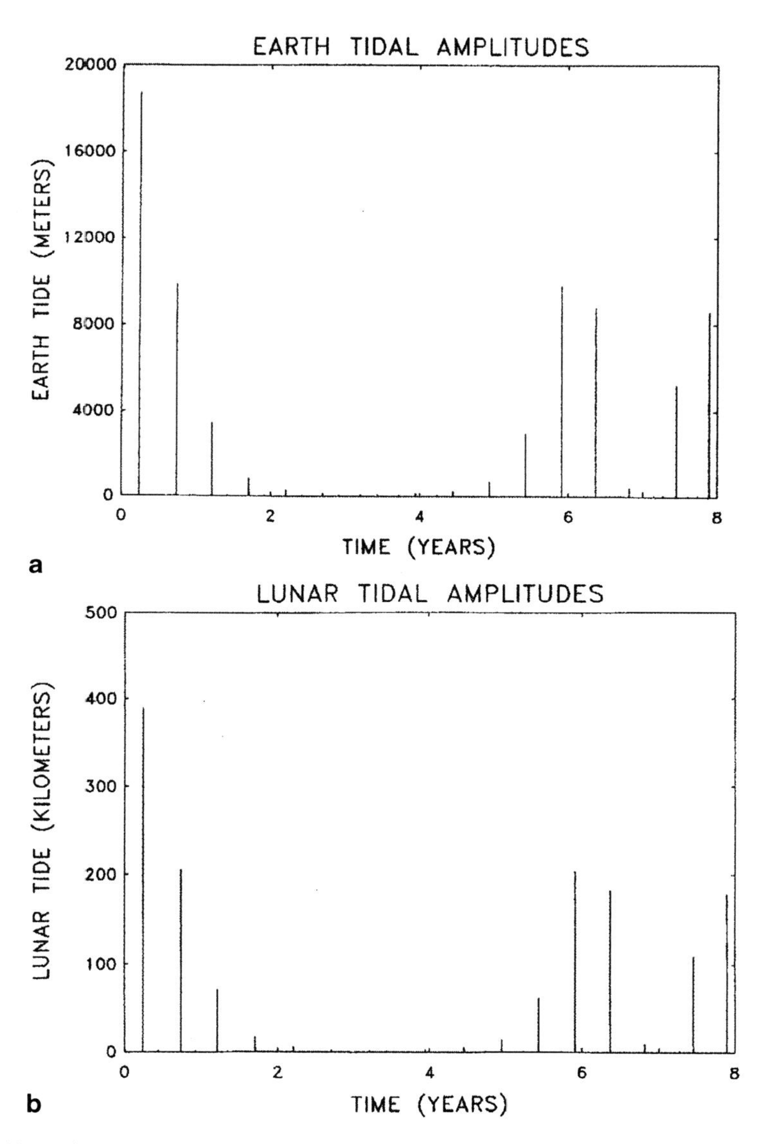

Fig. 5.63 **a** Diagram showing Earth tidal amplitudes vs. Eight Earth Years of time. Note that the initial encounter of the encounter sequence is the closest one at 1.43 earth radii. This causes rock tidal amplitudes on Earth up to 18 km. **b** Diagram showing Lunar tidal amplitudes vs. Eight Earth years of time for the lunar-like planetoid. Note that the initial encounter raises rock tides of about 400 km on the lunar-like planetoid

of TIDAL VORTICITY INDUCTION (TVI) from Bostrom (2000) (illustrated in Fig. 5.73). Figure 5.73a is a diagrammatic sketch of an equatorial cross-section of Earth showing the various zones of action. For this diagram the lunar-mass body is considered as a stationary body because it is revolving very slowly relative to

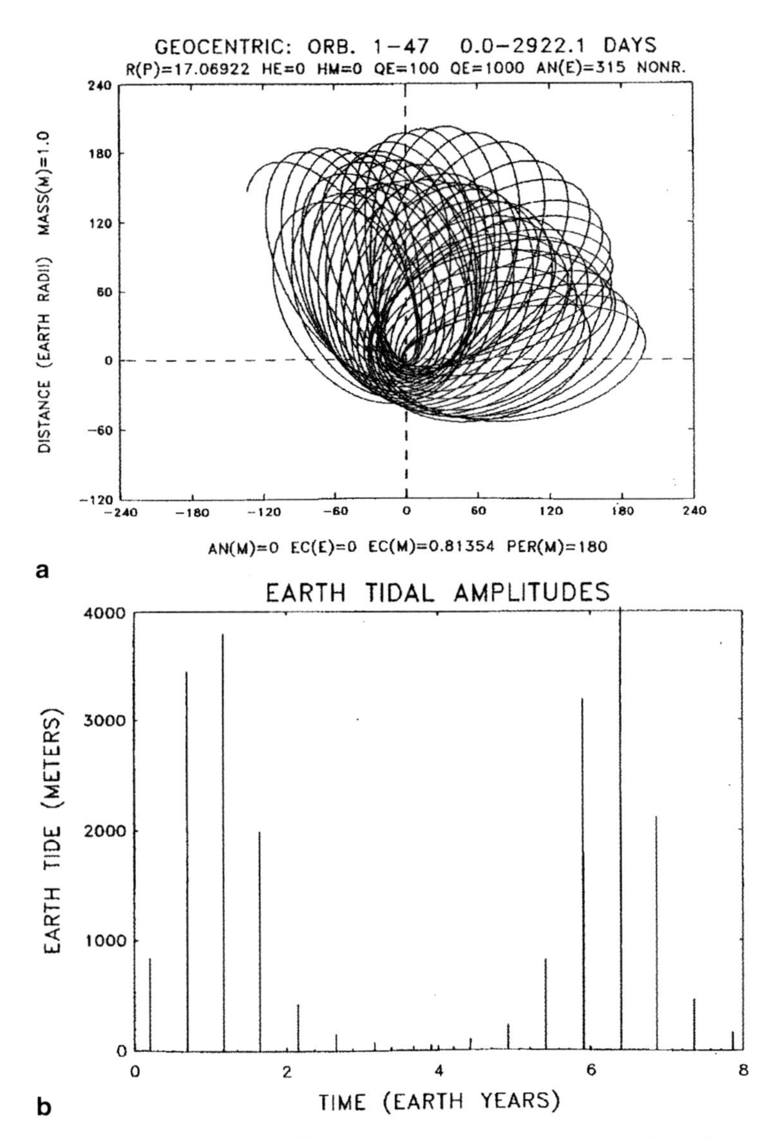

Fig. 5.64 The scene for Earth about 100 years after capture. **a** 8 years of orbits. **b** 8 years of tidal amplitudes

the rapidly rotating planet. The tidally "undisturbed" mantle is rotating in the same direction as the crustal complex but at a slower rate because of the tidal retardation of the crustal complex. This differential rotation of the crustal complex relative to the lower mantle causes TVI cells to rotate in a retrograde (clockwise direction) when viewed from the north pole. The down-going limbs of the convection cells relate to subduction zones in the presently operating plate tectonics model and primitive volcanic arcs are predicted to form above these subduction zones.

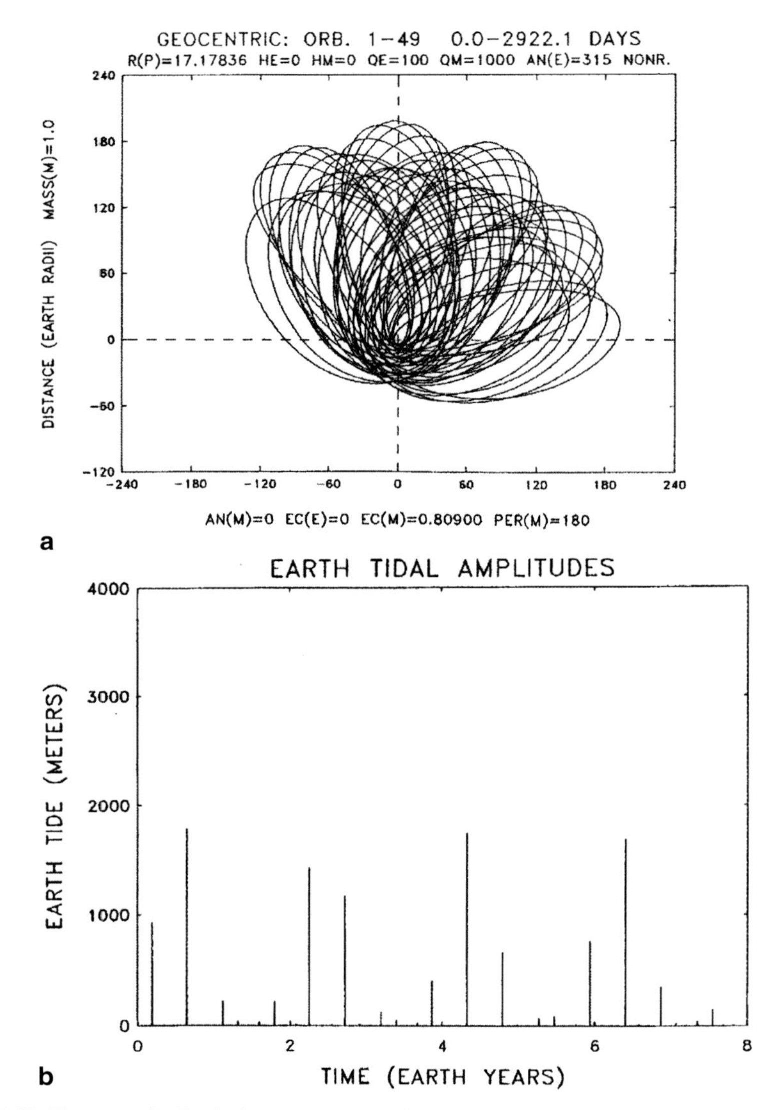

Fig. 5.65 The scene for Earth about 1000 years after capture. **a** 8 years of orbits. **b** 8 years of tidal amplitudes

Recycling of a Hadean-age primitive crust was also featured in a recent article by Nebel et al. (2014) in which they suggested that the Late Heavy Bombardment (LHB) may be the cause of the change in thermal signature of the zircon crystals about 3.9 Ga. I am suggesting that the forced subduction associated with the Gravitational Lunar Capture mechanism would give a "spiked" beginning at ~3.95 Ga and crustal recycling would continue in an exponentially decreasing mode until ~3.6 Ga. In contrast, the LHB mechanism could have a sharply defined beginning

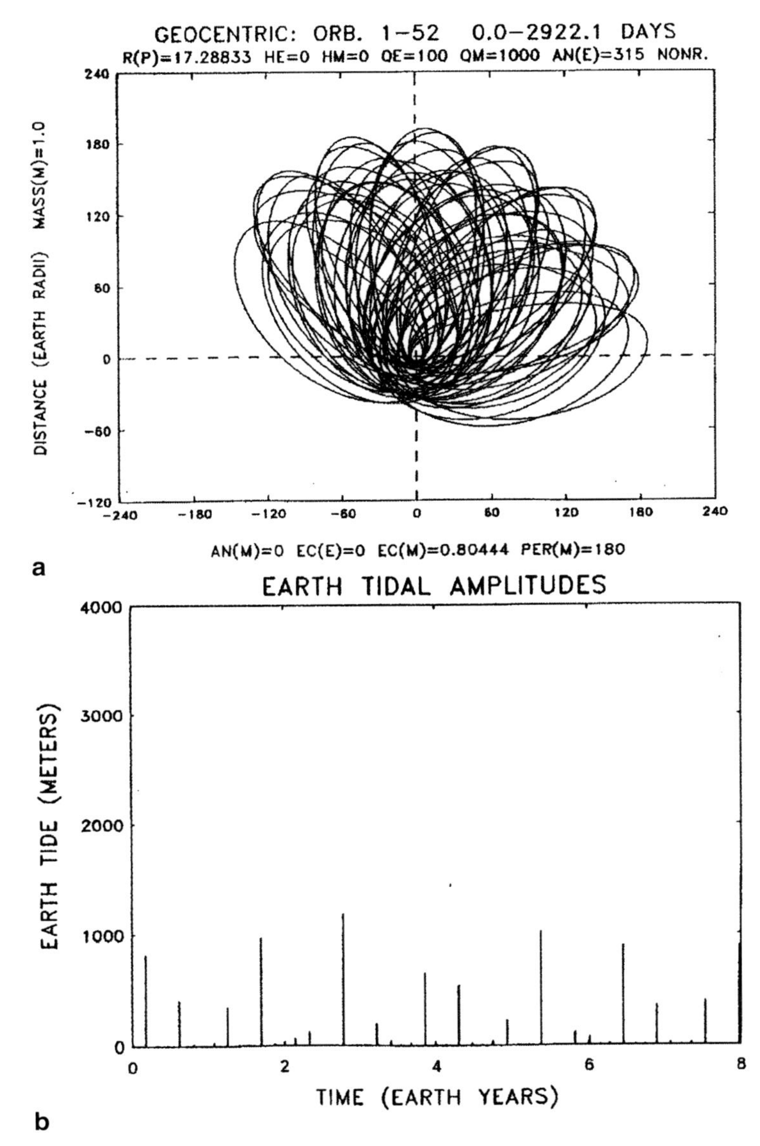

Fig. 5.66 The scene for Earth about 4000 years after capture. **a** 8 years of orbits. **b** 8 years of tidal amplitudes

but the process would be challenged to continue until ~3.6 Ga. In addition, the recycling for an LHB process would not be nearly as organized as a unidirectional rock-tide recycling system.

Figure 5.73b is a sketch of a map view of the equatorial zone of Earth. Note the position of the subduction zones relative to the volcanic arcs. The downward

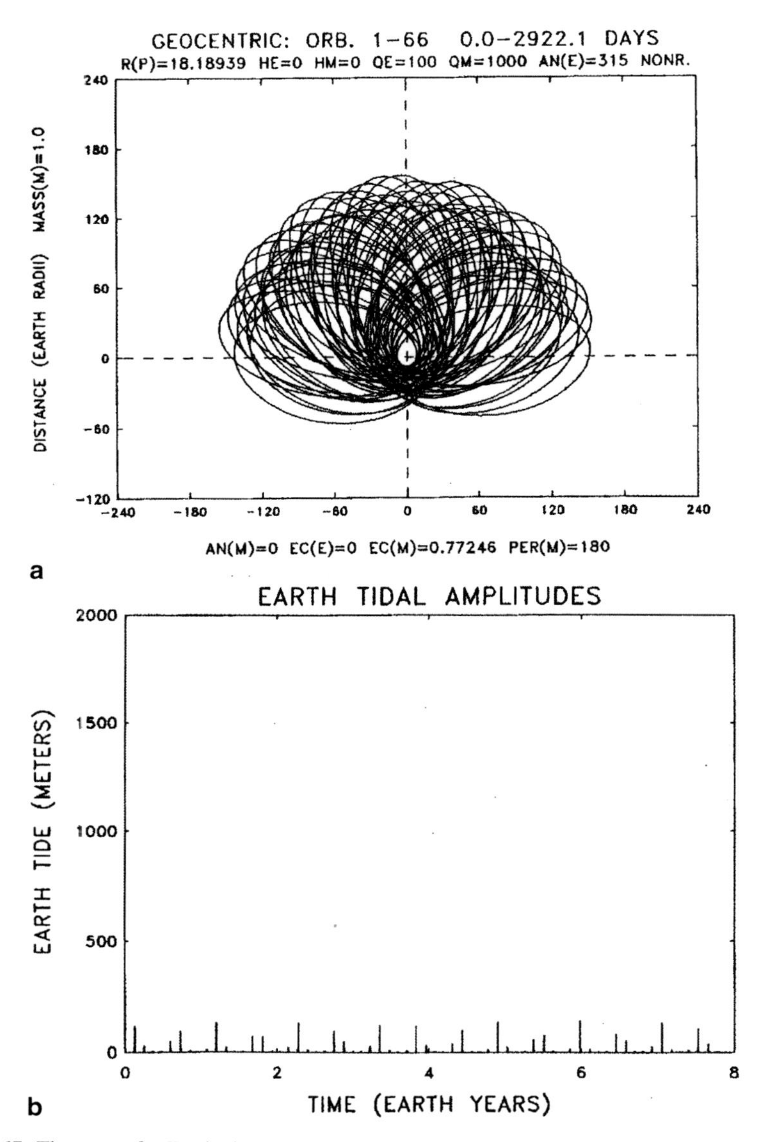

Fig. 5.67 The scene for Earth about 350,000 years after capture. **a** 8 years of orbits. **b** 8 years of tidal amplitudes. (Note change of scale on tidal amplitude plot)

portion of each TVI cell is a subduction zone with an associated volcanic arc. Some granitic material is generated by the volcanic arc operation but most of that would be recycled in time in the equatorial zone. Over time this recycling action decreases in intensity as the lunar orbit circularizes and the recycling activity eventually becomes very sluggish. Some of the primitive crust in the polar zones survives the subduction process because these regions are sheltered from the early severe rock

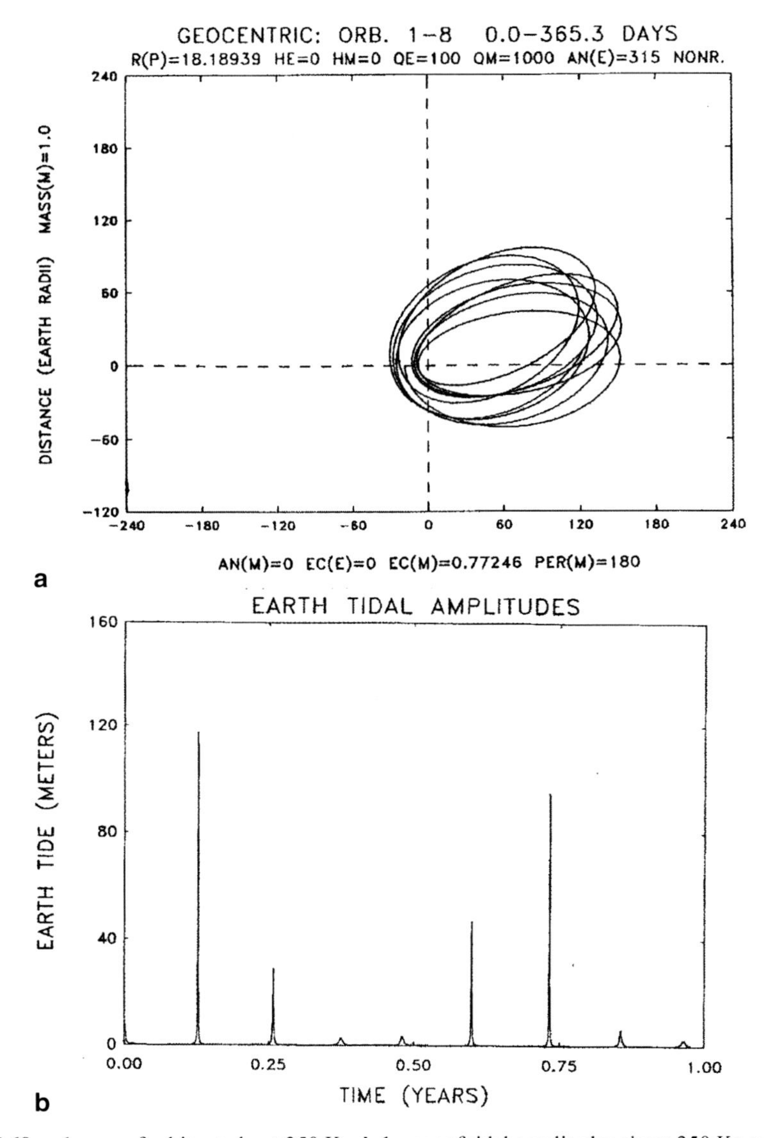

Fig. 5.68 **a** 1-year of orbits at about 350 Ka. **b** 1-year of tidal amplitudes about 350 Ka ago

tidal activity. The polar regions, then, can be a source of zircons for sedimentary processes (i.e., these polar areas can be a source for the sandstones, conglomerates, and associated zircon crystals for something like the Jack Hills Formation).

An appropriate quote from Bell et al. (2011, p. 4816) reinforces the conclusions of Zeh et al. (2008), which occurred earlier in this vignette, and is consistent with the timing of the GRAVITATIONAL CAPTURE SCENARIO.

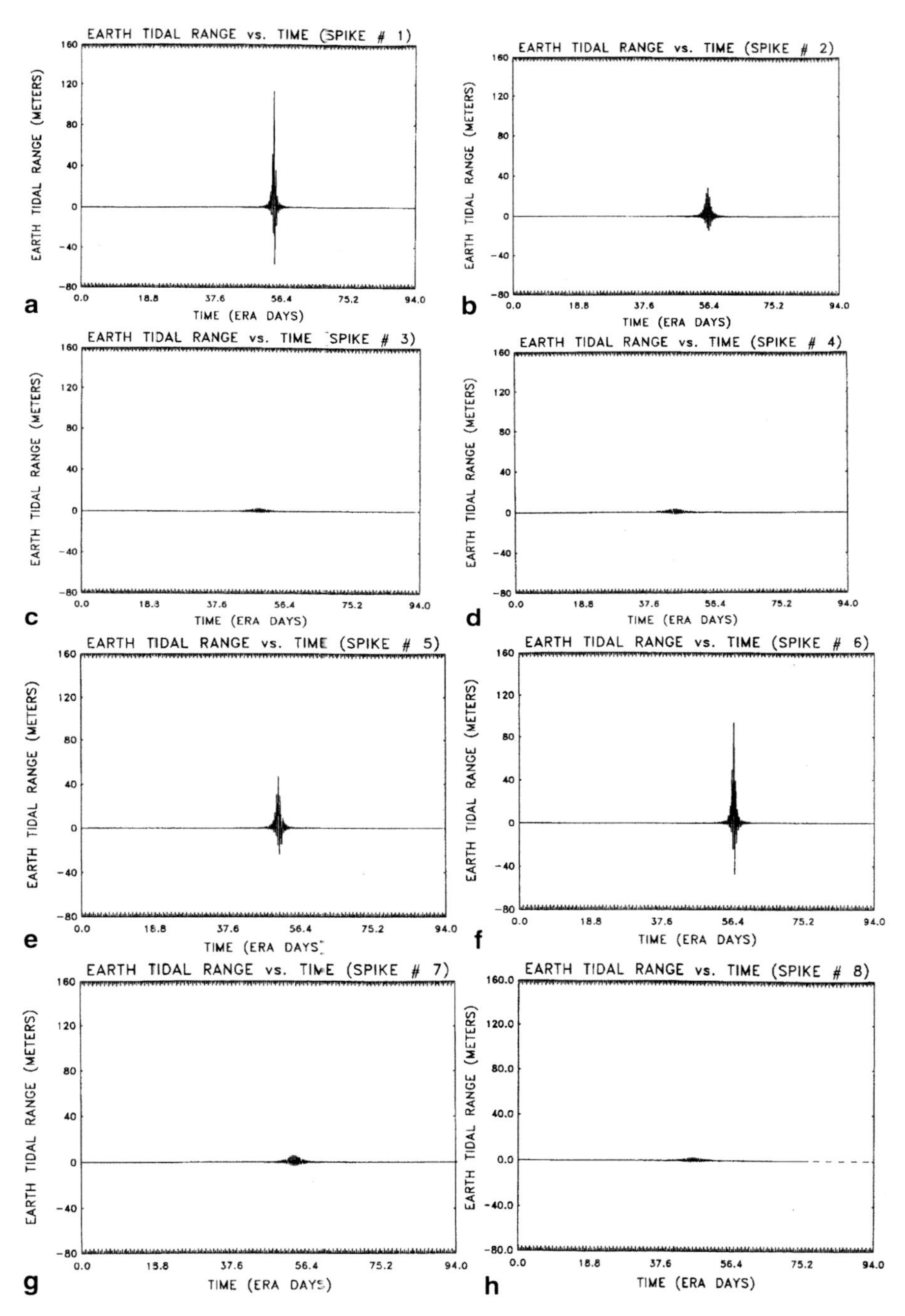

Fig. 5.69 Sequence of monthly rock tidal range diagrams for **a** typical year about 350 Ka ago. **a–h** shows the rock tidal range diagrams for each month. Note: The values for the ocean tidal ranges are about 1.4 times those of the rock tidal ranges

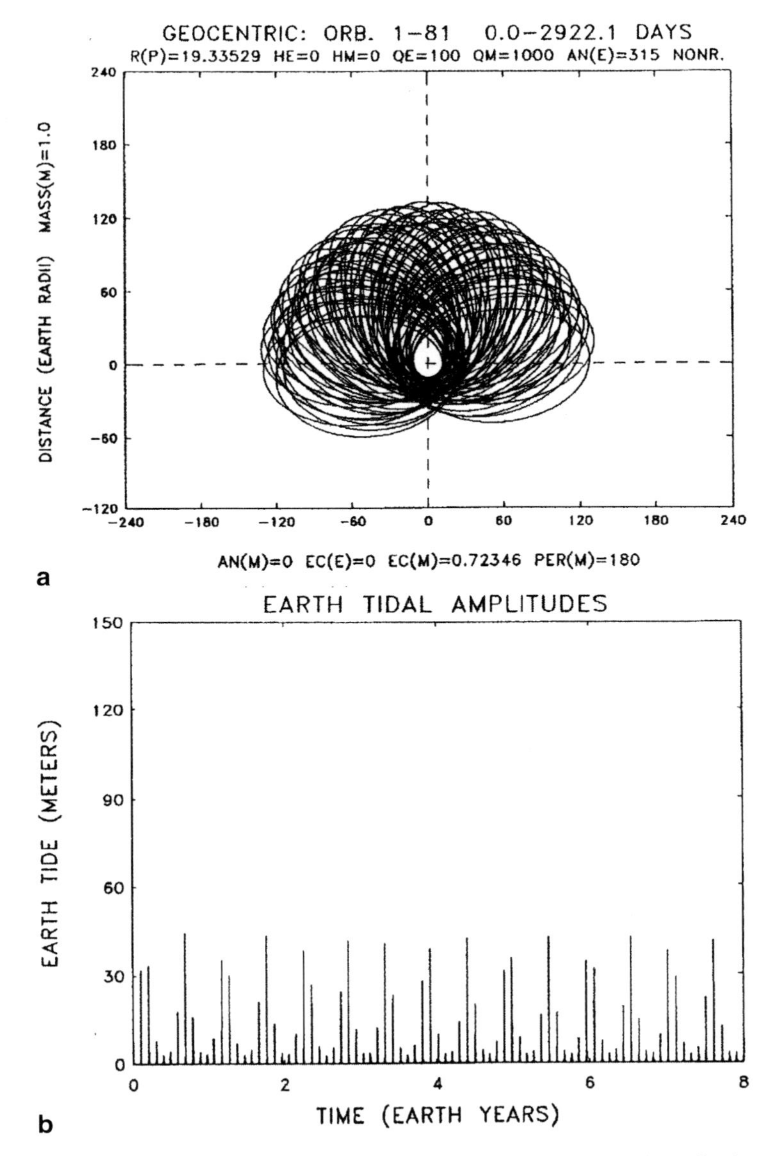

Fig. 5.70 **a** 8 years of orbits about 4 Ma after capture. **b** 8 years of rock tidal amplitudes

The shift in the Lu—Hf systematics together with a narrow range of mostly mantle-like delta ^{18}O values among the <3.6 Ga zircons (in contrast to the spread towards sedimentary delta ^{18}O among Hadean samples) suggests a period of transition between 3.6 and 4 Ga in which the magmatic setting of zircon formation changes and the highly unradiogenic low Lu/Hf Hadean crust ceased to be available for intracrustal reworking.

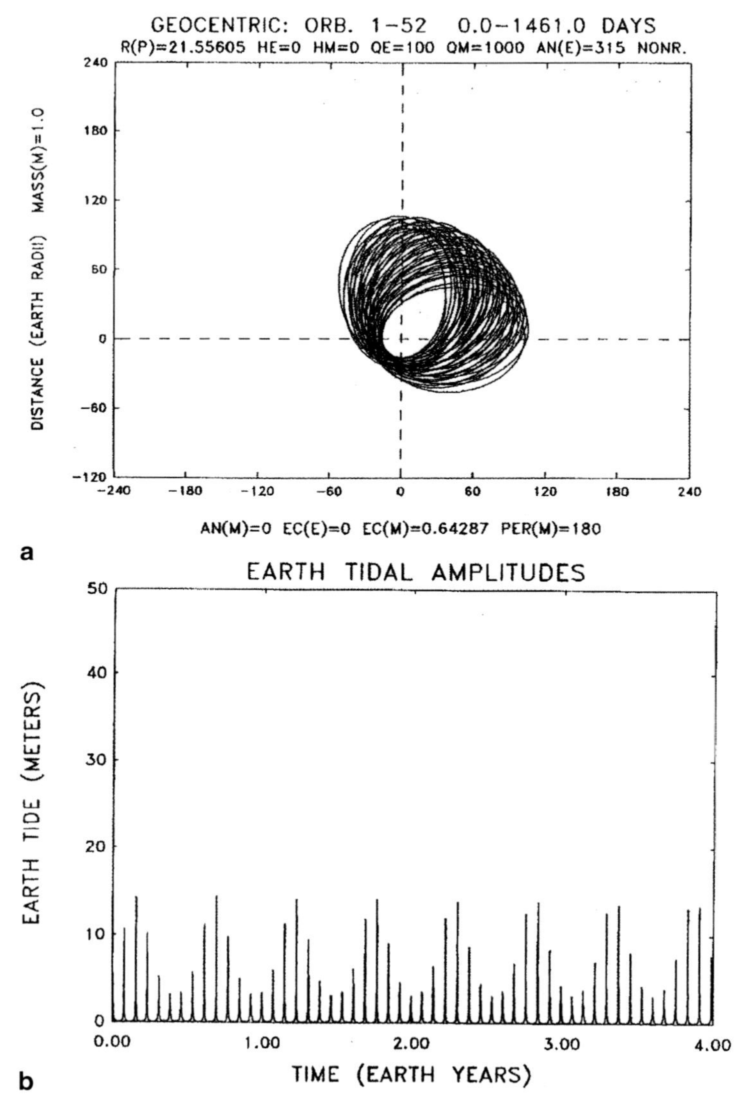

Fig. 5.71 **a** 4 years of orbits about 30 Ma after capture. **b** 4 years of rock tidal amplitudes. (Note change of scale on tidal amplitude plot)

5.6.6 Summary

1. There is substantial evidence for the concept of "A COOL EARLY EARTH".
2. There is substantial evidence for the development of an enriched crust throughout the Hadean eon.

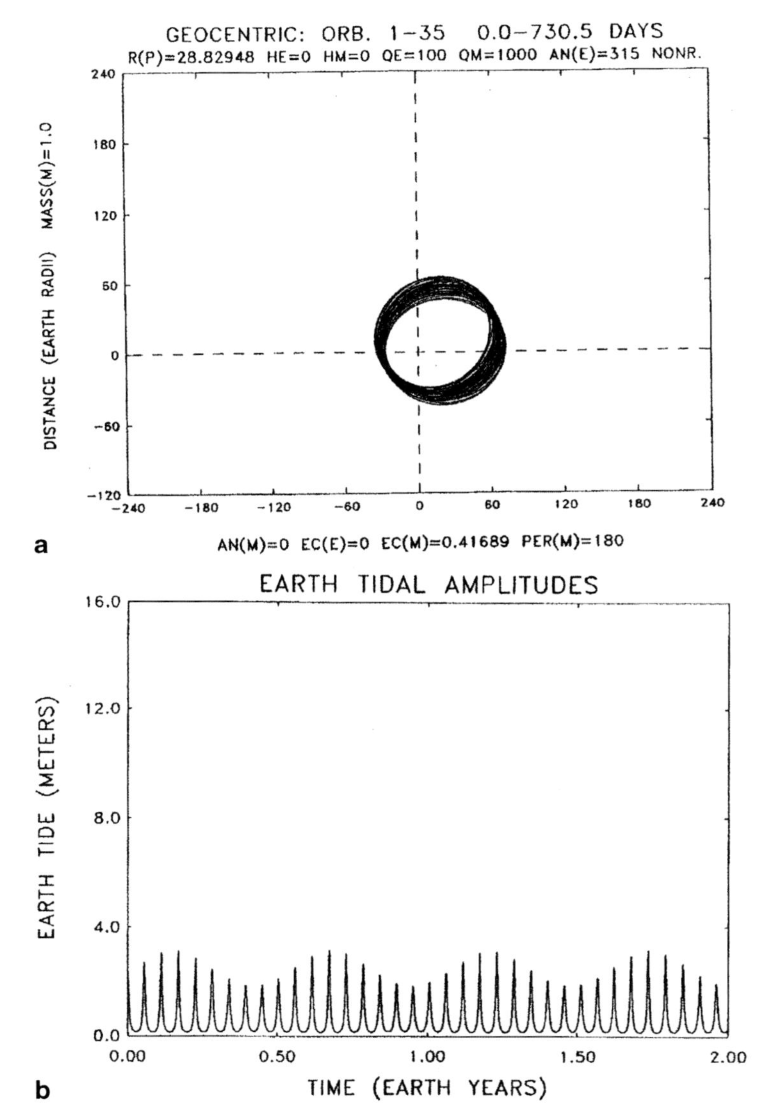

Fig. 5.72 **a** 2 years of orbits about 300 Ma after capture. **b** 2 years of rock tidal amplitudes. Note that the maximum rock tidal amplitude is down to about 3 m (rock tidal range about 4.5 m). (Note change of scale on tidal amplitude plot)

3. Tidally induced subduction appears to be an adequate mechanism for recycling an enriched primitive crust into the mantle in the equatorial zone (broadly defined) over a time period of about 300 Ma.
4. Portions of a felsic, enriched crust in the polar regions can be a source for Hadean-age zircons.

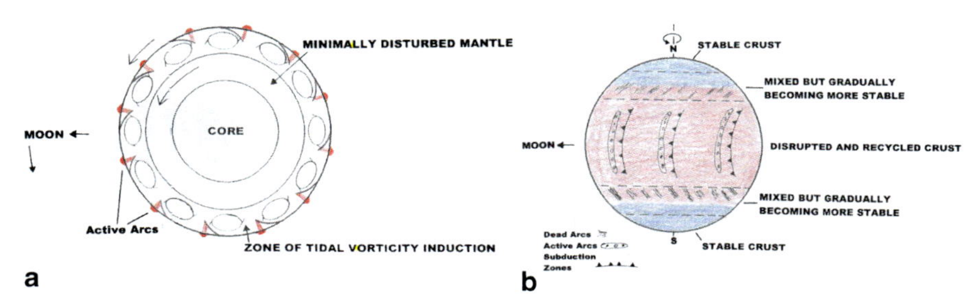

Fig. 5.73. **a** Diagrammatic sketch of an equatorial cross-section of Earth as viewed from the north pole. The zone of tidal vorticity induction, the location of subduction zones and volcanic arcs, and the direction of circulation of the TVI cells are shown. **b** Diagrammatic sketch of the surface of Earth is viewed from the equatorial plane of the planet. This is for an early stage in the orbit circularization sequence (perhaps about 350 Ka after capture). Note that in time the subduction zone and associated volcanic arc activity decreases and volcanic arc remnants are spalled off as dead (or fossil) volcanic arcs. More information concerning this diagram is in the text

5.6.7 Discussion

First let us consider some quotes from Amelin (2005) and from Naeraa et al. (2012). Then I will do a soliloquy on how all of these events and models may fit together for a story of "A COOL EARLY EARTH", the crust of which gets subducted in the broadly defined equatorial zone to be replaced by a basaltic complex at the dawn of the Archean Eon.

First, let us examine a quote concerning Earth's "dark age" (from Amelin 2005, p. 1914–1915).

> No rock or mineral record has been preserved from Earth's 'dark age'—the mysterious time after accretion of the planet about 4560 million years ago. Thanks to a continuous effort to find the oldest pieces of our planet, however, the duration of this unknown era is becoming shorter and shorter. Following the development of modern isotope dating, the extent of the dark age was established at about 800 to 1000 million years by the discovery of exceptionally old rocks in western Greenland (1). The discovery of still older grains of zircon (zirconium silicate, a common if not very abundant component of crustal rocks, and an extraordinarily resilient mineral) in archean sedimentary rocks of Western Australia reduced the dark age to 400 to 300 million years (2, 3) and recently to less than 200 million years (4), a mere 5 % of Earth's life span.

> Something reset the clock on a planetary scale. On the basis of the lunar record, a period of massive meteorite bombardment is thought to have occurred at that time. If so, it must have induced homogenization of mantle and crust on a scale vastly greater than can be explained by recent plate tectonic processes. We are indeed fortunate that the fragmented memory of an earlier time has survived in the form of zircon crystals.

Now let us consider a quote on the nature of the primitive crust relative to Archean-age crust on planet Earth (from Naeraa et al. 2012, p. 627).

> The observations suggest that 3.9–3.5-Gyr-old rocks differentiated from a >3.9-Gyr-old source reservoir with a chrondritic to slightly depleted hafnium isotope composition. In contrast, rocks formed after 3.2 Gyr ago register the first additions of juvenile depleted

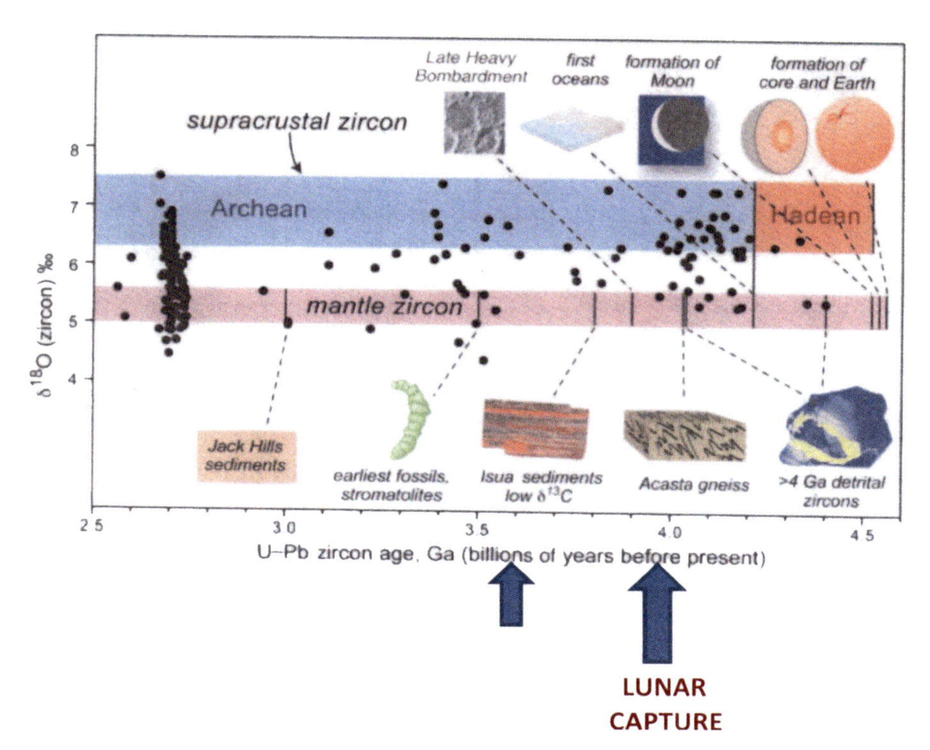

Fig. 5.74 Plot of crystallization age vs. Oxygen isotope ratios for zircons from the first 2.1 Ga of Earth history. This is the same diagram as in Fig. 5.60 but with the time of capture and the subsequent orbit circularization sequence indicated by the *arrows*. (Diagram adapted from Valley 2006, Fig. 2, with permission from Mineralogical Society of America)

> material (that is, new mantle-derived crust) since 3.9 Gyr ago, and are characterized by striking shifts in hafnium isotope ratios similar to those shown by Phanerozoic subduction-related orogens[10-12].
>
> These data suggest a transitional period 3.5–3.2 Gyr ago from an ancient (3.9–3.5 Gyr old) crustal evolutionary regime unlike that of modern plate tectonics to a geodynamic setting after 3.2 Gyr ago that involved juvenile crust generation by plate tectonic processes.

Some additional recent articles on the concept of the formation of an enriched primitive crust and/or recycling enriched primitive crust into the mantle are O'Neil et al. (2011, 2012, 2013a, b); Darling et al. (2013); Nutman et al. (2013); Rizo et al. (2013); Zeh et al. (2014), and Turner et al. (2014).

The following is a "soliloquy" that may relate to many elements of this geological puzzle. Figures 5.74 and 5.75 is a summary diagram relating to many of the major events.

In the Intact Capture Model the Earth accretes and differentiates into a metallic core and silicate mantle. Eventually a somewhat stable, mainly basaltic crust separates from the mantle within the first 200 Ma of earth history. This crust is enriched

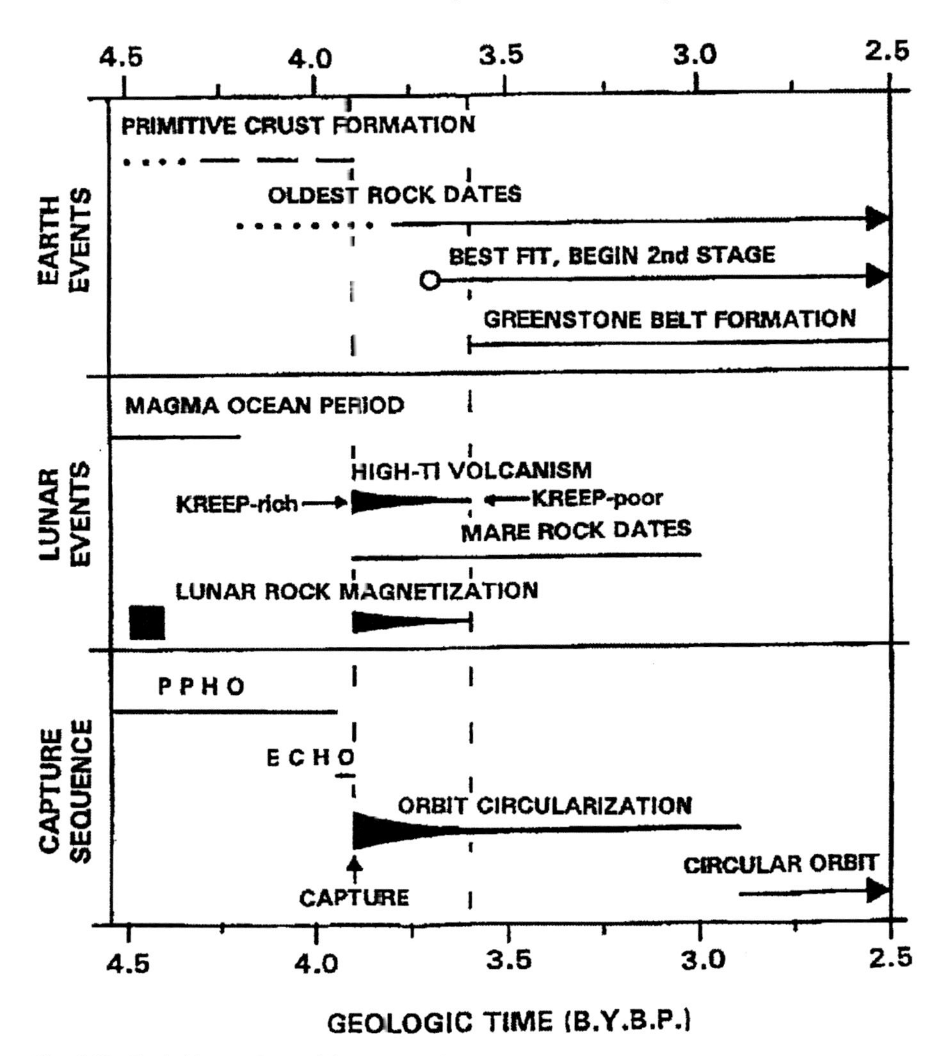

Fig. 5.75 Earth history, Lunar history, and the gravitational capture scenario. Diagram showing the possible relationships of a TIDAL CAPTURE MODEL and a number of terrestrial and lunar events. History of the diagram: First appearance in preliminary form was in Malcuit et al. (1977); a more advanced form was in Malcuit et al. (1992). And now the current form. Sources of data for EARTH EVENTS: Oldest rock dates (Black et al. 1971; Goldich and Hedge 1974; Baadsgaard et al. 1976; Windley 1984); Pb-Pb age of Swaziland Sequence (Saager and Koppel 1976); best fit for second stage of lead isotope evolution (Stacey and Kramers 1975; Albarede and Juteau 1984). More recent information on Earth Events (Valley et al. 2002; Valley 2006; Harrison et al. 2009; Bell et al. 2011; Mueller et al. 2009). Sources of information for LUNAR EVENTS: lunar-rock magnetization (Cisowski and Fuller 1983; Cisowski et al. 1983; Runcorn 1983); "pre-mare volcanism" (Ryder and Taylor 1976; Taylor 1982); mare-rock dates (Taylor 1982; Ryder 1990). Capture-sequence symbols: PPHO = Planet-Perturbed Heliocentric Orbit; ECHO = Earth-Crossing Heliocentric Orbit

with the incompatible elements and maintains its identity during modifications due to mantle upwellings and minor bolide bombardment. In the capture model there is no satellite until the time of capture. Thus, the only rock tides are weak solar gravitational tides. Once a stable "chill" crust is formed, then the era of a COOL EARLY EARTH begins as water vapor condenses from the primitive atmosphere and accumulates as liquid water on the lower elevations of the cooling planet. A basaltic crust is processed to include some granitic rocks by way of the mechanisms outlined by Bedard (2006) which he calls "a catalytic delamination-driven tectono-magmatic model".

Since the signature of TIDAL CAPTURE is high rock and ocean tides on Earth (up to 20 km amplitude) and extremely high rock tidal amplitudes on the Moon (up to 400 km amplitude), there is plenty of dynamical action in the crust-mantle region of both bodies. The bulk of the energy of capture (between 1 and 2×10^{28} J) must be dissipated within the interacting bodies. The smaller body, the Moon, must store and subsequently dissipate over 90 % of the energy for capture. As the post capture orbit circularizes, the rate of energy dissipation decreases but this tidal evolution scenario continues for several 100 Ma.

Since stable capture must occur near the plane of the planets, the equatorial zone of the planet (assuming a low obliquity) has the highest rock tidal amplitudes. These unidirectional rock tides cause convection cells to form in the upper mantle [tidal vorticity induction (TVI) cells as described by Bostrom (2000)]. These convection cells are systematically recycling the "enriched" crust into the mantle and the two entities are partially rehomogenized chemically. The "enriched" crust in the polar regions is protected from this recycling action and perhaps some survives. This polar crust, then, is a possible source rock for the Jack Hills zircons (and perhaps for zircons to be discovered in other areas in the future).

Since the captured planetoid must dissipate nearly all of the energy for capture and this energy must be dissipated within the body of the planetoid in about 2 h of time (*i.e.*, during the initial encounter of a successful capture sequence of events), a zone about 50 km thick is melted in the lunar upper mantle during the capture encounter and more is added during the close encounters associated with the early post-capture orbit evolution. Thus, it is inescapable that the captured planetoid goes through a major and sharply defined thermal episode.

This supra-tidal action and associated magmatism on Earth and Moon at about 3.95 Ga has the potential for explaining most, and perhaps all, of the features of the so-called "LATE HEAVY BOMBARDMENT" as being associated with a sequence of related events associated with a GRAVITATIONAL CAPTURE SCENARIO. In addition, a LUNAR MAGNETIC FIELD is generated within the remelted lunar magma ocean by convection cells (TVI cells) formed within the lunar magma ocean during the orbit circularization sequence between 3.95 and 3.60 Ga. The mechanism suggested for generating a lunar magnetic field is magnetic amplifier cells operating in the remelted lunar magma ocean, a mechanism suggested by Smoluchowski (1973a, b). The "seed" field for the lunar amplifier cells for this model is the Earth's magnetic field.

Since many of the KREEP basalts yield ages around 3.95 Ga and acquired a remanent magnetization from an internal lunar magnetic field and the time of disappearance of the Earth's crust is about 3.95 Ga, I think that these are all simultaneous events associated with the GRAVITATIONAL CAPTURE EPISODE. Thus, a GRAVITATIONAL CAPTURE SCENARIO may be a solution to the recent intriguing question posed by Amelin (2005) which can be paraphrased as: "WHERE DID ALL THE OLD CRUST GO?"

NOTE: See the APPENDIX FOR THE "COOL EARLY EARTH" VIGNETTE at the end of this book for a more complete treatment of the tidal amplitudes associated with the orbit circularization era.

5.7 Vignette F. On the Origin of Earth's Oceans of Water

As stated in Chap. 1, there are several special features of the Earth as a planet. The three outstanding special features are: (1) it is the only planet in the Solar System that has liquid water at the surface at the present time, (2) it is the only planet in the Solar System to have a significant partial pressure of oxygen in its atmosphere, and (3) it is the only planet in the Solar System that has a highly developed biological system. Most natural scientists agree that these three features are intimately intertwined. In addition to the features listed above, the Earth is the only planet with a very strong magnetic field relative to the other terrestrial planets and the mechanism for its generation is not well understood. Then there is this large satellite, the MOON. The mass ratio of the Moon to the Earth is 1:81. This is by far the largest mass ratio for any of the regular eight planets (i.e., excepting the Charon-Pluto system) in the Solar System. As the reader has probably noticed, the theme of this book is that our Moon is a very special object in our environment. Furthermore, in my view, all of the first four special features of the Earth are caused by, or have something to do with, the presence of this large satellite in orbit about the planet.

The subject of this section of the chapter is WATER! Planet Earth is commonly referred to as "THE WATER PLANET" especially after we humans got a view of the planet from space. We literally have oceans of water and until recently there was no reasonable explanation for the origin of this large quantity of water on the planet.

Models for the source of our water can be placed into three categories: (1) the Simplest Source, (2) the Far-Out Source, and (3) the Intermediate Source. A model for the Simplest Source for the water in the terrestrial oceans is presented by Drake and Righter (2002). Their concept is that the water on Earth was outgassed from the Earth's mantle over geologic time. The underlying assumption is that the water was associated with the mixture of meteoritic debris that accumulated (accreted) to form the planet. This is a commonly held view among geoscientists and it is the "simplest" view, but it does not relate very well to the "facts to be explained by a successful model".

Proponents of the Far-Out Source suggest that the water associated with Earth and neighboring planets was delivered by water-rich comets from the Kuiper Belt/Oort Cloud toward the later part of the planet accretion process (Owen and Bar-Nun 1998).

Table 5.1 The Deuterium/Hydrogen Ratio for Earth, neighboring planets, and other bodies. (values from Albarede 2009)

D/H ratio of the Sun and Jupiter	$20 \pm 4 \times 10^{-6}$
D/H ratio of Earth's Oceans	$149 \pm 3 \times 10^{-6}$
D/H ratio of Comets	$310 \pm 40 \times 10^{-6}$
D/H ratio of Carbonaceous Chondrites (clay)	$140 \pm 20 \times 10^{-6}$ (spread from 120–280)
D/H ratio of Mars (meteorites)	300×10^{-6} (perhaps a comet source)
D/H ratio of the Atmosphere of Mars	810×10^{-6}
D/H ratio of Interstellar Ice	up to 1×10^{-2}

Deuterium is "HEAVY HYDROGEN". When heat is applied to a batch of mixed hydrogen, the "light" hydrogen leaves and the "heavy" hydrogen is enriched in the batch of hydrogen that remains

Table 5.2 Potassium content normalized to uranium for the Moon, Earth, and Mars. (Values from Albarede 2009)

Moon ~ 3000	Moon lacks ~ 95 % of the nebular K inventory
Earth ~ 10,000	Earth lacks ~ 85 % of the nebular K inventory
Mars ~ 20,000	Mars lacks ~ 70 % of the nebular K inventory

Potassium is a fairly volatile element and Uranium is a refractory element. Thus the K/U ratio is a *Volatility Index*

The first problem with this source is the Deuterium to Hydrogen ratio of the Earth's oceans; it is *not* similar to comets. A second problem with this model is that the source region (Kuiper Belt/Oort Cloud) is very distant from Earth and it may take some time to transfer these "dirty snow balls" to the vicinity of Earth.

Now we come to the Intermediate Source the proponents of which advance the idea that the water for Earth and neighboring planets was delivered by water-rich carbonaceous chondrite meteorites from the outer reaches of the Asteroid Zone. A major fact explained by this model is that the Deuterium/Hydrogen (D/H) ratio of the water-rich carbonaceous chondrites is similar to the D/H ratio of the Earth's oceans (Robert 2001).

5.7.1 Some Facts to be Explained by a Successful Model for the Origin of Water on Earth and Neighboring Planets

a. Table 5.1 shows the D/H ratios for a variety of bodies in the Solar System. Cursory examination of this Table suggests that the D/H ratio of the Earth's oceans is very similar to that of the clay minerals in carbonaceous chondrites. It is also clear that the D/H ratio of the mantle rocks of Mars (obtained from SNC meteorites found on Earth and thought to be impact debris from Mars) is very similar to the D/H ratio of comets.

b. The Potassium to Uranium (K/U) ratio for various Solar System bodies needs to be explained. Table 5.2 gives some summary information about relative potassium content of Moon, Earth, and Mars from Albarede (2009, p. 1228). For Earth there is also a depletion from 92–98 % of the following volatile elements: Zn, Ag, As, Sb, Sn, Pb, and S. "Planets that lack such large fractions of moderately volatile elements cannot have been endowed with large amounts of water" (Albarede 2009, p. 1228): i.e., the water would have been removed with the moderately volatile elements.

c. The Lead-206/Lead-204 (i.e., the Pb/Pb) age of the Earth needs to be explained. The facts, in this case, are the experimentally determined quantities of the lead isotopes. The ages are interpretations of the lead isotope abundances in the substance. It turns out that the LEAD GEOCHRONOMETER for the Earth gives a younger date for the planet than do the other geochronometers. Albarede (2009) suggests that this material came in as a LATE VENEER along with the water for the oceans and other volatile elements.
d. The quantity of water in the Earth system needs to be explained. How much water is involved? A reasonable estimate is the following. An estimate for the volume of the oceans is about 1664 million km^3 Lodders and Fegley (1998). Clay minerals can contain up to 50 % water. Thus we need to multiply the volume of the oceans by a factor of two to get a reasonable minimum estimate. Some guesstimates of the water content of the mantle go up to five times that in the surface oceans. Thus we need about 16,640 million cubic kilometers of water-bearing asteroidal material to reach this maximum estimate.

5.7.2 The Asteroidal Source of Water as Proposed by Albarede (2009)

This model of Albarede was published in the journal *Nature* on 29 October 2009. He proposed that the source for the water and the associated volatile elements should be water-rich asteroids. Then on 29 April 2010, there were two articles published in the same issue of *Nature*, by two different teams of observational astronomers, who detected unusually high quantities of water ice and organics on Themis-type asteroids.

The general concept of the Albarede Model (a general concept also supported by Taylor 1998, 2001) is that the inner part of the Solar System was dehydrated in the very earliest history of the Solar System before the chondrules and matrix of the chondritic meteorites began to accrete into the various types of chondritic meteorites. The heat pulses of the early Sun (perhaps some combination of the X-Wind and the Disk-Wind models) would push the volatiles out to the vicinity of the Asteroid Belt and the orbit of Jupiter (the Frost Line concept mentioned in Chap. 2). Then, after the terrestrial planets and asteroids formed, the hydrous material began to migrate back into the inner part of the Solar System to add a LATE VENEER of hydrous material about 100 Ma (±50 Ma) after formation of the Earth and the other terrestrial planets (see Fig. 5.76 for a graphic representation of Albarede's model). A "missing link" in Albarede's model was that he could not identify any asteroids with sufficiently large quantities of water. That is why the spectroscopic analyses of the Themis-type asteroids by Rivkin and Emery (2010) and Campins et al. (2010), published about 6 months after Albarede's report, are so important.

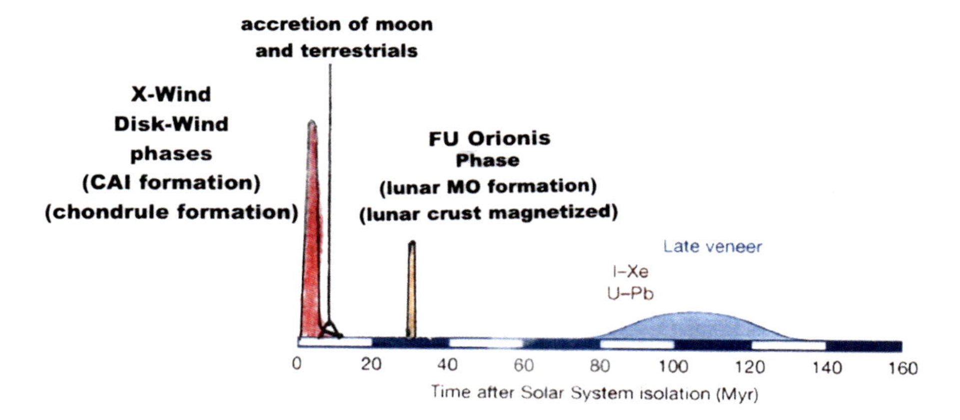

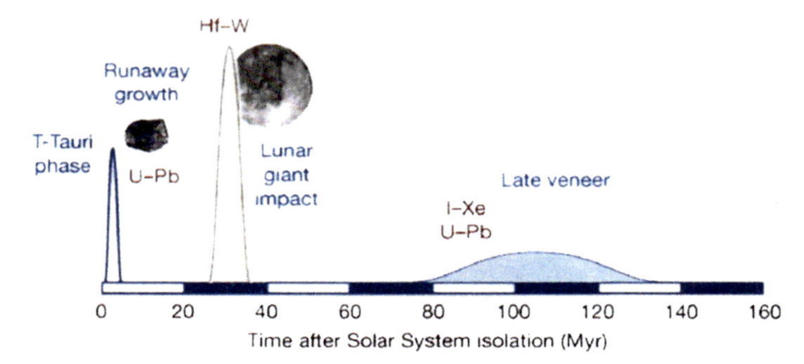

Fig. 5.76 **a** Diagram adapted from Albarede (2009, Fig. 5, with permission from nature publishing group) for the timing of the arrival of the "late veneer"; these modifications make the concept very compatible with the cool early Earth model (Valley et al. 2002) as well as the gravitational capture model which features capture at about 3.95 Ga. **b** The original diagram from Albarede (2009, Fig. 5). In the author's view, the Lunar Giant impact episode is not needed and really complicates the late veneer concept. In addition, the X-Wind and Disk-wind models appear to explain the origin of CAIs and chondrules much better than the generalized T-Tauri model. The FU Orionis phase of solar evolution is much more powerful than the subsequent T-Tauri phase and the former may relate to the microwave (joule) heating of the Vulcanoid asteroids

I think that the model of Albarede (2009) relates to the "facts to be explained" better than any other of the proposed models for the origin of Earth's water and associated volatiles. The remaining question is about the delivery system for getting this asteroidal water to Earth by impact processes without vaporizing and subsequently ionizing a large quantity of that water. The mechanism that I propose is the "soft" impact of water-bearing asteroids onto the surface of the planet. Such a "soft" impact process can only occur when the asteroidal body has a close (grazing) encounter with the earth from a near-earth-like orbit. Thus, it has to get transferred from an asteroidal orbit (in the Asteroid Belt) to an Earth-like heliocentric orbit.

And several of these water-rich asteroids must have a "soft" impact on Earth to explain the quantity of water in the oceans and interior of planet Earth.

My main suggestions for modification to the earliest history portion of the model of Albarede (2009) are the following: (1) The Giant-Impact Model for the origin of the Moon is not needed (simply eliminate that episode). A Gravitational Capture model for the origin of the Moon at about 3.95 is very consistent with the data; with a late capture, about 600 Ma after formation of the planet, the Earth can have a lengthy "COOL EARLY EARTH" stage and can have a much better chance to preserve the water it receives from the asteroidal source. (2) The X-Wind model works well for the origin of CAIs as well as for yielding the thermal power to clear the inner part of the Solar System of the more volatile elements at a stage where the largest bodies are probably of decimeter size or smaller. These modifications are illustrated in Fig. 5.76.

5.7.3 A Proposed Delivery Mechanism for the Water-Bearing Asteroids

Figure 5.77 shows the main characters in my proposed delivery system for the LATE VENEER water. The Sun is the most important character on the stage and Jupiter is the next because Jupiter controls most of the action in the Asteroid Belt via orbital resonance action. Some orbits in the Asteroid Belt are stable relative to the gravitational perturbations by Jupiter and some are not. The asteroids that are in stable orbits are still out there in the Asteroid Belt. We are interested in a few of the asteroids that get perturbed into unstable orbits and had gravitational interactions with other asteroids and Mars and eventually end up in a near earth-coincident orbit from which a "soft" impact can occur.

Figure 5.78 illustrates how a circular orbit of a Themis-type asteroid can get perturbed into an elliptical orbit from which it could have encounters with planet Mars. Figure 5.79 illustrates how such an asteroid orbit can become earth-crossing and eventually begin interacting with both Venus and Earth.

Figure 5.80 illustrates how Aquarioid asteroids can deliver water to Earth and Venus. A straight-in, high-speed, collision (Fig. 5.80b) would not add much water to the planet because of the heat generated by the impact. A tangential, low-speed, impact such as shown in Figs. 5.80a and c would lead to a significant deposition of water onto the surface of a terrestrial planet like Venus or Earth.

For additional background information on this fascinating subject of the origin of our planet's water and other volatile elements and the potential delivery systems I would suggest the following articles: Alexander et al. (2012), Cooper et al. (2014), Cry et al. (1998), Dauphas (2003), Dauphas and Marty (2002), Dauphas et al. (2000), Hutsemekers et al. (2009) Jacquet and Robert (2013), Libidi et al. (2013), Morbidelli et al. (2000), Raymond et al. (2013), and Wang and Becker (2013).

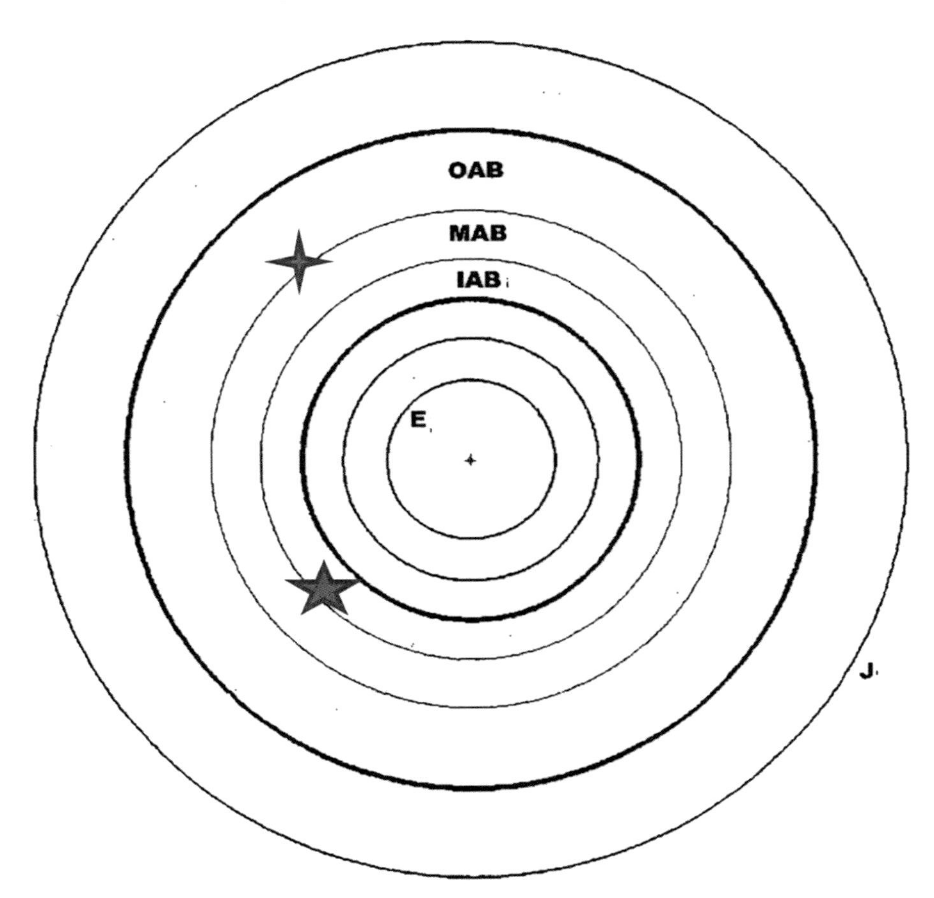

Fig. 5.77 Diagram of the some orbits of bodies in the Solar System, viewed from the north pole of the Solar System, showing the position of the Sun and the orbits of Jupiter, Earth, and the Outer Asteroid Belt (*OAB*), Middle Asteroid Belt (*MAB*), and Inner Asteroid Belt (*IAB*). The Martian orbit is shown but not labeled (between Earth's orbit and the *IAB*). The general position of asteroid 24 Themis (~3.2 AU) (4-point star symbol) (Rivkin and Emery 2010; Campins et al. 2010) as well as the theoretical asteroid "Aquarius" (5-point star symbol) are shown. The orbit of "Aquarius" (~2.3 AU) will be used in the orbital demonstrations for how a "soft" impact process may be used to deliver water to Earth

5.7.4 Summary and Conclusions

1. The original material that formed the Earth was apparently very water poor as well as poor in other volatile elements.
2. The DEUTERIUM to HYDROGEN (D/H) ratio appears to be a key fact to be explained in the search for a source for EARTH'S OCEANS OF WATER.
3. A special brand of WATER-RICH ASTEROIDS ("aquarioids") appear to be the best source of water for planets Earth (and Venus).

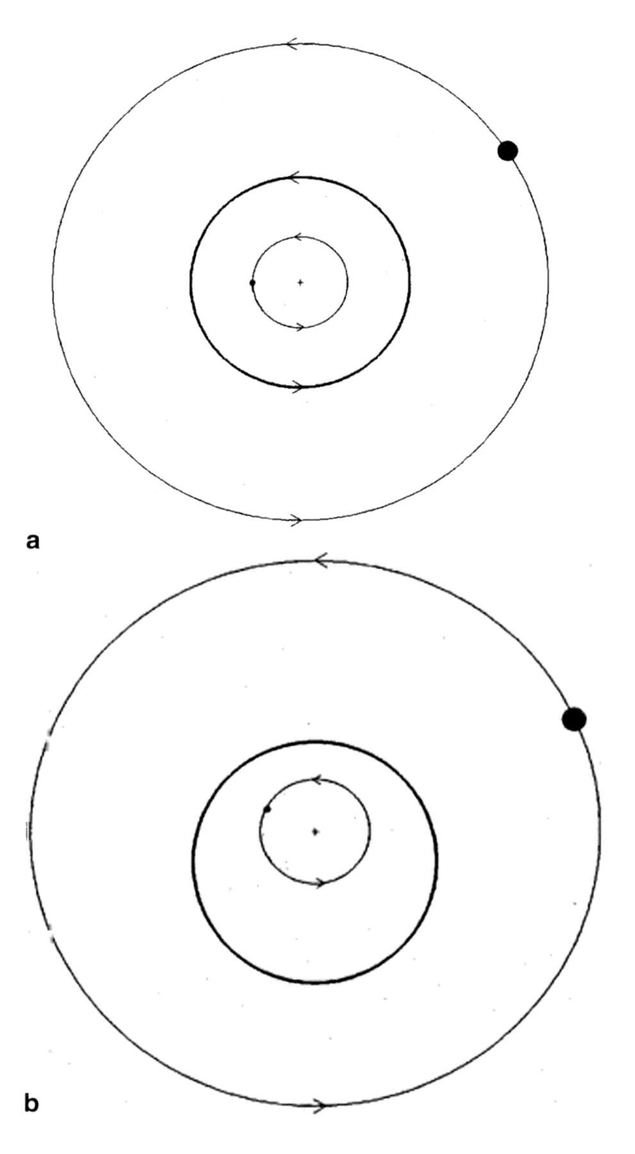

Fig. 5.78 Early stages of a gravitational perturbation sequence; planet Jupiter is in control of the perturbations. View is from the north pole of the Solar System. **a** Circular orbit of Aquarius at 2.3 AU relative to the orbits of Jupiter and Earth. **b** Eccentricity of the orbit of Aquarius is increased to 0.25

4. Planet Mars apparently got most of its water from a Jupiter-captured comet during a grazing collision with Mars (as strongly suggested by the D/H ratio for the mantle rocks of Mars).

5.7.5 Discussion

The origin of the water reservoirs on planet Earth is a serious and perplexing problem. The normal solution [i.e., that the water was associated with the

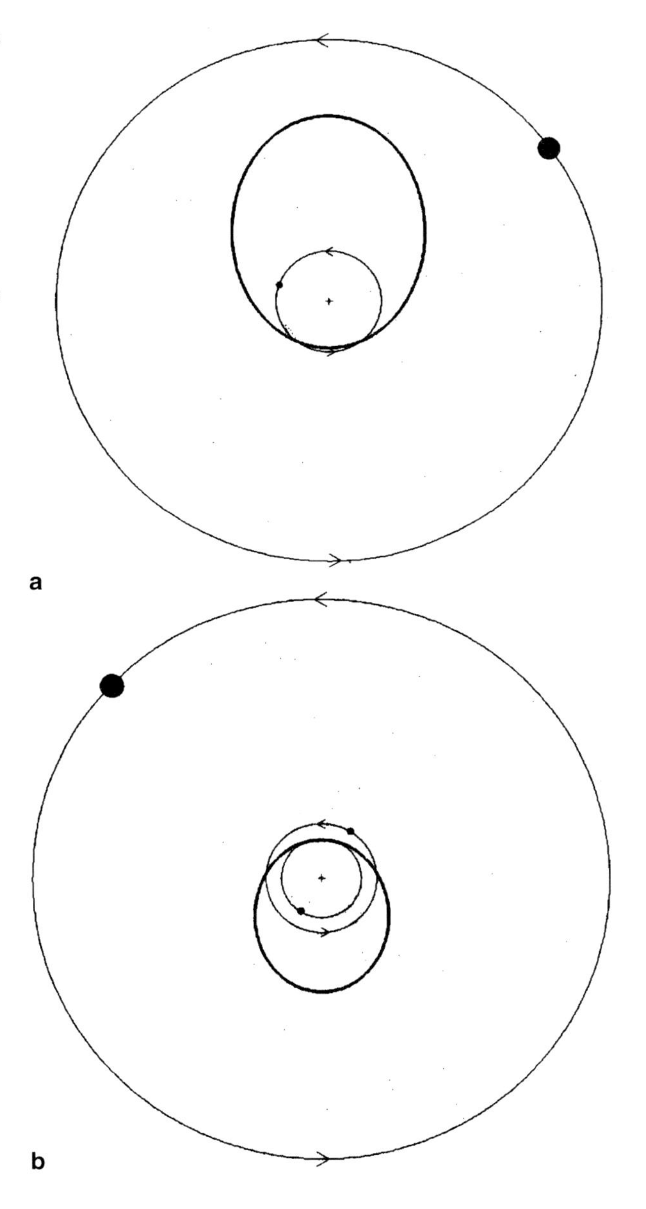

Fig. 5.79 Intermediate stages of a gravitational perturbation sequence. View is from the north pole of the Solar System. **a** Earth is now in control of the perturbations. Eccentricity of the orbit of Aquarius has increased to 0.5. **b** The orbit of Aquarius (now 1.4 AU for semi-major axis and eccentricity is 0.5) is intersecting the orbits of both Earth and Venus

meteoritic debris that accreted to make the Earth (Drake and Righter 2002)], has very little, if any, evidence in its favor. The Albarede (2009) Model is consistent with nearly all of the evidence but is a very complex model. As with most of the other "big picture" models in the natural sciences such as the CONTINENTAL DRIFT/PLATE TECTONICS MODEL for shifting land masses over time and the MILANKOVITCH MODEL for Earth's climate variations over time, the ORIGIN OF WATER problem appears to be complex.

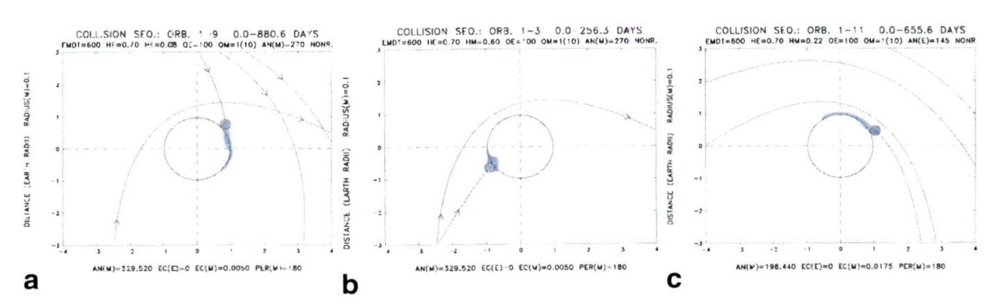

Fig. 5.80 Some suggestions for the latest stages of proposed delivery system. **a** Retrograde grazing collision. **b** Straight-in high-speed collision. **c** Prograde grazing collision

A brief summary of how this model fits into early Solar System development is presented here. The precursor of the Solar System is a cloud of dust and gas from previous stellar explosions. A central mass begins to condense and gravitational attraction causes an influx of material to a central area that eventually becomes the Sun. The accumulation of mass is so rapid that the Sun eventually transformed into a rapidly rotating nuclear furnace with massive bipolar outflow of material. The solar equatorial zone is characterized by a strong magnetic flux guiding the inflow and outflow of material. Major heat pulses are generated in the solar equatorial plane by the breaking and reforming of magnetic flux lines. This X-WIND (i.e., Shu et al. 2001) action gives rise to chemical and physical features of Calcium-Aluminum Inclusions as well as to the thermal pulses that propels some of the CAI material to regions beyond their zone of formation. The thermal pulses also dehydrate and devolatilize the surrounding small particles and remove a large portion of the gaseous material along the mid-plane of the Solar System. As the X-WIND wanes, the DISK WIND (Salmeron and Ireland 2012) continues the thermal processing of material in the solar mid-plane as material is still falling in from the proto-stellar cloud. This thermal processing by the DISK WIND, although less intense than the X-Wind, causes flash melting of the adjacent material by the snapping and reforming of magnetic field lines to form the various types of CHONDRULES. The previously formed material, the CAIs and the CHONDRULES interact with unused materials and newly infalling material to form various types of chondritic meteorites. This unused and newly infalling material becomes the matrix, or binding material, of the chondritic meteorites.

The precursor chemistry that eventually forms the chondrules is gradationally zoned from the thermal pulses associated with the X-Wind, which is very strong at first and then wanes in intensity and eventually gives way to the succeeding Disk Wind action. The main result of thermal processing is that the volatiles (water and other volatile elements and compounds) are progressively removed from the inner regions of the solar nebula and displaced progressively to the Asteroid Zone and to the vicinity of Jupiter's orbit. Accreting planets accumulated whatever material was in the immediate environment and the composition of the resulting planets reflects the composition of the material in their respective accretion tori. As a result Venus and Earth are more volatile-poor than Mars and Mars is more volatile-poor than the material in the Asteroid Zone.

Jupiter then takes over and redistributes some of the hydrous and volatile-rich material to the inner part of the Solar System from which it originally came. Jupiter is a giant "asteroid herder" and marshals them into orbital bands by orbital resonance action. Some asteroids are ushered into stable orbits relative to the combined gravitational action of the Sun and Jupiter and others end up in unstable orbits. The water and volatile-rich asteroids in the UNSTABLE orbits are the important ones here because some of them became THE MAJOR SOURCE OF WATER AND VOLATILES for planets Earth and Venus. Planet Mars probably gets some asteroidal water but the D/H ratio of the mantle chips from Mars suggest that it got most of its water from a comet or two. These comets apparently become "Jupiter-Captured Comets" meaning that they had a close encounter with Jupiter and Jupiter inserted the comets into somewhat circular prograde orbits in the zone of the asteroids.

Once the water and volatile bearing asteroids are in unstable orbits and eventually get worked into Mars-crossing and eventually Earth-crossing orbits, there is still a good bit of "luck" involved for getting them successfully guided to Earth. A high-speed impact (as illustrated in Fig. 5.80b) will probably deliver very little water and volatile material to the planet (most evaporates and thermally dissociates and goes off into space, never to return). Only "soft" impacts (*i.e.*, low speed and grazing impacts as illustrated in Figs. 5.80a and c) deliver much of the water-saturated clay minerals to the surface of the planet. The hydrous, volatile bearing material is heated by the impact but much ends up on or in the basaltic crust of the newly formed planet. The liquid water then fills-in the low spots to form the first mud-puddles on the nascent "Cool Early Earth". With several repeated "soft" impacts (and some that were not so "soft") the Earth eventually accumulates its inventory of ASTEROIDAL WATER. A similar story can be worked out for planet Venus because the D/H ratio of the atmosphere suggests that it had oceans of water in its geologic past (Donahue et al. 1982; Donahue 1999).

If the reader considers this WATER STORY to be TOO COMPLICATED, go ahead and try to simplify the story!!

5.8 Discussion of the Speculations by Harold Urey and Zdenek Kopal

"It is therefore more than probable that the principal period of cosmic bombardment which mutilated the lunar surface extended over less than the first billion years of the existence of our satellite, and was not far removed from the time of origin of the solar system as a whole. On the Earth and other terrestrial planets all landmarks of comparable age must have been completely obliterated aeons ago by the joint action of their atmospheres and oceans, not to speak of internal heat at deeper layers of their crusts. However, as any geological change on the Moon can proceed only at an exceedingly slow rate, its present face still bears scars and traces of many events which took place in the inner precincts of the solar system not long after its formation. If so, this should make our Moon the most important fossil of the solar system and a correct interpretation of its stony palimpsest should bring rich scientific rewards. (Zdenek Kopal, 1973, The Solar System: Oxford University Press, p. 82 (with permission of Oxford University Press)

> Urey felt that the Moon might well be a primitive body, formed in the early days of the solar system. If this were so, the record of its early history preserved on the surface of the Moon, would provide invaluable clues to the origin and evolution of other bodies of the solar system as well. (Homer and Newell 1973, Harold Urey and the Moon: The Moon, v. 7, p. 1)

In this chapter we have covered a number of topics under the rubric of "vignettes". Some "vignettes" are short and others are a bit lengthy but they are all related directly or indirectly with the Lunar Origin Problem. Most of them are concerned with the first billion years in the life of the Moon, the timeframe that concerned both Urey and Kopal.

First I did a critique of the rules of interpretation of the lunar surface that were presented in great detail by Wilhelms (1987, 1993). Of the 17 rules, there are three that are questionable and two others that are found to be a bit misleading.

Then we looked at directional properties of the maria and their features, a topic in my opinion that has been neglected for decades and contains much valuable information (many of these properties qualify as "stony palimpsest" features in Kopal's terminology).

Next we considered the subject of mass concentrations (mascons) on the lunar surface and found that the simplest interpretation has been passed over because it does not fit with the current paradigm that only solid body impact processes have affected the lunar surface.

A brief review of the so-called Late Heavy Bombardment reveals that the source of these external impacting bodies is still an unsolved problem some 40 years after its inception. And it is not clear that any bodies other than the Earth and Moon (and Sun) were involved in these high energy events that certainly left a record on the lunar surface and were centered on a timeframe of 4.0 to 3.8 Ga.

A discussion of the Cool Early Earth model strongly suggests that the Giant Impact model is incompatible with the basic tenants of the model. In contrast, a Lunar Capture model featuring the capture episode at about 3.95 Ga features partial disruption of the lunar body during a very close encounter soon after the capture encounter. This particular capture model relates to (1) the origin of lunar mascons, (2) the exponentially decreasing strength of the remanant magnetization of lunar mare rocks and breccias spanning the time from about 3.9 to 3.6 Ga, (3) the recycling of a primitive crust of the Earth which accumulates from ~4.4 to 3.95 Ga and then gets progressively recycled into the mantle from about 3.95 to 3.60 Ga.

The origin of water for planet Earth, at first glance, seems very tangential to the lunar origin story but, in the end, it is intimately involved. In the beginning the water story is heavily associated with the devolatilization/dehydration events in the very earliest history of the Solar System (i.e., before the formation of chondrules). The precursor particles of the lunar body (CAIs and related entities) are affected by the same "wind" events (X-Wind, Disk Wind, etc.) as the dust and gas located farther from the proto-sun. Some of the water and other volatiles that were removed from the inner part of the Solar System are then returned to the vicinity of Earth and Venus as water-rich asteroids about 100 Ma after the formation of those planets.

In summary, Urey and Kopal perhaps were correct in their speculations that the Moon may be a faithful recorder of Solar System events because, in this particular version of capture, Luna and siblings occupy spaces in their "birthplace" near the Sun before moving outward in the Solar System. Since Luna appears to be largest survivor, it could be a "ROSETTA STONE OF THE SOLAR SYSTEM" (as Urey suggested) and "A CORRECT INTERPRETATION OF ITS STONY PALIMPSEST SHOULD BRING RICH SCIENTIFIC REWARDS" (as Kopal suggested).

5.9 Summary Statement

The Giant Impact Model is currently the favored model for the origin of the Moon. A common expression is the following from a fairly recent paper:

> The histories of Earth and Moon are intrinsically linked, with a catastrophic giant impact considered to be the most likely mode of origin of the Earth-Moon system (1,2). Day et al. (2007, p. 217)

Since 1984, the Giant Impact Model has been considered a "default" model in that it was commonly thought that no version of the three traditional models would relate to the features of the Earth, Moon, and the Earth-Moon system. In my treatment of this chapter, we find that it is the GIM that generates many questions but very few answers and most of the suggestions for solutions are of an "*ad hoc*" nature. (1) There is no firm place of origin for Theia, the planetoid that collides with the proto-earth and purportedly leads to the formation of the Moon. (2) There is no firm source for the solid body impactors that are involved in the sharply defined Late Heavy Bombardment (all solutions are of an "*ad hoc*" nature). (3) There is no hint of a solution to the lunar rock magnetization mechanism(s). (4) There has been no reasonable solution for how the GIM can be made compatible with the Cool Early Earth Model. If no suggestions are forthcoming for these outstanding problems of lunar science, then perhaps it is time to search for another "default" model!

NOTE: I should note that there is no doubt in my mind that giant impacts occurred in the early history of the Solar System and that some small satellites may have resulted from the debris of the impacts. What has not been demonstrated is that a lunar-like body with many of the characteristics of the Earth's Moon can result from a giant impact between two planet-sized bodies in the vicinity of the orbit of Earth.

Perhaps it is time to give a new version of the Gravitational Capture Model a chance as a "default" model. This capture model has a place of origin for the precapture moon (Luna). This place of origin was first suggested by Cameron (1972, 1973) and the X-Wind model of Shu et al. (1997, 2001) gives a reasonable chemical and physical environment for the formation of Luna and sibling Vulcanoid planetoids. The proposed time of capture is ~3.95 Ga and numerical simulations of capture can be used to explain not only the stable capture episode but also the directional properties of many of the maria and other surface features. This capture model also provides a testable model for the emplacement of mascons in the near surface environment. Most of these features result from a very close encounter subsequent to the capture encounter that results in partial tidal disruption of the lunar body. Some of the disrupted material goes to the Earth but most falls back onto the lunar surface along a great-circle pattern of large circular maria. In addition, two major features, one magnetic and one geochemical, can be explained as episodes occurring simultaneously during the early phases of the post-capture orbit circularization sequence. These are (1) the generation of a lunar magnetic field via lunar rock tides due to periodic strong tidal interactions with the Earth and (2) the systematic subduction of an apparently all-enclosing, geochemically enriched crustal complex that developed on the planet during the Cool Early Earth era. This "highly unradiogenic low Lu/Hf Hadean crust" was apparently recycled into the upper mantle between 4 and 3.6 Ga (Bell et al. 2011). In this capture model, this crustal recycling operation is powered by lunar tides operating unidirectionally on a fairly rapidly rotating Earth (rotation rate from 10–13 h/day).

This new version of a Gravitational Capture Model for capture of a Vulcanoid planetoid may develop into a "grand unifying theory" (a term used by Taylor 2001) for the Earth and Planetary Sciences. It appears to explain a good bit about the Earth, Moon, and Earth-Moon system, but it also has the potential to explain the origin of 4 Vesta, Angra, and perhaps a few other "volcanic" asteroids and it can be used to explain the current situation of our sister planet Venus via retrograde capture of a Vulcanoid planetoid and the subsequent orbit evolution, a scenario suggested by Singer (1970). Thus, this model is more than a potential default model for the origin of the Moon and the Earth-Moon system. It may serve as an explanation for several other mysteries of the Solar System.

Appendix

The "Cool Early Earth" Vignette (Sect. 5.6.)

This appendix consists mainly of plots of orbits and tidal amplitudes for an earth-like planet during an orbit circularization sequence associated with the recycling mechanism for a primitive terrestrial crust.

NOTE: This is a full set of orbital and rock tidal amplitude diagrams for the orbit circularization sequence for the Condensed Orbital Evolution.

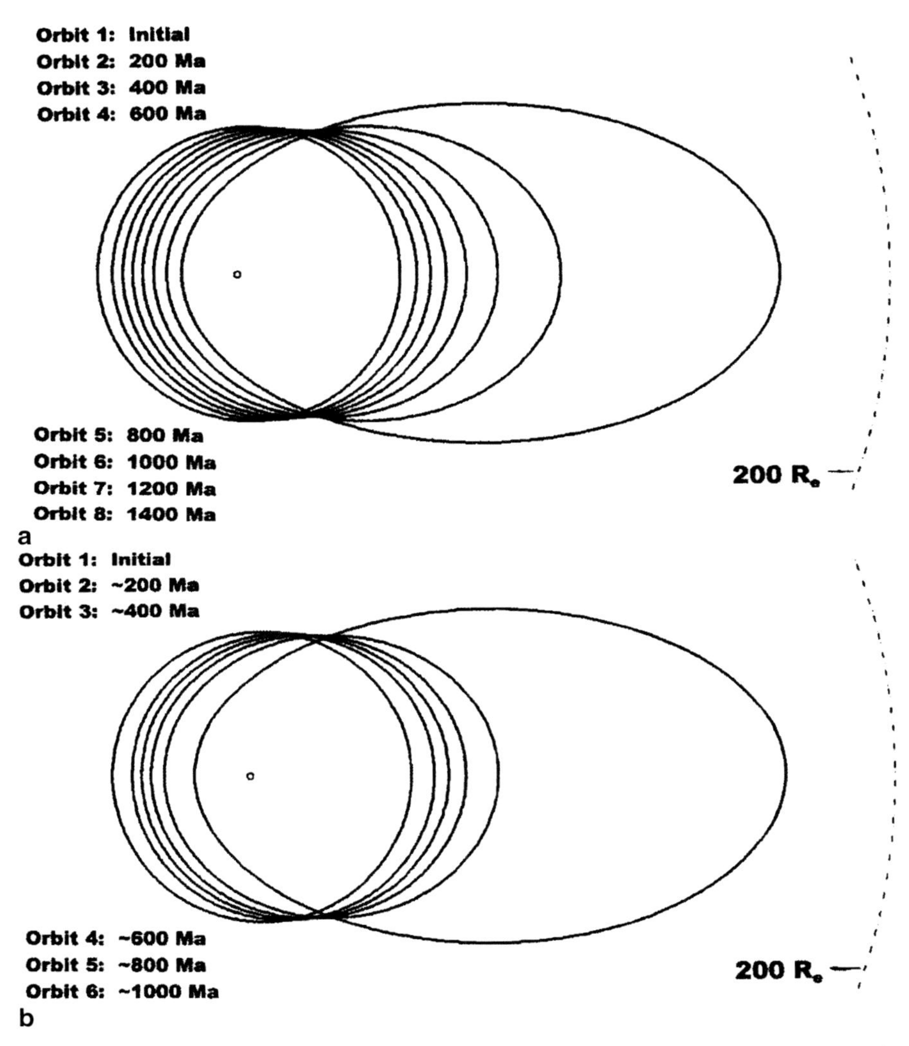

Fig. 5.81 Normal calculated two-body evolution timescale vs. Condensed three-body timescale. **a** Normal calculated two-body timescale, 200 Ma intervals. **b** Condensed three-body timescale, 200 Ma intervals

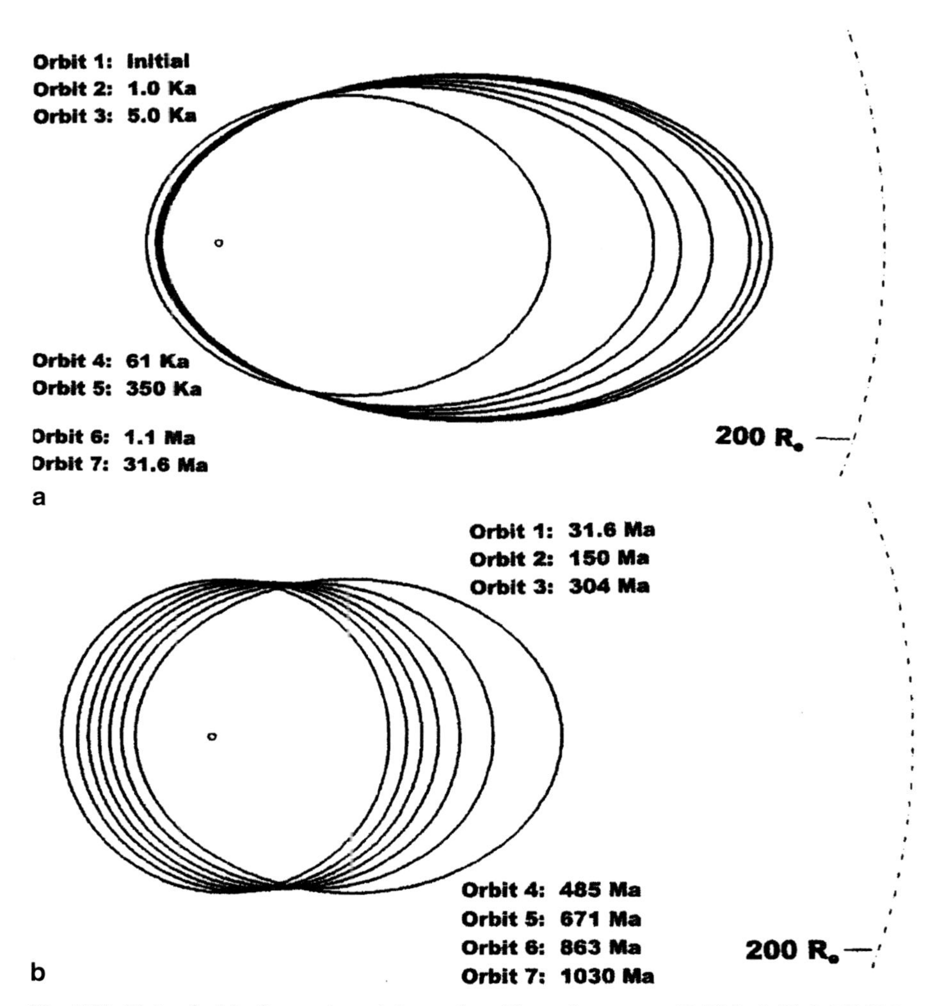

Fig. 5.82 Plots of orbits for condensed timescale. **a** Time of capture to 31.6 Ma. **b** 31.6–1030 Ma. (~10% eccentricity)

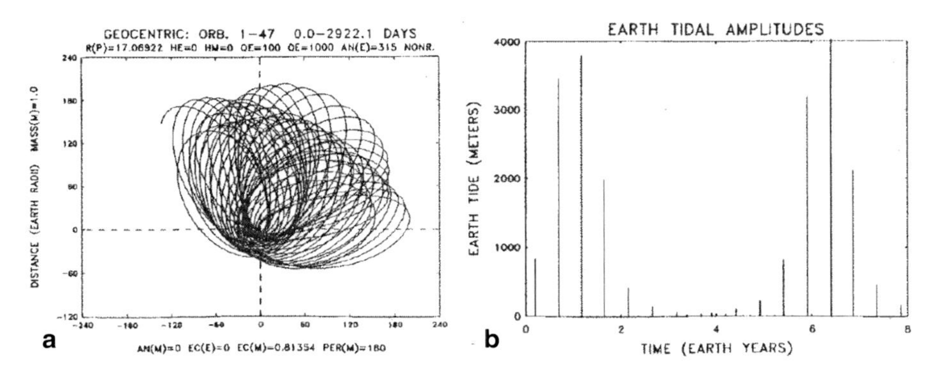

Fig. 5.83 **a**. 8 years of orbits soon after capture. **b** 8 years of rock tidal amplitudes on Earth soon after capture

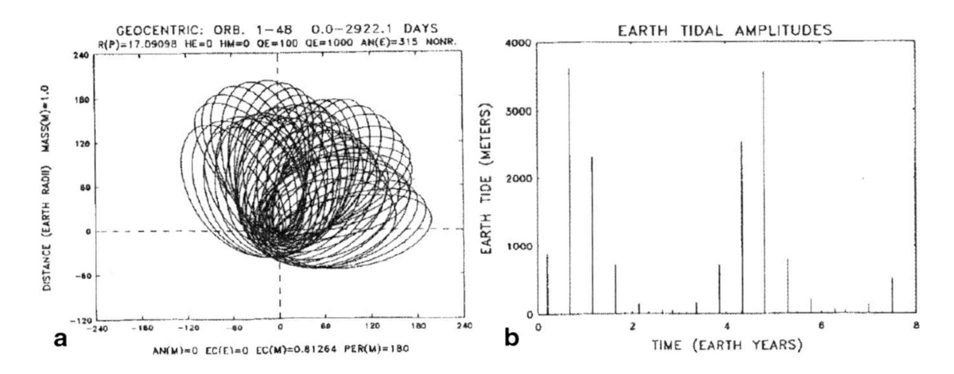

Fig. 5.84 8 years of orbits ~126 years after capture. **b** 8 years of rock tidal amplitudes on Earth ~126 years after capture

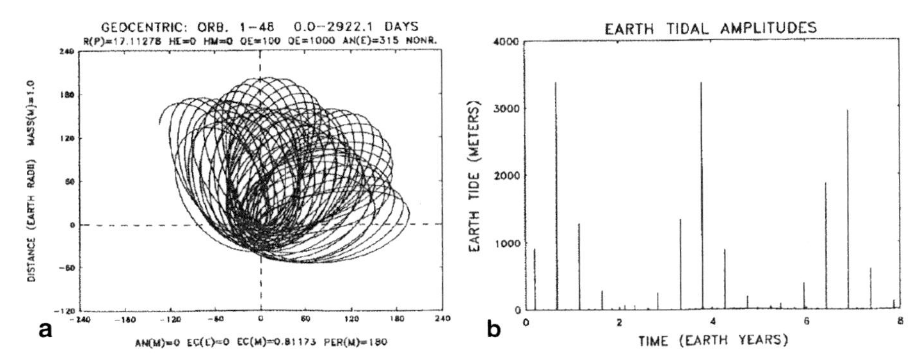

Fig. 5.85 **a** 8 years of orbits ~281 years after capture. **b** 8 years of rock tidal amplitudes on Earth ~281 years after capture

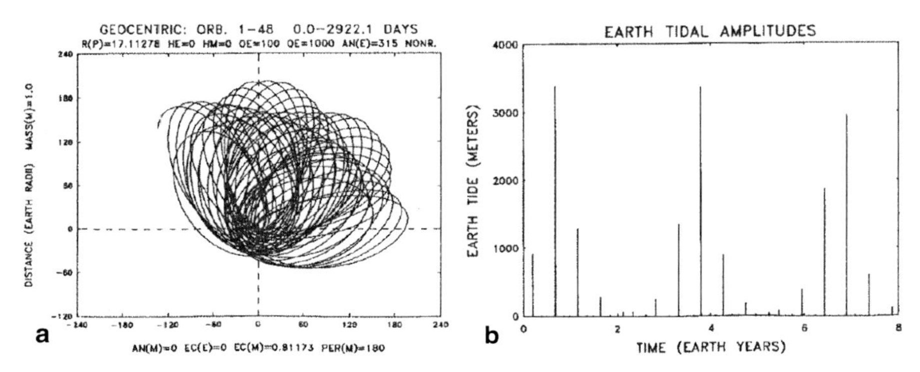

Fig. 5.86 **a** 8 years of orbits ~481 years after capture. **b** 8 years of rock tidal amplitudes on Earth ~481 years after capture

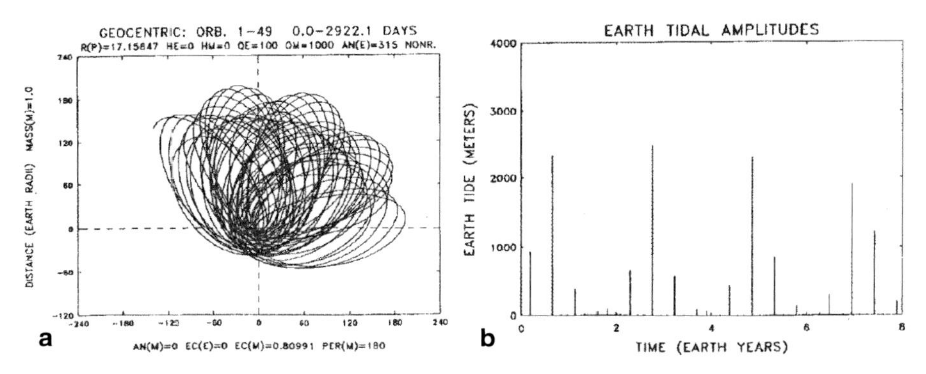

Fig. 5.87 **a** 8 years of orbits ~748 years after capture. **b** 8 years of rock tidal amplitudes on Earth ~748 years after capture

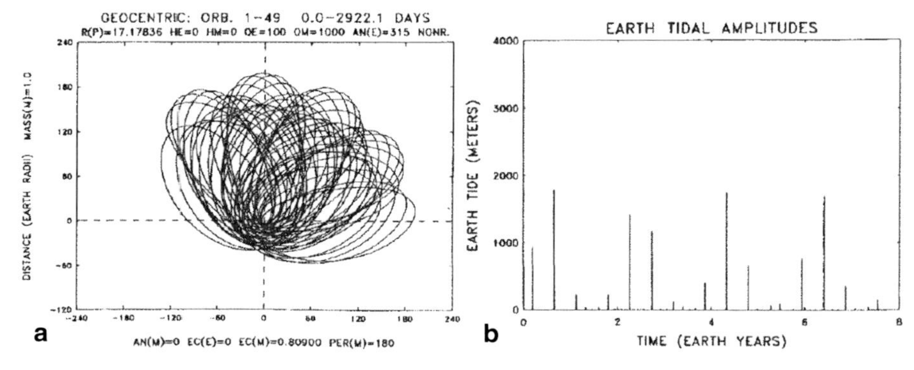

Fig. 5.88 **a** 8 years of orbits ~1085 years after capture. **b** 8 years of rock tidal amplitudes on Earth ~1085 years after capture

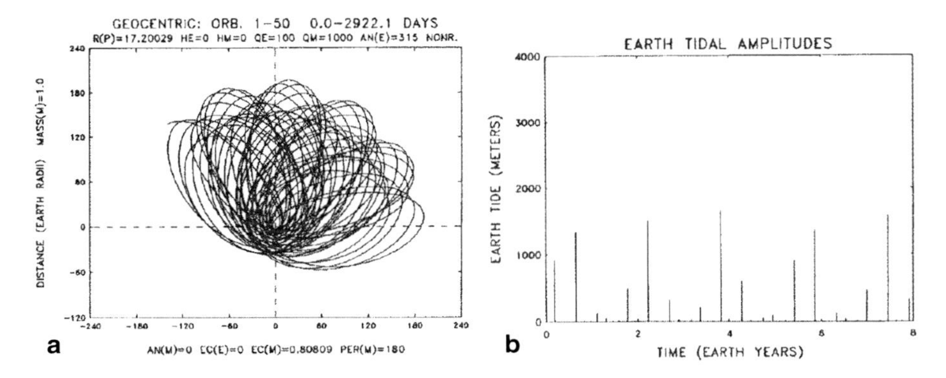

Fig. 5.89 **a** 8 years of orbits ~1515 years after capture. **b** 8 years of rock tidal amplitudes on Earth ~1515 years after capture

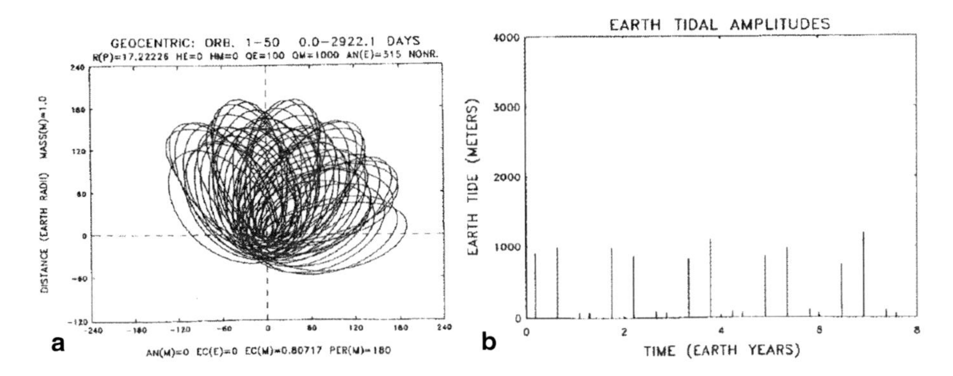

Fig. 5.90 **a** 8 years of orbits ~2033 years after capture. **b** 8 years of rock tidal amplitudes ~2033 years after capture

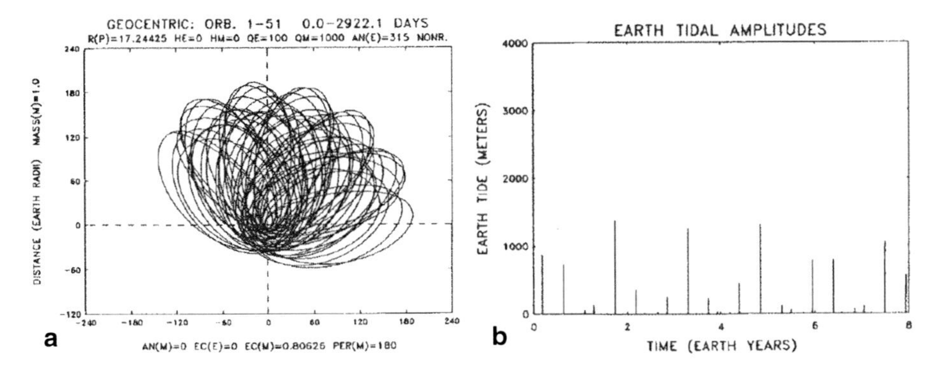

Figure 5.91 **a** 8 years of orbits ~2733 years after capture. **b** 8 years of rock tidal amplitudes on Earth ~2733 years after capture

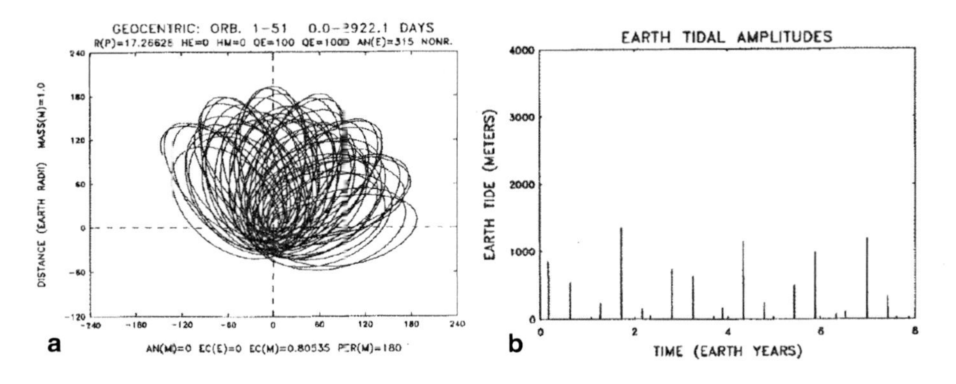

Fig. 5.92 **a** 8 years of orbits ~3467 years after capture. **b** 8 years of rock tidal amplitudes on Earth ~3467 years after capture

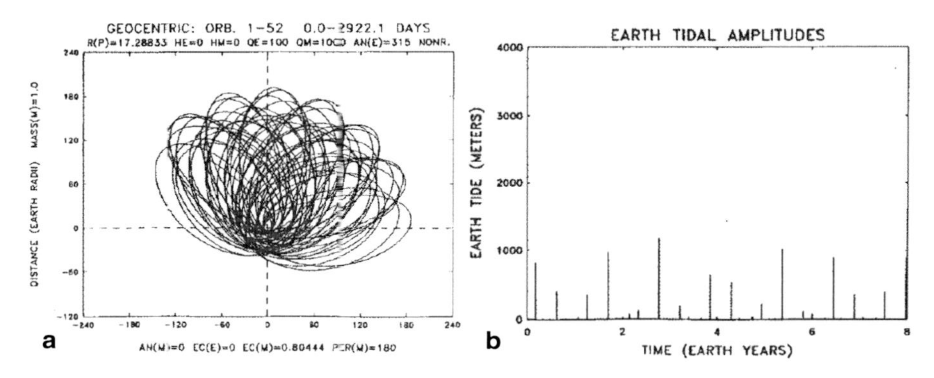

Fig. 5.93 **a** 8 years of orbits ~4347 years after capture. **b** 8 years of rock tidal amplitudes of Earth ~4347 years after capture

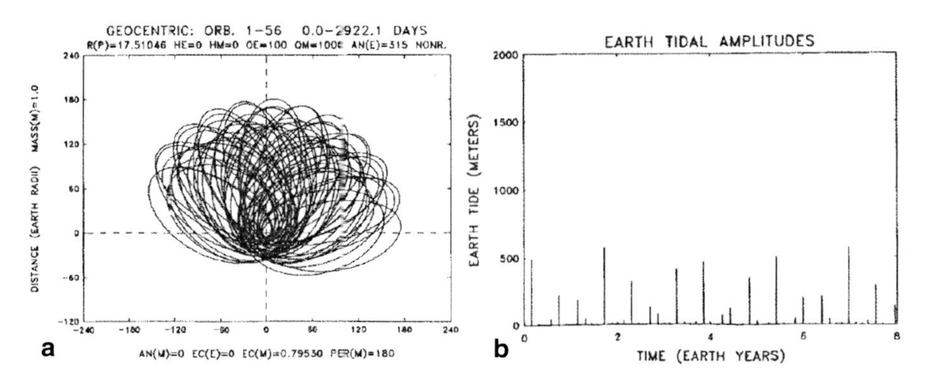

Fig. 5.94 **a** 8 years of orbits ~18.59 Ka after capture. **b** 8 years of rock tidal amplitudes on Earth ~18.59 Ka after capture

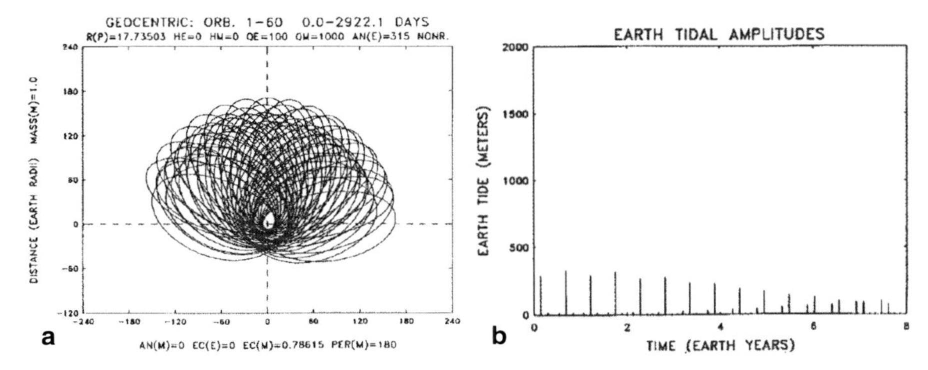

Fig. 5.95 8 years of orbits ~61.04 Ka after capture. **b** 8 years of rock tidal amplitudes on Earth ~61.04 Ka after capture

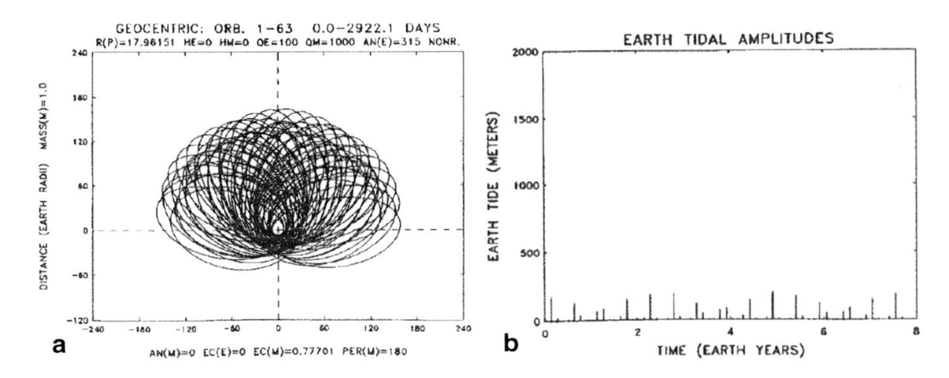

Fig. 5.96 **a**. 8 years of orbits ~160 Ka after capture. **b** 8 years of rock tidal amplitudes on Earth ~160 Ka after capture

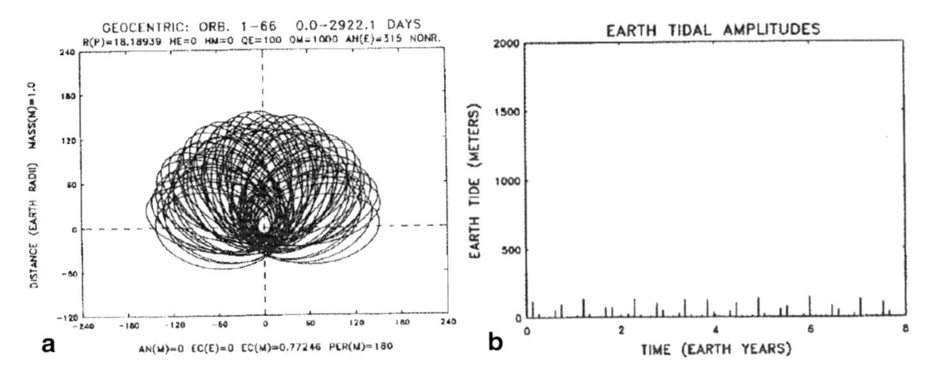

Fig. 5.97 **a** 8 years of orbits ~347 Ka after capture. **b** 8 years of rock tidal amplitudes on Earth ~347 Ka after capture

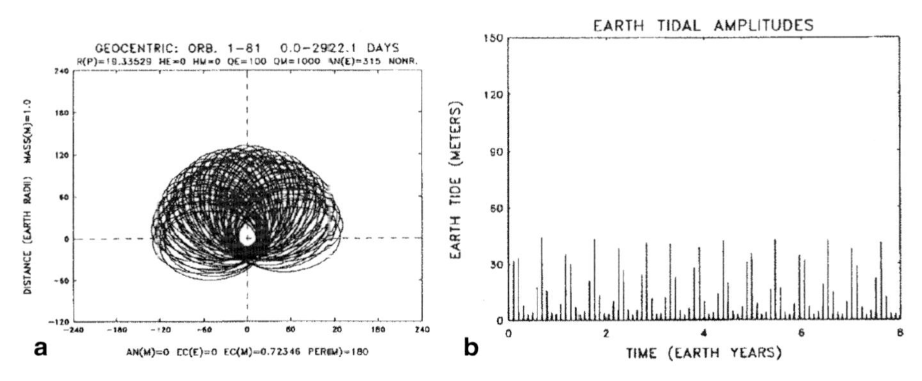

Fig. 5.98 **a** 8 years of orbits ~3.857 Ma after capture. **b** 8 years of rock tidal amplitudes ~3.857 Ma after capture

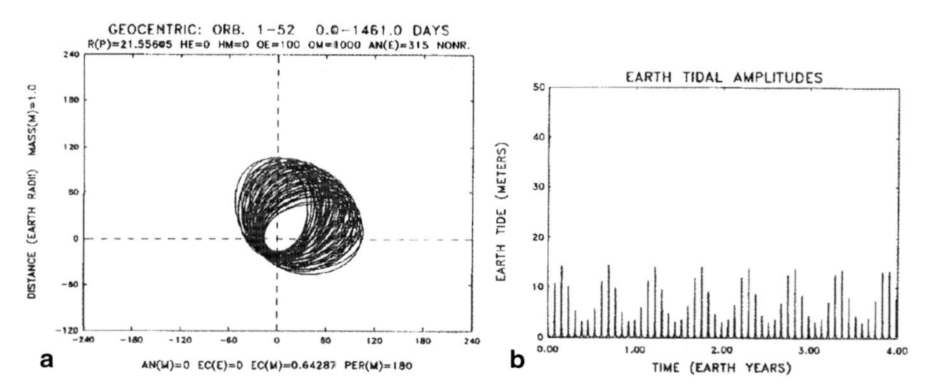

Fig. 5.99 **a** 4 years of orbits ~31.6 Ma after capture. **b** 4 years of rock tidal amplitudes on Earth ~31.6 Ma after capture

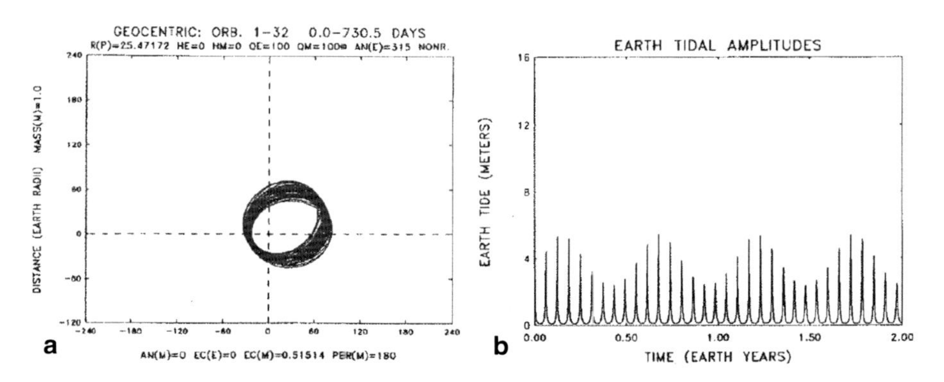

Fig. 5.100 **a** 2 years of orbits ~149 Ma after capture. **b** 2 years of rock tidal amplitudes on Earth ~149 Ma after capture

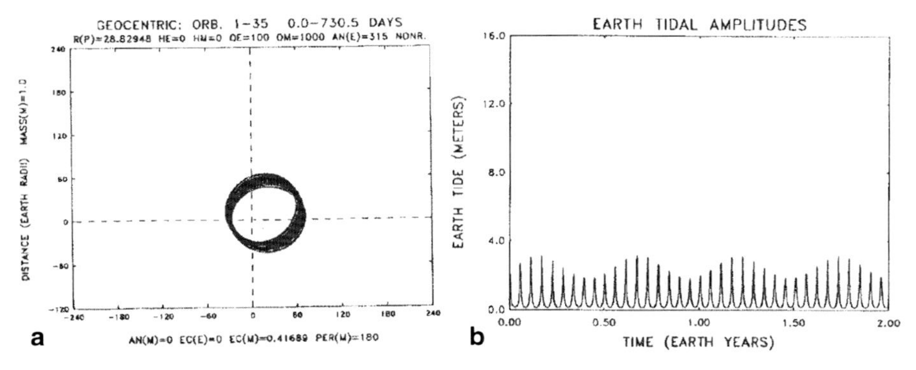

Fig. 5.101 **a** 2 years of orbits ~304 Ma after capture. **b** 2 years of rock tidal amplitudes on Earth ~304 Ma after capture

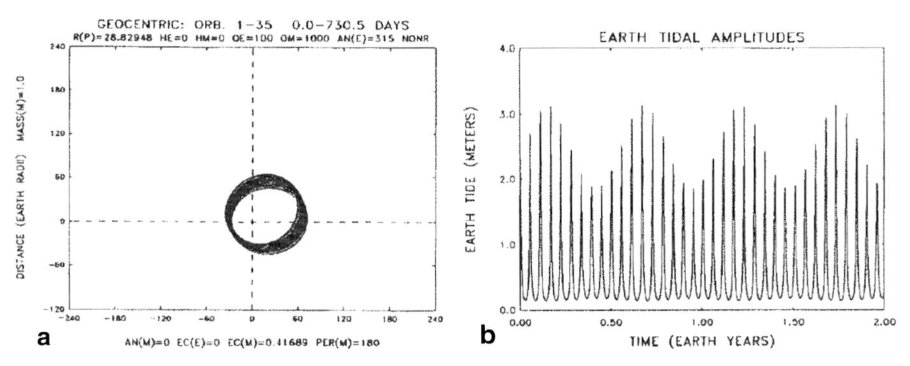

Fig. 5.102 **a** 2 years of orbits ~485 Ma after capture. **b** 2 years of rock tidal amplitudes on Earth ~485 Ma after capture

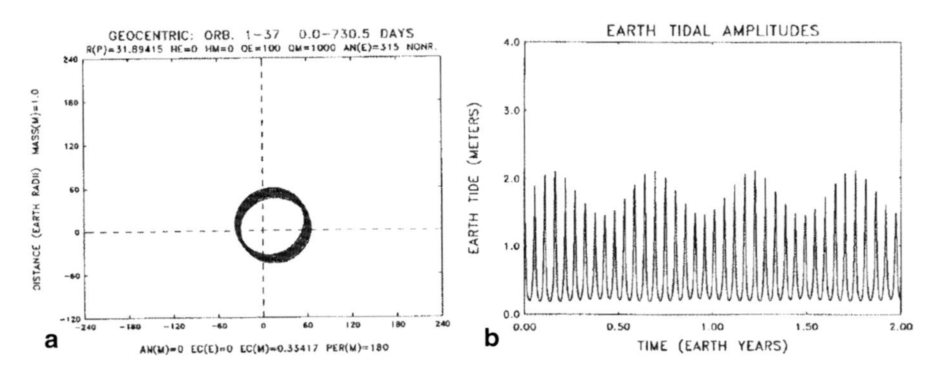

Fig. 5.103 **a** 2 years of orbits ~671 Ma after capture. **b** 2 years of rock tidal amplitudes on Earth ~671 years after capture

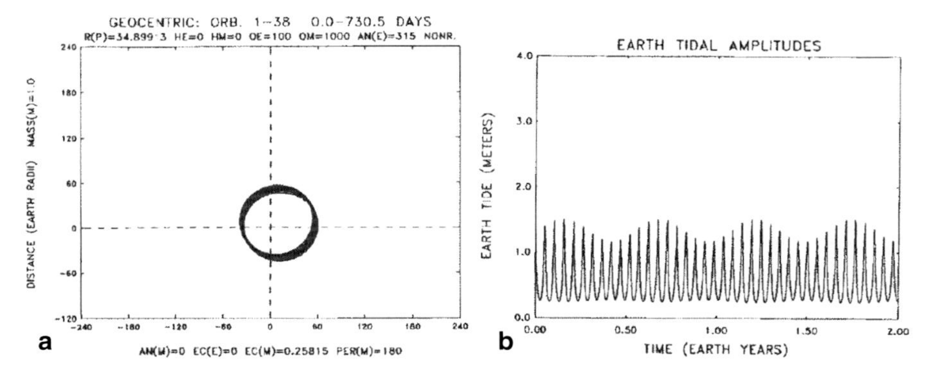

Fig. 5.104 **a** 2 years of orbits ~863 Ma after capture. **b** 2 years of rock tidal amplitudes ~863 Ma after capture

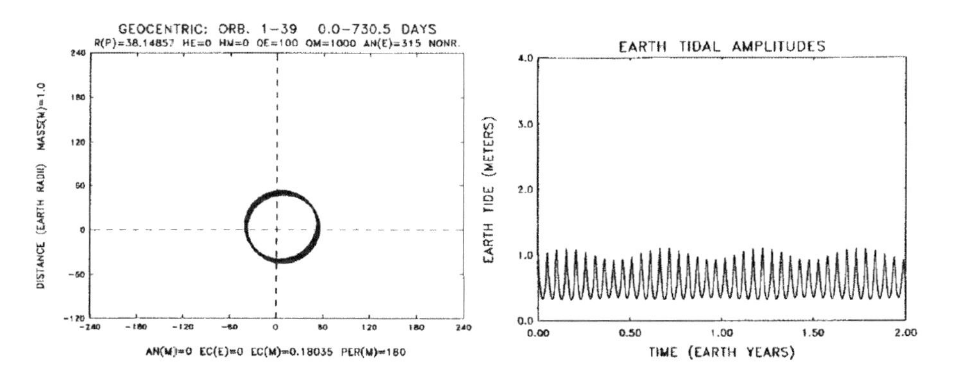

Fig. 5.105 **a** 2 years of orbits ~1030 Ma after capture. **b** 2 years of rock tidal amplitudes on Earth ~1030 Ma after capture

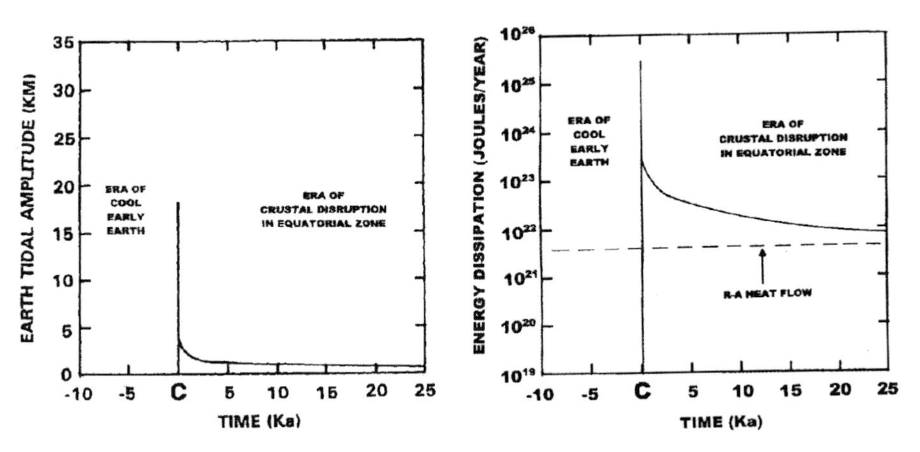

Fig. 5.106 **a** Plot of Earth tidal amplitudes (Capture to 25 Ka after capture). **b** Plot of energy dissipation in Earth. (Capture to 25 Ka after capture)

References

Albarede F (2009) Volatile accretion history of the terrestrial planets and dynamic implications. Nature 461:1227–1233
Albarede F, Juteau M (1984) Unscrambling the lead model ages. Geochim Cosmochim Acta 48:207–212
Alexander CM O'D, Bowden R, Fogel ML, Howard KT, Herd CDK, Nittler LR (2012) The provenances of asteroids, and their contributions to the volatile inventories of the terrestrial planets. Science 337:721–723
Amelin YA (2005) A tale of early earth told in zircons. Science 310:1914–1915
Anderson DL (1972) The origin of the Moon. Nature 239:263–265
Anderson DL (1973) The Moon as a high-temperature condensate. The Moon 8:33–57
Anderson DL (1975) On the composition of the lunar interior. J Geophys Res 80:1555–1557
Baadsgaard H, Lambert R St J, Krupicka J (1976) Mineral isotope age relationships in the polymetamorphic Amitsoq gneisses, Godthaab district, West Greenland. Geochim Cosmochim Acta 40:513–537
Bedard JH (2006) A catalytic delamination-driven model for coupled genesis of Archaean crust and sub-continental lithospheric mantle. Geochim Cosmochim Acta 70:1188–1214
Bell EA, Harrison TM, McCulloch MT, Young ED (2011) Early Archean crustal evolution of the Jack hills zircon source terrane inferred from Lu-Hf, $^{207}Pb/^{206}Pb$ and delta ^{18}O systematics of Jack hills zircons. Geochim Cosmochim Acta 75:4816–4829
Besse S, Sunshine JM, Gaddis LR (2014) Volcanic glass signatures in spectroscopic survey of newly proposed lunar pyroclastic deposits. J Geophys Res Planet 119:355–372
Black LP, Gale NH, Moorboth S, Panhurst RJ, McGregor VR (1971) Isotopic dating of very early Precambrian amphibolite facies gneisses from the Godthaab district, West Greenland. Earth Planet Sci Lett 12:245–259
Bostrom RC (2000) Tectonic consequences of earth's rotation. Oxford University Press, London, p 266
Bowin C, Simon B, Wollenhaupt WR (1975) Mascons: a two-body solution. J Geophys Res 80:4947–4955
Cadogan PH (1974) Oldest and largest lunar basin?. Nature 250:315–316
Campins H, Hargrove K, Pinilla-Alonso N, Howell ES, Kelly MS, Licandro J, Mothe-Diniz T, Fernandez Y, Ziffer J (2010) Water ice and organics on the surface of the asteroid 24 Themis. Nature 464:1320–1321
Canup RM (2004a) Dynamics of lunar formation. Annu Rev Astron Astrophys 42:441–475
Canup RM (2004b) Origin of the terrestrial planets and the Earth-Moon system. Phys Today 57(4):56–62
Canup RM (2008) Lunar-forming collisions with pre-impact rotation. Icarus 196:518–538
Chapman CR, Cohen BA, Grinspoon DH (2007) What are the real constraints on the existence and magnitude of the late heavy bombardment? Icarus 189:233–245
Charnoz S, Morbidelli A, Dones L, Salmon J (2009) Did Saturn's rings form during the Late Heavy Bombardment? Icarus 109:413–428
Cisowski SM, Fuller M (1983) Lunar sample magnetic stratigraphy. Abstracts of the 14th lunar and planetary science conference. Lunar and Planetary Institute, Houston, p 115–116
Cisowski SM, Collinson DW, Runcorn SK, Stephenson A, Fuller M (1983) A review of lunar paleointensity data and implications for the origin of lunar magnetism. Proceedings of the 13th lunar and planetary science conference. J Geophys Res Supplement 88:A691–A704
Clark JT, Prange R, Ballester GE, Trauger J et al (1995) HST far-ultraviolet imaging of Jupiter during the impacts of Shoemaker-Levy 9. Science 267:1302–1307
Cloud PE (1968) Atmospheric and hydrospheric evolution on the primitive Earth. Science 160:729–736
Cloud PE (1972) A working model of the primitive earth. Am J Sci 272:537–548

Conel JE, Holstrom GB (1968) Lunar mascons: a near-surface interpretation. Science 162:1403–1404
Conway BA (1986) Stability and evolution of primeval lunar satellite orbits. Icarus 66:324–329
Cooper G, Horz F, Spees A, Chang S (2014) Highly stable meteoritic organic compounds as markers of asteroidal delivery. Earth Planet Sci Lett 385:206–215
Cyr KE, Sears WD, Lunine JI (1998) Distribution and evolution of water ice in the solar nebula: implications for solar system body formation. Icarus 135:537–548
Dalrymple GB, Ryder G (1993) Ar-40/Ar-39 age spectra of Apollo 15 impact melt rocks by laser step-heating and their bearing on the history of lunar basin formation. J Geophys Res 98:13085–13095
Dalrymple GB, Ryder G (1996) Ar-40/Ar-39 age spectra of Apollo 17 highlands breccia samples by laser step-heating and age of the Serenitatis basin. J Geophys Res 101:26069–26084
Darling JR, Moser DE, Heaman LM, Davis WJ, O'Neil J, Carlson R (2013) Eoarchean to Neoarchaen evolution of the Nuvvuagittuq supracrustal belt: new insights from U-Pb zircon geochronology. Am J Sci 313:844–876
Dauphas N (2003) The dual origin of the terrestrial atmosphere. Icarus 165:326–339
Dauphas N (2013) Sulpher from heaven and hell. Nature 501:175–176
Dauphas N, Marty B (2002) Inferences on the nature and mass of Earth's late veneer from noble metals and gases. J Geophys Res 107(E12):5129–5135
Dauphas N, Robert F, Marty B (2000) The late asteroidal and cometary bombardment of earth as recorded in water deuterium to protium ratio. Icarus 148:508–512
Day JMD, Pearson DG, Taylor LA (2007) Highly siderophile element constraints on accretion and differentiation of the Earth-Moon system. Science 515:217–219
Dietz RS (1946) The meteoritic impact origin of the Moon's surface features. J Geol 54:359–375
Donahue TM (1999) New analysis of hydrogen and deuterium escape from Venus. Icarus 141:226–235
Donahue TM, Hoffman JH, Hodges RR Jr, Watson AJ (1982) Venus was wet: a measurement of the ratio of deuterium to hydrogen. Science 216:630–633
Drake MJ, Righter, K (2002) Determining the composition of the Earth. Nature 416:39–44
Farinella P, Gonczi R, Froeschle Ch, Froeschle C (1993) The injections of asteroid fragments into resonances. Icarus 101:174–187
Franz J (1912) Der Mond, 2 edn. Leipzig, BG Teubner, pp 120
Franz J (1913) Nova Acta, Abh. Der Kaiserl. Leop. Deutschen Akademie der naturforscher, 99, 1. Halle
Frei R, Rosing MT (2005) Search for traces of the late heavy bombardment on Earth—Results from high precision chromium isotopes. Earth Planet Sci Lett 236:28–40
Gaddis LR, Staid MI, Tyburczy JA, Hawke BR, Petro NE (2003) Compositional analyses of lunar pyroclastic deposits. Icarus 161:262–280
Gast PW (1972) The chemical composition and structure of the Moon. The Moon 5:121–148
Gault DE, Widekind JA (1978) Experimental studies of oblique impacts. Proceedings of the 9th Lunar and Planetary Science Conference. Pergamon Press, London, pp 3843–3875
Goldich SS, Hedge CE (1974) 3,800-Myr granitic gneiss in south-western Minnesota. Nature 252:467–468
Gottlieb P, Muller PM, Sjogren WL (1969) Nonexistence of large mascons at Mare Marginis and Mare Orientale. Science 166:1145–1147
Griffin WL, Belousova EA, O'Neill C, O'Reilly SY, Malkovets V, Person NJ, Spetsius S, Wilde SA (2014) The world turns over: Hadean-Archean crust-mantle evolution. Lithos 189:2–15
Gustafson JO, Bell JF III, Gaddis LR, Hawke BR, Giguere TA (2012) Characterization of previously unidentified lunar pyroclastic deposits using Lunar Reconnaissance Orbiter camera data. J Geophys Res 117:21. doi:10.1029/2011JE003893-21
Harrison TM (2009) The Hadean crust: evidence from ≥ 4 Ga zircons. Annu Rev Earth Planet Sci 37:479–505
Hartmann WK (1977a) Cratering in the solar system. Sci Am 236(1):84–99
Hartmann WK (1977b) Relative crater production on planets. Icarus 31:260–276
Hartmann WK, Wood CA (1971) Moon: origin and evolution of multi-ring basins. Moon 3:4–78

Head JW (1982) Lava flooding of ancient planetary crusts: geometry, thickness, and volumes of flooded lunar impact basins. Moon Planet 26:61–88

Humayun M, Clayton, RN (1995) Potassium isotope cosmochemistry: genetic implications of volatile element depletion. Geochim Cosmochim Acta 59:2131–2148

Hutsemekers D, Manfroid J, Jehin E, Arpigny C (2009) New constraints on the delivery of cometary water and nitrogen to Earth from the $^{15}N/^{14}N$ isotopic ratio. Icarus 204:346–348

Holsapple KA, Michel P (2006) Tidal disruptions. A continuum theory for solid bodies. Icarus 183:331–348

Holsapple KA, Michel P (2008) Tidal disruptions: II. A continuum theory for solid bodies with strength, with applications to the solar system. Icarus 193:283–301

Jacquet E, Robert F (2013) Water transport in protoplanetary disks and the hydrogen isotope composition of chondrites. Icaurs 223:722–732

Janle P (1981) A crustal gravity model for the Mare Serenitatis—Mare Crisium area of the Moon. J Geophys 49:57–65

Kaula WM (1969) Interpretation of lunar mass concentrations. Phys Earth Planet Inter 2:123–137

Kerr RA (2000) Beating up on a young earth and possibly life (News Focus). Science 290:1677

Koeberl C (2006) Impact processes on the early earth. Elements 2:211–216

Kring DA, Cohen BA (2002) Cataclysmic bombardment throughout the inner solar system 3.9–4.0 Ga. J Geophys Res 107:6. doi:E2. 10.1029/2001JE001529–6

Kunze AWG (1974) Lunar mascons: another model and its implications. Moon 11:9–17

Kunze AWG (1976) Evidence for isostasy in lunar mascon maria. Moon 15:415–419

Labidi J, Cartigny P, Moreira M (2013) Non-chrondritic sulphur isotope composition of the terrestrial mantle. Nature 501:208–211

Lawrence SJ, Stopar JD, Hawke BR, et al (2013) LRO observations of morphology and surface roughness of volcanic cones and lobate lava flows in the Marius Hills. J Geophys Res: Planets 118:615–634

Lena R, Fitzgerald B (2014) On a volcanic construct and a lunar pyroclastic deposit (LPD) in northern Mare Vaporum. Planet Space Sci 92:1–15

Lepland A, van Zuilen MA,Arrhenius G, Whitehouse MJ, Fedo CM (2005) Questioning the evidence for earth's earliest life—Akilia revisited. Geology 33:77–79

Lipskii YuN (1965) Zond-3 photography of the moon's far side. Sky Telesc 30:338–341

Lipskii YuN, Rodionova ZhF (1977) Antipodes on the Moon, in the Soviet-American Conference on Cosmochemistry of the Moon and the Planets. pp 755–761

Lipskii YuN, Pskovskii Yu P, Gurshtein AA, Shevchenko VV, Pospergelis MM (1966) Current problems of the morphology of the moon's surface. Kosmicheskie Issledovaniya 4(6):912–922 (Translated to English at a later date)

Lodders K, Fegley B Jr (1998) The planetary scientist's campanion. Oxford University Press, Oxford, pp 371

Mackin JH (1969) Origin of lunar maria. Geol Soc Am Bull 80:735–748

Malcuit RJ, Byerly GR, Vogel TA, Stoeckley TR (1975) The great-circle pattern of large circular maria: product of an Earth-Moon encounter. Moon 12:55–62

Malcuit RJ, Winters RR, Mickelson ME (1977) Is the Moon a captured body? Abstracts Volume, Eight Lunar Science Conference, Houston, p 608–610

Malcuit RJ, Mehringer, DM, Winters RR (1988) Computer simulation of "intact" gravitational capture a lunar-like body by an Earth-like body: Abstract Volume, Lunar and Planetary Science XIX, Lunar and Planetary Institute, Houston, pp 718–719

Malcuit RJ, Mehringer DM, Winters RR (1989) Numerical simulation of gravitational capture of a lunar-like body by earth. Proceedings of the 19th Lunar and Planetary Science Conference. Lunar and Planetary Institute, Houston, pp 581–591

Malcuit RJ, Mehringer DM, Winters RR (1992) A gravitational capture origin for the Earth-Moon system: implications for the early history of the earth and moon. In: Glover JE, Ho SE (eds) The Archaean. Terrains, processes and metallogeny. Proceedings volume, 3rd international archaean conference, vol 22. Univeristy of Western Australia Publication, Crawley, pp 223–235

Matter A, Guillot T, Morbidelli A (2009) Calculation of the enrichment of the giant planet envelopes during the "late heavy bombardment". Planet Sp Sci 57:816–821
McSween HY Jr (1999) Meteorites and their parent bodies, 2nd edn. Cambridge University Press, Cambridge, p 310
Melosh HJ (1989) Impact cratering: a geologic process. Oxford University Press, London, p 245
Mizuno H, Boss AP (1985) Tidal disruption of dissipative planetesimals. Icarus 63:109–133
Moore HJ (1976) Missile impact craters (White Sands Missile Range, New Mexico) and applications to lunar research. U. S. Geol Surv Prof Pap 812–B:47
Morbidelli A, Chambers J, Lunine JI, Petit JM, Robert F, Valsecchi GB, Cyr KE (2000) Source regions and time scales for delivery of water to earth. Meteorit Planet Sci 35:1309–1320
Mueller PA, Wooden JL, Nutman AP (2009) U-Pb ages (3.8–2.7 Ga) and Nd isotope data from the newly identified Eoarchean Nuvvuagittuq supracrustal belt, Superior Craton, Canada. Geol Soc Am Bull 121:150–163
Muller PM, Sjogren WL, Wollenhaupt WR (1973) Lunar shape via the Apollo laser altimeter. Science 179:275–278
Mutch TA (1972) Geology of the moon: a stratigraphic view. Princeton University Press, Princeton, p 391
Naeraa T, Schersten A, Rosing MT, Kemp AIS, Hoffman JE, Kokfelt TF, Whitehouse MJ (2012) Hafnium isotope evidence for a transition in the dynamics of continental grouth 3.2 Gyr ago. Nature 485:627–630
Nash DB (1963) On the distribution of lunar maria and the synchronous rotation of the Moon. Icarus 1:372–373
Nebel O, Rapp RP, Yaxley GM (2014) The role of detrital zircons in Hadean crustal research. Lithos 190–191:313–327
Nuth JA (2008) What was the volatile composition of the planetesimals that formed the earth? Earth Moon Planets 102:435–445
Nutman AP, Bennett VC, Friend CRL, Hidaka H, Yi K, Ryeollee S, Kamiichi T (2013) The Itsaq gneiss complex of Greenland: episodic 3900 to 3660 Ma juvenile crust formation and recycling in the 3660 to 3600 Ma Isukasian orogeny. Am J Sci 313:877–911
Nyquist LE, Shih CY (1992) The isotopic record of lunar volcanism. Geochim Cosmochim Acta 56:2213–2234
O'Neil J, Francis D, Carlson RW (2011) Implications of the Nuvvuagittuq greenstone belt for the formation of earth's early crust. J Petrol 52:985–1009
O'Neil J, Carlson RW, Paquette J-L, Francis D (2012) Formation age and metamorphic history of the Nuvvuagittuq greenstone belt. Precambrian Res 220–221:23–44
O'Neil C, Debaille V, Griffin W (2013a) Deep earth recycling in the Hadean and constraints on surface tectonics. Am J Sci 313:912–932
O'Neil J, Boyet M, Carlson RW, Paquette J-L (2013b) Half a billion years of reworking of Hadean mafic crust to produce the Nuvvuagittuq eoarchean felsic crust. Earth Planet Sci Lett 379:13–25
Owen T, Bar-Nun A (1998) From the interstellar medium to planetary atmospheres via comets: Faraday Discussions No. 109:453–462
Pearce JA (2014) Geochemical fingerprinting of the Earth's oldest rocks. Geology 42:175–176
Pidgeon RT, Nemchin AA, van Bronswijk W, Geisler T, Meyer C, Compston W, Williams JS (2007) Complex history of a zircon aggregate from lunar breccia 73235. Geochimi Cosmochim Acta 71:1370–1381
Raymond SN, Schlichting HE, Hersant F, Selsis F (2013) Dynamical and collisional constraints on a stochastic late veneer on the terrestrial planets. Icarus 226:671–681
Rivkin AS, Emery JP (2010) Detection of ice and organics on an asteroidal surface. Nature 464:1322–1323
Rizo H, Boyet M, Blichert-Toft J, Rosing MT (2013) Early mantle dynamics inferred from ^{142}Nd variations in Archean rocks from southwest Greenland. Earth Planet Sci Lett 377–378:324–335
Robert F (2001) The origin of water on earth. Science 293:1056–1058
Runcorn SK (1983) Lunar magnetism, polar displacements and primeval satellites in the Earth-Moon system. Nature 304:589–596

Russell CT (1980) Planetary Magnetism. Rev Geophys Space Phys 18:77–106
Ryder G (1990) Lunar samples, lunar accretion and the early bombardment of the Moon. Trans Am Geophys Union 71(10):313, 322–323
Ryder G (2002) Mass flux in the ancient Earth-Moon system and benign implications for the origin of life on earth. J Geophys Res. 107:14. doi:(E-4) 10.1029/2001KE 001583
Ryder G, Taylor GJ (1976) "Pre-mare" volcanism: Abstracts of the 7th lunar and planetary science conference, Lunar and Planetary Institute, Houston, pp 755–757
Saager R, Koppel V (1976) Lead isotopes and trace elements from sulfides of Archean greenstone belts in South Africa—a contribution to the knowledge of the oldest known mineralizations. Econ Geol 71:44–57
Salmeron R, Ireland TR (2012) Formation of chondrules in magnetic winds blowing through the proto-asteroid belt. Earth Planet Sci Lett 327-328:61–67
Schoenberg R, Kamber BS, Collerson KD, Moorbath S (2002) Tungsten isotope evidence from ~3.8 Gyr metamorphosed sediments for early meteorite bombardment of the Earth. Nature 418:403–405
Schultz PH (1976) Moon morphology. University of Texas press, Austin, p 626
Shultz PH, Gault DE (1990) Prolonged global catastrophes from oblique impacts. In: Sharpton VL, Ward PD (eds) Global catastropies in earth history. Geological Society of America Special Paper 247, Boulder, CO, pp 239–261
Schultz PH, Gault DE (1991) Impact decapitation from laboratory to basin scales: Abstracts, 22nd lunar and planetary science conference, Lunar and Planetary Institute, Houston, pp 1195–119
Schultz PH, Lutz-Garahan AB (1982) Grazing impacts on Mars: a record of lost satellites: Proceedings of the 13th lunar and planetary science conference, Lunar and Planetary Institute, Houston, pp A84–A96
Shu FH, Shang H, Glassgold AE, Lee T (1997) X-rays and fluctuating X-Winds from protostars. Science 277:1475–1479
Shu FH, Shand H, Gounelle M, Glassgold AE, Lee T (2001) The origin of chondrules and refractory inclusions in chondritic meteorites. Astrophys J 548:1029–1050
Singer SF (1970) How did Venus loose its angular momentum? Science 170:1196–1198
Sjogren WL, Wollenhaupt WR (1973) Lunar shape via the Apollo laser altimeter. Science 179:275–278
Smoluchowski R (1973a) Lunar tides and magnetism. Nature 242:516–517
Smoluchowski R (1973b) Magnetism of the moon. Moon 7:127–131
Spudis PD, McGovern PJ, Kiefer, WS (2013) Large shield volcanoes on the Moon. J Geophys Res: Planets 118:1063–1081
Stacey JS, Kramers JD (1975) Approximation of terrestrial lead isotope evolution by a two-stage model. Earth Planet Sci Lett 26:207–221
Stuart-Alexander DE, Howard KA (1970) Lunar maria and circular basins—a review. Icarus 12:440–456
Taylor SR (1975) Lunar science: A post-Apollo view. Pergamon Press, New York, p 372
Taylor SR (1982) Planetary science: a lunar perspective. Lunar and Planetary Institute, Houston, p 481
Taylor SR (1998) Destiny or chance. Our solar system and its place in the cosmos. Cambridge University Press, Cambridge, p 229
Taylor SR (2001) Solar system evolution: a new perspective, 2nd edn. Cambridge University Press, Cambridge, p 460
Tera F, Papanastassiou DA, Wasserburg GJ (1974) Isotopic evidence for a terminal lunar cataclysm. Earth Planet Sci Lett 22:1–21
Trail D, Mojzsis SJ, Harrison TM (2007) Thermal events documented in Hadean zircons by ion microprobe depth profiles. Geochim Cosmochim Acta 71:4044–4065
Turner S, Rushmer T, Reagan M, Moyen J-F (2014) Heading down early on? Start of subduction on earth. Geology 42:139–142
Ulrich GE, Hodges CA, Muehlberger WR (eds) (1981) Geology of the Apollo 16 area, Central Lunar Highlands. USGS Professional Paper 1048, 539 p. (12 plates)

Valley JW (2006) Early earth. Elements 2:201–204
Valley JW, Peck WH, King EM, Wilde SA (2002) A cool early Earth. Geology 30:351–354
Wang Z, Becker H (2013) Ratios of S, Se and Te in the silicate earth require a volatile-rich late veneer. Nature 499:328–331
Wasserburg GJ, Papanastassiou DA, Tera F, Huneke JC (1977) Outline of a lunar chronology. Philos Trans R Soc Lond A 285:7–22
Wetherill GW (1981) Nature and origin of basin-forming projectiles, In: Schultz PH, Merrill RB (eds) Multi-ring basins: proceedings of the lunar and planetary science conference. 12A:1–18
Wichman RW, Schultz PH (1994) The Crisium basin. Implications of an oblique impact for basin ring formation and cavity collapse, In: Dressler BO, Greive RAF, Sharpton VL (eds) Large meteorite impacts and planetary evolution. Geological Society of America Special Paper 293, pp 61–72
Wilhelms DE (1987) The geologic history of the moon: U. S. Geological Survey Professional Paper 1348, p 302
Wilhelms DE (1993) To a rocky moon: a geologist's history of lunar exploration. The University of Arizona Press, Tucson, p 477
Windley BF (1984) The evolving continents, 2nd edn. Wiley, New York, p 399
Zeh A, Gerdes A, Klemd R, Barton JM Jr (2008) U-Pb and Lu-Hf isotope record of detrital zircon grains from the Limpopo belt—evidence for crustal recycling at the Hadean to early-Archean transition. Geochim Cosmochim Acta 72:5304–5329
Zeh A, Stern RA, Gerdes, A (2014) The oldest zircons of Africa—their U-Pb-Hf-O isotope and trace element systematics and implications for Hadean to Archean crust-mantle evolution. Precambrian Res 241:203–230

Lunar Geologic Maps Cited

Lucchitta BK (1978) Geologic map of the north side of the moon. United States Geological Survey, Miscellaneous Investigations Series, Map I-1062
Scott DH, McCauley JF, West MN (1977) Geologic map of the west side of the moon. United States Geological Survey, Misccellaneous Investigations Series, Map I-1034
Stuart-Alexander D (1978) Geologic map of the central far side of the moon. United State Geological Survey, Miscellaneous Investigations Series, Map I-1047
Titley SR (1967) Geologic map of the Mare Humorum Region of the moon. Unites States Geological Survey, Miscellaneous Investigations Series, Map I-495
Wilhelms DE, El-Baz F (1977) Geologic map of the east side of the moon. United States Geological Survey, Miscellaneous Investigations Series, Map I-948
Wilhelms DE, McCauley JF (1971) Geologic map of the near side of the moon. United States Geological Survey, Miscellaneous Investigations Series, Map I-703

Lunar Charts Cited

Lunar Chart (1970) LPC-1, 1st edn.
Lunar east side chart (1970) LMP-1, 2nd edn.
Lunar far side chart (1970) LMP-2, 2nd edn.

Chapter 6
Origin and Evolution of the Venus-Adonis System: A Retrograde Gravitational Capture Model

> *"... we postulate capture of a moonlike object from an initially retrograde orbit: it would despin Venus... and transform the planet's rotational kinetic energy into internal heat, which would lead to volcanism and the liberation of large amounts of volatiles. The moon would disappear by crashing into the surface of Venus." From Singer 1970, p. 1196.*

The *history* of a terrestrial planet, in my opinion, is very important in determining the habitability of that planet. But even more important is the history of the planet-satellite system because interactions with a large satellite is a major control over (1) *the rotation rate of the planet* and (2) *the tidal regime of the planet*. If a planet has some water on the surface, then the tidal regime can be very important for the evolution of life forms because of the systematic movements of ocean water especially in the equatorial zone of the planet.

In Chap. 4 of this book I concentrated on *a prograde gravitational capture model of a lunar mass planetoid by an Earth-like planet and the subsequent orbit evolution of the system*. I refer to this as *A BENIGN ESTRANGEMENT SCENARIO*. The resulting system is a planet-satellite system very similar to our own—a very habitable planet. In the case of a Venus-like planet, the satellite is captured in the opposite direction to the rotation of the planet and the resulting terrestrial planet has basic physical characteristics of planet Venus (i.e., mass, density, present surface and atmospheric conditions, and present rotation rate). This model features *retrograde capture of a 0.5 moon-mass satellite and the subsequent evolution of the system*. I intend to show that the resulting planet-satellite pair could yield a habitable planet for about 3.5 Ga (from about 4.2 to 0.7 Ga). But the satellite is forever orbiting closer and closer to the planet as well as despinning the planet. Eventually the satellite orbits within the Roche limit for a solid body, breaks up in orbit, and coalesces with the surface of the planet. I refer to this as *A FATAL ATTRACTION SCENARIO*.

In general, habitable conditions for any terrestrial planet-satellite combination would have a limited lifetime. Depending on the details, the habitable conditions on planet Venus could have been more or less favorable than those on Earth today.

Robert J. Malcuit, *The Twin Sister Planets Venus and Earth*,
DOI 10.1007/978-3-319-11388-3_6

Details and happenstance play a part in determining the favorability rating. We must keep in mind that the solar radiation output after the Sun reaches the Main Sequence of evolution is estimated to be about 20% lower than at present. This condition should be taken into consideration when modeling the possibilities for habitable conditions on the twin sister planets, Venus and Earth.

6.1 Origin of the Concept of Retrograde Capture of a Lunar-Like Body by Planet Venus

The basic concept of retrograde capture of a satellite for Venus was published in a paper by Singer (1970) (see also the introductory quote of this chapter). The title of the article is: "HOW DID VENUS LOSE ITS ANGULAR MOMENTUM". The abstract is important for the critique that follows.

Abstract

Venus now has a retrograde and negligible spin, but it very likely started with a typical planetary spin: prograde and with a 10- to 20-h period. The usual assumed mechanism of solar tidal friction is quite insufficient to remove this angular momentum. Instead, we postulate capture of a moonlike object from an initially retrograde orbit: it would despin Venus and suddenly transform the planet's rotational kinetic energy into internal heat, which would lead to volcanism and the liberation of large amounts of volatiles. The moon would disappear by crashing into the surface of Venus.

My critique of Singer's model is: (1) he did not demonstrate that gravitational capture of a moon-like satellite is physically possible, (2) a moon-mass satellite causes the planet-satellite system to evolve too rapidly [there is a need to explain how Venus got completely resurfaced about 1.0–0.5 Ga ago (but that "fact" was not known by Singer in 1970)], and (3) the long timescale is important and a 0.5 moon-mass object gives a long timescale for the tidal interaction with Venus as well as for the eventual coalescence of the satellite with planet Venus. Nonetheless, Singer (1970) outlined the essential features of a retrograde satellite capture model.

McCord (1966, 1968) did calculations on the evolution of retrograde satellites orbiting planets with prograde rotation. He outlined a generalized scenario for despinning planet Venus as well as for the post-capture evolution of the retrograde orbit of Triton around planet Neptune. McCord (1966, 1968) did not speculate on the mechanism(s) of the capture process. Counselman (1973) also presented a generalized treatment of the tidal evolution of planet-satellite pairs. His conclusion was that the final result depends mainly on the planet-satellite mass ratio, the initial rotation rate of the planet, and the orbital angular velocity of the satellite.

6.2 Some Facts to be Explained by a Successful Model

Listed below are a set of facts and generally accepted interpretations of facts that need to be explained by a successful model for the evolution of planet Venus to its present condition.

1. the retrograde sidereal rotation of Venus (−243 earth days/rotation)
2. the very low obliquity of planet Venus (~3°)
3. the basaltic resurfacing of the planet somewhere between 1.0 and 0.5 Ga ago (Herrick 1994)
4. the origin of Beta Regio and other highland areas on Venus
5. the somewhat uniform cratering pattern on the surface of the planet (i.e., a spherical distribution of impact craters with all surfaces about the same age)
6. the dense carbon dioxide atmosphere of planet Venus [~90 atmospheres of pressure; there is about 1.6 times more carbon dioxide in the atmosphere of Venus than there is in a combination of the oceans, atmosphere, and crustal rocks on Earth at the present time (Cloud 1972, 1974)].

Proposed Name for a Phantom Satellite for Planet Venus

Adonis was the product of an incestuous relationship between a father and daughter. He turned out to be a beautiful child and a very attractive young man. Once Aphrodite (Venus) saw him she fell in love with him instantaneously. Well, another women, Persephone, was interested in Adonis also. Zeus then held a hearing about who gets Adonis. Since both Persephone and Aphrodite wanted Adonis, Zeus decided that they should share Adonis, each getting Adonis for one-half of the year. Aphrodite (Venus) was not happy with this arrangement and essentially crowded the other women out of the relationship.

Adonis died young as a result of a boar-hunting accident and Aphrodite was very near to the scene when it happened. So Adonis got killed and Aphrodite (Venus) was forever saddened by the death of Adonis (Summarized from Rosenzweig 2004).

It does sound a bit like a *FATAL ATTRACTION SCENARIO*!

6.3 Place of Origin of Adonis and Sibling Planetoids and the Original Rotation Rate of Planet Venus

Adonis is just another Vulcanoid planetoid that formed from accreted CAI material that resulted from the processing of proto-solar cloud material during the X-Wind era. Venus is formed from whatever material was in the accretion torus of Venus. I accept the analysis of MacDonald (1963, 1964) for the Angular Momentum Density vs. Mass for the planets of the Solar System (see Chap. 1, Fig. 4). In this analysis the initial rotation rate of Venus is ~ 13.5 h/day. Lissauer and Kary (1991) concluded that these rotations rates can be justified if moderate sized planetesimals accreted from somewhat elliptical orbits to form the terrestrial planets. This is referred to as the "orderly component of planetary rotation". Since Earth and Venus are considered twin sisters, let us treat their early histories similarly. I am assuming "cool early earth" conditions on Venus before Adonis comes onto the scene. Thus, after accretion, a fairly thick "stagnant lid" basaltic crust with TTG enclaves developed on Venus over a period of about 300 Ma (i.e., from 4.56 to 4.26 Ga).

6.4 Migration History of Adonis and Sibling Planetoids

A migration history for the Vulcanoids was presented in Chap. 4. Adonis was included in this group. For model purposes my guesstimate is that Adonis is in a Venus-like heliocentric orbit about 300 Ma after formation. It probably undergoes a differentiation history very similar to Luna and probably accumulates a multi-kilometer thick anorthositic crust that may have had a magnetic signature of some sort. The anorthositic crust may be important later in this purported history of the Venus-Adonis system. Once Adonis is in a near Venus-like orbit then the stage is set for close encounters that may result in gravitational capture.

6.5 Gravitational Capture of Adonis and the Subsequent Orbit Circularization—A Two-Body Analysis

For a two-body analysis of capture, prograde and retrograde capture have an equal chance of occurrence. However, in the three-body numerical simulations, prograde capture is nearly impossible because of the geometry of Stable Capture Zones for the Venus-Adonis system. So for this two-body analysis, I will use a retrograde capture scheme.

6.5.1 Retrograde Capture of a 0.5 Moon-Mass Planetoid from a Co-Planar, Venus-Like Orbit

Figure 6.1 shows a scale sketch of retrograde capture of a planetoid by Venus. The maximum size of a stable post-capture orbit is about 150 venus radii, the approximate limit of the Hill sphere for Venus. Equation 6.1 shows the factors involved in the capture of a planetoid from a venus-like heliocentric orbit into a venocentric orbit. Table 6.1 shows the numerical results for energy dissipation for placing a 0.5 moon-mass planetoid into venocentric orbits for major axes of different dimensions.

Equation 6.1 is for capture of a planetoid from a heliocentric orbit into a planetocentric orbit. The first term relates to the kinetic energy of the planetoid, in this case Adonis, when the orbit of Adonis is not coincident with the orbit of the Venus.

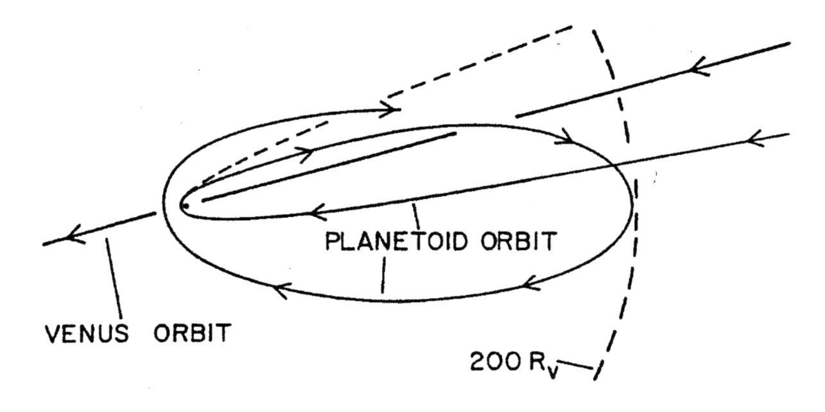

Fig. 6.1 Scale sketch of retrograde capture of a 0.5 moon-mass planetoid into the largest co-planar venocentric orbit possible

Table 6.1 Results of calculations for capture into co-planar venocentric orbits of different dimensions when v_∞ is set to 0

Major axis (R_v)	Energy necessary for capture
200	$E_{capture}$ = 0 + (0.987 E28 Joules) = 0.987 E28 Joules
150	$E_{capture}$ = 0 + (1.316 E28 Joules) = 1.316 E28 Joules
100	$E_{capture}$ = 0 + (1.973 E28 Joules) = 1.973 E28 Joules
50	$E_{capture}$ = 0 + (3.946 E28 Joules) = 3.946 E28 Joules

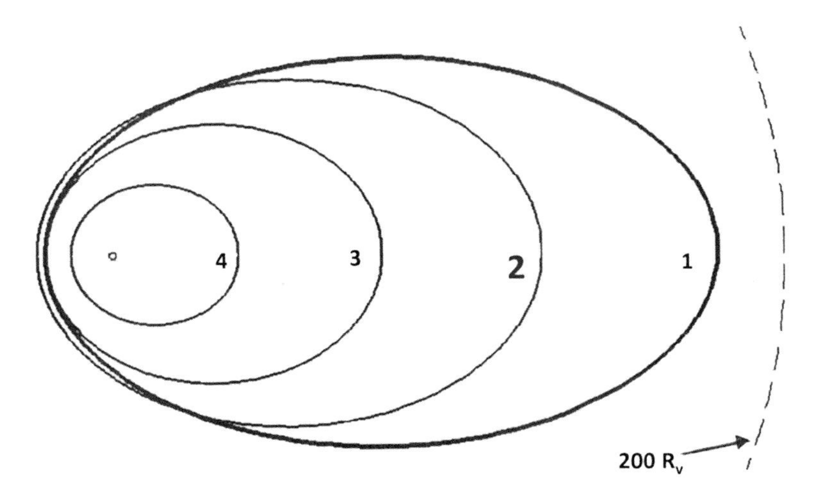

Fig. 6.2 Plot of potential venocentric orbits resulting in capture that relate to values in Table 6.1

Table 6.2 Energy dissipation necessary for capture from a heliocentric orbit that is smaller than the orbit of the planet when v_∞ has a value but the venocentric orbit is the largest stable major axis possible (~150 R_v)

Ecc. (%)	v_∞ (km/s)	Energy for capture
~0.015	0.25	$E_{capture}$ = 0.115 E28 Joules—(−1.316 E28 Joules) = 1.431 E28 Joules
~0.026	0.50	$E_{capture}$ = 0.459 E28 Joules—(−1.316 E28 Joules) = 1.775 E28 Joules
~0.040	0.75	$E_{capture}$ = 1.034 E28 Joules—(−1.316 E28 Joules) = 2.350 E28 Joules
~0.054	1.00	$E_{capture}$ = 1.838 E28 Joules—(−1.316 E28 Joules) = 3.154 E28 Joules

The second term is the potential energy of the resulting venocentric orbit. Figure 6.2 show a graphic representation of the orbits for Table 6.1,

Equation 6.1:

$$E_{capture} = 0.5M_{pl}v_{¥}^{2} - \left(\frac{-G\,M_v M_{pl}}{2a}\right) \tag{6.1}$$

Where M_{pl} is the mass of the planetoid, v_∞ is the velocity of the planetoid at infinity, G is the gravitational constant, M_v is the mass of Venus, and 2a is the major axis of the planetocentric orbit.

If the planetocentric orbit of the planetoid is not coincident with the heliocentric orbit of the planet, then the first term of Eq. 6.1 is non-zero. Table 6.2 gives some values for v_∞ and the associated pre-encounter heliocentric orbit eccentricities of the planetoids. Figure 6.3 gives an orbital representation of pre-encounter heliocentric orbits of the encountering bodies.

I note here that all four PARADOXES associated with the gravitational capture model that were discussed in Chap. 4 apply to the Venus-Adonis analysis. The main paradox is that the encountering planetoid must dissipate well over 90 % of the energy for its own capture. Discussion of Eqs. 6.2 and 6.3 will illustrate this point.

Equation 6.2 is *for energy storage in Venus* during one close encounter. The energy necessary for capture into an orbit with a major axis of 150 venus radii is ~1.316 E28 Joules. As demonstrated in Chap. 4 the closeness of the encounter is important here because of the $1/r^6$ dependence for energy dissipation. Using h (venus) = 0.7 and Q (venus) = 100, the energy that is stored and subsequently dissipated is *grossly insufficient* for capture. Table 6.3 gives some numerical results for single encounters at four different distances of closest approach. With the deformation parameters given above even a slightly grazing encounter, when body deformation is considered, at 1.25 venus radii dissipates only about 1/1000 of the energy necessary for capture. So capture by energy dissipation within the planet, as Singer (1970) suggested for his model, is very improbable.

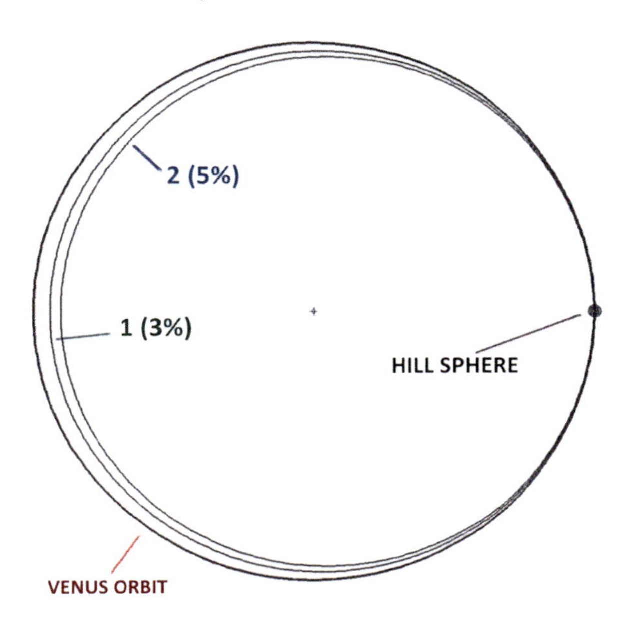

Fig. 6.3 Graphical representation of some encounter conditions in the range of values listed in Table 6.2. The energy necessary for capture increases rapidly as the v_∞ increases. That is why successful gravitational capture by tidal energy dissipation can only occur from orbits that are within ~3 % difference in eccentricity from that of the planet. (Examples above: 3 % difference in eccentricity results in v_∞ ~ 0.53 km/s velocity at infinity; 5 % difference in eccentricity results in v_∞ ~ 0.93 km/s velocity at infinity)

Table 6.3 Values for energy dissipated in a Venus-like planet during a single encounter

Pericenter radius	Energy stored
For 1.25 venus radii (a grazing encounter)	= 1.406 E25 Joules
For 1.30 venus radii (a non-grazing encounter)	= 1.110 E25 Joules
For 1.35 venus radii (a non-grazing encounter)	= 0.886 E25 Joules
For 1.40 venus radii (a non-grazing encounter)	= 0.712 E25 Joules

Equation 6.2:

$$E_{stored\text{-}venus} = \frac{3\,G\,M_{pl}^{\;2} R_v^{\;5} h_v}{5 r_p^{\;6}} \tag{6.2}$$

where G is the gravitational constant, M_{pl} is the mass of the planetoid, R_v is the radius of Venus, h_v is the displacement Love number of Venus, and r_p is the distance of separation of the centers of Venus and the planetoid (Peale and Cassen 1978).

Equation 6.3 is for energy storage and subsequent dissipation *within the planetoid* during one close encounter. The energy necessary for capture into an orbit with a major axis of 150 venus radii is ~ 1.316 E28 Joules. Using h (Adonis) = 0.5 and Q (Adonis) = 1, a slightly grazing encounter at 1.25 venus radii is sufficient for capture. Table 6.4 gives some numerical results for single encounters at four different distances of closest approach. However, an encounter with a periven passage of about 1.40 venus radii will dissipate only about 50 % of the energy necessary for capture. So the case for a two-body analysis of gravitational capture of a reasonably deformable planetoid for planet Venus is very similar to the case for gravitational

Table 6.4 Values for energy dissipation in Adonis during a single encounter

Pericenter radius	Energy stored
For 1.25 venus radii (a grazing encounter)	= 1.277 E28 Joules
For 1.30 venus radii (a non-grazing encounter)	= 1.009 E28 Joules
For 1.35 venus radii (a non-grazing encounter)	= 0.805 E28 Joules
For 1.40 venus radii (a non-grazing encounter)	= 0.647 E28 Joules

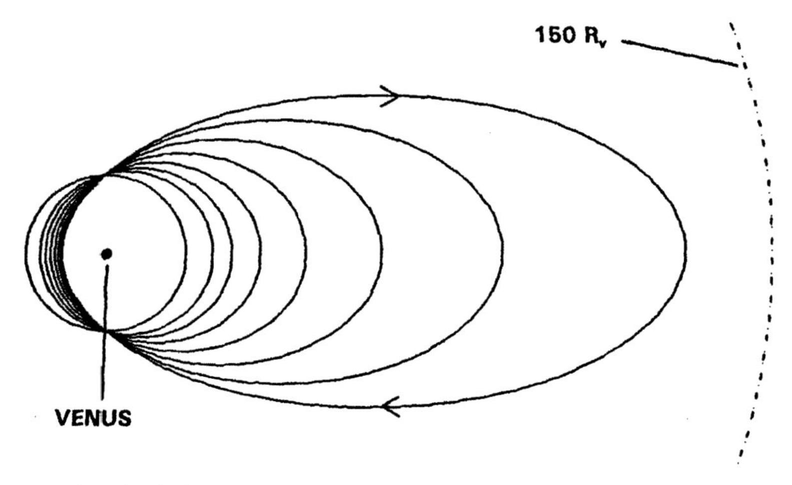

Fig. 6.4 Series of elliptical orbits illustrating the orbit circularization sequence for the Venus-Adonis system. Since no angular momentum is exchanged during this orbit circularization, the angular momentum of the large elliptical orbit and the final circular orbit are identical

capture of a reasonably deformable planetoid for Earth; the numbers are marginal but not very convincing.

Equation 6.3:

$$E_{\text{stored-planetoid}} = \frac{3\,G\,M_v{}^2 R_{pl}{}^5 h_{pl}}{5 r_p{}^6} \tag{6.3}$$

where G is the gravitational constant, M_v is the mass of Venus, R_{pl} is the radius of the planetoid, h_{pl} is the displacement Love number of the planetoid, and r_p of the distance of separation between the centers of the planetoid and the Venus-like planet.

6.5.2 *Post-Capture Orbit Circularization Era*

Figure 6.4 shows a series of venocentric orbits for a post-capture orbit circularization sequence assuming (for simplicity) no change in angular momentum during circularization. The result is that the small circular orbit has the same angular momentum content as the large elliptical orbit. The calculated time-scale for circularization is relatively short and the tidal activity associated with the orbit circularization is significant during periven passages and probably causes recycling of the planetary crust in the equatorial zone of the planet.

6.5.3 *Circular Orbit Evolution*

McCord (1966, 1968) and Counselman (1973) present some generalized calculations for the evolution of satellites of various masses in retrograde orbits tidally interacting with planets with prograde rotation. In this section I am interested mainly in the effects of this circular orbit evolution on the crust-mantle system of the planet.

Once the elliptical orbit is circularized, then a lengthy circular orbit evolution ensues. The planet is rotating in the prograde direction and the satellite is revolving in the retrograde direction. Thus, there is a cancellation of angular momentum and the result is that the prograde planetary rotation rate decreases as the retrograde orbit of the satellite decreases. Figure 6.5 shows the early stages of evolution from 18 venus radii to 10 venus radii and Fig. 6.6 shows the circular orbit evolution from 10 venus radii to 2 venus radii. The tidal regime for the planet gets more severe as the satellite gets closer to the planet in a circular orbit (see Table 6.5 for the tidal and orbital parameters of the system). At 2 venus radii the orbit of the satellite is well within the classical Roche limit where a liquid body would begin to disintegrate by necking along the sub-venus point and the anti-sub-venus point. But a solid body would not begin to lose mass by gravitational disintegration until it is well within the Roche limit for a Solid Body at about 1.6 venus radii. *But this sketch of the orbital evolution is not yet over*. The satellite will eventually separate into three major parts, each of which is in its own orbit. The three newly born satellites will be chipping away at each other producing a multitude of missiles for future impacts.

In the meantime, the surface of the planet has become a basaltic cauldron because of energy dissipation associated with the extremely high rock tides as the planet continues to lose prograde angular momentum. The final act is for the three large satellites to crash onto the planet via atmospheric drag and then the smaller bodies fall onto the planet in due course. (Note: There will be more detail on this scenario in the next section of this chapter.)

Figure 6.6a is a plot for the approximate values of the semi-major axis for the satellite orbit in geologic time and Fig. 6.6b is a plot of the approximate energy dissipation in the Venus-like planet over time.

6.6 Numerical Simulations of Retrograde Planetoid Capture for Venus and a 0.5 Moon-Mass Planetoid

Now that we have looked at the two-body version of a retrograde gravitational capture scenario for the history of Venus, let us now consider some numerical simulations of the capture process as well as certain aspects of the subsequent orbital evolution that can be treated with a three-body interaction code.

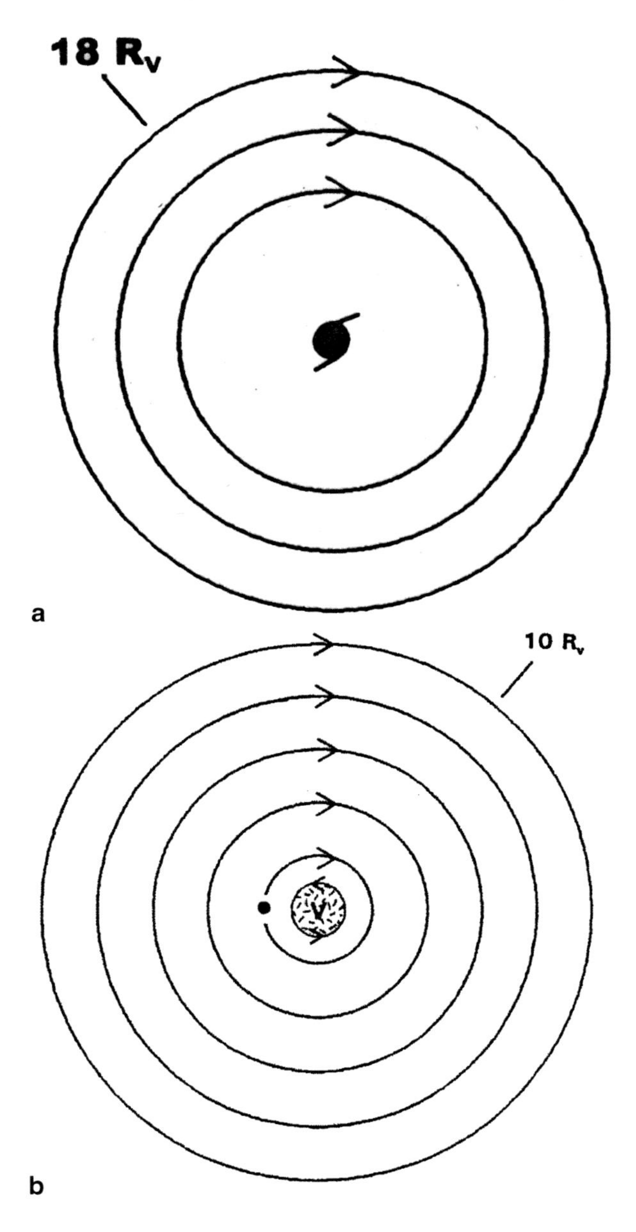

Fig. 6.5 **a** Circular orbit evolution from 18 to 10 venus radii. **b** Circular orbit evolution from 10 venus radii to 2 venus radii

6.6.1 Coordinate System for Plotting the Results

Figure 6.7 shows the coordinate system for plotting the results of the numerical simulations. The plots are similar to those in Fig. 4.21, but the numbers are different. Planet Venus orbits the sun in ~226 earth days. For these co-planar simulations the Venus orbit is at 0.72 AU and has zero eccentricity.

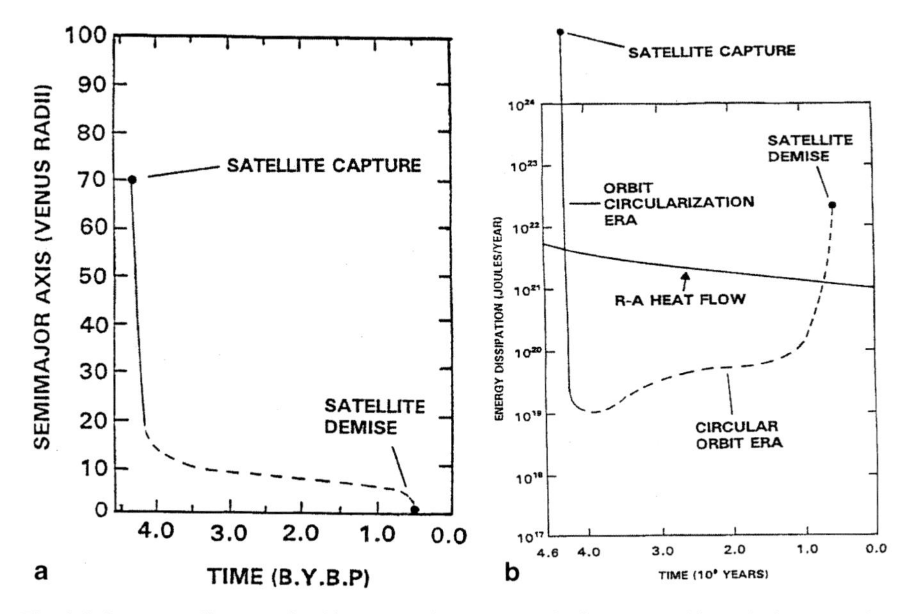

Fig. 6.6 Summary diagrams for this retrograde capture and subsequent orbit evolution scenario. **a** Semi-major axis of Adonis versus time. **b** Tidal energy dissipation in Venus versus time

Table 6.5 Equilibrium rock tides, day length, and sidereal months/Venus year for this two-body orbit evolution scenario

Orb. Rad (Rv)	Tidal ampli. (Venus) (m)	Hours/Venus Day	Months/Venus Yr.
18	5.48	21.0	−49.1
14	11.7	25.2	−71.6
10	32.0	32.9	−118.7
8	62.5	39.8	−165.8
6	148.0	52.4	−255.3
4	500.0	84.0	−469.1
2	4000.0	386.5	−1326.7

6.6.2 *A Sequence of Orbital Encounter Scenarios Leading to Stable Retrograde Capture*

Figure 6.8 shows a heliocentric orientation for an encounter that could lead to retrograde capture if sufficient energy is dissipated in the encountering planetoid for capture. In this case no energy is dissipated and the planetoid returns to a heliocentric orbit of the same dimensions after the two-orbit encounter.

Figure 6.9 shows the results of a five-orbit escape scenario. Since insufficient energy is dissipated for capture, then the planetoid escapes into a somewhat changed heliocentric orbit (For calculation procedures consult Malcuit et al. 1989, 1992 and Chapter 4 of this book).

Figure 6.10 shows the results of a 12-orbit collision scenario. On the 12th orbit the planetoid collides with the planet and the simulation is ended.

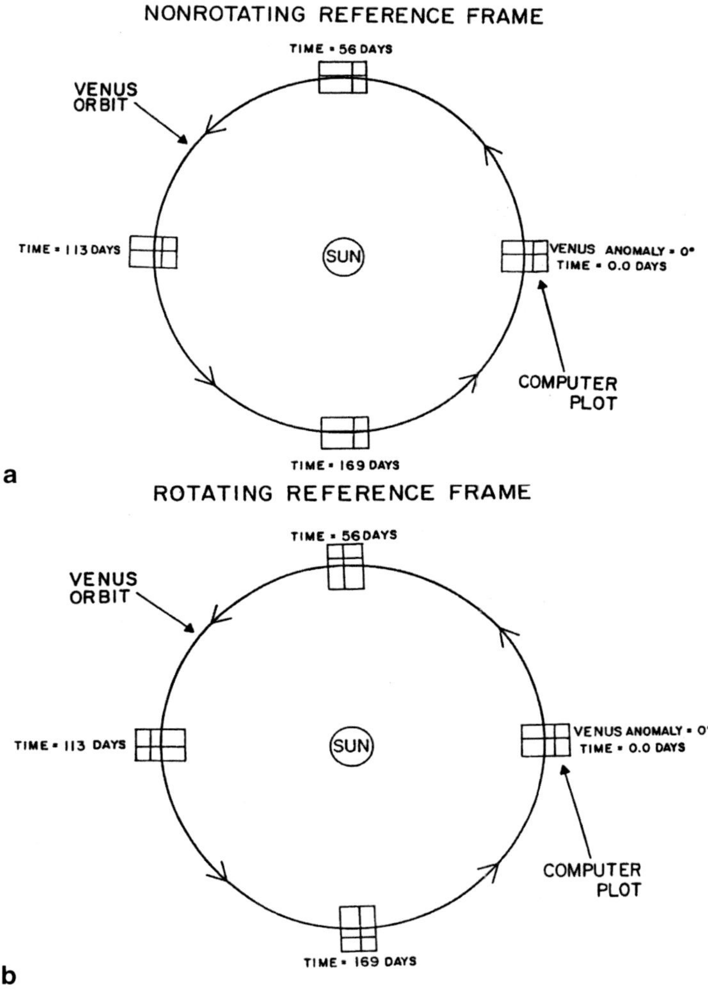

Fig. 6.7 **a** Non-rotating coordinate system for the Venus-Adonis planetocentric plots. **b** Rotating coordinate system for the Venus-Adonis planetocentric plots. (Note: The rotating coordinate system will not be used in this chapter)

Figure 6.11 shows the results of a stable capture scenario. Figure 6.12 gives the tidal amplitudes for Venus and Adonis for this simulation. Table 6.6 is a summary of the encounter results as the h of the planetoid is sequencially increased.

There were a few surprises from the numerical simulations of the Venus-Adonis system: *Surprise # 1* was that about 50 % of the energy estimated from the two-body system was needed to capture the 0.5 Moon-mass planetoid from a heliocentric orbit into a stable venocentric orbit. *Surprise # 2* was that *retrograde* stable capture zones were reasonably large but that the *prograde stable capture zones for the Venus-Adonis system were extremely small.*

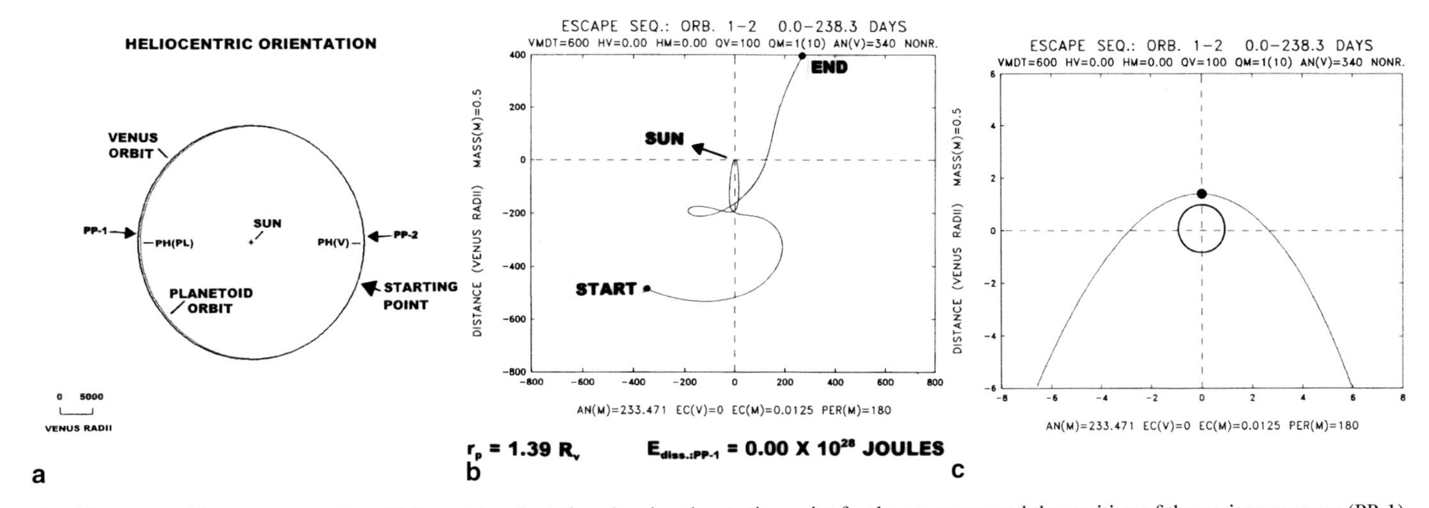

Fig. 6.8 A one-orbit escape scenario. **a** Heliocentric orientation showing the starting point for the encounter and the position of the periven passage (PP-1). **b** Venocentric orientation for the numerical simulation. Note that the second encounter is in the *prograde* direction and then the planetoid escapes. **c** This is a close-up of the initial encounter. The surfaces of the two bodies are separated by about 1000 km

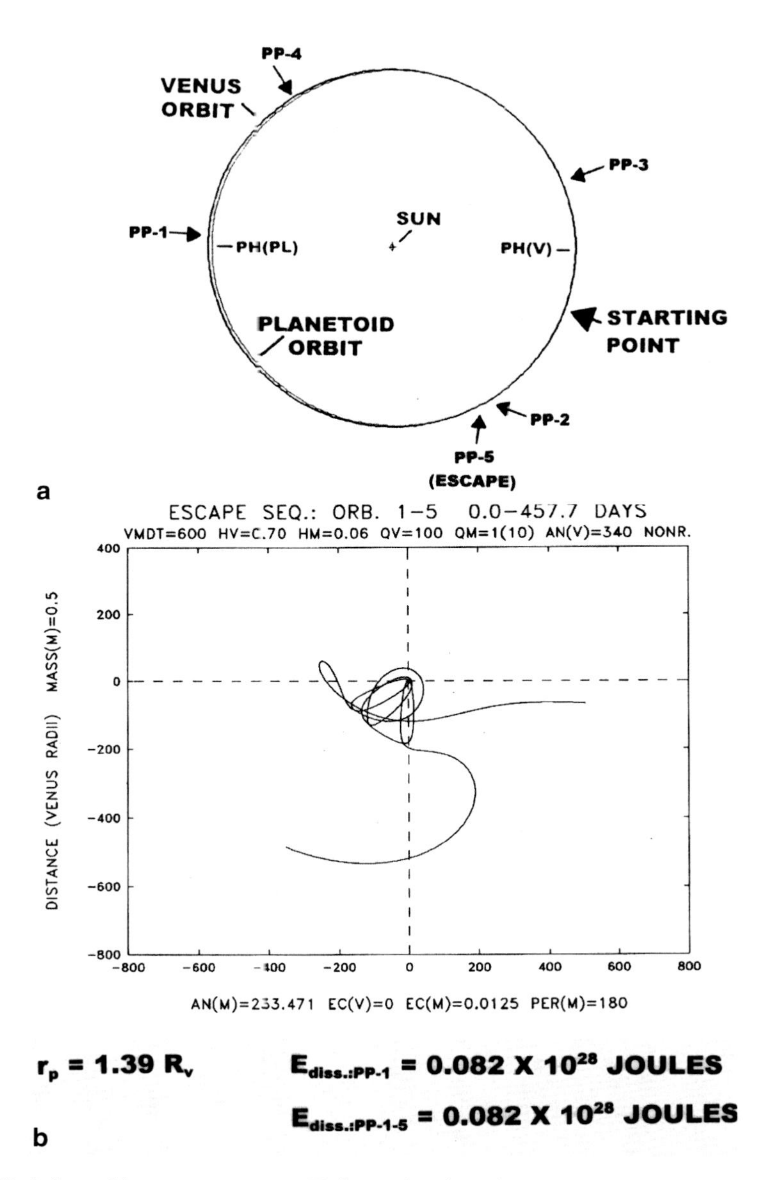

Fig. 6.9 A five-orbit escape scenario. **a** Heliocentric orientation showing the positions of the five periven passages. **b** Venocentric plot showing the geometry of the orbits before the escape. Note that on the fifth periven passage the planetoid again encounters the planet in the *prograde* direction

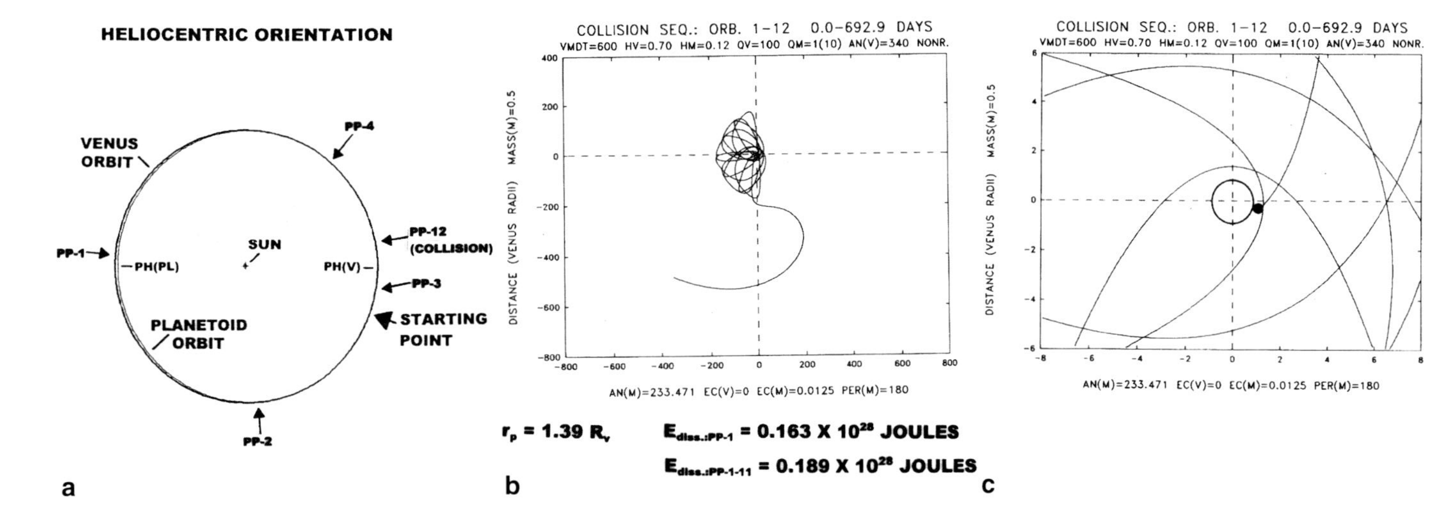

Fig. 6.10 A 12-orbit collision scenario. **a** Heliocentric orientation showing the positions of the first 4 periven passages and that of PP-12. **b** Venocentric plot of the simulation. On the 12 periven passage, the planetoid collides with the planet at a distance of 0.84 venus radii in the retrograde direction. This is a solid collision and the planetoid would not survive. **c** Close-up of the collision scene

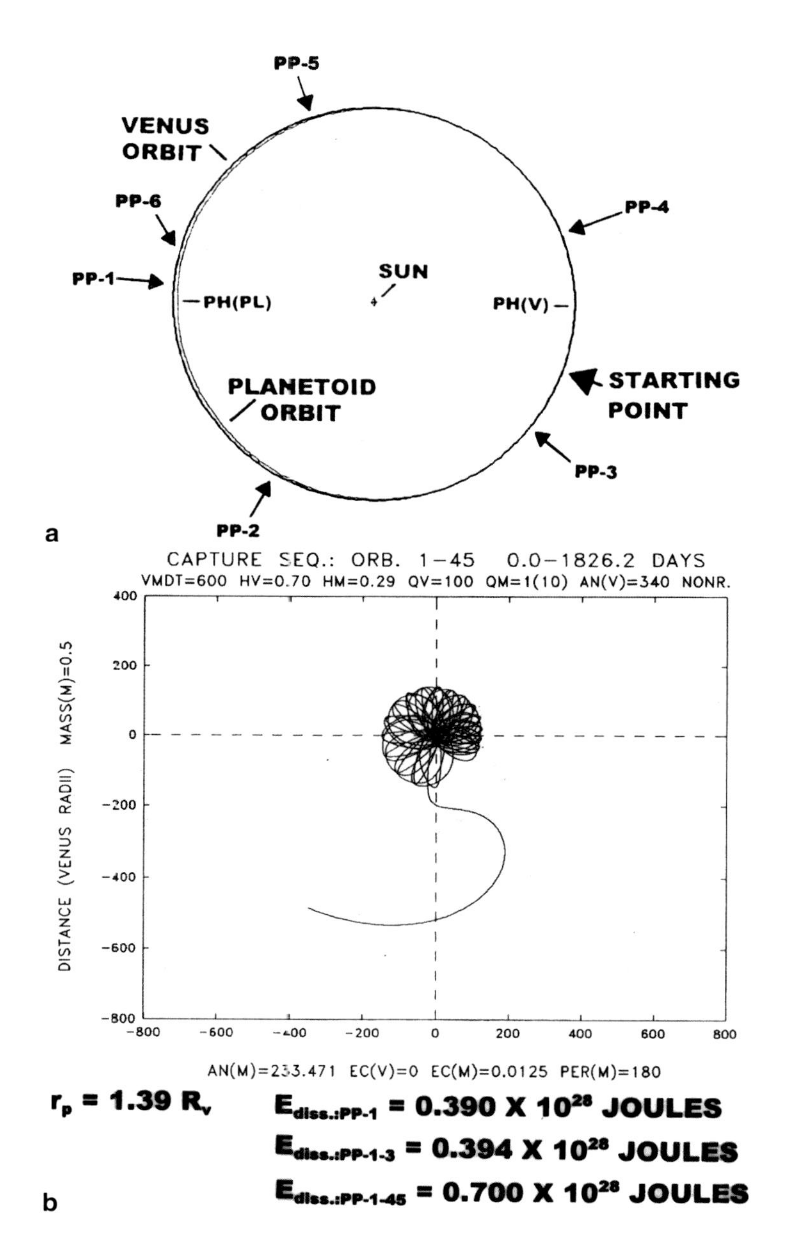

Fig. 6.11 Results of a stable capture scenario. **a** Heliocentric orientation for the simulation showing the positions of the first 6 periven passages. **b** Venocentric orientation of the first 45 orbits (8 venus years, 5 earth years) for this stable capture scenario

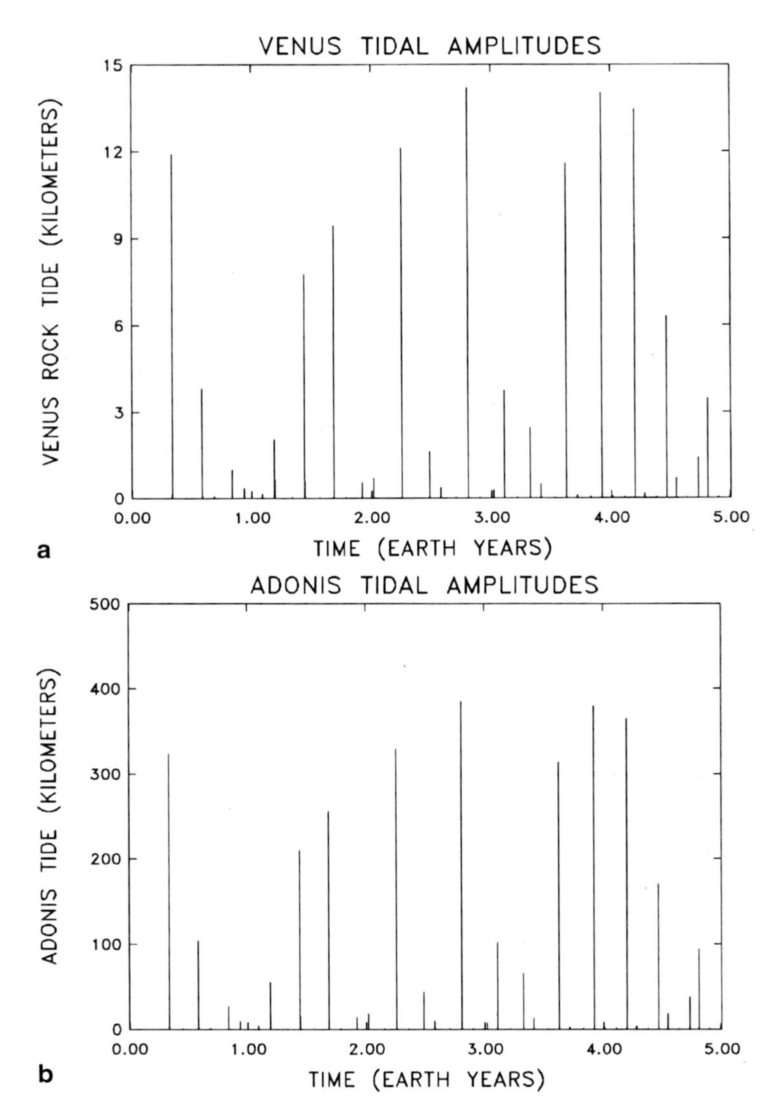

Fig. 6.12 Plots of tidal amplitudes for the two bodies for the stable capture scenario shown in Fig. 6.11; there are 5 earth years of tidal amplitudes shown on the plots (=8 Venus years of tidal amplitudes). **a** Tidal amplitudes for the venus-like body. Note that the initial encounter raises tides of ~ 12 km on the planet. **b** Tidal amplitudes for the planetoid. Note that the initial encounter raises tides over 300 km on the planetoid and that subsequent encounters raise even higher tides than the initial encounter

Figure 6.13 shows the geometry of the stable capture zones for the Venus-Adonis system as well as the position of the encounter parameters of the scenarios shown in Figs. 6.8, 6.9, 6.10, 6.11, and 6.12. Note that the useful zone for *prograde* gravita-

Table 6.6 Summary of encounter results as the displacement Love number of Adonis is sequentially increased to a value where stable capture can occur. Note that the h value of Venus is constant at 0.7

h (Adonis)	Result	h (Adonis)	Result
0.0	2-orbit escape		
0.01	3-orbit escape	0.21	3-orbit collision
0.02	3-orbit escape	0.22	3-orbit collision
0.03	3-orbit collision	0.23	3-orbit collision
0.04	3-orbit collision	0.24	3-orbit collision
0.05	17-orbit collision	0.25	3-orbit collision
0.06	5-orbit escape	0.26	14-orbit collision
0.07	4-orbit collision	0.27	18-orbit collision
0.08	4-orbit collision	0.28	20-orbit collision
0.09	7-orbit collision	0.29	100-orbit capture
0.10	31-orbit collision	0.30	100-orbit capture
0.11	9-orbit collision	0.31	100-orbit capture
0.12	12-orbit collision	0.32	100-orbit capture
0.13	200(+)-orbit capture	0.33	100-orbit capture
0.14	45-orbit collision	0.34	100-orbit capture
0.15	31-orbit collision	0.35	100-orbit capture
0.16	36-orbit collision	0.36	100-orbit capture
0.17	47-orbit collision	0.37	100-orbit capture
0.18	21-orbit collision	0.38	100-orbit capture
0.19	16-orbit collision	0.39	100-orbit capture
0.20	3-orbit collision	0.40	100-orbit capture

SUMMARY: 4 escapes, 24 collisions, 1 quasi-capture, 12 stable captures

tional capture by tidal dissipation processes is very small; this makes *prograde* capture for a Venus-Adonis combination highly improbable. In contrast, the retrograde SCZs are much larger and capture from either an orbit slightly larger or slightly smaller than the orbit of Venus with planetoid eccentricity values between 0.3 and 0.6 % yield a reasonable probability if the body deformation parameters are within a certain range of values. It is fairly clear from the SCZ diagrams that if stable capture is to occur for the Venus-Adonis combination, it will probably happen in the retrograde direction (Malcuit and Winters 1996).

6.6.3 Post-Capture Orbit Circularization Era

A post-capture orbit circularization code was used to calculate a time-scale for the two-body evolution. Without any angular momentum exchange via the rock tide processes, the orbit evolved to 10 % eccentricity in a mere 70 million years. With three-body corrections, the time scale is even shorter because of the large percentage of close periven passages that are characteristic of high eccentricity retrograde orbits; the instability of these orbits is caused by solar perturbations. Here, again, someone in the future can do a continuous three-body simulation from the time of capture until orbit circularization.

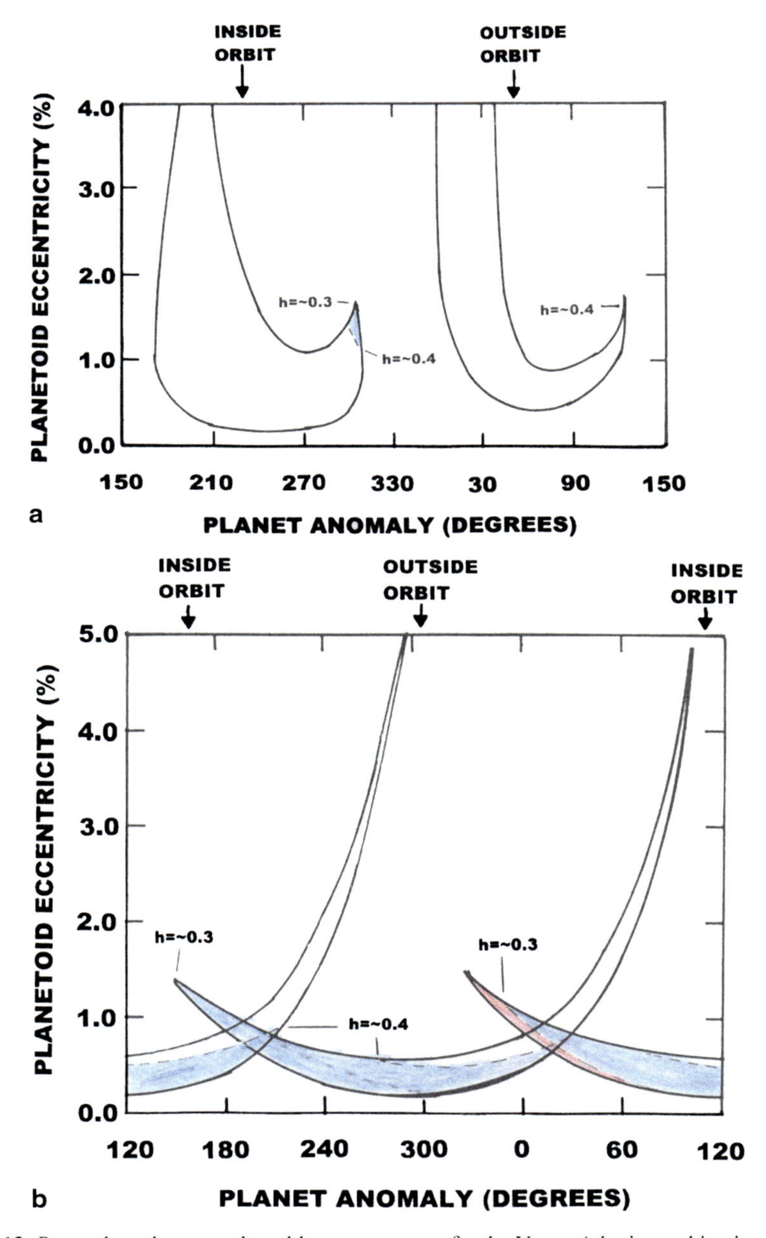

Fig. 6.13 Prograde and retrograde stable capture zones for the Venus-Adonis combination. **a** Plot of prograde stable capture zones. **b** Plot of retrograde stable capture zones. Discussion of the relative probability of stable capture is in the text. The position of the simulations used in Figs. 6.8, 6.9, 6.10, 6.11, and 6.12 are planet anomaly = 340 and planetoid eccentricity = 0.0125

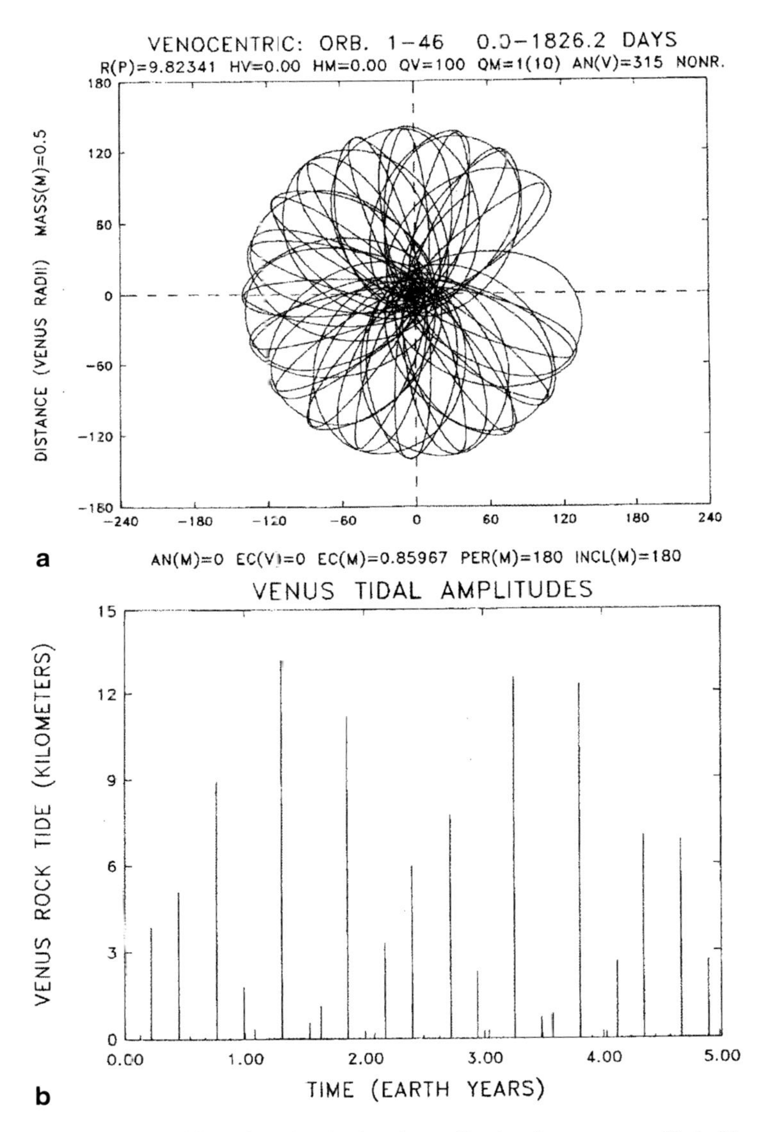

Fig. 6.14 Orbit and rock-tide regime for the time immediately after capture. **a** Eight Venus years of retrograde orbits. **b** Eight venus years of rock tidal amplitudes on Venus (5 earth years). Note that the rock tidal amplitudes are high and irregular

The rock tides associated with a stable capture scenario would probably lead to the recycling of the primitive crust of Venus in the equatorial zone (broadly defined) of the planet. The five orbit and tidal amplitude diagrams in Figs. 6.14, 6.15, 6.16, 6.17, and 6.18 will give the reader a firm concept of the tidal regime experienced

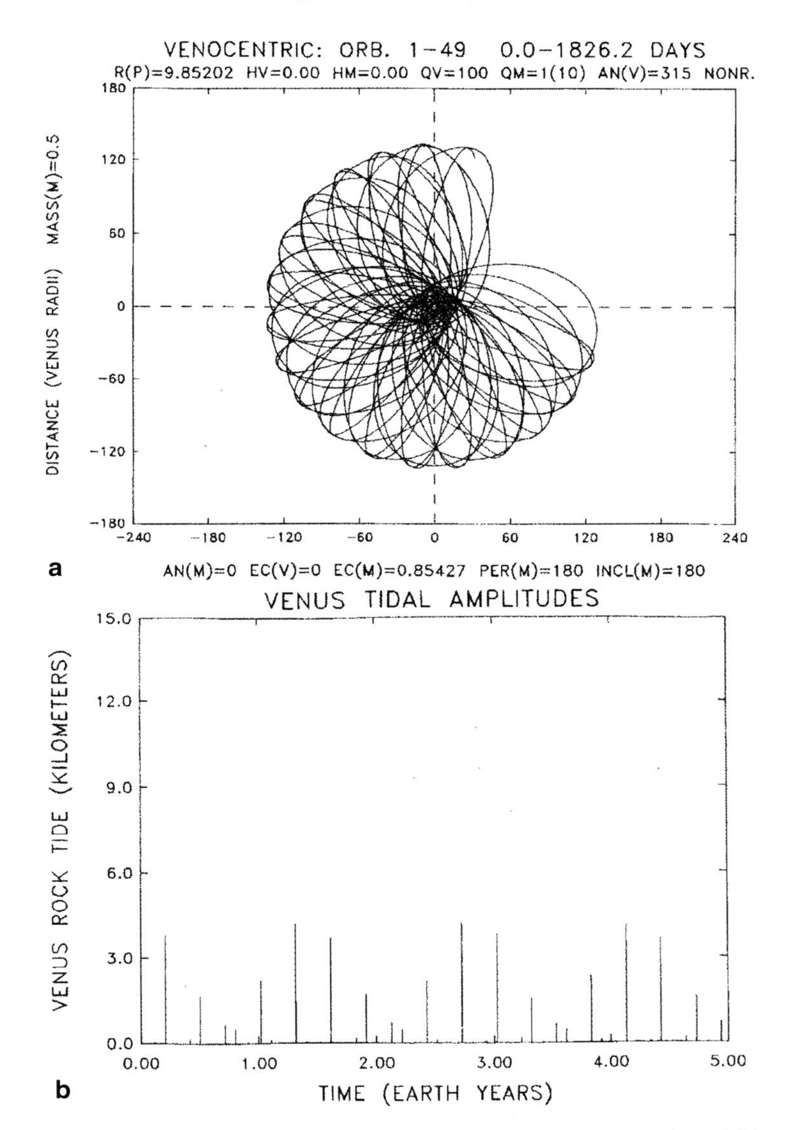

Fig. 6.15 Orbit and rock tide regime on Venus about 1.0 Ka after capture (about 5 Ma on two-body timescale). **a** Eight venus years of retrograde orbits. **b** Eight venus years of rock tidal amplitudes on Venus. Note that in 1000 years, the tidal amplitudes have decreases to a maximum of about 5 km

by planet Venus in this scenario. After the crustal recycling associated with the orbit circularization sequence is completed, crustal stability returns to Venus which eventually becomes a stagnant-lid (one-plate) planet.

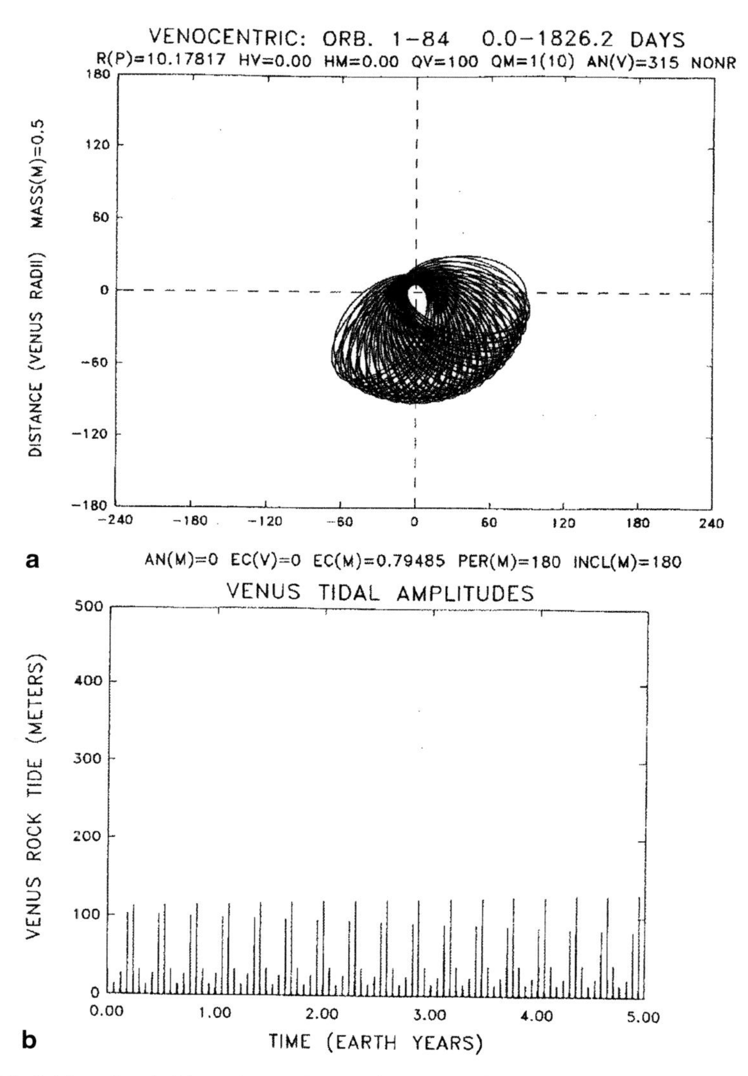

Fig. 6.16 Orbit and rock tide regime on Venus about 5 Ka after capture (about 10 Ma on 2-B timescale). **a** Eight venus years of orbits **b** Eight venus years of rock tidal amplitudes on Venus. Note that the rock tidal amplitudes have settled down to a regular pattern and that the maximum rock tide in the equatorial zone is a bit over 100 m

6.6.4 Sequence of Diagrams Showing the Possible Surface and Interior Effects on Planet Venus for Retrograde Capture of Adonis and Subsequent Orbit Circularization

Figures 6.19 and 6.20 show my concept of the general conditions before the capture episode and "snapshots" of three times during the post-capture orbit circularization sequence.

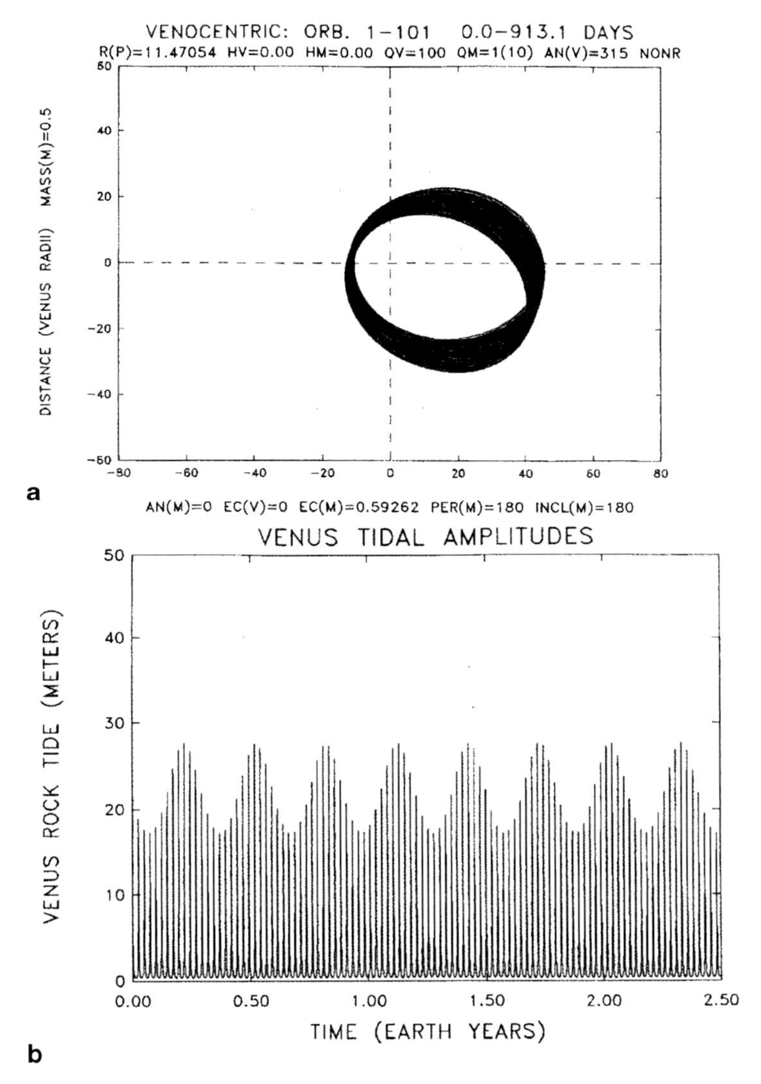

Fig. 6.17 Orbit and rock tide regime on Venus about 7 Ka after capture (about 30 Ma on 2B timescale). **a** Four venus years of orbits. **b** Four venus years of rock tidal amplitudes on Venus. Note the large number of tidal cycles/venus year and the regularity of the rock tidal regime. The maximum tide at this time is just under 30 m

6.6.5 Diagrams Showing Possible Surface Effects During the Circular Orbit Era

Figures 6.21 and 6.22 show "snapshots" of my view of surface condition on planet Venus during the circular orbit evolution era. This is the time following the orbit circularization era and before the severe tidal consequences of rapid contraction of

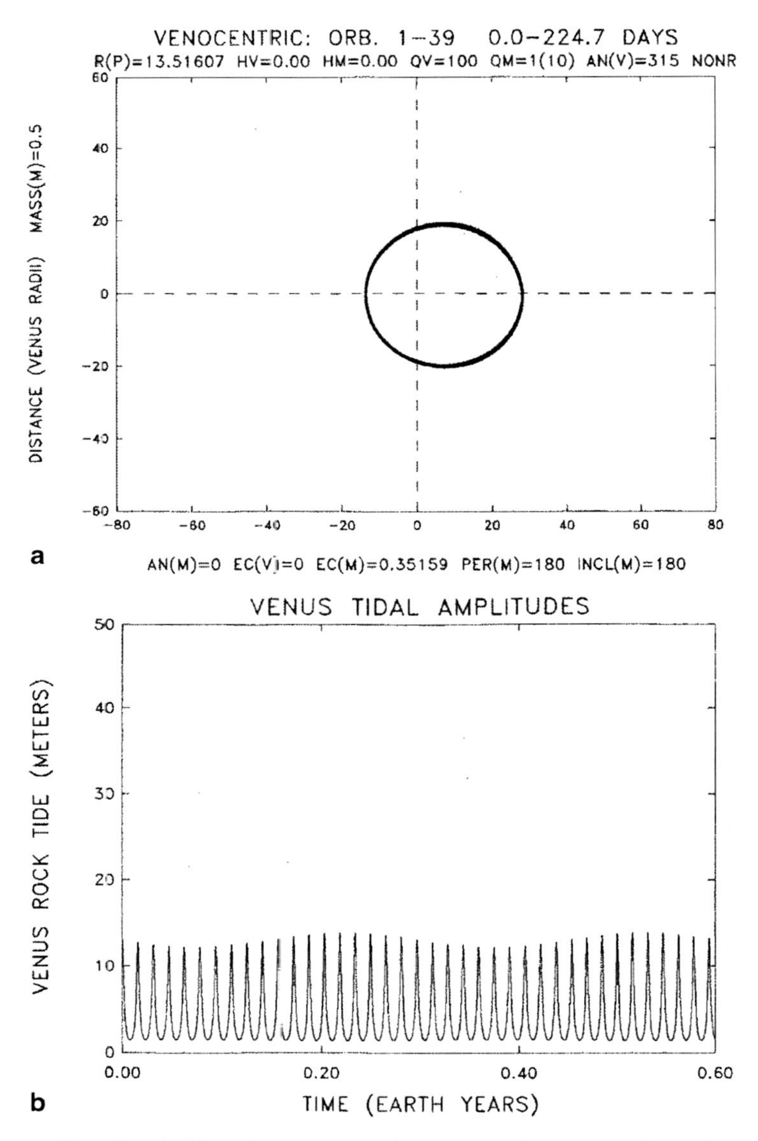

Fig. 6.18 Orbit and rock tide regime on Venus about 10 Ka after capture (about 50 Ma on 2-B timescale). **a** One venus year of orbits (on two-body timescale). **b** One venus year of rock tides on Venus. Note that there are about 40 tidal cycles/year on Venus at this time but the maximum tidal amplitude is about 13 m. The tidal amplitudes and ranges are lower in mid-latitude and polar regions

the orbit of Adonis. Conditions during this era *could be very conducive to the origin and evolution of life forms*. This placid time of circular orbit evolution is estimated to be about 3.0 billion years.

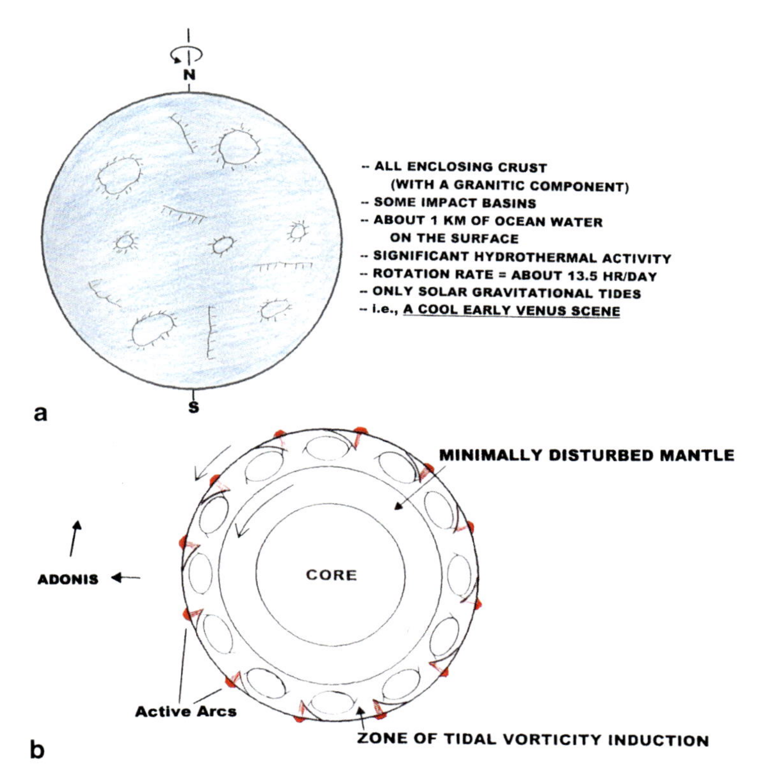

Fig. 6.19 **a** Hemispherical sketch of the probable condition of planet Venus before the capture of Adonis (~4.26 Ga ago). Solar radiation is about 15–20% less than at present. The crust is all enclosing and is being thermally processed via vertical tectonics via the Bedard (2006) model. There is an original atmospheric composition resulting from the accretion process and a major thermal pulse due to metallic core formation. The atmosphere is a combination of nitrogen and carbon dioxide with other gasses as minor constituents. There probably is some combination of liquid water and water ice on the surface at various latitudes. The rotation rate is about 21 h/day. In my view, both Venus and Earth are characterized by features of the "Cool Early Earth Model" of Valley et al. (2002): i.e., the "twin sisters" begin their planetary lives in a very similar manner. **b** Equatorial cross-section of Venus ~1 Ka after capture (~5 Ma on 2-B timescale). View is from the north pole of the planet. Note that Adonis is moving in the opposite direction to the rotation of the planet. Note also that the Tidal Vorticity Induction (TVI) cells are rotating in the retrograde direction. Concept of the TVI cells is from Bostrom (2000)

6.6.6 Summary and Commentary on Conditions during this 3.0 Billion Year Era

Adonis is in a circular orbit at a comfortable distance. Although plate tectonic action may be operating in the equatorial zone of Venus during the orbit circularization era, any rock tidal amplitudes are much lower than they were during the orbit circularization era. Thus the rock tides *may not* be sufficient to cause organized and sustained subduction activity.

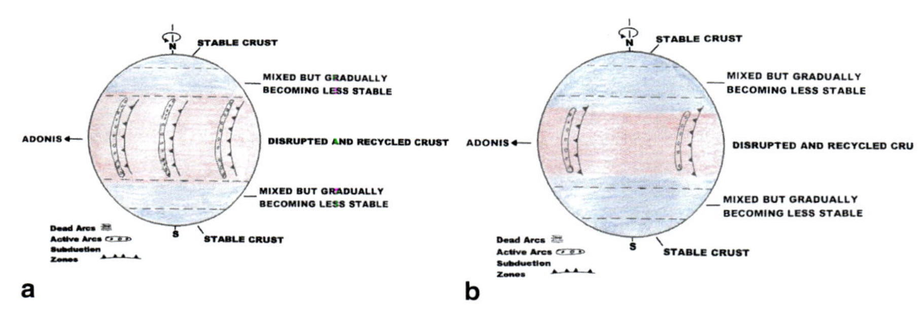

Fig. 6.20 Hemispherical sketches of possible surface conditions at two times during the post-capture orbit circularization era. **a** Probable surface scene ~1 Ka after capture (~5 Ma after capture on 2-B timescale). Maximum rock tides on Venus at this time are about 5000 m; there is a continuous cover of ocean water on the surface but the water is sloshing around a bit during close periven passages; tidally induced plate tectonics is operating during this era and causes partial recycling of the primitive crust. **b** Probable surface condition about ~10 Ka after capture (~50 Ma on 2-B timescale). Maximum rock tides on Venus are now about 13 m; there is a continuous cover of ocean water on the surface; the ocean tide as well as the rock tide are very regular. Surface volcanism is exponentially decreasing in time. The rotation rate of Venus is about 21 h/day and there are about 40 months/venus year. Note: The subduction zones on the illustrations are associated with the downward movement of the TVI cells and the upward movement is on the backside of the volcanic arcs

The planet, in my view, may go back to a one-plate (stagnant lid) condition. Fairly large volcanoes could accumulate on the thickened crust (i.e., large shield volcanoes similar to what we see on planet Mars). There is local thermally induced upwarps and local volcanism but no organized pattern of volcanic arcs and subduction zones. Ocean tides tend to stir the ocean water systematically.

Depending on the obliquity of the planet, there could be significant seasonality. With Adonis in orbit in the plane of the planets, the obliquity, whatever the value, should be stabilized. Let us assume a value for obliquity between 20° and 25°.

The water source for Venus mimics that of Earth and is an asteroidal source as outlined by Albarede (2009). This asteroidal water arrives in the first 100 Ma or so: i.e., well before the capture episode. Water subducted and degassed from the mantle of Venus during accretion, capture, and subsequent orbit circularization yields a quantity of liquid water with an average thickness of 1.0 km on the surface of Venus. But after capture and the subsequent orbit circularization, the topography is complex and the shoreline morphology is complex also. Erosion is rapid and many sedimentary rocks result from the erosion and deposition cycles.

Depending on the carbon dioxide content of the atmosphere, there can be "greenhouse" or "icehouse" conditions on the surface of the planet. The carbon dioxide would tend to be fixed as carbonate rocks in shallow ocean basins. Polar ice caps and pack ice could accumulate seasonally or more permanently in polar areas. Bacterial and algal life forms are expected to form and evolve under these conditions.

Note on the Rotation Rate of Venus During this Era A reasonable original prograde rotation rate for Venus is about 13.5 h/day (MacDonald 1963; Singer 1970; Lissauer

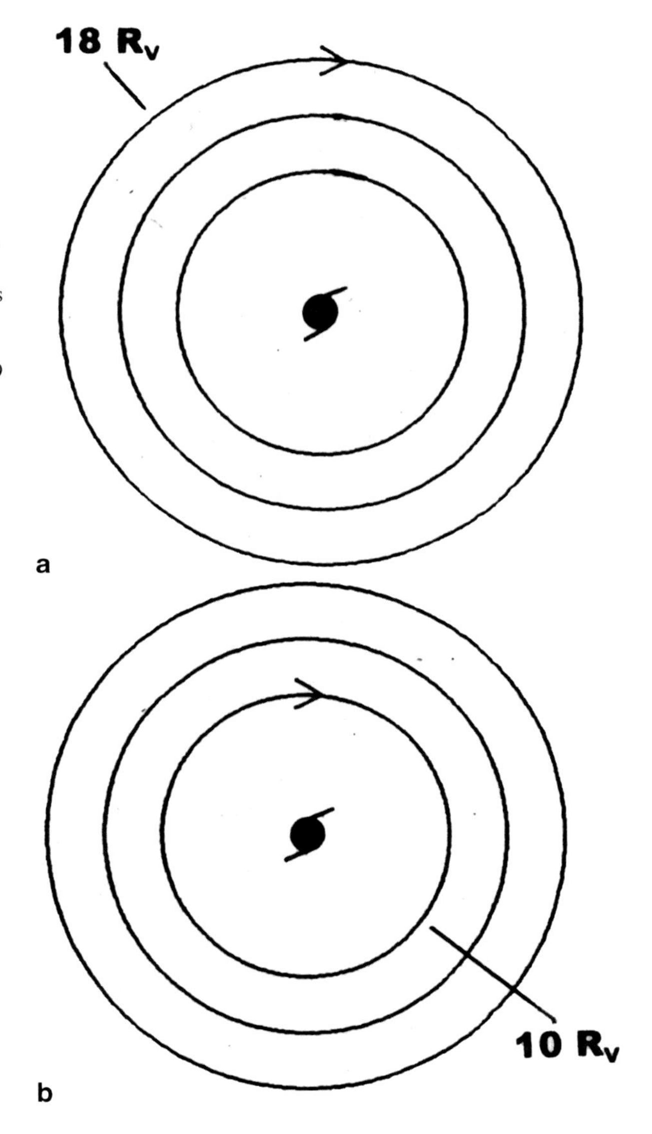

Fig. 6.21 Beginning of the Circular Orbit Era. **a** Scale sketch of the orbit of Adonis at 18 venus radii (view from north pole of Solar System). **b** Hemispherical view of what the surface conditions on Venus may have been like about 4.0 Ga ago. Conditions at this time are: equilibrium rock tidal amplitude on Venus ~5.5 m; Venus rotation rate ~21 h/day *prograde*; ~257 sidereal days/venus year; ~49 sidereal months/venus year *retrograde*

and Kary 1991, Lissauer 1993; Lissauer et al. 1997). But in order for Adonis to despin Venus to a very slow retrograde condition, only the angular momentum of a Venus rotating at 21.0 h/day is needed. So, what is up??? Well, solar tides have despun planet Mercury and they do effect planet Earth in a minor way (Goldriech and Soter 1966; Burns 1973). In general, it is difficult to justify the concept that solar tides could possibly account for this apparent angular momentum loss for Venus, a loss equivalent to several hours/venus rotation (i.e., hours/venus sidereal day). A suggestion is that a few collisional scenarios, either retrograde collisions during capture attempts or high speed retrograde collisions by the Vulcanoid planetoid

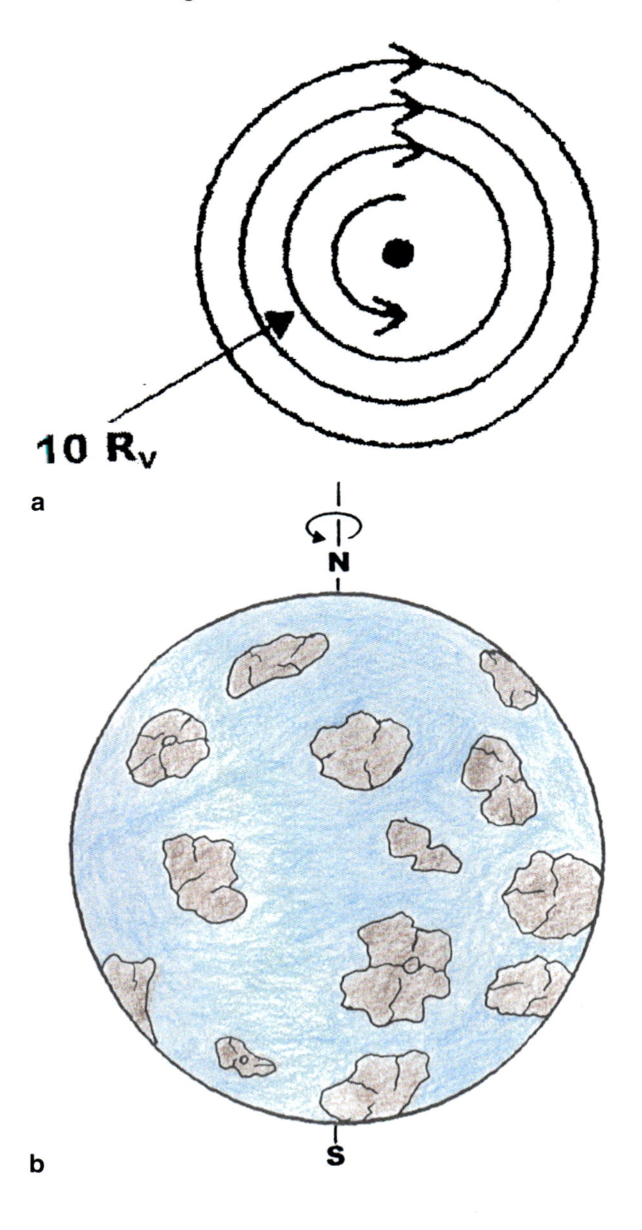

Fig. 6.22 End of the "comfortable" time of the Circular Orbit Era: **a** scale sketch of the orbit of Adonis at 10 Venus radii (view from north pole of Solar System). **b** Hemispherical view of what the surface conditions on Venus may have been like about 1.0 Ga ago. Conditions at this time are: equilibrium rock tidal amplitude on Venus ~ 32 m, Venus rotation rate ~ 32.9 h/day *prograde*; ~ 164 sidereal days/venus year; ~ 119 sidereal months/venus year *retrograde*. Note that the rock tides could begin to cause a "forced subduction" via the operation of the TVI convection cells

population could cause a significant loss of prograde rotational angular momentum for Venus. Since planet Venus is nearer to the source of the Vulcanoid planetoids, it would have a greater chance to interact with Vulcanoid planetoids that are passing through the orbital space of Venus.

Figure 6.23 gives an orbital summary of the later stages of circular orbit evolution.

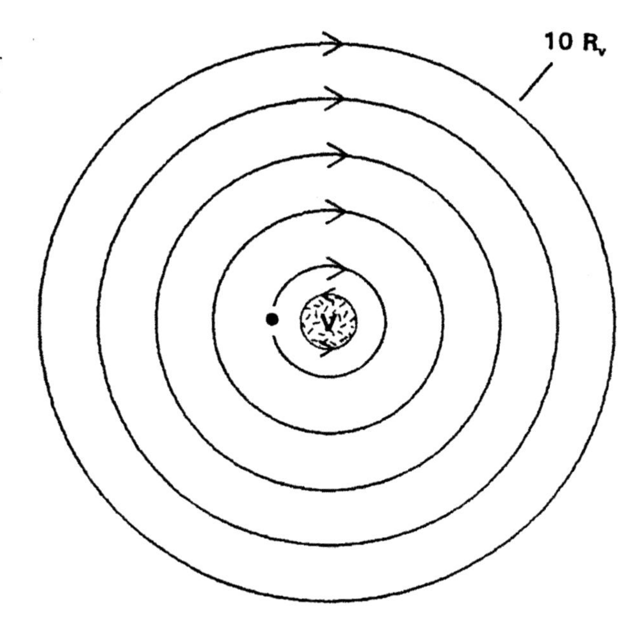

Fig. 6.23 Scale diagram of the later stages or the circular orbital contraction of Adonis. The rock tidal amplitudes at 10 venus radii are 32 m (48 m for tidal range). View is from the north pole of the Solar System

Figure 6.24 shows "snapshots" of my view of the possible effects of rock tides on the surface and upper mantle of Venus as the circular orbit decreases from 10 venus radii to 2 venus radii. I am not showing the TVI cells in cross-sectional views; I am showing only the hemispherical surface view of the effects of the TVI cell circulation. In general, the crust that accumulated during the "placid era" on Venus is systematically recycled into the venusian mantle by the rock tidal action of Adonis. The severe reworking of the surface and upper mantle of Venus probably begins when Adonis is in a retrograde circular orbit at about 10 venus radii. At that distance, the rock tides have an equilibrium amplitude of about 32 m (range = 48 m) in the equatorial zone of the planet. The number of months/venus year is about 193 at that point and this number increases to about 2000 months/venus year when Adonis is at 2.0 venus radii at which time the equilibrium rock tide is ~4000 m. In general, the retrograde orbital motion of Adonis for this model is much more important for the TVI action than the prograde rotation rate of Venus.

During this era when the orbit of Adonis is decreasing in orbital radius from 10 venus radii to 2 venus radii, the surface water systematically evaporates from the surface and carbon dioxide gas from the mantle accumulates in the atmosphere to begin a super-greenhouse heating which is still present today. The crust is systematically recycled into the upper mantle of Venus. As Adonis gets ever closer to Venus, the TVI cells extend their influence poleward. The last crust to survive would be at the poles and eventually even it would be recycled. *BUT THE SHOW IS NOT OVER!!*

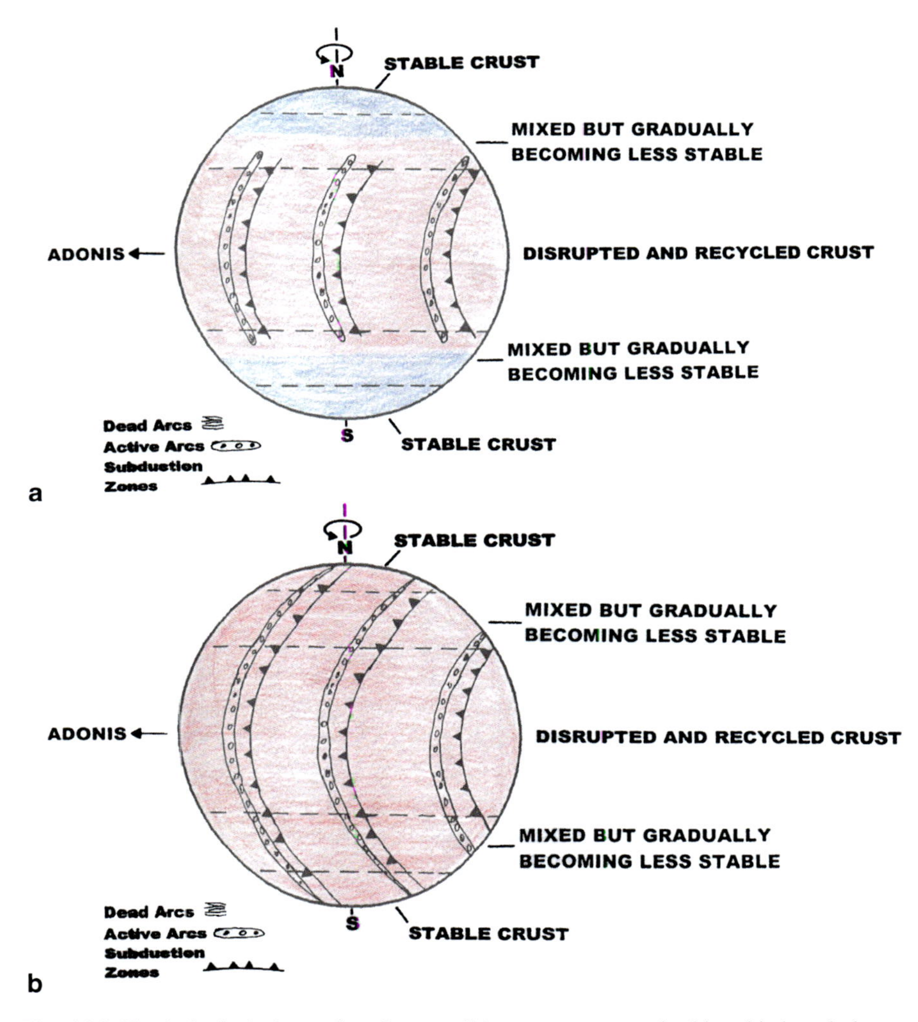

Fig. 6.24 Hemispherical views of surface conditions at two stages in this orbital evolution. **a** Surface activity when Adonis is at 6 venus radii. Equilibrium tidal amplitude is ~148 m; Venus rotation rate is ~52.4 h/venus sidereal day, *prograde*; sidereal days/venus year is ~103; sidereal months/venus year is ~255, *retrograde*. **b** Surface activity when Adonis is at 2 venus radii. Equilibrium tidal amplitude is ~4000 m; Venus rotation rate is ~386 h/venus sidereal day, *prograde*; sidereal days/venus year are ~14; sidereal months/venus year are ~1327, *retrograde*

6.6.7 A Model for the Final Demise of Adonis from the Roche Limit for a Solid Body to Breakup in Orbit and Eventual Coalescense with Planet Venus

Before we get to the final stage of the demise of Adonis we need some additional discussion of the Roche Limits for Satellites of Terrestrial Planets: The classical Roche Limit (Jeffreys 1947) is the radial distance at which a liquid (molten rock)

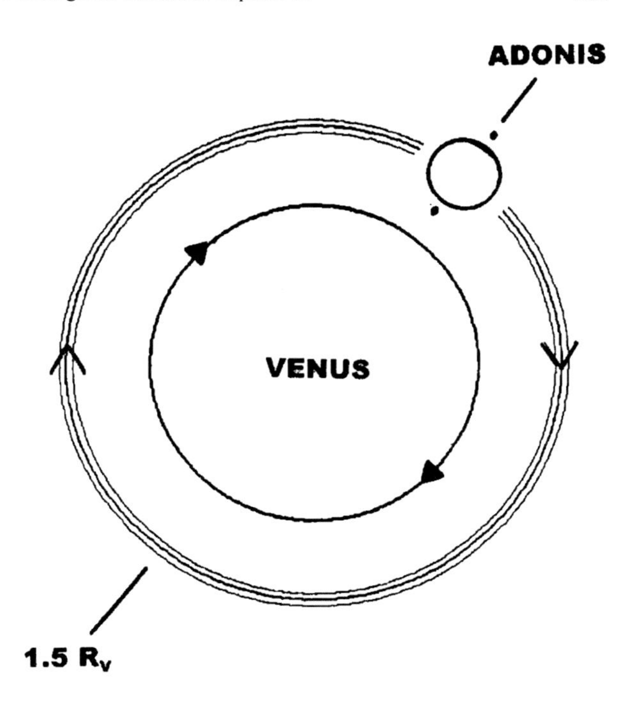

Fig. 6.25 Scale sketch of Adonis slightly within the Roche limit for a solid body. Equilibrium tidal deformation of Adonis ~332.5 km (h=0.5); this is about 24% radial tidal deformation; Venus rotation rate=2334 h/venus day, *retrograde*. Adonis is orbiting Venus at about 2043 orbits/Venus year, *retrograde*, near the equatorial plane of the planet. Both the front-side and back-side masses are about to "peel" away from the mantle mass

satellite in a circular orbit becomes unstable because of stretching along the line of centers connecting the planet and the satellite. The satellite separates into two or more masses at about 2.89 venus radii.

The Roche Limit for a solid body is the radial distance at which a solid body in a circular orbit becomes unstable due to tidal deformation and begins to break up due to ductile faulting (cracking) into two or more parts (Jeffreys 1947; Aggarwal and Oberbeck 1974; Holsapple and Michel (2006, 2008). Any magma in a subcrustal magma zone facilitates this disintegration process. This Roche limit for a solid body is at about 1.6 venus radii. Figure 6.25 shows a stage in the demise of Adonis.

Figure 6.26a is a graphic depiction of the orbital evolution of the Adonian parts after the tripartite breakup of the satellite. After the tripartite breakup the parts undergo mutual gravitational perturbations. These interactions consist of close encounters, grazing collisions, and occasional solid collisions. Because of the very low encounter velocities during collisions, there is not much vaporization but the collisions produce many smaller (10's of meters to multi-kilometer-sized) objects. Mutual gravitational interactions lead to much scattering of the orbits of the particles so that they occupy much of the retrograde orbit space about Venus. Some particles get perturbed into prograde orbit space as well. Figure 6.26b may give the reader some idea of what this chaotic orbital motion looks like in two dimensions. Your imagination must put it in three dimensions.

Figures 6.27 and 6.28 will give the reader some idea of how particles, large and small, can interact with the very dense atmosphere of Venus and subsequently impact on the surface of Venus.

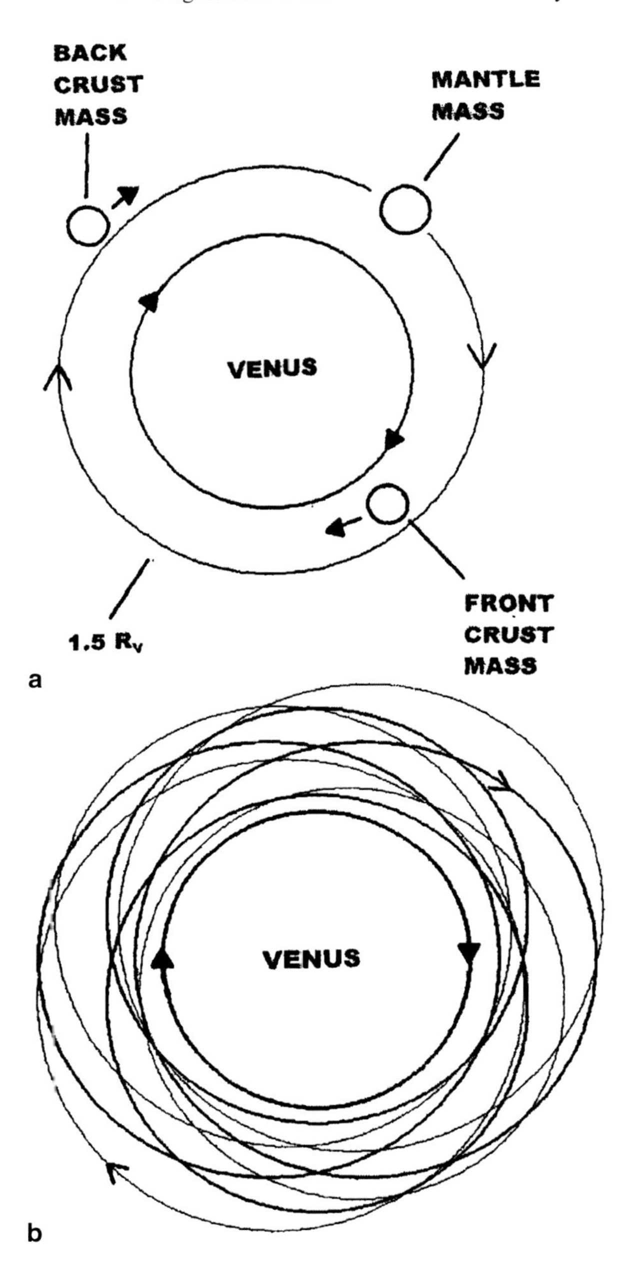

Fig. 6.26 **a** Initial tripartite break-up of Adonis in the equatorial plane of Venus. The front-side crust and the back-side crust separate from the Adonian mantle complex and all three go into their individual retrograde orbits. Venus rotation is very slow, retrograde; mantle mass velocity is ~6.0 km/s; back-side crustal mass velocity is ~5.5 km/s; front-side crustal mass velocity is ~6.4 km/s; closing time for mantle mass and frontside crustal mass is ~4.8 h; closing time for mantle mass and backside crustal mass is ~6.0 h. **b** Scale sketch of a two-dimensional version of a swarm of orbiting objects near the equatorial plane about planet Venus. Venus rotation is very slow in retrograde direction. Adonian mantle and crustal "chips" are in an array of orbits about planet Venus. In this schematic diagram the darker lines represent mantle parts and the lighter lines represent crustal parts or fragments. Most, if not all, of these fragments are eventually perturbed so that they enter the dense atmosphere of Venus and impact onto the basaltic surface of the planet

6.7 Summary for the Chapter

1. In a two-body analysis for capture of a sizeable satellite (0.5 moon-mass or greater), prograde capture appears to be as likely as retrograde capture although both are just marginally physically possible.

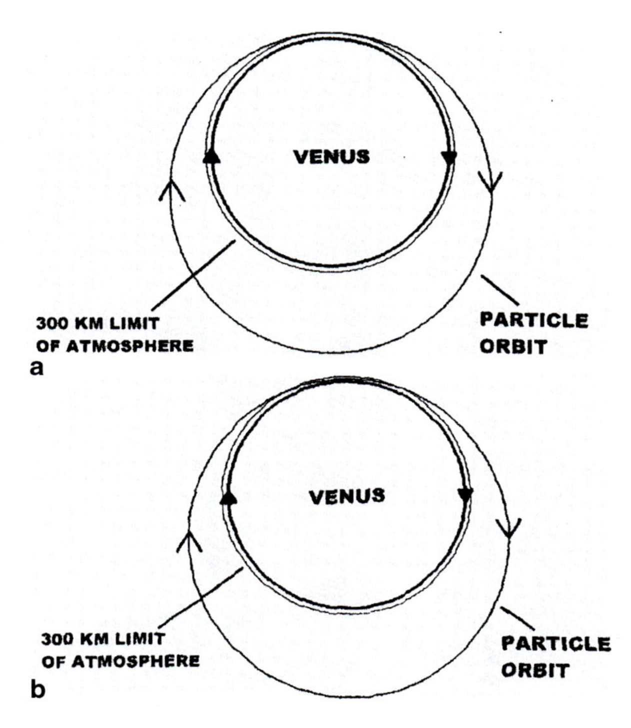

Fig. 6.27 Two scale sketches for particles in elliptical orbits in the equatorial plane of Venus. Venus rotation is very slow in retrograde direction. Both particles have a major axis of 1.4 venus radii. Velocity at periven for both is ~2 km/s. **a** Particle eccentricity for this diagram is 0.25 which places it just above the dense atmosphere of Venus. **b** Particle eccentricity for this diagram is 0.27 which has it penetrating the atmosphere of Venus

2. In a three-body numerical simulation, the energy dissipation requirements for retrograde capture are decreased over 50 % relative to a two-body estimate for planetary eccentricity values ~0.5 %. Furthermore, the useful portion of the Stable Capture Zone for retrograde capture for planet Venus is reasonably large.

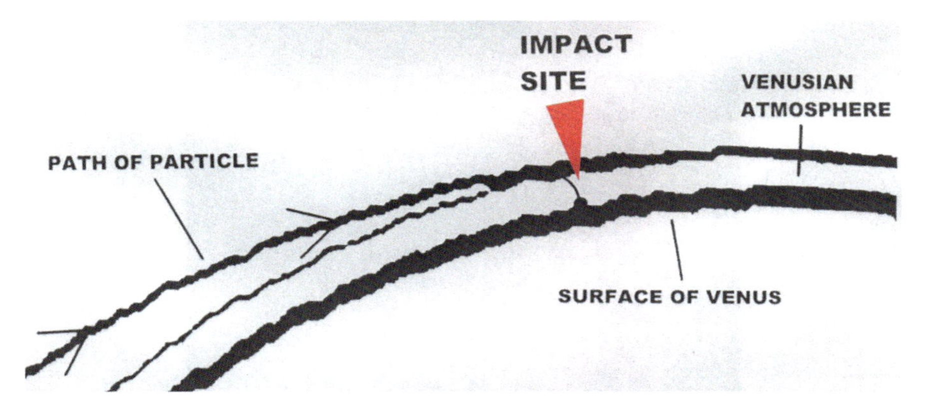

Fig. 6.28 The author's view of a typical low velocity particle impact on the surface of Venus. Since the present atmosphere of Venus is about 92 times as dense as the atmosphere of Earth, the particle would begin to decelerate and burn at about 200 km above the surface and would probably be falling nearly vertically at the time of impact

3. In a three-body numerical simulation for prograde capture from a heliocentric orbit, solar perturbations are not very effective and the useful portion of the prograde Stable Capture Zone is very limited making prograde capture of a sizeable satellite for planet Venus highly improbable.
4. After a successful retrograde capture of Adonis by Venus, the satellite orbit circularized over a timescale of about 100 Ma to a circular orbital radius of about 18 venus radii.
5. The circular orbit of Adonis then undergoes an orbital evolution, via tidal activity, in which the satellite orbit decreases in radius as the planetary rotation decreases due to cancellation of the angular momentum of the system. This era is estimated to be about 3 Ga long.
6. As the orbital radius decreases to ~10 venus radii, then the rate of energy dissipation within Venus begins to increase exponentially and the rate of orbital evolution increases exponentially as well.
7. Adonis begins to disintegrate at the Roche limit for a solid body which is located at about 1.6 venus radii.
8. After severe tidal stretching along the line of centers of the two major bodies, Adonis separates into three major parts: a front-side crustal mass, a backside crustal mass, and a mantle mass.
9. These three masses have different orbital periods and are gravitationally perturbing each other fairly regularly.
10. In a short period of time the bodies are colliding with each other and chipping off both major and minor particles, each of which becomes another satellite of Venus.
11. In this scenario the rotation of Venus changes from prograde to retrograde when Adonis is about to break into three parts. This change from prograde to retrograde rotation has no significant effect on the subsequent events.
12. The chips are expected to be perturbed to all latitudes around the planet but retrograde impact vectors are expected to be much more numerous.
13. The impact velocities for the smaller orbits are low (about 2.0 km/s). Larger particles with more eccentric orbits have higher velocities when entering the atmosphere of Venus. [All of these atmospheric entry velocities are low relative to a body falling onto Venus from infinity (~7.3 km/s)].
14. All bodies entering the atmosphere of Venus would undergo much frictional drag and are slowed dramatically before impacting on the surface of the planet. Even some of the larger bodies in low Venus orbit could impact fairly "softly" onto the surface.

THE END OF THE VENUS-ADONIS STORY!!!

References

Aggarwal HR, Oberbeck VR (1974) Roche limit for a solid body. Astrophys J 191:577–588

Albarede F (2009) Volatile accretion history of the terrestrial planets and dynamic implications. Nature 461:1227–1233

Bedard JH (2006) A catalytic delamination-driven model for coupled genesis of Archaean crust and sub-continental lithospheric mantle. Geochim Cosmochim Acta 70:1188–1214
Bostrom RC (2000) Tectonic consequences of Earth's rotation. Oxford University Press, London, p 266
Burns JA (1973) Where are the satellites of the inner planets? Nat Phys Sci 242:23–25
Cloud PE (1972) A working model of the primitive Earth. Am J Sci 272:537–548
Cloud PE (1974) Rubey conference on crustal evolution (Meeting Report). Science 183:878–881
Counselman CC III (1973) Outcomes of tidal evolution. Astrophys J 180:307–314
Goldriech P, Soter S (1966) Q in the solar system. Icarus 5:375–389
Herrick RR (1994) Resurfacing history of Venus. Geology 22:703–706
Holsapple KA, Michel P (2006) Tidal disruptions. A continuum theory for solid bodies. Icarus 183:331–348
Holsapple KA, Michel P (2008) Tidal disruptions. II. A continuum theory for solid bodies with strength, with applications to the solar system. Icarus 193:283–301
Jeffreys H (1947) The relation of cohesion to Roche's limit: monthly notices. R Astron Soc 3:260–262
Lissauer JJ (1993) Planet formation. Annu Rev Astron Astrophys 31:129–174
Lissauer JJ, Kary DM (1991) The origin of the systematic component of planetary rotation. 1. Planet on a circular orbit. Icarus 94:126–159
Lissauer JJ, Berman AF, Greenzweig Y, Kary DM (1997) Accretion of mass and spin angular momentum by a planet on an eccentric orbit. Icarus 127:65–92
MacDonald GJF (1963) The internal constitutions of the inner planets and the moon. Space Sci Rev 2:473–557
MacDonald GJF (1964) Tidal friction. Rev Geophys 2:467–541
Malcuit RJ, Winters RR (1996) Geometry of stable capture zones for planet Earth and implications for estimating the probability of stable gravitational capture of planetoids from heliocentric orbit. Abstracts Volume, XXVII Lunar and Planetary Science Conference, Lunar and Planetary Institute, pp 799–800
Malcuit RJ, Mehringer DM, Winters RR (1989) Numerical simulation of gravitational capture of a lunar-like body by Earth. In: Proceedings of the 19th Lunar and Planetary Science Conference, Lunar and Planetary Institute, Houston, pp 581–591
Malcuit RJ, Mehringer DM, Winters RR (1992) A gravitational capture origin for the Earth-Moon system: implications for the early history of the Earth and Moon. In: Glover JE, Ho SE (eds) Proceedings Volume, 3rd International Archaean Symposium, vol 22. The University of Western Australia, Crawley, pp 223–235
McCord TB (1966) Dynamical evolution of the Neptunian system. Astron J 71:585–590
McCord TB (1968) The loss of retrograde satellites in the solar system. J Geophys Res 73:1497–1500
Peale SJ, Cassen P (1978) Contributions of tidal dissipation to lunar thermal history. Icarus 36:245–269
Rosenzweig R (2004) Worshipping aphrodite: art and cult in classical athens. The University of Michigan Press, Ann Arbor, p 157
Singer SF (1970) How did Venus loose its angular momentum? Science 170:1196–1198
Valley JW, Peck WH, King EM Wilde SA (2002) A cool early Earth. Geology 30:351–354

Chapter 7
A Retrograde Gravitational Capture Model for the Earth-Moon System

An Alternate Reality Scenario That May Yield Some Insights on the Habitability of Terrestrial Planets

A prograde capture scenario featuring a one moon-mass planetoid appears to explain a number of outstanding features of the Earth-Moon system and a retrograde capture scenario appears to explain a number of the outstanding features of planet Venus (Saunders 1999). A simple question is: COULD THESE SCENARIOS BE REVERSED? The answer is YES! BUT the probabilities of occurrence would be different for the two opposite scenarios.

Much of the probability issue is determined by the geometry of SCZs for Venus and Earth. Since Venus is closer to the Sun, solar gravitational assistance during capture is much lower for prograde capture than for retrograde capture. Stable prograde capture is essentially impossible from an orbit either outside or inside the orbit of Venus. On the other hand, stable retrograde capture of a 0.5 lunar-mass body can be accomplished with about equal facility from either inside orbits or outside orbits using reasonable h and Q values for the encountering planetoid. The bottom line is that for the Venus situation retrograde capture is heavily favored over prograde capture.

The case for Earth is different. The probability of stable capture of a lunar-mass body into a prograde orbit is not much different from the probability of stable capture into a retrograde orbit. When considered in detail, *retrograde* capture is favored because of the size of the SCZs as well as the lower energy dissipation requirements for stable *retrograde* capture. So we can ask the question: WOULD EARTH HISTORY BE DIFFERENT IF A LUNAR-MASS BODY WAS CAPTURED BY EARTH IN A RETROGRADE DIRECTION? There is no doubt in my mind that our world would be different, MUCH DIFFERENT!! The Earth could still have a few billion years of continuously habitable conditions even with a retrograde capture scenario but eventually the result would be a “FATAL ATTRACTION SCENARIO” for planet Earth. Then the twin sister planets would look very similar with each planet having consumed its satellite and in the process becoming a BASALTIC INFERNO.

Robert J. Malcuit, *The Twin Sister Planets Venus and Earth*,
DOI 10.1007/978-3-319-11388-3_7

7.1 Purpose

The purpose of this chapter is to show (1) that a retrograde capture scenario could leave a distinct signature on the lithosphere, hydrosphere, and atmosphere of an earth-like planet as well as on the rotation rate of the planet, (2) that a retrograde capture model is compatible with the development of primitive life forms on an Earth-like planet for a significant portion of time in the history of the planet, and (3) that when we are searching for exoplanets with habitable conditions, the presence or absence (or evidence of a former presence) of a large satellite may be an important factor.

7.2 Overview of a Retrograde Capture Scenario for Earth: A Two-Body Analysis

The early history of Adonis Magnus would be similar to the history of Adonis in Chap. 6 (Malcuit and Winters 1995; Malcuit 2009). The major difference is size. Adonis Magnus is a *lunar-mass body* and it goes through an early history identical to that of Luna in Chap. 4. The early history of Earth is the same as in Chap. 4. It undergoes a COOL EARLY EARTH SCENARIO. After about 600 Ma of independent planet evolution, the Earth and Adonis Magnus find themselves in essentially identical heliocentric orbits and begin to have some orbital "sashays" while in heliocentric orbits.

Figure 7.1 is a scale diagram of a retrograde gravitational capture of a moon-mass planetoid by Earth. Once Adonis Magnus is securely captured, a post-capture orbit circularization occurs (Fig. 7.2). The main trends in this evolution are: (1) there are high and irregular tides affecting both bodies in the early stages of this orbit evolution and (2) the tidal amplitudes on both bodies exponentially decrease as the orbit of Adonis Magnus circularizes. An internal magnetic field is generated within the partially remelted magma ocean of Adonis Magnus and severe tidal action causes partial recycling via subduction processes of the "stagnant lid" crust on the earth-like planet. Once the orbit is circularized, then the crustal complex on the planet tends to return to something like a "stagnant lid" condition.

The early stages of the circular orbit era (Fig. 7.3) are on a long time scale because the tidal friction processes are in a diminished state. Conditions for life forms are good at this time with (1) ocean water on the surface, (2) a fairly rapid rotation rate for the planet, (3) significant ocean tidal action affecting the shoreline areas in a broadly defined equatorial zone, and (4) a healthy meteorological regime. As time goes on and Adonis Magnus orbits closer to the planet, there is more rock and tidal zone action at the surface of the planet: the tidal ranges get higher, the tidal phenomena occur more often and, in general, the surface conditions are more active. But when the tidal amplitudes and ranges get beyond a threshold level, the rock tides initiate basaltic volcanism via crustal breakup, subduction, and associated volcanic arc activity. At this point in time the ocean water begins to evaporate to the atmosphere and the near surface conditions begin to become detrimental for biological entities.

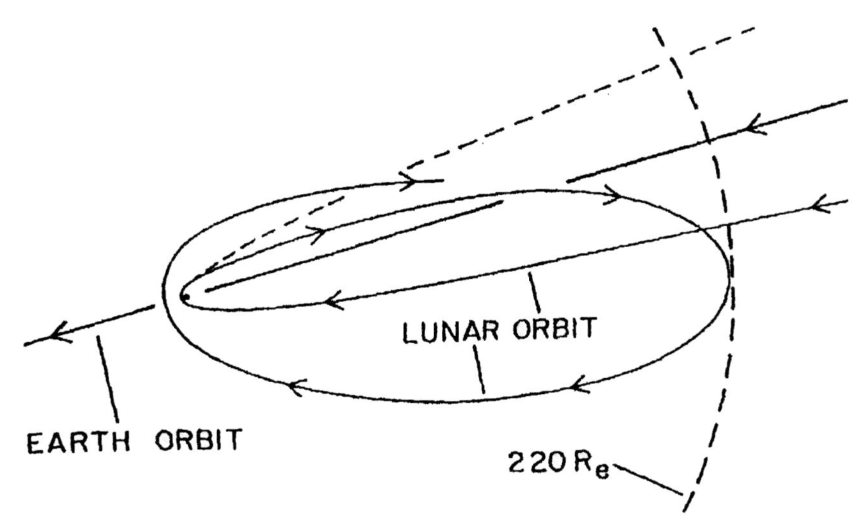

Fig. 7.1 Scale sketch of retrograde gravitational capture of a moon-mass planetoid by Earth. View is from the north pole of the Solar System. Before the encounter both Earth and Adonis Magnus are in very similar heliocentric orbits. Adonis Magnus encounters Earth in the retrograde (*clock-wise*) direction. If no energy is dissipated during the close encounter, then Adonis Magnus escapes from Earth along the dashed line and returns to a heliocentric orbit that is very similar to what it had before the encounter. If sufficient energy is dissipated during the timeframe of the close encounter, then the lunar-like body is inserted into a highly elliptical, but stable, retrograde geocentric orbit

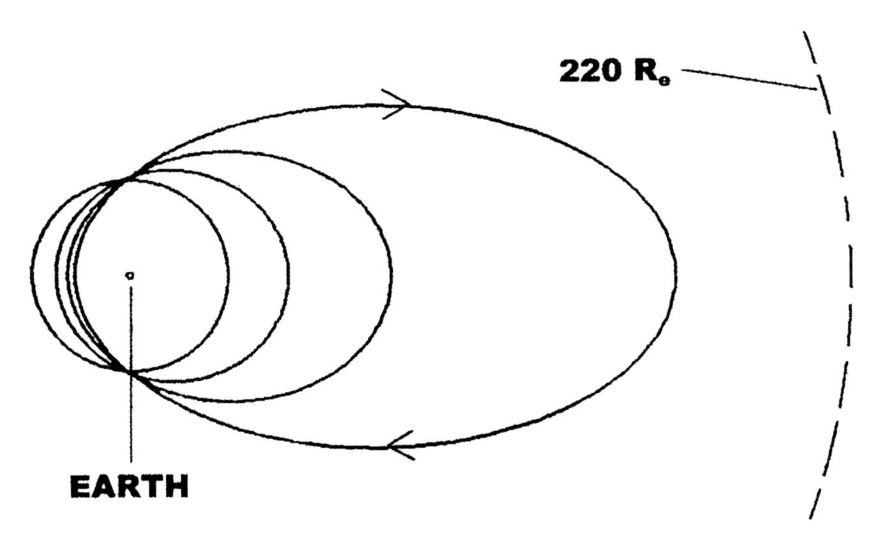

Fig. 7.2 Post-capture orbit evolution. Earth rotation rate at the time of capture is 10 h/day prograde. The lunar-mass satellite is revolving in the retrograde direction. For this two-body calculation the tangential component of the Earth tide is turned off. Thus the orbital angular momentum does not change during the calculation: i.e., the angular momentum of the large elliptical orbit is the same as that of a circular orbit of 30 earth radii. Note also that the calculated two-body evolution timescale is about twice as long as a three-body calculation. View is from the north pole of the Solar System

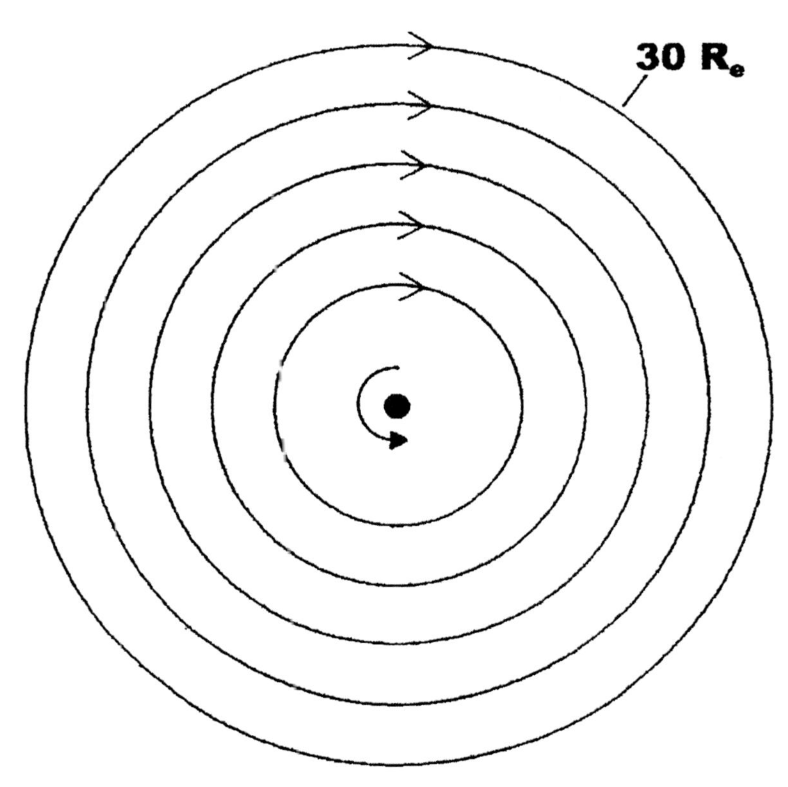

Fig. 7.3 Stages of orbital contraction of the circular lunar orbit in this scenario. Earth rotation rate changes from 10 to 25.8 h/day as the lunar orbital radius changes from 30 to 10 earth radii

Table 7.1 gives a summary of the orbital and tidal parameters for the early part of the circular orbit era. There is a dramatic increase in the tidal amplitudes and ranges in the equatorial zone. The number of hours/day increases from 10 to 25.8 and the months per year increase five-fold. Nonetheless, living conditions are reasonable in the middle to high latitudes of the planet.

Before we proceed to the later part of the orbital contraction scenario, let us look at some three-body simulations of retrograde capture of Adonis Magnus by an earth-like planet.

7.3 Numerical Simulations of Retrograde Planetoid Capture for Earth and a Moon-Mass Planetoid and the Subsequent Circularization of the Post-Capture Orbit

The non-rotating coordinate system as well as the terminology for the three body calculation are presented in Chap 4 and in Malcuit et al. (1989, 1992). Here I will start with a set of retrograde encounters of Adonis Magnus and Earth that can result

Table 7.1 Equilibrium rock tides, day length, and sidereal months per Earth year for this two-body orbit evolution scenario

Orbit rad (R_e)	Tidal amplitude (earth) (m)	Hours/earth day	Months/earth year
30	2.03	10.0	−38.1
25	3.51	11.4	−50.1
20	6.86	13.6	−70.0
15	16.25	17.4	−107.8
10	54.85	25.8	−198.1

in stable retrograde capture if sufficient energy is dissipated within the interacting bodies. Figures 7.4, 7.5, 7.6, 7.7 show a sequence of four simulations with the same starting positions within a retrograde Stable Capture Zone but with different displacement Love numbers. This sequence of simulations will give the reader some feeling for the range of variation of the outcomes.

Figure 7.8 shows the tidal amplitudes for both the earth-like planet and Adonis Magnus.

Figure 7.9 is a plot of the geometry of retrograde SCZs for the Earth-Adonis Magnus system. Table 7.2 gives a summary of the results of a progressive change in h values leading to stable capture.

Figures 7.10, 7.11, 7.12, 7.13, 7.14 show a sequence of diagrams illustrating possible surface and interior effects on an earth-like planet resulting from retrograde capture of a moon-mass satellite and the subsequent orbit circularization. Figure 7.10 is a depiction of the "Cool Early Earth" as deduced by Valley et al. (2002), Zeh et al. (2008), Bell et al. (2010), and others. After a hot accretion and core formation, a chill crust forms over the planet. Some of the early crust founders but eventually a basaltic crust with pockets of granitic rock forms. Additional processing via a petrologic model proposed by Bedard (2006) adds granitic rock masses and extracts some mafic components which delaminate and founder into the mantle. Throughout this "Cool Early Earth" era the earth-like planet has many characteristics similar to planet Mars today: i.e., it would be a one-plate planet featuring a "stagnant-lid" with some volcanic structures on a mainly basaltic crust. In contrast to present day Mars, this earth-like planet has a rotation rate of about 10 h/day, a fairly dense atmosphere of carbon dioxide and nitrogen, shallow oceans of water, and a vigorous global atmospheric circulation system. A very active hydrologic system would cause erosion of the conic volcanic features and granitic domes that are the predicted products of the Bedard (2006) model. This "Cool Early Earth" model features an interesting medley of igneous, sedimentary, and metamorphic rocks.

A Project for Modeling the Cool Early Earth

Modeling this "cool early earth" operation would be an interesting project. The atmospheric circulation system modeling would be a challenge for an earth-like planet with a 10-h rotation rate. Vertical and longitudinal

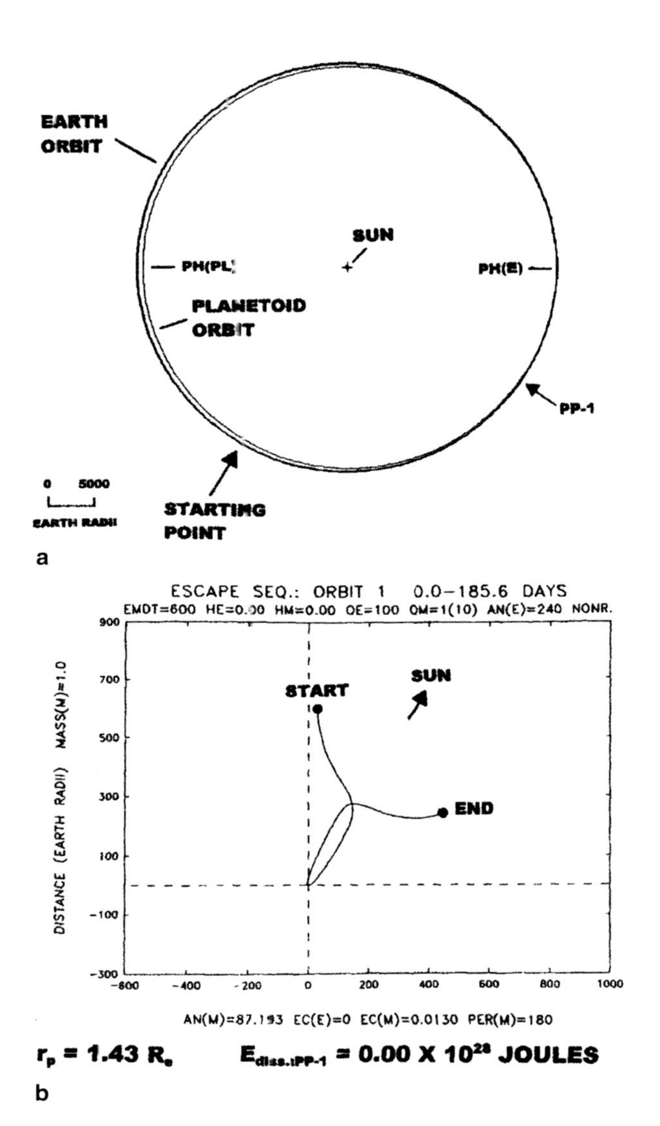

Fig. 7.4 A one-orbit escape scenario. **a** Heliocentric orientation for a retrograde encounter that is within a Stable Capture Zone, in this case for an orbit that is slightly smaller than the orbit of Earth. The starting point of the simulation as well as the position of the close encounter are shown on the plot. **b** Geocentric plot of a numerical simulation that has a favorable orientation for capture but in this simulation no energy is dissipated within the bodies. The position of the Sun is shown for the beginning of the simulation; also shown are the beginning and ending points for the lunar-mass body during the simulation

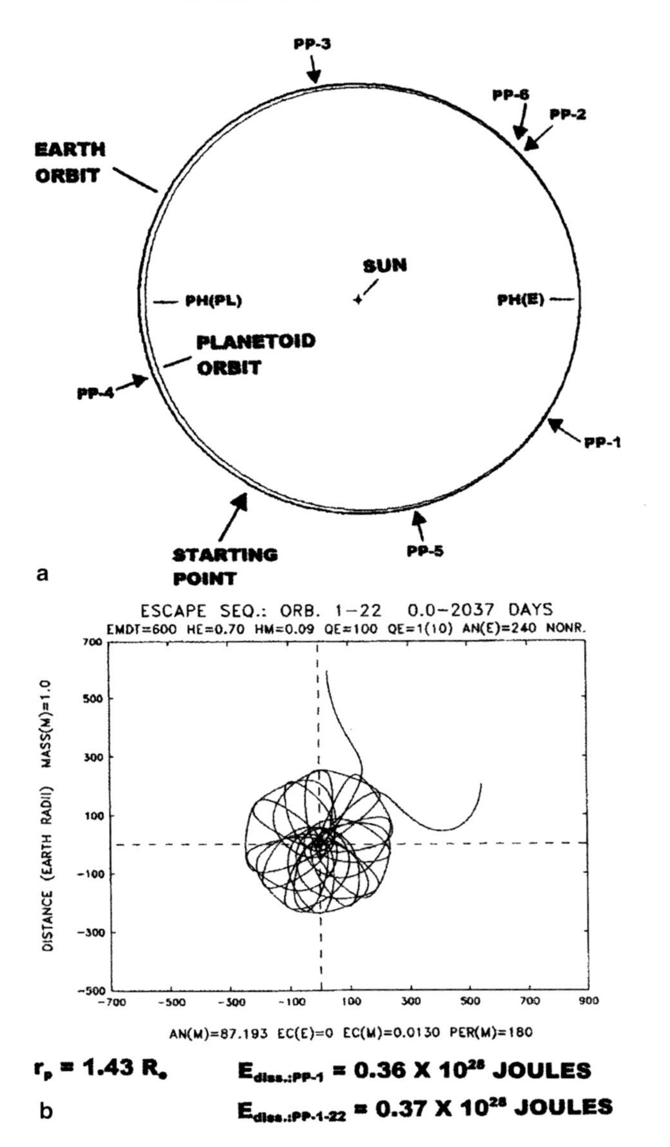

Fig. 7.5 A 22-orbit escape scenario. **a** Heliocentric orientation showing the starting point of the simulation and the positions of the first six perigee passages. **b** Geocentric plot showing the 22 orbits and the subsequent escape after about 5 years of orbits. During the first encounter about 36 % of the energy necessary for capture was dissipated and that quantity of energy was enough to keep the planetoid orbiting the earth-like planet for a while. No subsequent encounter was close enough to dissipate much additional energy. The lunar-mass planetoid escaped from Earth and attained a slightly different heliocentric orbit from which it can have a future encounter with the Earth

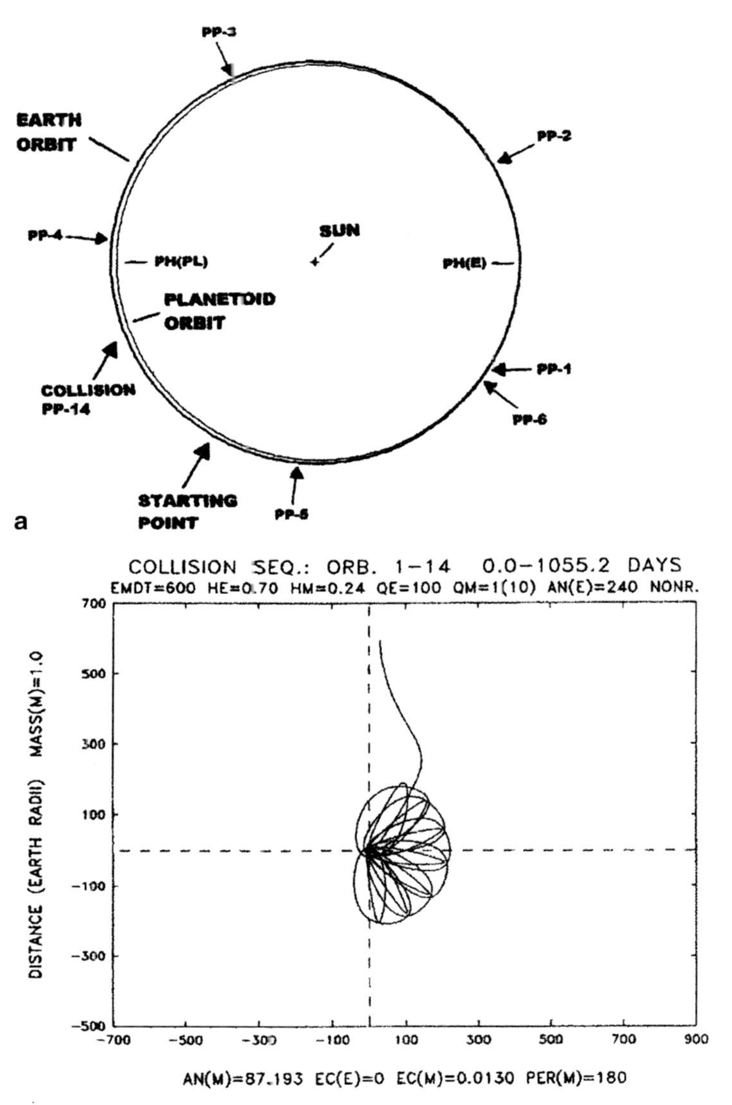

Fig. 7.6 A 14-orbit collision scenario. **a** Heliocentric orientation showing the positions of the first six perigee passages as well as the position of the collision encounter. **b** Geocentric plot showing the simulation of the 14 orbits. About 96 % of the energy dissipation necessary for capture was dissipated during PP-1. This lunar-mass planetoid would be consumed by the Earth

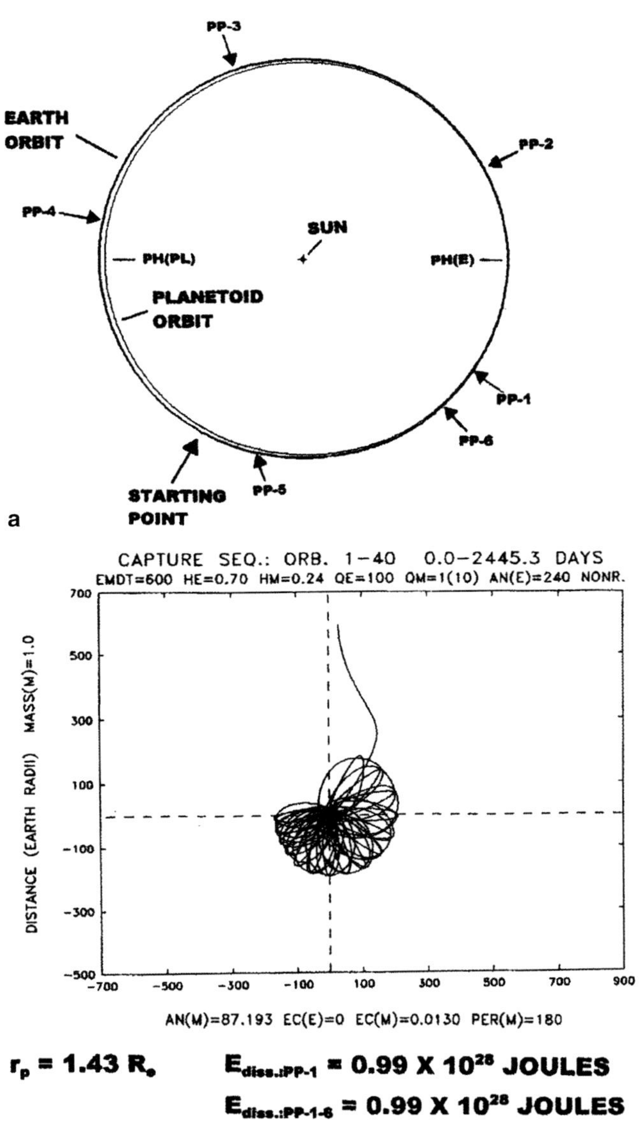

Fig. 7.7 A stable capture scenario. **a** Heliocentric orientation showing the positions of the first 6 perigee passages. **b** Geocentric plot of the first 40 simulated orbits. The planetoid is securely captured after the first perigee passage and eventually other close encounters cause the major axis of the orbit to decrease making the orbit even more stable

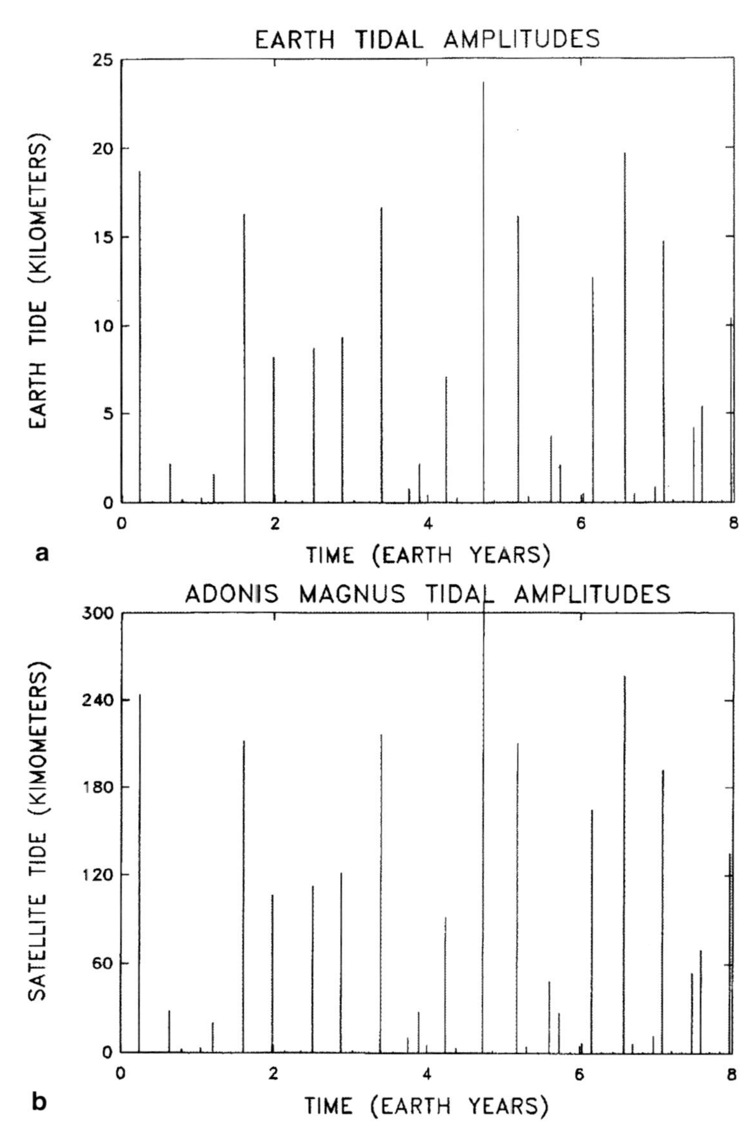

Fig. 7.8 Equilibrium tidal amplitudes for the interacting bodies. **a** Equilibrium rock tides for the earth-like body. **b** Equilibrium rock tides for Adonis Magnus. Note that the highest tides for the planet are in the range of 20–25 km; those for Adonis Magnus are in the range of 240–300 km. Also note that retrograde capture scenarios, in general, are characterized by many more close encounters following stable capture than are prograde capture scenarios. This increased energy dissipation results in the post-capture orbit undergoing much more rapid orbit circularization than in the case of stable prograde capture (e.g., compare and contrast the tidal amplitudes plots above with those in Fig. 4.26)

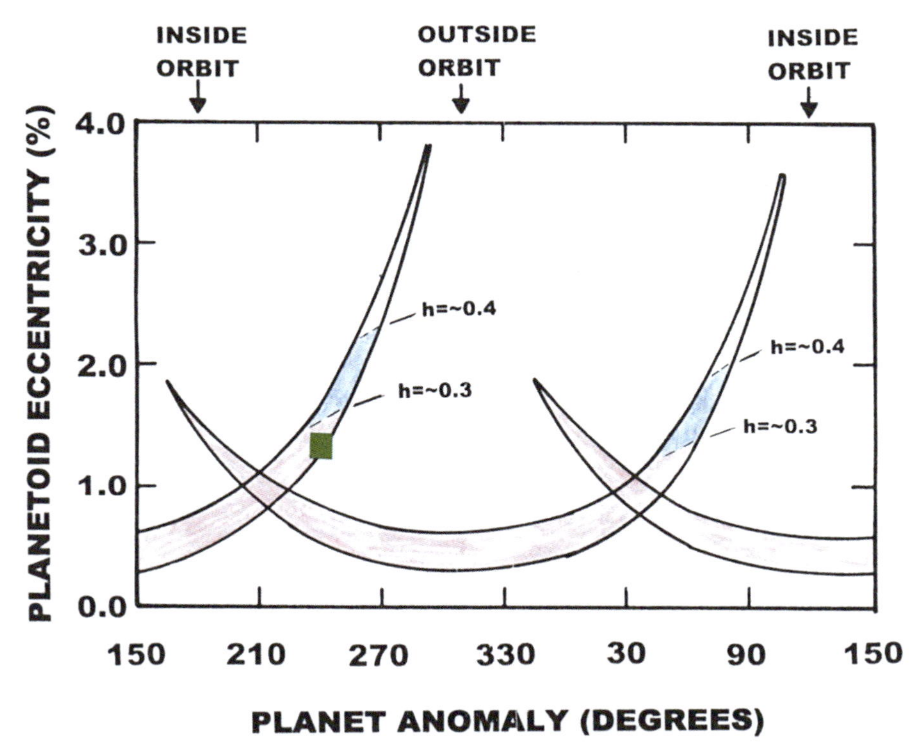

Fig. 7.9 Diagram of retrograde stable capture zones for an earth-like planet and a lunar-mass planetoid showing the position of the parameters for this series of simulations. Note that when examining the h trials in Table 7.2, stable capture does not occur until the planetoid has an h of 0.25. The trials up to $h = 0.25$ are either escapes or collisions with the vast majority being collisions. This propensity for collisions for encounters that are within the Stable Capture regions of the diagram is associated only with retrograde encounters. This factor *decreases* the probability for successful retrograde capture and may have something to do with the ultimate outcome of prograde capture of Luna by Earth

atmospheric zonation as well as composition and partial pressures of the atmospheric gasses would be important factors. Solar radiation would be ~15–20% less than today and the modelers would have some liberty to choose various obliquity angles. The Milankovitch cycles would be operating but the details of their effectiveness depends on the obliquity and other features. The calculations by Laskar et al. (1993) and Laskar and Robutel 1993 give some information that relates to obliquity cycles on a moon-less earth. We must keep in mind that without a large satellite, this "Cool Early Earth" system could have continued for all of geologic time and our world would be much different. The implications for biological systems, at all levels, challenge the imagination. A climate simulation model such as that used by Charnay et al. (2013) for investigating the "faint young Sun problem" would be a place to start with such a project.

Table 7.2 Summary of encounter results as the displacement Love number of Adonis Magnus is sequentially increased to a value where stable capture can occur. The Q value for Adonis Magnus is 1 for the initial encounter and is increased to 10 for all succeeding encounters. For this set of simulations the h value of the earth-like planet is 0.7 and the Q value is 100

h (Adonis M.)	Result	h (Adonis M.)	Result
0.00	1-orbit escape	0.16	3-orbit collision
0.01	1-orbit escape	0.17	3-orbit collision
0.02	1-orbit escape	0.18	3-orbit collision
0.03	1-orbit escape	0.19	3-orbit collision
0.04	1-orbit escape	0.20	3-orbit collision
0.05	2-orbit escape	0.21	3-orbit collision
0.06	3-orbit escape	0.22	3-orbit collision
0.07	3-orbit escape	0.23	3-orbit collision
0.08	3-orbit collision	0.24	14-orbit collision
0.09	22-orbit escape	0.25	100-orbit capture
0.10	4-orbit escape	0.26	100-orbit capture
0.11	4-orbit collision	0.27	100-orbit capture
0.12	4-orbit collision	0.28	100-orbit capture
0.13	4-orbit collision	0.29	100-orbit capture
0.14	4-orbit collision	0.30	100-orbit capture
0.15	5-orbit collision	0.31	100-orbit capture

This "Cool Early Earth" scene comes to an abrupt end when a lunar-mass body gets stably captured into a retrograde orbit. Since stable capture can only occur near the plane of the planets and the obliquity of the planet would probably be low (about 5–25°), the very high rock and ocean tides would affect the equatorial zone (broadly defined) more than the polar regions. The era of extremely high tides would occur for only a few thousand years but elevated tides, especially on an earth-like planet with a rotation rate between 10 and 12 h/day yields a very dynamic rock and ocean tidal regime. Figure 7.10b shows the author's concept of the process of systematic tidally-induced recycling of the Cool Early Earth crust into the mantle of the planet. The process was discussed in some detail in Vignette E of Chap. 5.

Figure 7.12 shows a sequence of orbital states for the evolution of the post-capture orbit circularization era. The calculated timescale for a two-body simulation with h (earth) = 0.7, Q (earth) = 100 and h (satellite) = 0.5, Q (satellite) = 1000 and only radial tides operating (no tangential tidal component) on the earth-like planet is 600 million years. Corrections via the three-body code would decrease this timescale considerably mainly for the early part of the orbit evolution. My guess is that the orbit would be circularized to 10 % eccentricity in 300 million years or less. The short timescale for the tidally induced tectonic activity following capture as well as for the subsequent orbit circularization sequence has severe consequences for the lithosphere inherited from the "Cool Early Earth" time.

Figure 7.11b shows my interpretation of the surface view of the convection cells that would be set in motion as a result of retrograde capture and the subsequent orbit evolution. This general pattern of Tidal Vorticity Induction cells (Bostrom 2000) would persist in the mantle in some form throughout the orbit circularization era.

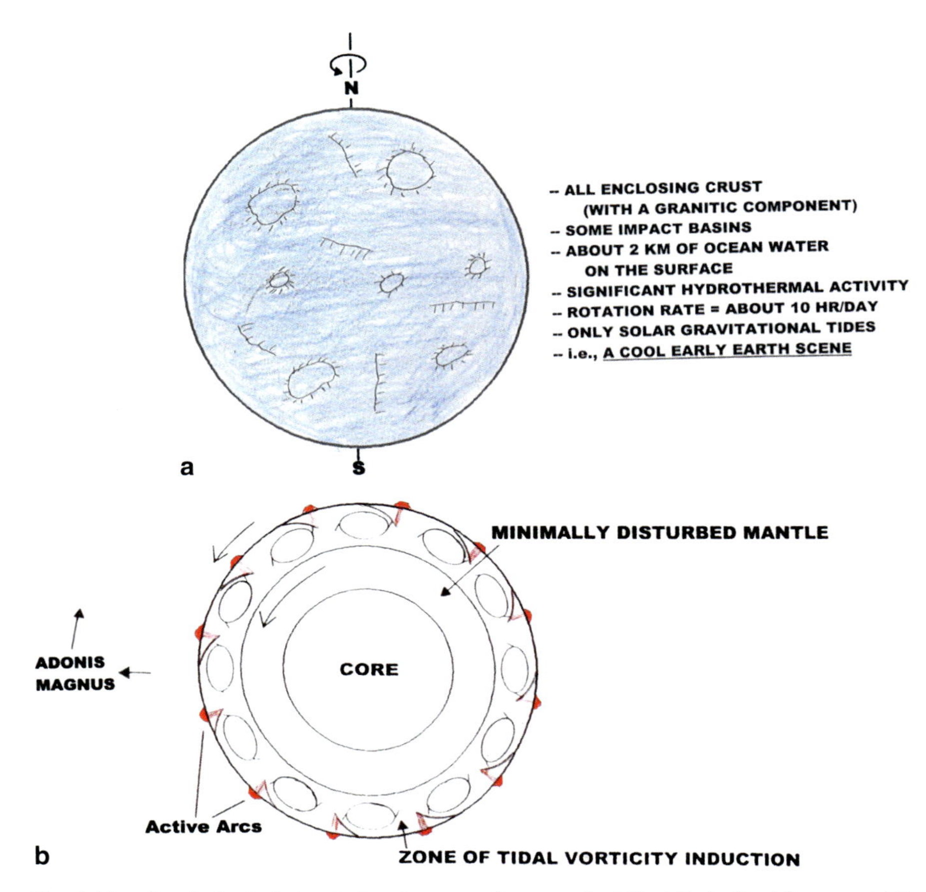

Fig. 7.10 **a** Hemispherical sketch showing some features of a "Cool Early Earth" scene. Additional information is in the text. **b** Equatorial cross-section of an earth-like planet about 1 Ka (5 Ma on the two-body timescale) after capture. Note that Adonis Magnus is revolving in the opposite direction to the rotation of the planet. Note also that the mantle circulation cells rotate in the opposite direction to the rotation of the planet. The concept of Tidal Vorticity Induction is from Bostrom (2000)

The cells would decrease in velocity and change morphology as the rock tidal action decreases and these changes would be manifested by the surface features. Since the lunar-like body dissipates nearly all of the energy for its own capture and a significant quantity of energy for the orbit circularization sequence, the earth-like body, although deformed significantly, stays relatively "cool" throughout this process: i.e., cool enough to permit ocean water to be present at the surface of the planet throughout this tumultuous era in the history of the planet.

During the latter part of the circularization sequence, the TVI cells are widening (flattening) and decreasing in velocity. As a result the subduction zones and associated volcanic arcs are more widely spaced and decreased in length. As the orbit circularizes the tidal energy dissipation within the planet decreases, something like a "stagnant-lid" condition would persist on the planet.

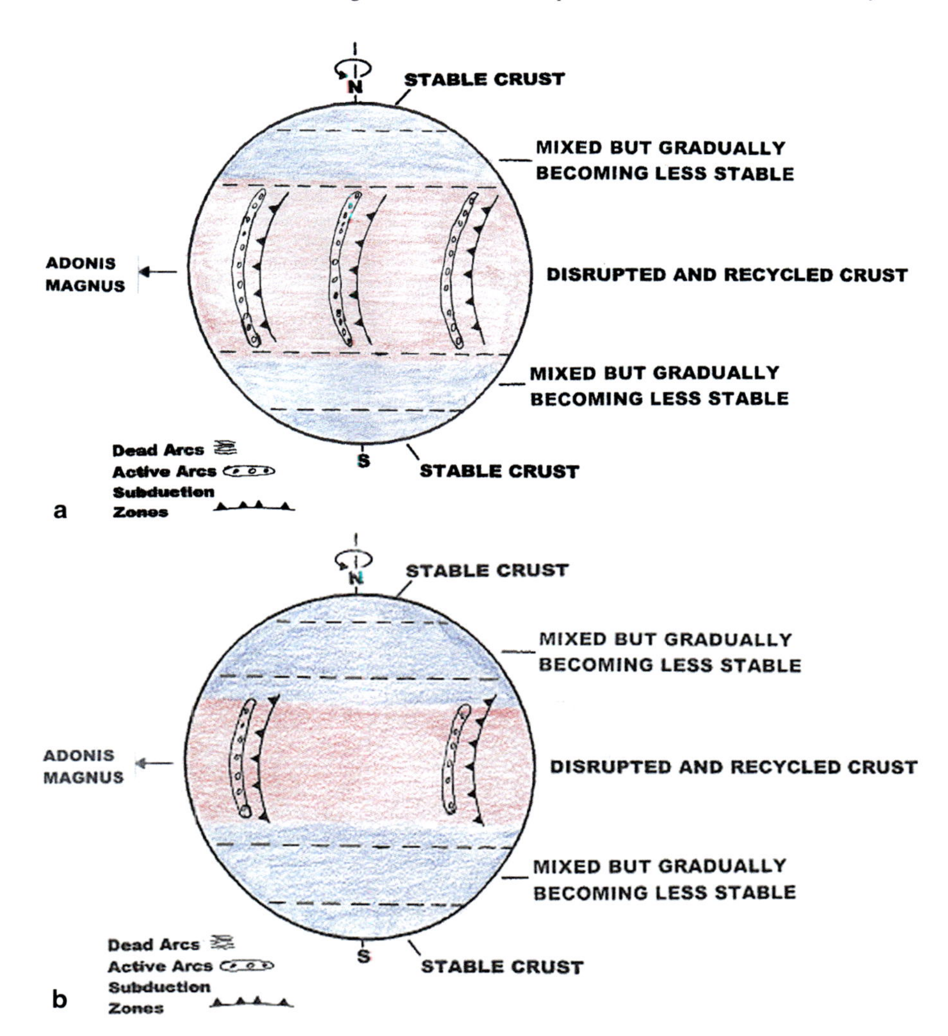

Fig. 7.11 Hemispherical sketch of the earth-like planet showing the zonation of probable effects on the surface at two times during the orbit circularization sequence. The subduction zones on the diagrams are the down-going part of the TVI cell and the volcanic arc results from the forced subduction of this circulation system. The upward moving part of the cell is coming up in front of the volcanic arc. These TVI cells have a 3-D morphology of curved tapered roller bearings and the size of the cell as well as the motion would decrease to essentially zero in the polar zones. **a** The scene soon after capture. Some dynamical characteristics for this time are: minimum distance of closest approach ~4.1 earth radii, maximum rock tidal amplitude ~1000 m, months/year ~8, day ~10 h, continuous cover of water on the surface in the polar zones and in the equatorial zone when the lunar-like body is away from the earth-like body. **b** The scene during the latter part of the circularization sequence. Some dynamical characteristics for this time are: minimum distance of closest approach ~13.2 earth radii, maximum rock tide on the planet ~23.8 m, day ~10 h, continuous cover of water over the surface of the planet

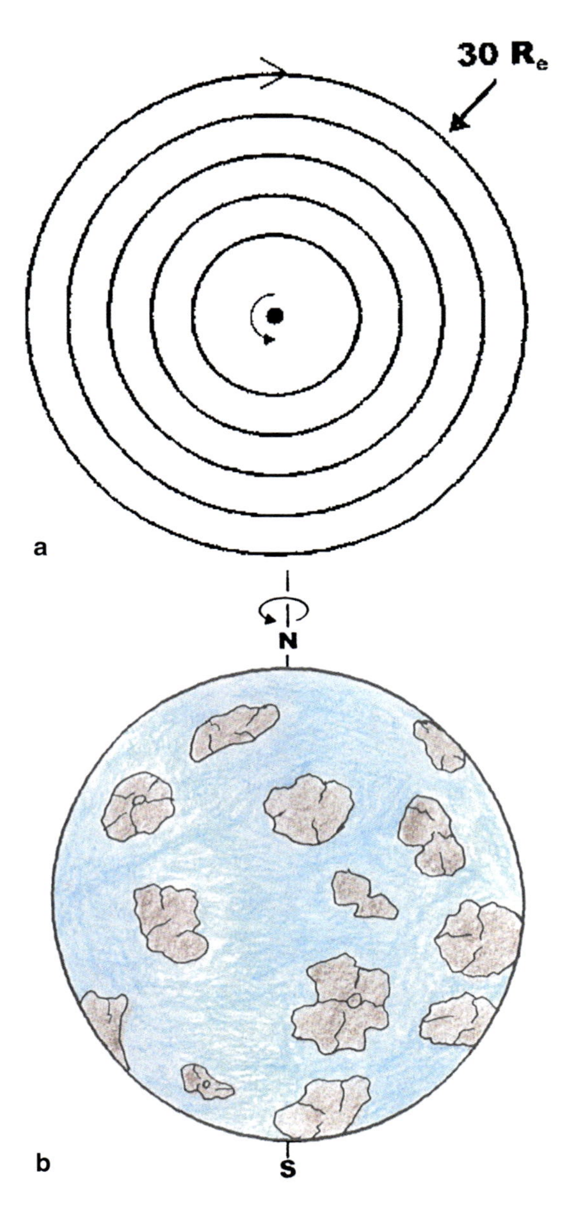

Fig. 7.12 **a**. Scale sketch of the geometry of orbits for the early phase of the circular orbit era. View is from the north pole of the Solar System. **b**. Hemispherical sketch of the probable surface conditions of the earth-like planet during this era. Orbital and tidal information is in Table 7.1

7.4 Circular Orbit Era

Figure 7.12a is a sketch of some orbital states for a circular orbit era. The surface diagram in Fig. 7.12b is the author's concept of what an earth-like planet may look like during a portion of this era. There may or may not be periodic episodes of

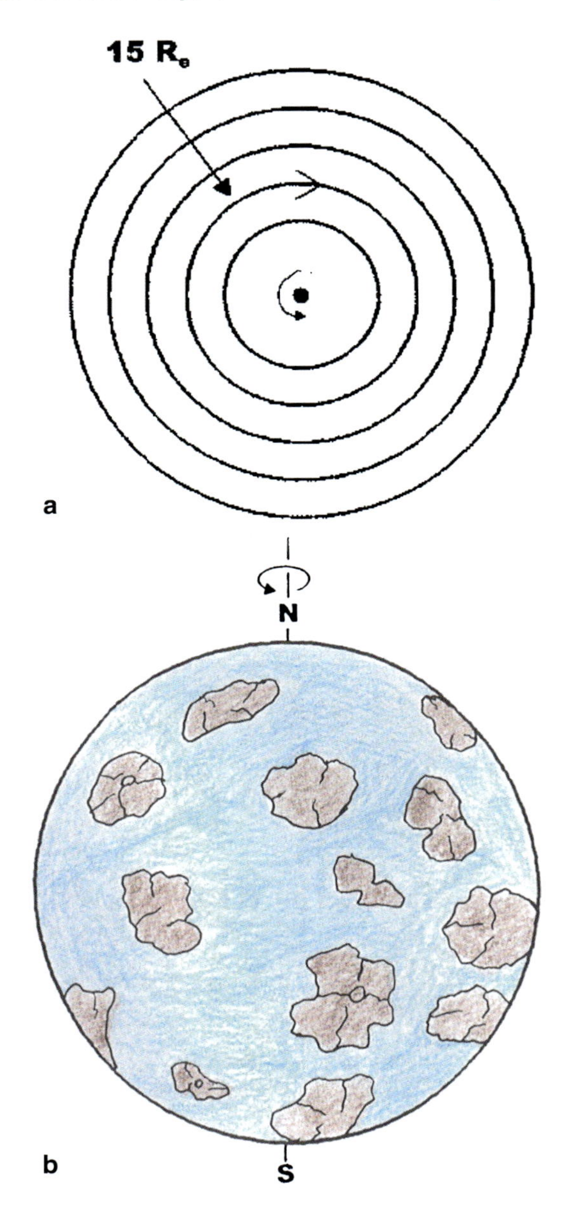

Fig. 7.13 **a** Scale sketch of the geometry of orbits for the late phase of the circular orbit era. View is from the north pole of the Solar System. **b** Hemispherical sketch of the probable surface conditions of the earth-like planet during this era. Orbital and tidal information is in Table 7.1

rock-tide-driven subduction. Certainly the ocean tidal regime would cause much agitation along the shorelines especially in the equatorial zone. The equilibrium rock tidal range in the equatorial zone at 30 R_e is ~3 m.

Figure 7.13 shows a possible scene at 15 R_e when the rotation rate is ~17 h/day and the equilibrium rock tidal range in the equatorial zone is ~24 m. If the mantle of the planet is cool enough, tidally assisted "slab-pull" plate tectonics can be expected

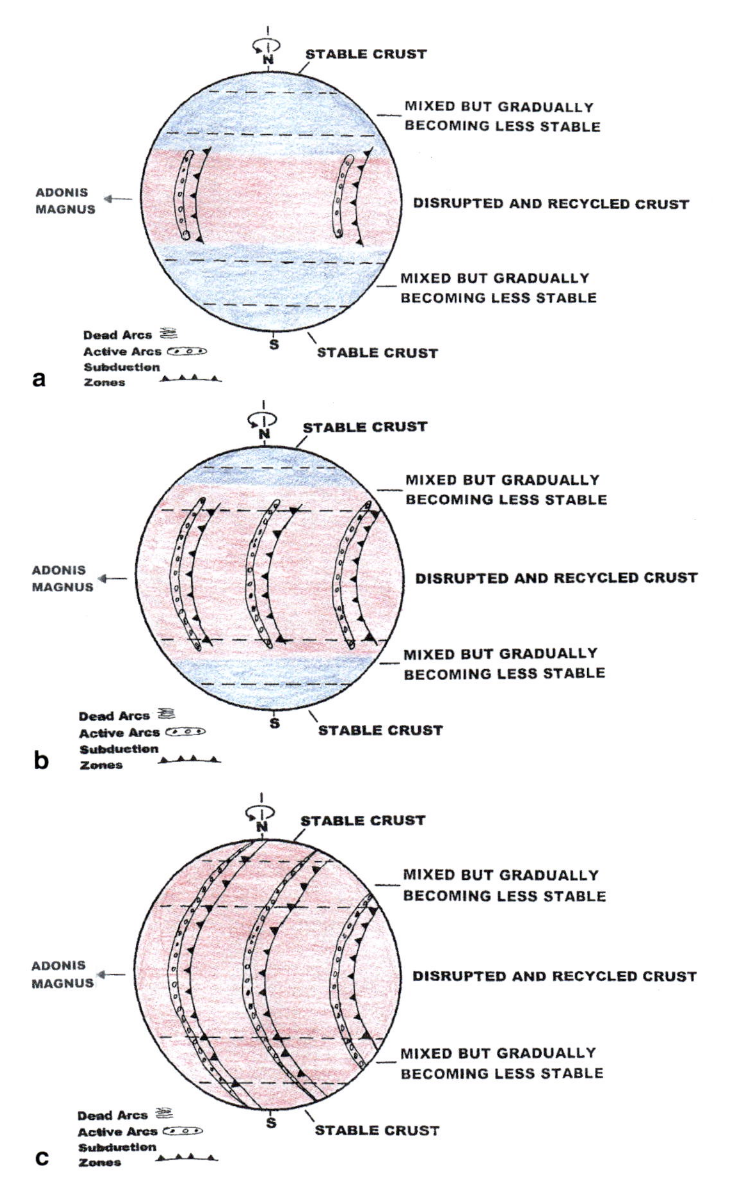

Fig. 7.14 **a** Hemispherical sketch of the probable surface conditions of the earth-like planet when Adonis Magnus is at about 10 earth radii. Note that TVI cells are beginning to operate in the mantle of the planet and that the "stagnant-lid" crust from an earlier era is beginning to be recycled into the mantle. **b** Hemispherical sketch of the probable surface conditions of the earth-like planet when Adonis Magnus is at about 6 earth radii. Note that the crust in the polar regions has not yet been recycled. **c** Hemispherical sketch of surface conditions when Adonis Magnus is at about 2 earth radii. Note that all of crust is now being recycled, at somewhat different rates depending on latitude, by the Tidal Vorticity Induction cells. Although the satellite is well within the classical Roche limit, it is not quite to the Roche limit for a "solid" body

Table 7.3 Equilibrium rock tides, day length, and sidereal months per Earth year for this two-body orbit evolution scenario

Orb. rad (R_e)	Tidal ampli. (earth) (m)	Hours/earth day	Months/earth year
10	54.9	25.8	− 198.1
8	107.1	33.5	− 276.8
6	254.0	50.5	− 426.2
4	857.4	125.7	− 782.9
2	6856.8	− 132.2	− 2214.7

in the equatorial zone of the planet. In general, fairly placid conditions persist on the earth-like planet during this era. For long periods of time early in this era the days would be short (i.e., 10–20 h/day). A shorter day is favorable for life forms because of a more equal distribution of solar energy during the day/night cycle. The number of months per year is more than we humans are accustomed to but since the satellite is in a circular orbit the tidal "beat" is fairly regular. There are about 550–800 days per year but this simply adds more tidal cycles to the environmental mix. In general, the conditions for life forms are reasonable until the rock tides reach about 40-plus meters in the equatorial zone. At this point the rock tides cause significant tectonic action but the life forms can migrate pole-ward where the tidal action is not so severe.

7.5 Late Phase of the Circular Orbit Evolution Era

This later phase of the evolution of the circular orbit transforms the earth-like planet from an abode for life forms to a basaltic inferno very similar to the present state of planet Venus. During this era the orbital radius of Adonis Magnus decreases from 10 earth radii to 2 earth radii and eventually breaks up in orbit and is consumed by the earth-like planet. Table 7.3 gives a summary of a variety of features associated with this sequential orbital evolution.

Figure 7.14 shows three "snapshots" of how the rock tide operation can cause the crustal complex of the earth-like planet to be systematically recycled into the mantle by the rock tide action of Adonis Magnus via the process of Tidal Vorticity Induction (Bostrom 2000). In this model the severe reworking of the surface and upper mantle of the earth-like planet begins when Adonis Magnus is in a circular orbit at about 10 earth radii. At this distance, the rock tides have an equilibrium amplitude of about 54.8 m (range = 82.2 m) in the equatorial zone of the planet. It does seem ironic that these very energetic tidal conditions in the retrograde capture scenario are associated with a planet rotation period near 24 h prograde. In the prograde capture model at a time that the rotation rate is similar, the tidal conditions on Earth are very placid.

Returning to the evolution of the retrograde orbit evolution of the Earth-Adonis Magnus system at 10 earth radii, the number of sidereal months per year is about 198.1 at that point but this number is increasing to over 2200 sidereal months/year at 2.0 earth radii. In general, the orbital motion of Adonis Magnus is much more important for the TVI action than the rotation rate of the earth-like planet. The number of sidereal months per year has been increasing rapidly and eventually the angular momentum exchange between the orbiting satellite and the planet causes a decrease in prograde rotation rate and the planet is eventually forced to rotate very slowly in the retrograde direction when Adonis Magnus is at 2 earth radii.

Now we proceed to the final phase of this FATAL ATTRACTION MODEL featuring the lunar-mass satellite, Adonis Magnus, and an earth-like planet. (See Chap. 6 for more details on the classical Roche limit and the Roche limit for a solid body.) Figure 7.15 shows the final orbital states of this scenario.

After the tripartite break-up of Adonis Magnus, the parts would undergo mutual gravitational perturbations. These interactions consist of close encounters, grazing collisions, and occasional solid collisions. Because of the very low encounter velocities during collisions, there is not much vaporization but the collisions produce many smaller (10's of meters to multi-kilometer-sized) bodies. Mutual gravitational interactions lead to much scattering of the orbits of the particles so that they probably occupy much of the retrograde orbit space. Some are perturbed into prograde orbits as well. The following diagram (Fig. 7.16) will give the reader some idea of what this chaotic orbital motion would look like in two dimensions. Your imagination must put it in three dimensions.

Figure 7.17 has summary diagrams for this scenario for both the time evolution of the semi-major axis of the orbit as well as for the energy dissipation within the earth-like body during this scenario.

7.6 Summary for the Retrograde Capture and Subsequent Orbital Evolution of an Earth-Like Planet and a Lunar-Mass Satellite System

A Vulcanoid planetoid (Adonis Magnus) is gravitationally captured into a retrograde orbit by an earth-like planet about 3.95 Ga ago; the earth-like planet has a prograde rotation rate of about 10 h/day at the time of capture. The planet was *satellite-free* for about 600 Ma and acquired a water inventory from an asteroidal source.

The orbit of the satellite undergoes an orbit circularization by way of mutual rock tidal interactions between the earth-like planet and the newly captured satellite.

The circular orbit era is about 3 billion years long and the conditions are conducive for the origin and evolution of a biological system.

As the circular orbit decreases in radius, the ocean and rock tides get higher and the conditions for the biological system become less favorable and a dense atmosphere of mainly carbon dioxide develops.

Adonis Magnus begins to disintegrate at the Roche limit for a solid body which is at about 1.6 earth radii.

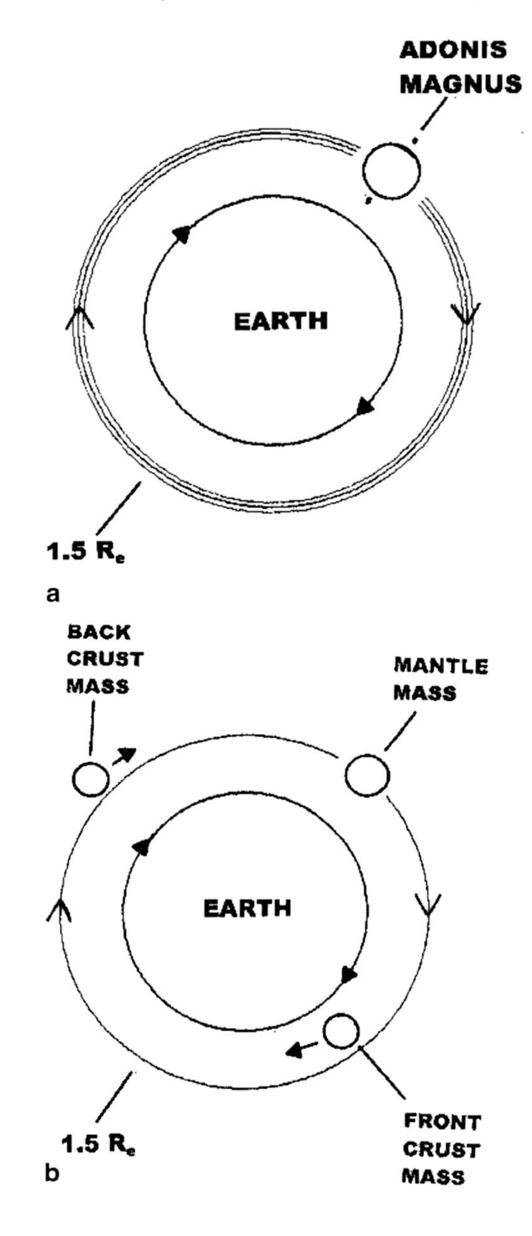

Fig. 7.15 The final orbital states featuring the demise of Adonis Magnus. **a** Adonis Magnus at 1.5 earth radii is slightly within the Roche limit for a solid body and is gravitationally locked on the earth-like planet. The *dots* along the line of centers show the extent of radial tidal deformation of the satellite (with $h = 0.5$), about 20 % radial tidal deformation. This tidal deformation will increase until the satellite breaks into three parts. **b** The initial break-up of Adonis Magnus in the equatorial plane of the earth-like planet. The front-side crust and the back-side crust separate from the mantle mass and all three go into their individual retrograde orbits

After severe tidal stretching along the line of centers of the two bodies, Adonis Magnus separates into three major parts: a front-side crustal mass, a back-side crustal mass, and a mantle mass.

These three masses would have different orbital periods and would be gravitationally perturbing each other fairly regularly.

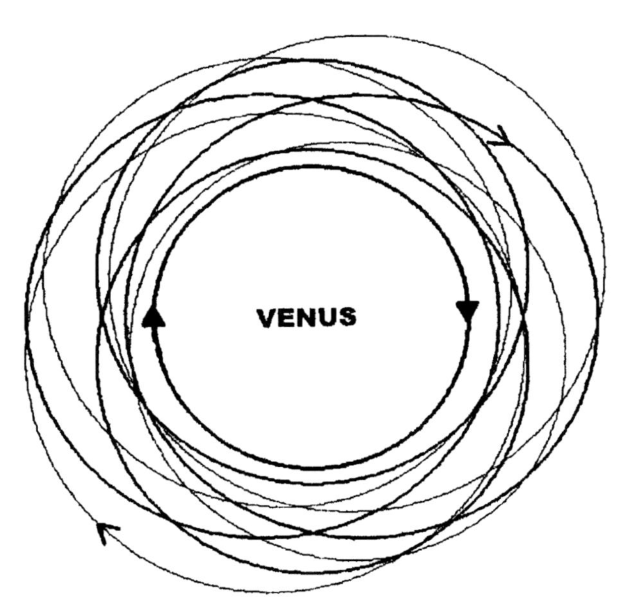

Fig. 7.16 Scale sketch of a two-dimensional version of a swarm of orbiting objects in the equatorial plane about the earth-like planet. The rotation rate of the earth-like planet is slow in the retrograde direction (~79.8 h/day) and the mantle of Adonis Magnus and crustal "chips" are in an array of orbits about the earth-like planet. In this schematic diagram the darker lines represent mantle parts and the lighter lines represent crustal parts or fragments. Most (if not all) of these fragments would eventually be perturbed so that they would enter the dense atmosphere of the earth-like planet and impact onto the basaltic surface of the planet

In a short period of time the bodies would be colliding with each other and chipping off both major and minor particles, each of which becomes another satellite of the earth-like planet.

Eventually the particles in orbit, large and small, interact with the very dense atmosphere of the planet and impact on the surface with a low impact speed.

The resulting planet in this model is a basaltic cauldron rotating very slowly in the retrograde direction and is characterized by a severe "greenhouse" surface temperature of 450–600 °C. The conclusion is that the VULCANOID planetoid partner, the late ADONIS MAGNUS, has left a distinctive and recognizable signature on the planet (see Fig. 7.18).

AGAIN, THIS IS A FATAL ATTRACTION SCENARIO

7.7 Discussion and Implications for the Search for Habitable Exoplanets

Lesson I A large satellite may be a very important factor for developing habitable conditions on a terrestrial planet. The Plate Tectonics mechanism may be all important for keeping a planetary ecosystem operating within certain limits. The very regular and unidirectional rock tides on the Earth may be aiding in the operation of plate tectonics (Nelson and Temple 1972; Scoppola et al. 2006).

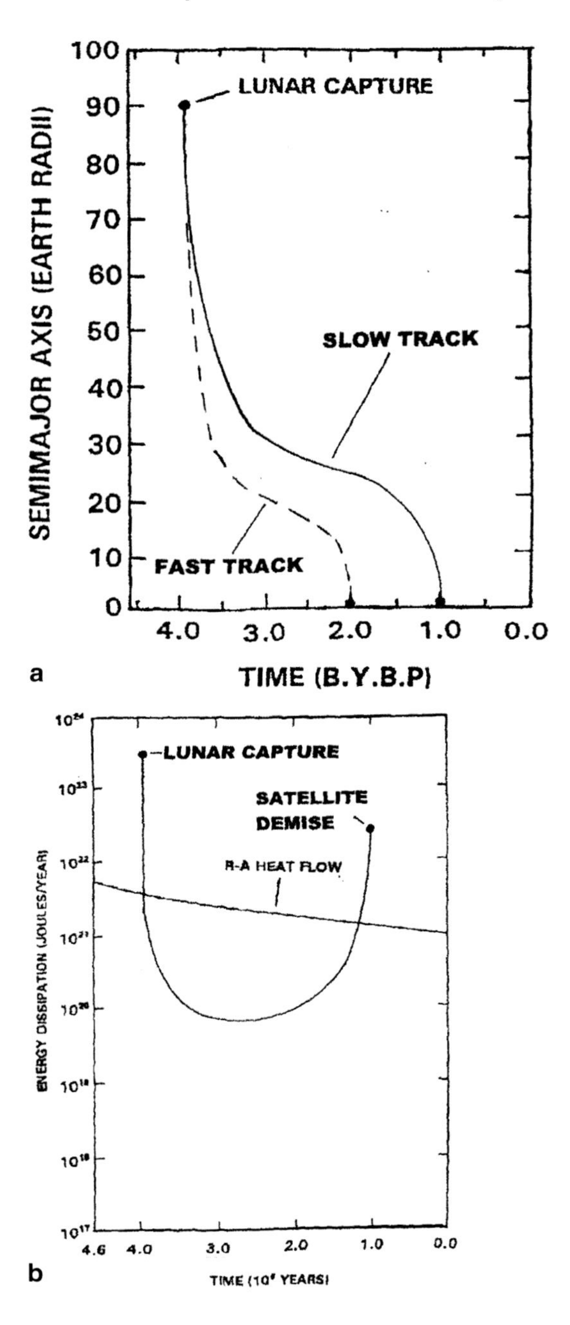

Fig. 7.17 Summary diagram for the evolution of an earth-like planet that captures a lunar-mass satellite in a retrograde orbit. **a** Plot of Semi-major Axis of Adonis Magnus vs. Geologic Time. Note that the "*Slow Track*" and "*Fast Track*" curves depend on the rate of dissipation within the bodies (i.e., it depends on the "Q" of the bodies). **b** Plot of Energy Dissipation (within the earth-like body) vs. Geologic Time

Lesson II The gravitational capture mechanism is physically possible, under certain specified conditions, for inserting large planetoids into a planetocentric orbit. The direction of capture may *not* be important for determining the habitability of a

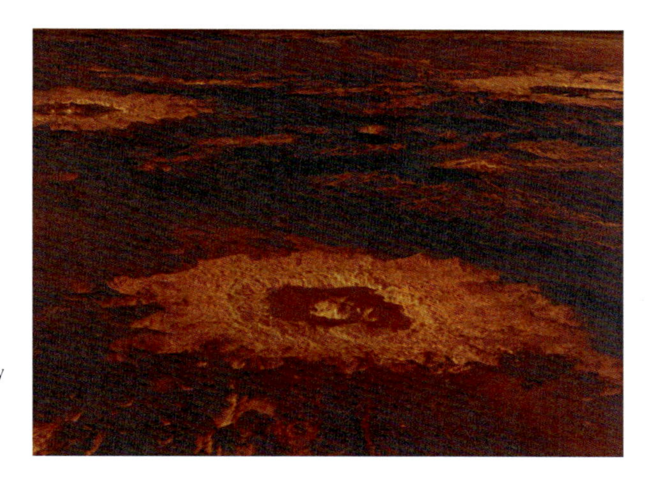

Fig. 7.18 A view of the alternate Earth. The earth-like planet is characterized by (1) a dense atmosphere rich in carbon dioxide, (2) an atmospheric pressure about 100 times the present 1 atmosphere of pressure of planet Earth, (3) a surface temperature about 450–600 °C compared to the present global average of about 15 Celsius, and (4) a rotation rate about 54.1 h/day *retrograde* compared to our present 24 h/day prograde. (Courtesy of NASA/JPL)

terrestrial planet. Our large satellite is in a prograde orbit, but the point of this chapter (and the previous chapter) is that retrograde capture could also yield habitable conditions for several 100 Ma of geologic time.

Lesson III Liquid water is a very important ingredient for developing habitable conditions on a terrestrial planet. But the presence of a large satellite may be of nearly equal or greater importance for determining habitability.

References

Bedard JH (2006) A catalytic delamination-driven model for coupled genesis of Archaean crust and sub-continental lithospheric mantle. Geochim Cosmochim Acta 70:1188–1214

Bell EA, Harrison TM, McCulloch MT, Young E (2010) Early Archean crustal evolution of the Jack hills zircon source terrane inferred from Lu-Hf, $^{207}Pb/^{206}Pb$, and delta ^{18}O systematics of Jack hills zircons. Geochim Cosmochim Acta 75:4816–4829

Bostrom RC (2000) Tectonic consequences of earth's rotation. Oxford University Press, New York, p 266

Charney B, Forget F, Wordsworth R, Leconte J, Millour E, Codron F, Spiga A (2013) Exploring the faint young sun problem and the possible climates of the Archean Earth with a 3-D GCM. J Geophys Res Atm 118:10414–10431

Laskar J, Robutel P (1993) The chaotic obliquity of the planets. Nature 361:608–612

Laskar J, Joutel F, Robutel P (1993). Stabilization of the Earth's obliquity by the moon. Nature 361:615–617

Malcuit RJ (2009) A retrograde planetoid capture model for planet Venus: implications for the Venus oceans problem, an era of habitability for Venus, and a global resurfacing event about 1.0–0.5 Ga ago. Geol Soc Am Abstr Progr 41(7):266

Malcuit RJ, Winters RR (1995) Numerical simulation of retrograde gravitational capture of a satellite by Venus: implications for the thermal history of the planet. Abstracts Volume, 26th Lunar and Planetary Science Conference. Lunar and Planetary Institute, pp 829–830

Malcuit RJ, Mehringer DM, Winters RR (1989) Numerical simulation of gravitational capture of a lunar-like body by Earth. In: Proceedings of the 19th Lunar and Planetary Science Conference, Lunar and Planetary Institute, Houston, pp 581–591

Malcuit RJ, Mehringer DM, Winters RR (1992) A gravitational capture origin for the Earth-Moon system: implications for the early history of the Earth and Moon. In: Glover JE, Ho SE (eds) Proceedings Volume, 3rd International Archaean Symposium. The University of Western Australia Publication 22, Crawley, pp 223–235

Nelson TH, Temple PG (1972) Mainstream mantle convection: a geological analysis of plate motion. Am Assoc Petrol Geol Bull 56:226–246

Saunders RS (1999) Venus. In: Beatty JK, Petersen CC, Chaikin A (eds) The new solar system, 4th edn. Cambridge University Press, Cambridge, pp 97–110

Scoppola B, Boccaletti D, Bevis M, Carminati E, Doglioni C (2006) The westward drift of the lithosphere: a rotational drag? Geol Soc Am Bull 118:199–209

Valley JW, Peck WH, King EM, Wilde SA (2002) A cool early earth. Geology 30:351–354

Zeh A, Gerdes A, Klemd R, Barton JM Jr (2008) U-Pb and Lu-Hf isotope record of detrital zircon grains from the Limpopo Belt—evidence for crustal recycling at the Hadean to early Archean transition. Geochim Cosmochim Acta 72:5304–5329

Chapter 8
Planet Orbit—Lunar Orbit Resonances and the History of the Earth-Moon System

Some Special Perks for Earth by Having Jupiter and Venus in the Neighborhood

"Another resonance of the above type is also stable. This keeps the perigee in step with jupiter's orbital motion at $a/a_{earth} = 53.4$. However, variations in the Earth's orbital eccentricity seem to make capture into this latter resonance unlikely (Yoder, private communication, 1977). Should capture have occurred, however, the effects on orbital history would be profound and significant heating of the Moon would occur if the eccentricity were driven to sufficiently large values."
From Peale and Cassen (1978, p. 260).

"If we choose starting positions of Venus and the earth that correspond to an inferior conjunction of Venus with the sun as observed from the earth, eight years later there would be a similar conjunction at which the sun and Venus would occupy positions in the sky very close to their starting positions. Consequently, conjunctions of the type discussed in the previous section should recur at eight-year intervals. In the same way, other phenomena of Venus such as superior conjunctions, eastern elongations, and western elongations should exhibit eight year cycles."

"After five conjunctions or eight years, they end up where they started."
From Chapman (1986, pp. 340–341).

After reviewing a number of geology and astronomy textbooks, a reader gets the feeling that the Moon is not all that important in the development of our habitable planet. The Moon raises ocean tides on the planet and it serves as a "night lantern" and these features of the Moon have been important for some human endeavors (e.g., in production of food crops in coastal areas and, in some special circumstances, in military campaigns at specific times in human history). But the question posed here is: HAS THE MOON BEEN IMPORTANT IN THE DEVELOPMENT OF PLANET EARTH INTO THE ONLY PLANET THAT WE KNOW OF THAT IS HABITABLE TODAY, AFTER NEARLY 4.6 BILLION YEARS OF GEOLOGICAL EVOLUTION?

My answer is an unqualified "YES". For starters, let us consider the rotation rate of Earth. According to MacDonald's (1963) analysis, the initial rotation rate of

Robert J. Malcuit, *The Twin Sister Planets Venus and Earth*,
DOI 10.1007/978-3-319-11388-3_8

Earth was about 10 h/day and it is now 24 h/day. Most scientists that are interested in this problem of Earth rotation will agree that tidal interactions with the Moon over geologic time have decreased the rotation rate of our planet to the present, very comfortable 24 h/day.

But is there more to the tidal evolution of the Earth-Moon system than the monotonic decrease in the rotation rate as the lunar orbital radius increases over time? I think that there is and I think that some of the crucial events in Earth History, such as the "GREAT OXIDATION EVENT" (about 2.45–2.32 Ga ago) (Holland 2002) and the "SECOND OXIDATION EVENT" (about 1.0–0.6 Ga ago) (Shields-Zhou 2011) are related to special events in the history of the lunar orbit. In my view, these "geological episodes" may be associated with PLANET ORBIT—LUNAR ORBIT resonances that were caused by gravitational perturbations by planets JUPITER and VENUS. Planet Jupiter can exert its special influence only when the lunar orbit has a radius of about 53.4 earth radii (Peale and Cassen 1978). But planet Venus has three opportunities for affecting life on Earth: the *first* is at ~47 earth radii (centered on ~2.45 Ga), the *second* is at ~55 earth radii (centered on ~0.5 Ga), and the *third* is at ~68 earth radii (centered on ~600 Ma in the future: i.e., in the Futurozoic).

The concept of PLANET ORBIT—LUNAR ORBIT RESONANCES has been mentioned by a few planetary physicists (e.g., Peale and Cassen 1978; Touma and Wisdom 1994, 1998; Cuk 2007) and falls under the heading of EVECTION RESONANCES. But to my knowledge no one has applied this concept to any specific episodes in Earth history, lunar history, or Earth-Moon system history although recently Cuk and Stewart (2012) and Canup (2013) have attempted to apply this concept to a certain phase in the evolution of the Earth-Moon system following a giant impact scenario.

The key factor in a PLANET ORBIT—LUNAR ORBIT RESONANCE model is the period of the perigean cycle of the Earth-Moon system. The present period of the perigean cycle is ~8.85 years in the prograde direction. As the lunar orbit *decreases* in size, the period of the perigean cycle *increases*.

8.1 Purpose

The purpose of this chapter is (1) to demonstrate in significant graphic detail how this special gravitational interaction of planets Jupiter and Venus can influence the lunar orbit and (2) to demonstrate how the resulting rock and ocean tidal activity may have helped to shape the chemical, physical, and biological world into the special habitable planet what we know as PLANET EARTH.

8.2 The Perigean Cycle for the Earth-Moon System

There are two long-term cycles associated with the evolution of the Earth-Moon system that are of significance to earth scientists:

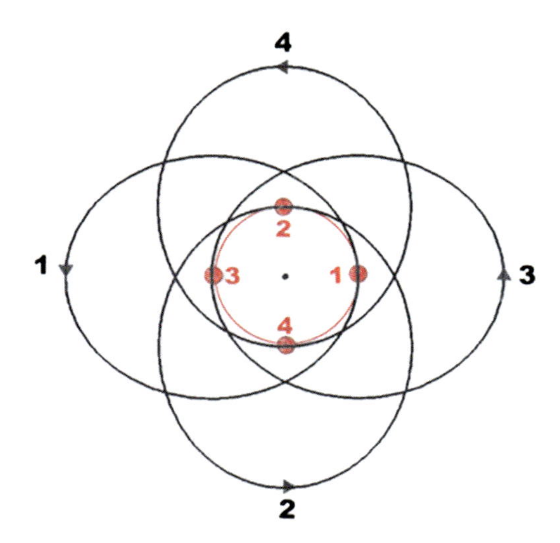

Fig. 8.1 Scale diagram of four sidereal months of lunar orbits illustrating the prograde progression of orbits in the perigean cycle. View is from the north pole of the Solar System. Apogee positions are marked in black; perigee positions are marked in red. The perigean cycle is the time for the perigee positions to complete a full circle

1. The NODAL CYCLE in which the nodes (intersection of the lunar orbital plane and the ecliptic plane) regress (i.e., retrograde motion) at the rate of 18.6 years/cycle (at present).
2. The PERIGEAN CYCLE in which the position of the perigee (monthly closest approach distance) moves in a prograde direction around the planet in 8.85 years/cycle (at present).

The NODAL CYCLE is recorded in some ancient tidal sequences [e.g., the Upper Mississippian Pride Formation (Miller and Eriksson 1997)] but I do not recall any report on evidence in the rock record for the PERIGEAN CYCLE. Nonetheless, the important cycle for this model is the *PERIGEAN CYCLE*. Figure 8.1 shows a simplified view of the perigean cycle.

In general, the length of the perigean cycle increases as the semi-major axis of the lunar orbit decreases. For example, the perigean cycle at present, with the lunar orbit semi-major axis at ~60.3 earth radii, is 8.85 years and at 47 earth radii the length of the perigean cycle is ~15 years. Figure 8.2 is a plot showing the present position of the perigean cycle as well as the positions of the perigean cycle values for the four orbital resonances discussed in this chapter.

Jupiter and Venus are two planets in the Solar System that can affect planet Earth in significant ways by way of interaction with the Moon in orbit about the Earth. Jupiter has the strongest influence because it is so massive and our twin sister, Venus, has about one-half as much influence but it is very close. Figure 8.3 shows the relevant geometry. The critical gravitational interactions take place when the Earth is near or at the sub-Jupiter or sub-Venus point in its orbit. These resonant episodes can only occur when the lunar orbit is at certain orbital radii. The duration of each of these episodes, in my models, occur over a few 100 million years of geologic time and result in periods of elevated rock and ocean tides that should leave a testable signature in the rock record of our planet.

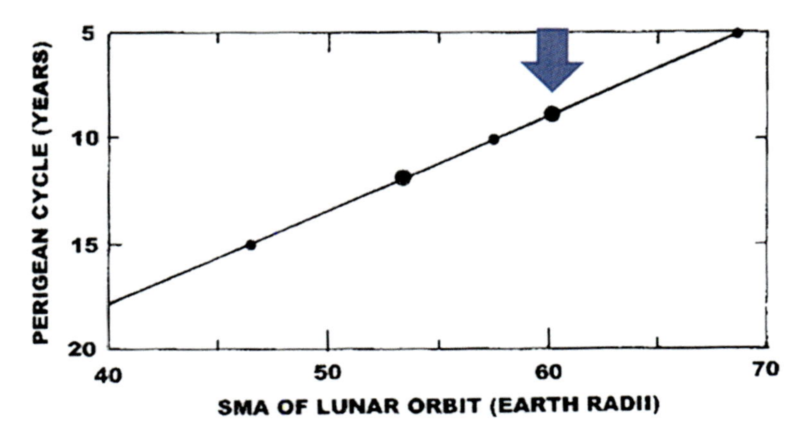

Fig. 8.2 Plot of Lunar Orbital Radius versus Length of the Perigean Cycle for lunar orbital radii between 40 and 70 earth radii. The small solid circles represent times in the history of the lunar orbit when planet Venus can interact with the lunar orbit via the perigean cycle. The large solid circle with the blue arrow (~60 earth radii) represents the present situation (perigean cycle ~8.85 years) and the other large solid circle at 53.4 earth radii represents a time at which the lunar orbit can be in resonance with planet Jupiter [perigean cycle ~11.8 years (Peale and Cassen 1978)]

My general scheme of presentation is to go with the strongest orbital resonance first. This is the Jupiter Orbit—Lunar Orbit resonance. This orbital resonance was identified by Peale and Cassen (1978) when they were interested in developing models for the thermal history of the Moon. I do think that there is evidence that the Moon was affected thermally during the era of interest, but my main interest is in the effects of this orbital resonance on the ocean and rock tidal regimes of planet Earth.

8.3 A Jupiter Orbit—Lunar Orbit Resonance

The Late Proterozoic Era is viewed by many geologists as an unusual time in Earth History. Knoll (1991), Windley (1995), and Conway-Morris (1990) present significant evidence that both the biological and tectonic worlds were involved.

Some of the outstanding events are:

1. There is evidence for two apparently global glaciations (and perhaps a few more) in which glacial ice was present in the equatorial zone at sea level (Hoffman et al. 1998).
2. There is evidence for significant excursions in the carbon-12 to carbon-13 isotope ratios (see Fig. 8.4).
3. There is evidence for low strontium-87 to strontium-86 ratios in marine carbonates implying enhanced hydrothermal activity (probably indicating enhanced ocean floor formation) during specific geological eras (Fig. 8.4 and Halverson and Shields-Zhou 2011).

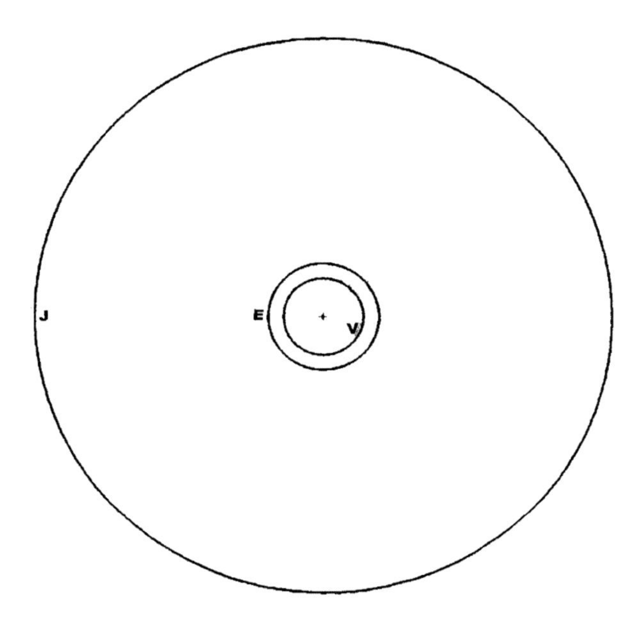

Fig. 8.3 Orbital geometry for the gravitational perturbers that cause the purported orbital resonances. The best index for the effects is the M_{planet}/R^2 at inferior or superior conjunction. M is the mass of the planet and R is the distance of separation of Earth and the perturbing planet at closest approach. An *Inferior Conjuction* occurs when a planet like Venus is between Earth and the Sun. A *Superior Conjunction* occurs when Earth is between a planet like Jupiter and the Sun. The M/R^2 value for Jupiter at superior conjunction is about 18 and the value for Venus at inferior conjunction is about 9. Thus, Jupiter has about *twice* the gravitational influence on the Earth, and the Moon in a geocentric orbit, as does Venus. (View is from the north pole of the Solar System)

4. There is an abundance of evidence for the development of continental rift zones: southern Africa, southern Australia, northwest Canada, western United States (Utah and Idaho), and eastern United States (Mount Rogers, Mitchum River, and Grandfather Mountain complexes in Virginia and North Carolina) (Bailey et al. 2007; Link et al. 1994; Miller 1994; Powell et al. 1994; Priess 1987; Rankin 1975, 1993).

Somehow, subsequent to all of this jumbling of events, or as a result of them, we have the development of the Ediacaran and metazoan life forms (Shields-Zhou and Och 2011). Figure 4 is a pictogram representing some of these events.

8.3.1 Geometry of a Jupiter Orbit—Lunar Orbit Resonance

In this resonance the critical alignment of planet Jupiter and the lunar orbit is associated with a perigean cycle that is 12 years long. But as the resonance state is approached, the forced eccentricity episodes occur for about 4 monthly cycles/year.

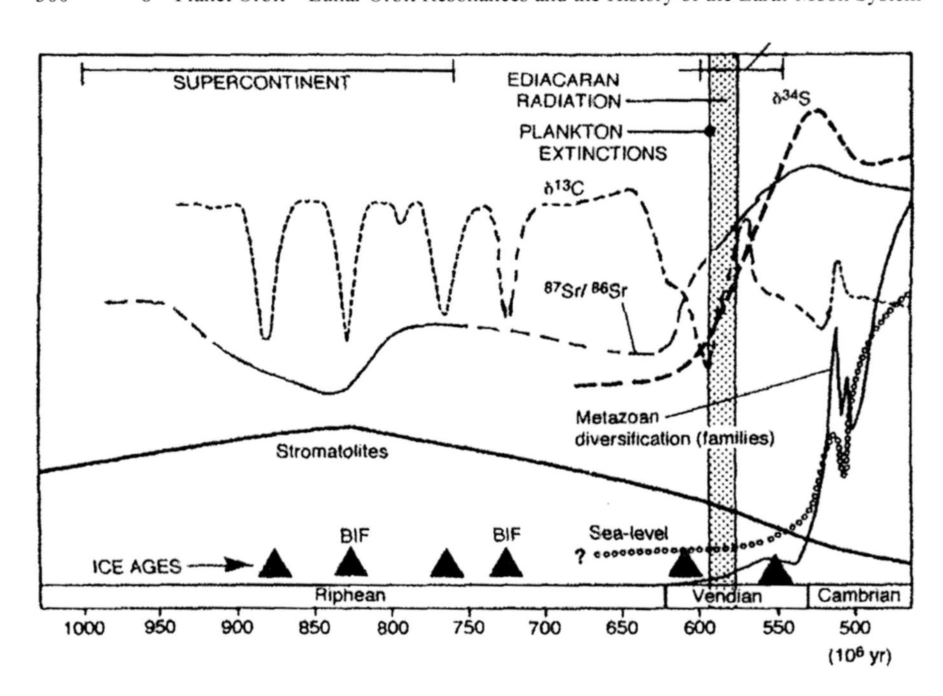

Fig. 8.4 Diagram showing several of the features of the Late Proterozoic that are mentioned in the text. (Diagram from Windley 1995, Fig. 13.1, with permission from John Wiley and Sons)

Figure 8.5 shows the geometry for about one and one-twelfth years. Jupiter is orbiting the Sun at the rate of about one orbit every 12 years (really 11.86 years/cycle). Figure 8.6 shows the orientation of the most favored lunar orbit for gravitational forcing for a Jupiter orbital cycle of about 12 earth years (11.86 years). (Note: All orbits in this section are viewed from the north pole of the Solar System unless stated otherwise.)

Figure 8.7 shows a general summary of the predictable stages in a FORCED ECCENTRICITY SCENARIO. Figure 8.7a shows the initial orbit of low eccentricity before the gravitational perturbations of the orbital resonance (orbit state 1). Both the semi-major axis and eccentricity are increased by the resonant motion to a maximum value represented by orbit state 2. This forced eccentricity is estimated to occur over a short period of geologic time (about 100 million years); thus the orbital angular momentum of orbit state 2 is only slightly higher from that of orbit state 1 because the SMA of the lunar orbit increases much more rapidly than the transfer of angular momentum from the rotating Earth to the lunar orbit. Figure 8.7b shows an intermediate orbit state (orbit state 3) where the semi-major axis is maintained at the resonant value of 53.4 earth radii but the angular momentum of the lunar orbit is increased relative to orbit stage 2 because of angular momentum transfer from the rotating Earth to the lunar orbit. Orbit state 4 is a circular orbit of low eccentricity representing a lunar orbit that is beyond the influence of the orbital resonance.

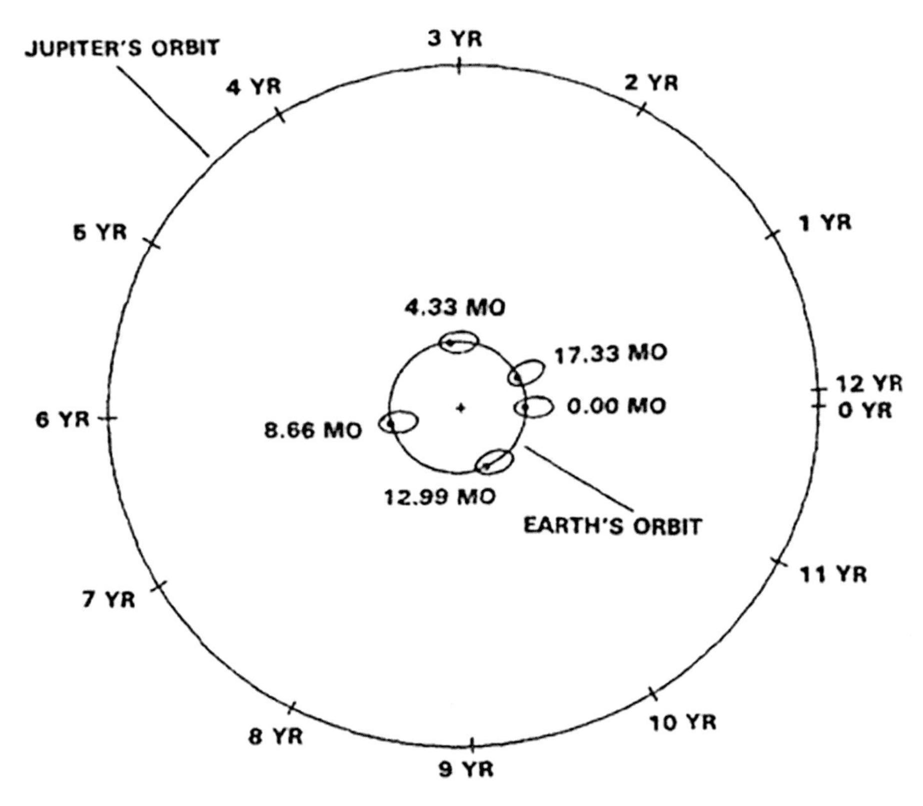

Fig. 8.5 Diagram summarizing one and one-twelfth years of lunar orbits when the perigean cycle is 12 earth years long. Note that throughout the year, the apogee of the lunar orbit is pointing toward planet Jupiter. With a semi-major axis of 53.4 earth radii, there are ~16 sidereal months/year. Thus after 17.33 months (1 plus 1/12 earth years) the apogee of the lunar orbit is still pointing directly at Jupiter. The only significant gravitational "tugs" can occur when the earth's gravitational influence is weakest (i.e., when the Moon is at apogee near the sub-Jupiter point). Although the lunar orbit is pointing toward Jupiter throughout the annual cycle, the strongest influence would occur only for a 4-month period when the lunar orbit attains this relative position. This near annual cycle of gravitational "tugs" occurs for this 4-month interval, year after year, until the resonance is either broken or simply "fades away." (Note: Lunar orbital eccentricity is greatly exaggerated in this illustration as well as in Fig. 8.6 in order to clarify the orientation of the lunar orbit)

8.3.2 *Orbital Geometry and Tidal Regime for a Forced Eccentricity Scenario*

Figure 8.8 shows the orbit and tidal scene at 50 earth radii before the resonance takes effect. Figure 8.9 represents the orbit and tidal scene after the resonance action has increased both the eccentricity and the semi-major axis to a maximum value. Figure 8.10 shows the orbit and tidal scene after the rotating Earth has transferred some angular momentum to the lunar orbit. The increase in angular momentum

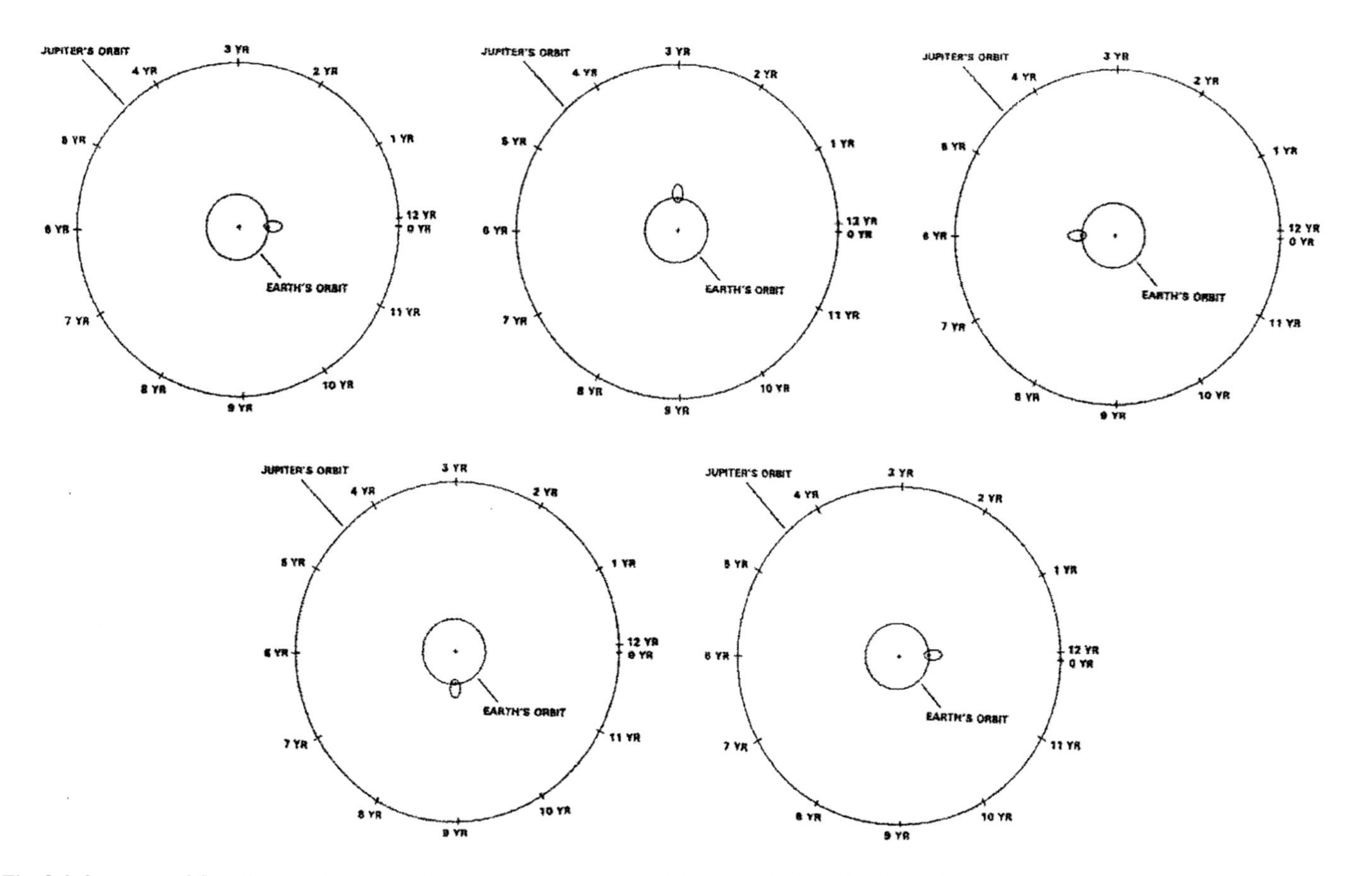

Fig. 8.6 Sequence of five diagrams illustrating the essential features of a JUPITER ORBIT—LUNAR ORBIT resonance for one Jupiter year. The apogee (greatest distance of Earth-Moon separation) is always pointing toward Jupiter. Since Jupiter is a very massive planet, the additive gravitational "tugs" are significant when the Earth is in the sub-Jupiter phase of its orbit for about 4 months/one and one-twelfth earth years. These additive effects result in a forced eccentricity of the lunar orbit (i.e., the apogee distance increases in time and, concomitantly, the perigee distance decreases in time)

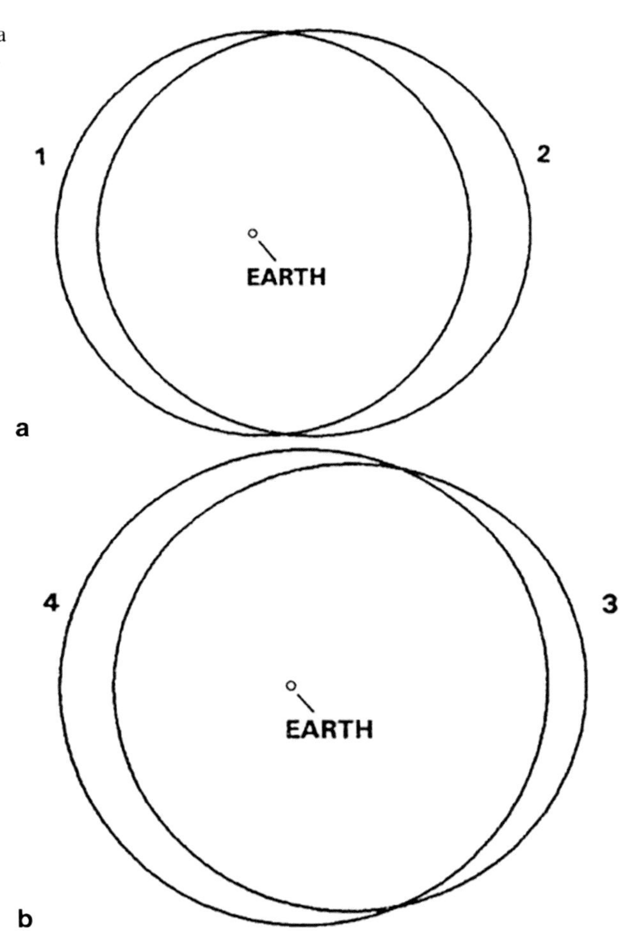

Fig. 8.7 Four orbit states in a typical orbital resonance scenario. **a** Orbit State 1 (about 50 earth radii = semi-major axis; orbital ecc. = 0.0550); Orbit State 2 (about 53.4 earth radii = semi-major axis; ecc. = 0.2856; angular momentum of the lunar orbit is only slightly higher than that of the initial orbit of the scenario). **b** Orbit State 3 (semi-major axis remains at the resonant value of 53.4 earth radii; eccentricity = 0.2519 is decreasing as the orbit increases in angular momentum because of the very slow transfer of angular momentum by the rotating Earth to the lunar orbit). Orbit State 4 (semi-major axis remains at 53.4 earth radii; ecc. = 0.100). The process of orbit circularization is truncated by the 10:1 Venus Orbit—Lunar Orbit resonance at this point in the orbital evolution. (Note: This minor complication is explained in the text.) The angular momentum of the lunar orbit at orbit stage 4 is that of a circular lunar orbit of 52.0 earth radii

gradually decreases the eccentricity of the lunar orbit but the semi-major axis remains at the resonant value of 53.4 earth radii. Figure 8.11 shows the orbit and tidal scene just before the lunar orbit very gradually transitions to the 10:1 Venus Orbit—Lunar Orbit resonance which gradually increases the SMA of the lunar orbit to 55.0 earth radii. (Note: Solar gravitational tides are *not* included in any of the calculations for the tidal amplitude or tidal range diagrams in Figs. 8.8, 8.9, 8.10, 8.11; the solar gravitational tide is very small relative to the tidal effects during the orbital resonance.) Figure 8.12 is a summary for the orbital and tidal parameters for the forced eccentricity scenario.

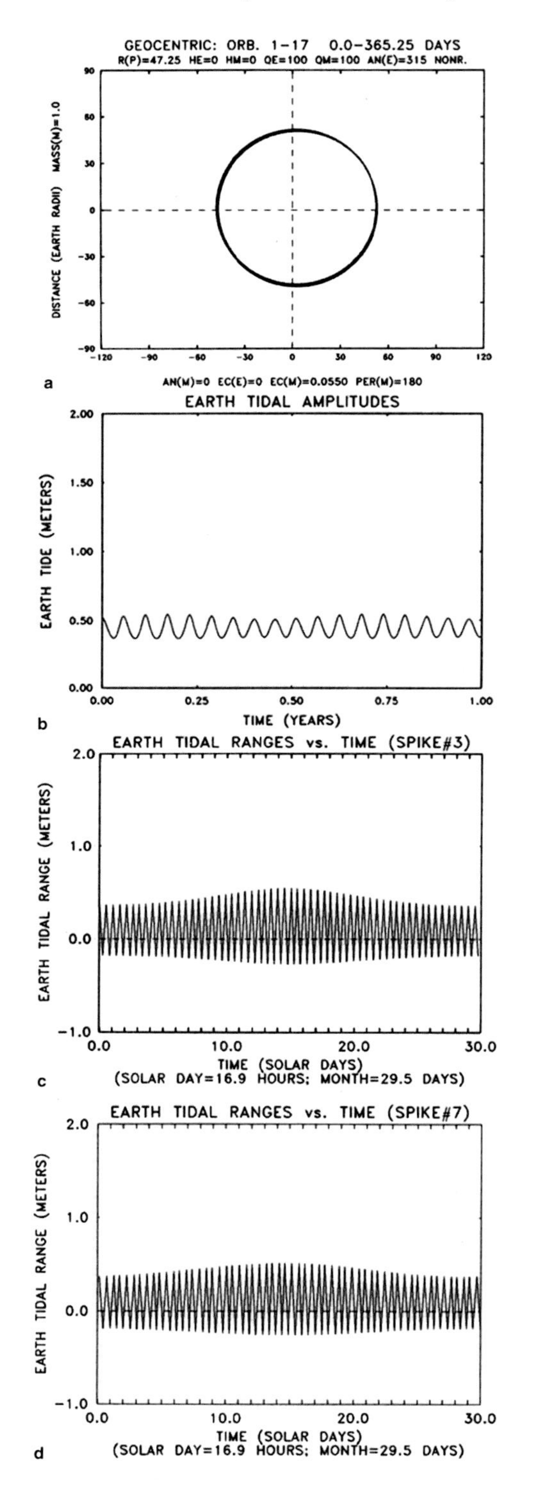

Fig. 8.8 Orbit and tidal setting before the resonance takes effect. **a** One year of orbits for a near circular orbit with eccentricity = 0.0550 and a semi-major axis of 50 earth radii. **b** One year of tidal amplitudes for this 18-month year. **c** One month of rock tidal ranges (month # 3) for this near circular orbit. Note that there is not much variation in tidal ranges over the month. Note also that there are two tidal cycles/day. The era day is ~16.9 h and there are ~28 era days/sidereal month. **d** Another month of rock tidal amplitudes (month # 4). Note that each month of the year for a near circular orbit looks very similar. (Note: Ocean tidal amplitude and ranges are ~1.4 times the rock tidal values)

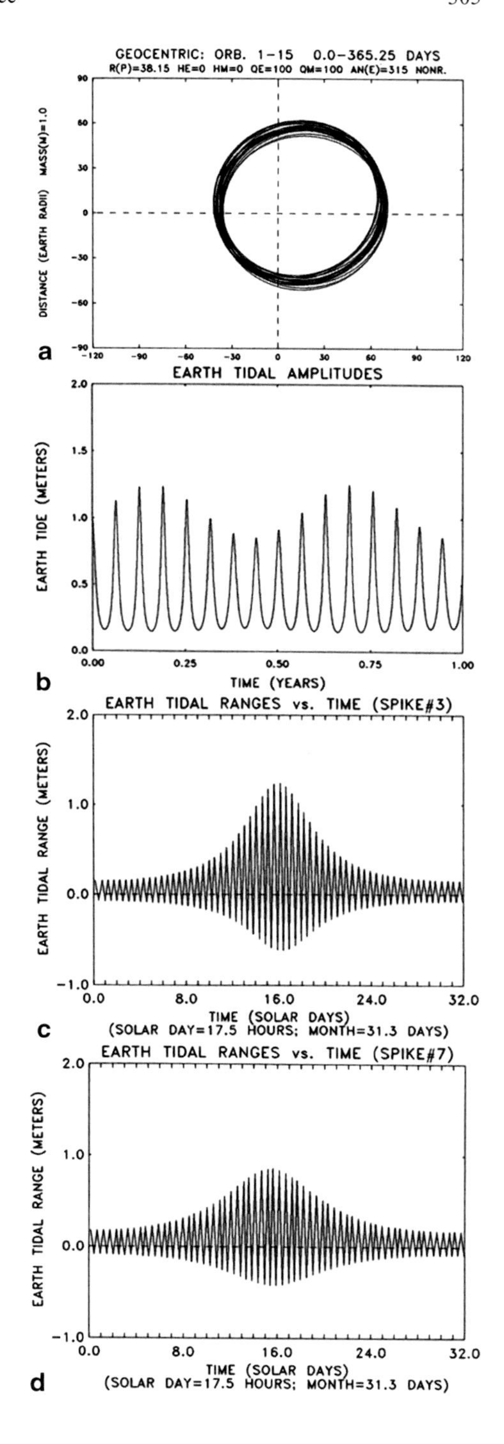

Fig. 8.9 Orbit and tidal setting for the purported maximum eccentricity state for the resonance. **a** One year of orbits for an orbit with ecc. = 0.2856 and semi-major axis of 53.4 earth radii. **b** One year of tidal amplitudes for this 15.9-month year. Note that the tidal amplitudes are alternately much higher and much lower than those in Fig. 8.8. **c** One month of tidal ranges for the highest value for the year. Note that there are fairly high tidal ranges for only about 12 days of the ~32-day month. **d** One month of tidal ranges for the lowest tidal range values for a typical year. Again note that there are fairly high tidal ranges for only about 12 days (~24 tidal cycles) during the ~32-day month

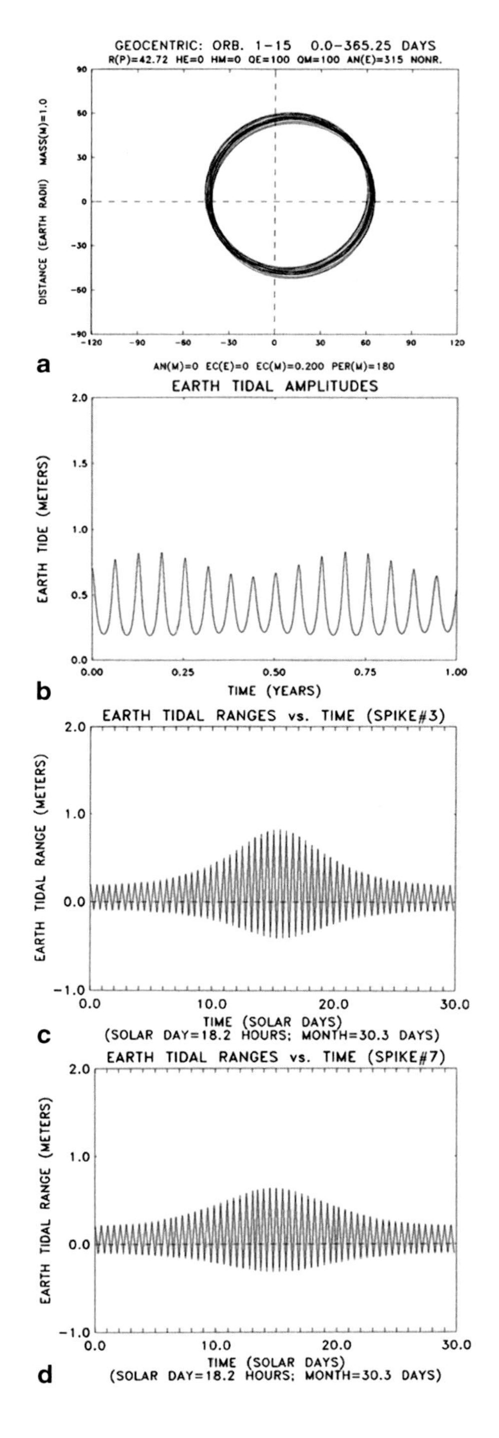

Fig. 8.10 Orbit and tidal setting during an intermediate stage in the history of the resonance. **a** One year of orbits for an orbit with ecc. = 0.2519 and semi-major axis of 53.4 earth radii. **b** One year of tidal amplitudes for this 15.9-month year. **c** One month of tidal ranges for the highest tidal month of the scenario. Note that the month now has 30 days because the rotation rate of the planet has decreased. **d** One month of tidal ranges for the lowest tidal month of the scenario

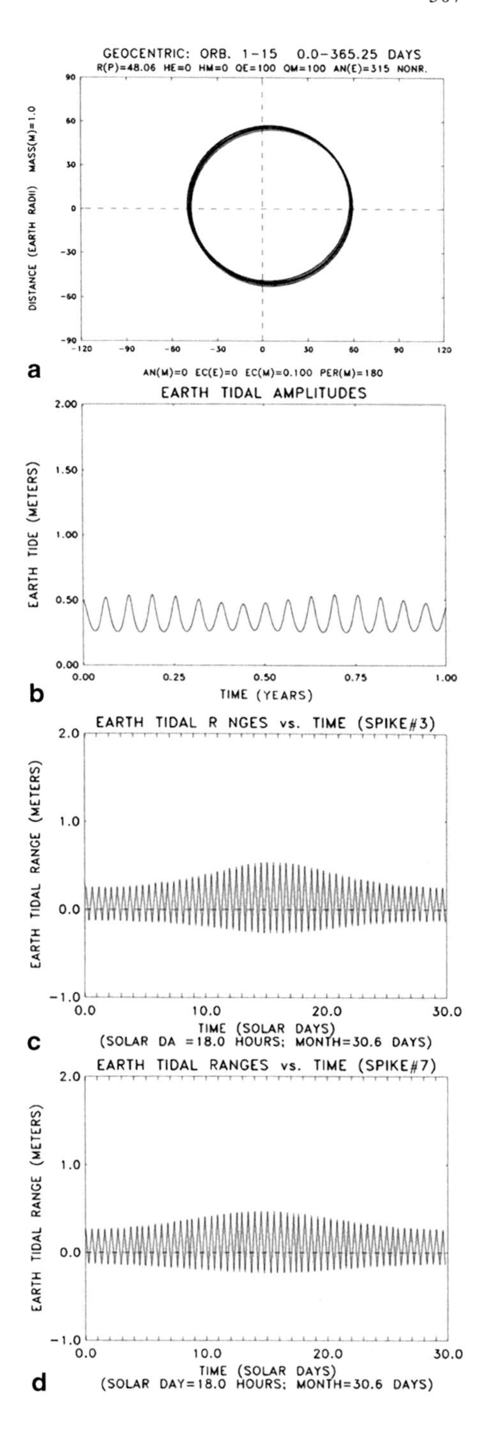

Fig. 8.11 Orbit and tidal setting near the end of the JO-LO resonance and just before the 10:1 VO-LO resonance takes effect. The orbit has an eccentricity of 0.100 and a semi-major axis of 53.4 earth radii. The angular momentum content of the lunar orbit is equal to a circular lunar orbit of ~ 52.0 earth radii. **a** One year of orbits for an elliptical orbit of 0.100 and a semi-major axis of 53.4 earth radii. **b** One year of ocean tidal amplitudes for the 15.9-month year. **c** One month of the highest ocean tidal ranges (month # 4) for this elliptical orbit. **d** One month of the lowest ocean tidal ranges (month # 7) for this year of tides. (Note: Solar gravitational tides are not included in any of the tidal amplitude and range diagrams in this section of this chapter)

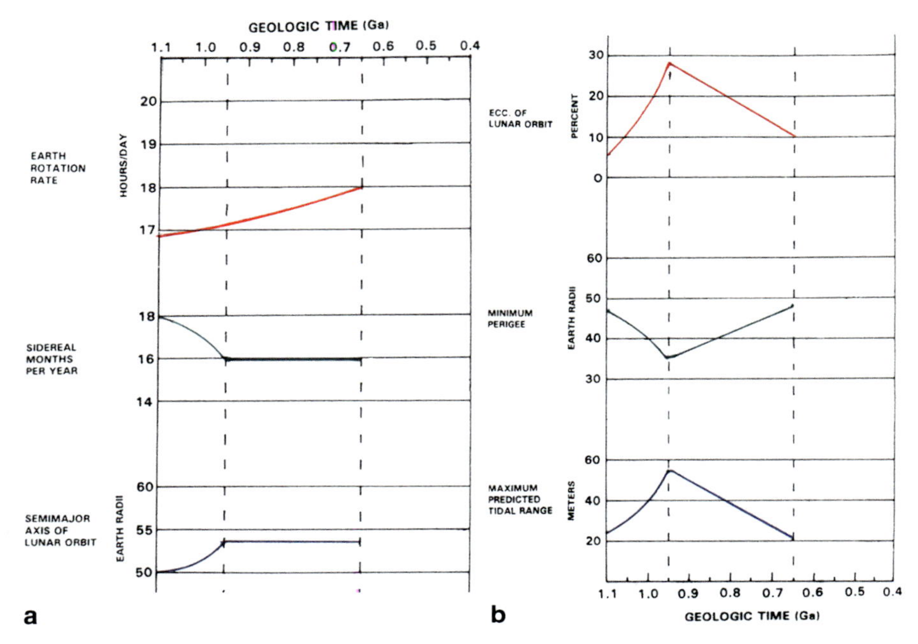

Fig. 8.12 Summary diagrams for some orbit and tidal parameters for this forced eccentricity scenario. **a** Earth rotation rate simply decreases as angular momentum of rotation is transferred to the lunar orbit. The number of months/year changes over 100 Ma or so to the resonant value of ~16 sidereal months/year and remains there until the lunar orbit decreases in eccentricity and drifts beyond the orbital resonant value of 53.4 earth radii (an episode that is prevented in this scenario). When the months/year=16, the semi-major axis remains constant at 53.4 earth radii. **b** The eccentricity of the lunar orbit increases fairly rapidly to the maximum value while the lunar body is cool. The Q value of the Moon controls the maximum value of the eccentricity. After the maximum value of eccentricity is attained, then the eccentricity decreases systematically while the semi-major axis remains at the resonant value and as angular momentum is transferred systematically from the rotating earth to the lunar orbit. The minimum perigee controls the height of the equilibrium tidal amplitudes and ranges. The effective maximum equilibrium rock and ocean tides are characteristic of a circular orbit of ~35 earth radii rather than 53.4 earth radii. Likewise, the maximum predicted tidal range, when using the Bay of Fundy tidal amplification factor of ~20, yields a maximum predicted tidal range of about 55 m in favored tidal embayments compared to near 20 m for the near circular orbits on both sides of the forced eccentricity scenario

8.3.3 Some Testable Predictions from this Forced Eccentricity Scenario

The main documents for testing the orbital and tidal features of the model are tidal rhythmite sequences and there are several of these in the rock record of the Late Proterozoic. From examination of the diagrams in Fig. 8.12, the two predictable constants are (1) that the semi-major axis of the lunar orbit should be near 53.4 earth radii for a few hundred million years and (2) that the number of sidereal months/year should be about 16 throughout this era. Assuming that the Earth rotation rate at 1.1 Ga (before the forced eccentricity episode) is ~16.9 h/day, the number of days/

year would decrease from ~519 d/yr at ~1.1 Ga to ~490 d/yr at the end of this orbital resonance era when the rotation rate has slowed to 17.9 h/d.

Other predictions are: (1) There should be an enhancement of tidal zone deposition, globally, in the geological record of Late Proterozoic relative to both Middle Proterozoic and Phanerozoic time: (2) There should be enhanced tectonic-magmatic activity on Earth due to enhanced rock tide activity during Late Proterozoic time relative to both earlier and later geological time periods, and (3) there should be enhanced tectonic-magmatic activity on the Moon in Late Proterozoic time relative to both earlier and later time periods. Information relating to these predictions is discussed below.

8.3.4 Summary and Discussion

Cursory examination of the essential features of a Jupiter Orbit—Lunar Orbit Resonance Model suggests a solution for at least some of the enigmatic features recorded in the Late Proterozoic rock record. Such an orbital resonance event would be a UNIQUE EVENT in the history of the Earth-Moon system when the lunar orbital radius is near 53.4 earth radii. This forced eccentricity scenario could occur only after the Moon had cooled significantly because the thermal state of the Moon is the ultimate control over the magnitude of eccentricity forcing. The body of the Moon should have been fairly cool by 1.2 Ga before present.

Prediction 1: Reports by Deynoux et al. (1993), Sonett et al. (1996), Sonett and Chan (1998), Williams (1989a, b), Banks (1973), Ehlers and Chan (1999), Kessler and Gallop (1988), and Saeed and Evans (1999) are about ocean tidal complexes of this age on four different continents (Africa, Australia, Europe, and North America). Information from these reports is consistent with the prediction of higher ocean tidal amplitudes and ranges in the broadly defined equatorial zone in Late Precambrian time.

Prediction 2: This prediction of enhanced tectonic-magmatic activity on Earth is consistent with the strontium isotope curves from marine carbonates deposited during Late Proterozoic time as summarized by Knoll (1991), Windley (1995; also see Fig. 8.4), and Halverson and Shields-Zhou (2011). It is also consistent with the evidence of major continental rift zones developed during the Late Proterozoic Eon. Reports by Eyles and Young (1994), Eyles and Januszczak (2004), and Powell et al. (1994) make a reasonable case for enhanced tectonic activity. Davies (1992) and Stern (2005) present a strong case for the beginning of the modern style of "slab-pull" plate tectonics in Late Proterozoic time. The predicted earth tide activity for this JO-LO resonance could enhance the subduction and associated rifting activity in this era via processes described by Nelson and Temple (1972), Uyeda and Kanamori (1979), Bostrom (2000), Scoppola et al. (2006) and others. This tidally driven activity is not "tidal drag", as pointed out by Jordan (1974), but is caused by a combination of the irregular vertical motion of the crust caused by earth tidal activity and the unidirectional tangential tidal component of the earth tide, a process well illustrated by both Bostrom (2000) and Scoppola et al (2006). An important part of this predicted increase in earth (and ocean) tidal activity associated with a planet orbit-lunar orbit resonance is that the tidal activity associated with an elliptical orbit is much more effective than that associated with a near circular orbit of low eccentricity like we have at the present time in earth history.

Prediction 3: The three-body numerical simulations for the maximum eccentricity state in this scenario suggest that energy dissipation in the body of the Moon, via radial tidal pumping, can equal or exceed the maximum heat flow through the lunar crust. Thus, it is possible to cause partial melting within the lunar upper mantle during such a forced eccentricity episode. It is interesting to note that in a summary paper on the time-frame of lunar mare volcanism, Schultz and Spudis (1983) cite evidence of young (relatively uncratered) lava flows on the lunar surface. Could these flows be of Late Proterozoic age? Perhaps future robotic missions to the Moon could be of value for this particular issue.

Some Calculations to be Done With a Four-Body Code for Planet Orbit–Lunar Orbit Resonances

The geocentric orbit code that I am using for these simulations will calculate the four-body system with reasonable accuracy and precession for up to 30 years for nearly circular lunar orbits. What is needed is a code that will calculate the four-body simulation for millions of years of time (really hundreds of millions of years). Again, this would be an interesting project for a physics/computer science student.

The increase in eccentricity is controlled by the Q factor of the Moon. If the Moon is cool, the orbital eccentricity will increase. As the lunar body gets warmer, the Q value increases and eventually the eccentricity will peak out and the major axis of the lunar orbit will be maintained at the resonant value as angular momentum is transferred from the rotating planet to the lunar orbit.

8.4 A Venus Orbit—Lunar Orbit Resonance Associated with a Perigean Cycle of 15 Earth Years (24 Venus Years) (A 15:1 VO-LO Resonance)

The Great Oxidation Event as described by Holland (2002, 2009), Reinhard et al. (2009), and Bekker and Holland (2012) is paraphrased as "A sharp rise in the concentration of atmospheric oxygen during the Paleoproterozoic between 2.45 and 2.32 Ga ago." Most investigators agree that the cause of the oxidation event is associated with (1) an increase in the rate of outgassing by volcanoes and (2) significant green (plant) photosynthesis.

The main question of this section is: CAN THE "GREAT OXIDATION EVENT" BE SOMEHOW ASSOCIATED WITH THE EVOLUTION OF THE EARTH-MOON SYSTEM? An answer is: YES, MAYBE, if the perigean cycle is in resonance with the orbit of planet Venus.

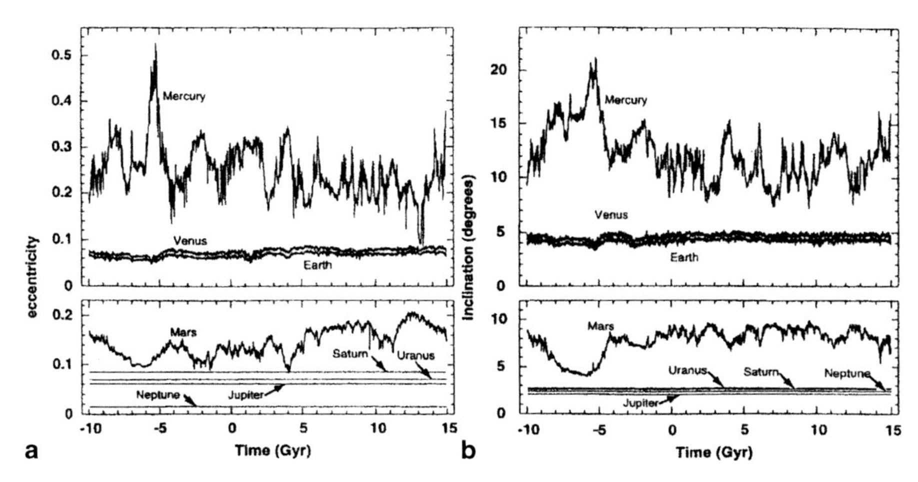

Fig. 8.13 Orbit stability diagrams from Laskar (1994, Fig. 1, with permission from Springer-Verlag) for planetary orbits in the Solar System for 10 Ga into the past and 15 Ga into the future. **a** Variation in orbital eccentricity during this time interval. Note the stability of orbits of Venus and Earth relative to Mercury and Mars. **b** Variation in orbital inclination during this time interval. Again, note the stability of the orbits of Venus and Earth relative to those of Mercury and Mars

Since Earth and Venus are in an approximate 5:8 orbital resonance (Chapman 1986; Bazso 2010), the critical numbers for the perigean cycle would be multiples of 5. The perigean cycle is 15 years when the lunar orbital radius is at ~47 earth radii. The essential condition for an effective resonance is for the apogee (the far distance) of the lunar orbit to be pointing toward Venus every 15 earth years (24 venus years). Using an orbital traceback of the Earth-Moon system similar to that of Hansen (1982) and Webb (1982), the Moon is at ~47 earth radii and the earth rotation rate is ~15.4 h/d about 2.7–2.5 Ga ago.

The main effect when moving forward in time from ~40 earth radii to 47 earth radii is a slow, but unidirectional, forced eccentricity of the lunar orbit from a nominal value of 0.0550 to perhaps 0.3 or higher. In this scenario the semi-major axis of the lunar orbit is gradually increased to the resonance value but the angular momentum of the lunar orbit lags behind. The highest rock and ocean tides occur at the end of this forced eccentricity scenario (i.e., equilibrium rock tidal amplitudes increase from 0.8 m at 0.0550 eccentricity to 2.5 m at 0.358 eccentricity). After the semi-major axis of the lunar orbit reaches the resonance value of 47 earth radii, then the period of the perigean cycle remains constant for several 100 Ma. The tidal amplitudes gradually decrease as the angular momentum of the lunar orbit increases, via tidal friction processes, and the eccentricity of the lunar orbit gradually decreases as the lunar orbit returns to a circular morphology with ~0.0550 eccentricity.

Another major question about this orbital resonance scenario is: How stable are the orbits of Earth and Venus relative to each other and how stable are these orbits relative to those of the other planets in the Solar System? Laskar (1994) did a series of numerical integrations of the averaged equations of motion for various Solar System bodies for 10 Ga into the past and 15 Ga into the future. The results of his calculations are shown in Fig. 8.13. The most noticeable features of the plots are

(1) that the orbits of the outer planets are stable throughout this long time interval and (2) that planets Venus and Earth (the Twin Sisters) appear to be in "lock-step" order throughout this long time interval while the orbits of both Mercury and Mars vary considerably in both orbital eccentricity and inclination. Laskar (1995) suggests that there is a "linear coupling" or an "angular momentum conservation constraint" between Venus and Earth and that this mechanism tends to confine their orbits to their present positions and that these mechanisms have been operating from the beginning of the Solar System. More recently Bazso et al. (2010) did extensive calculations on this "linear coupling" between Venus and Earth and found that the two planets follow each other in eccentricity cycles through the variations associated with the Milankovitch Model (i.e., 0–6 %) (Bazso et al 2010, Fig. 1). The conclusion, then, is THAT THE HELIOCENTRIC ORBIT RELATIONSHIPS OF VENUS AND EARTH APPEAR TO HAVE CHANGED LITTLE OVER GEOLOGIC TIME. Although several of the orbits of bodies in the Solar System are only marginally stable (Laskar (1996)).

8.4.1 Geometry of a Venus Orbit—Lunar Orbit Resonance when the Perigean Cycle is at 15 Earth Years (24 Venus Years)

Figure 8.14 shows the geometry and timing of an Inferior Conjunction for planet Venus relative to Earth. An INFERIOR CONJUNCTION is defined as the passage of planet Venus between planet Earth and the Sun. With a 5 to 8 resonance between the two planets, inferior conjunctions occur every 1.67 Earth years (2.67 Venus years). The time of significant gravitational interaction between Venus and the Moon in its orbit is about 60 earth days. Since the lunar orbital period at this time is about 20 earth days, this period is about 3 sidereal months long. The next significant interaction will be at the next inferior conjunction if the orientation of the lunar orbit is favorable.

Figure 8.15 shows the initial conditions for a VO–LO resonance model when the perigean cycle is at 15 earth years. Figure 8.16 is a series of diagrams illustrating the positions of the Inferior Conjunctions for one perigean cycle. (Note: All orbits in this section of the chapter are viewed from the north pole of the Solar System unless stated otherwise.)

Note that of the nine inferior conjunction positions in Fig. 8.16, only 3 have a favorable orientation for increasing the eccentricity of the lunar orbit. Only the final position gives the full gravitational effect and the other two are about 87 % effective (cos 30°).

8.4.2 Tidal Regime of this Venus Orbit—Lunar Orbit Resonance

Figure 8.17 shows four orbital states for a typical forced eccentricity scenario associated with a Venus Orbit—Lunar Orbit resonance when the perigean cycle of the lunar orbit is 15 years long (a 15:1 VO-LO resonance).

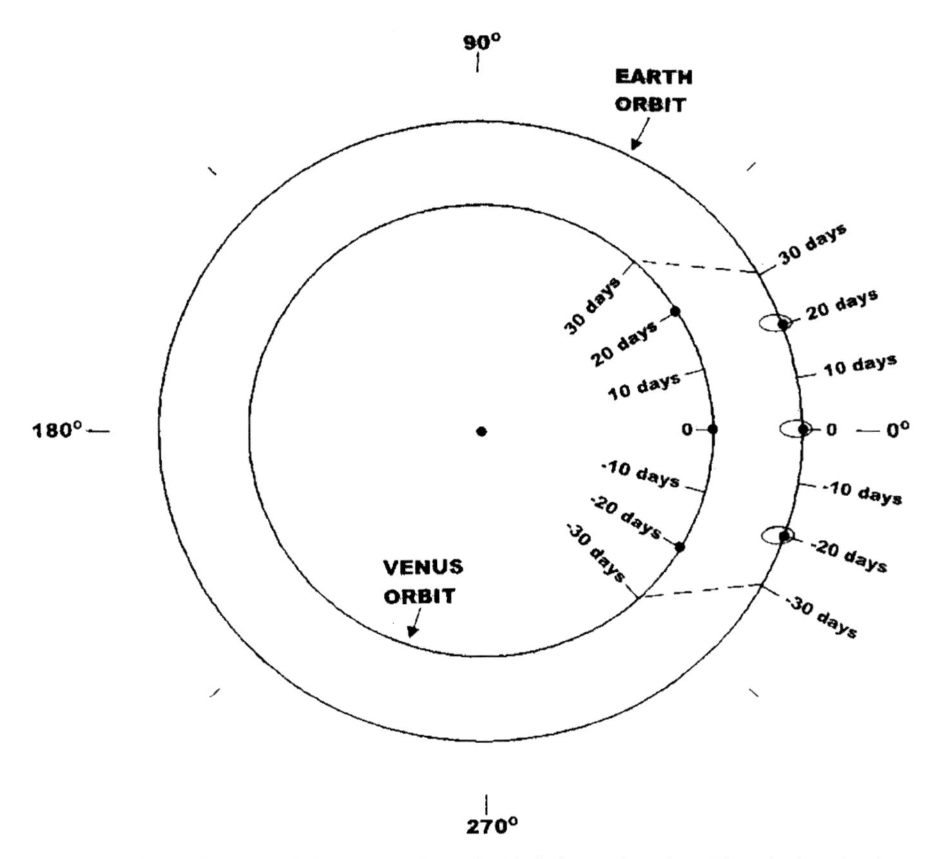

Fig. 8.14 Timing for an orbital passage of a typical inferior conjunction. There is time for three lunar orbits (three sidereal months) during which significant gravitational interaction can take place. Planet Venus is moving faster than the Earth. This diagram is centered on the time of the inferior conjunction. The reader will get a good feeling for what can happen when we examine the nine inferior conjunctions associated with this 15 earth year (24 venus year) scenario

Figure 8.18 shows the orbit and tidal scene with the semi-major axis at 40 earth radii before the resonance takes effect. Figure 8.19 represents the orbit and tidal scene after the resonance action has increased in both eccentricity and semi-major axis to a maximum value. Figure 8.20 shows the orbit and tidal scene after the rotating Earth has transferred some angular momentum to the lunar orbit. The increase in angular momentum decreases the eccentricity of the lunar orbit but the semi-major axis remains at the resonant value of 47 earth radii. Figure 8.21 shows the orbit and tidal scene after the lunar orbit very gradually exits the resonance value. These values are for a lunar orbit of 48 earth radii and for an ambient value of 0.0550 for the eccentricity (Fig. 8.22).

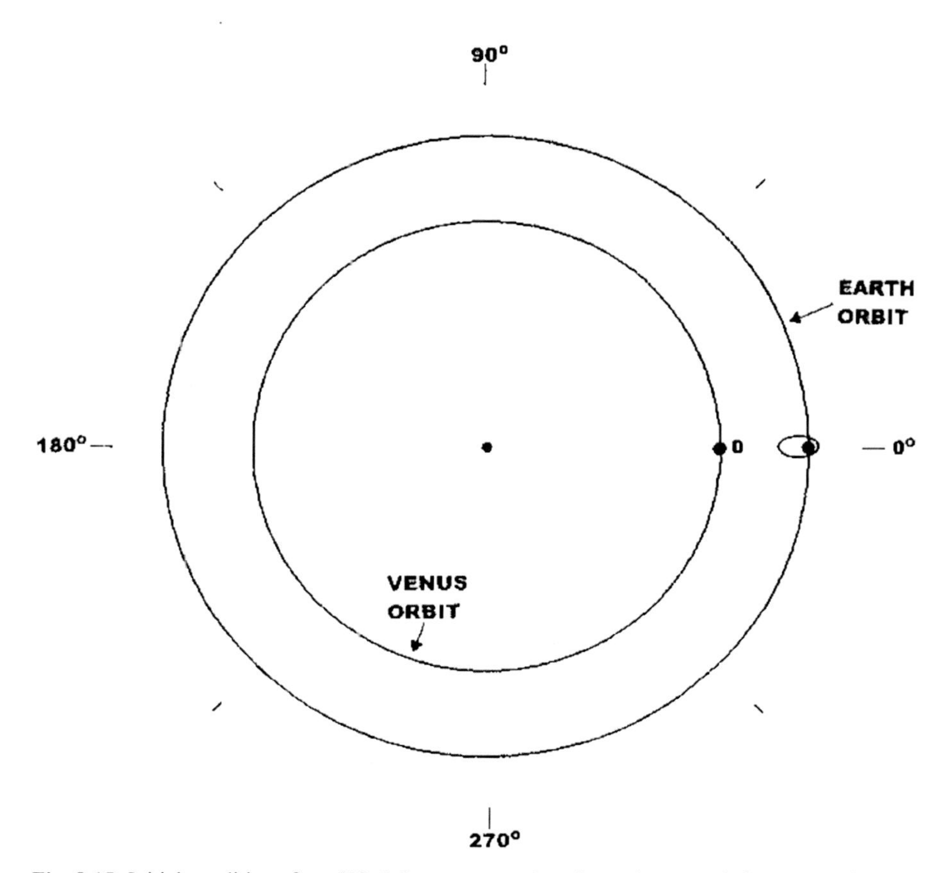

Fig. 8.15 Initial conditions for a VO–LO resonance when the perigean cycle is at 15 earth years (24 venus years). Venus is at the inferior conjunction position and the apogee of the lunar orbit is pointing directly toward planet Venus. Fifteen earth years later the apogee of the lunar orbit will have the same orientation. Every five earth years there will be an inferior conjunction at this location on the diagram but the apogee of the lunar orbit will *not* be pointing directly at planet Venus

8.4.3 Some Testable Predictions from this Forced Eccentricity Scenario

The main geological documents for testing this particular Venus Orbit—Lunar Orbit resonance are (a) the tidal rhythmite sequences and related features in sedimentary rocks of that era and (b) polar wandering curves for the continents in existence in the Early Proterozoic. Assuming a lunar orbital radius of 40 earth radii and a rotation rate of 12.6 h/day at 2.6 Ga, the two predictable constants for this era are (1) the semi-major axis of the lunar orbit should be near 47 earth radii (the resonant value) for a few hundred million years and (2) the number of sidereal months/year should

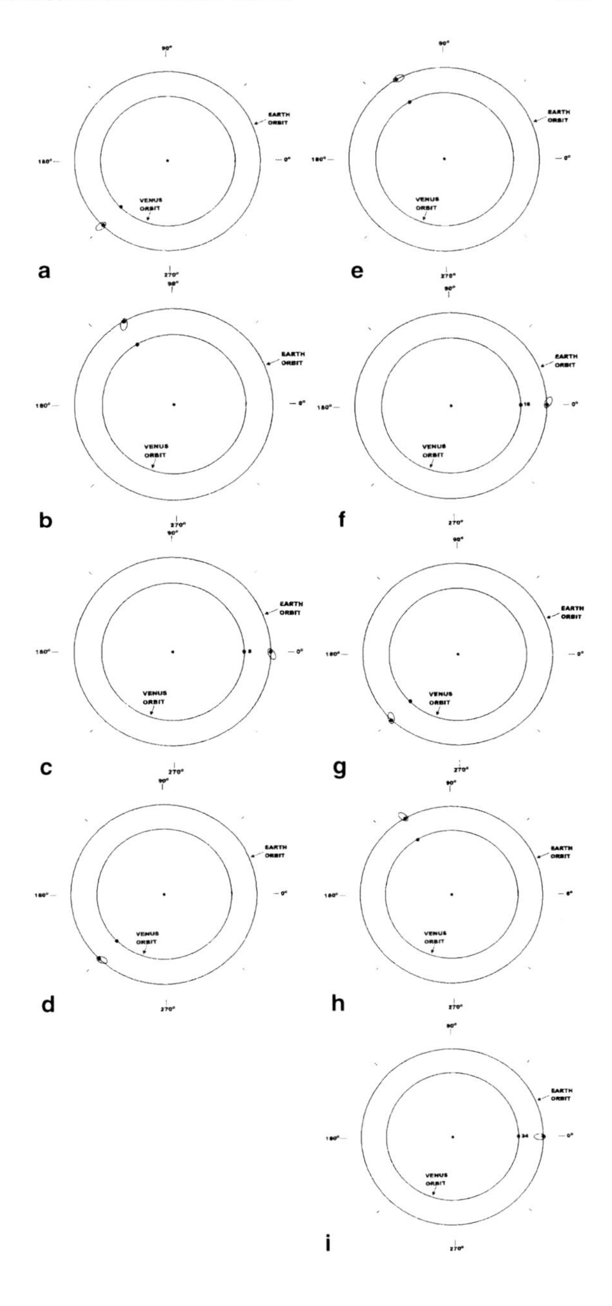

Fig. 8.16 Series of diagrams showing the nine inferior conjunction positions for one perigean cycle for the lunar orbit. Lunar orbital eccentricity is greatly exaggerated for clarity. **a** First IC (inferior conjunction). **b** Second IC **c** Third IC (Note that the Earth-Moon system is back to the starting position but that the perigean cycle is only one-third completed.) **d** Fourth IC. **e** Fifth IC. **f** Sixth IC (Note that the Earth-Moon system is back to the starting position but that the perigean cycle is only two-thirds completed.) **g** Seventh IC. **h** Eighth IC. **i** Ninth IC (Note that the Earth-Moon system has returned to the starting position and the perigean cycle has been completed)

be about 19 throughout this era. Under these conditions and assuming very little angular momentum exchange between the rotating Earth and the Sun, the number of era days/year would change from ~696 days/yr at 2.6 Ga to ~598 near the end of the orbital resonance era. Concomitantly, the rotation of the Earth would change from ~12.6 h/day to ~14.9 h/day during this time.

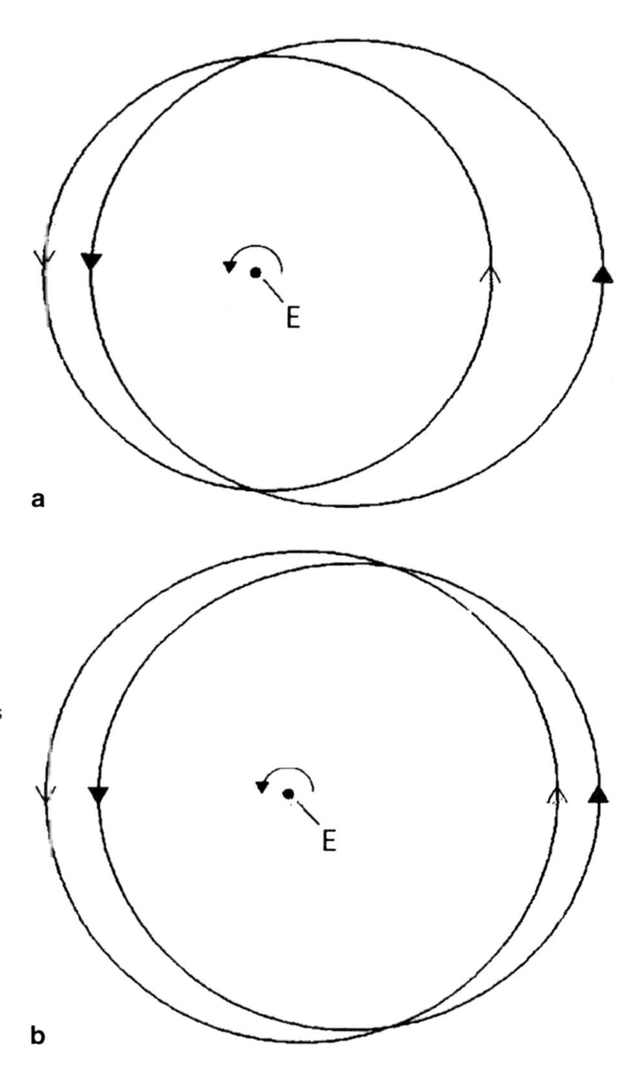

Fig. 8.17 Four orbit states in a typical VO–LO resonance scenario. **a** Orbit state 1 (about 40 earth radii = semi-major axis; orbital ecc. = 0.0550); Orbit state 2 (about 47 earth radii = semi-major axis; ecc. = 0.3580; angular momentum of the lunar orbit is only slightly higher than that of the initial orbit of the scenario). **b** Orbit state 3 (Semi-major axis remains at the resonant value of 47 earth radii; ecc. = 0.2470 is decreasing as the orbit increases in angular momentum because of the very slow transfer of angular momentum from the rotating Earth to the lunar orbit.) Orbit state 4 (ecc. is back to the ground state of 0.0550; the resonance is broken because the transfer of angular momentum has increased the semi-major axis of the lunar orbit to slightly beyond the resonant value of 47 earth radii)

Other predictions are very similar in nature to the Jupiter Orbit—Lunar Orbit resonance discussed in the earlier section:

1. There should be an enhancement of tidal zone deposition in the geological record of the Early Proterozoic relative to the Late Archean and Middle Proterozoic time.
2. There should be enhanced tectonic-magmatic activity on Earth due to enhanced rock tidal activity during Early Proterozoic time relative to both earlier and later geologic time periods.
3. There could be some tectonic-magmatic activity on the Moon in Early Proterozoic time because of the higher rock tides experienced by the Moon.

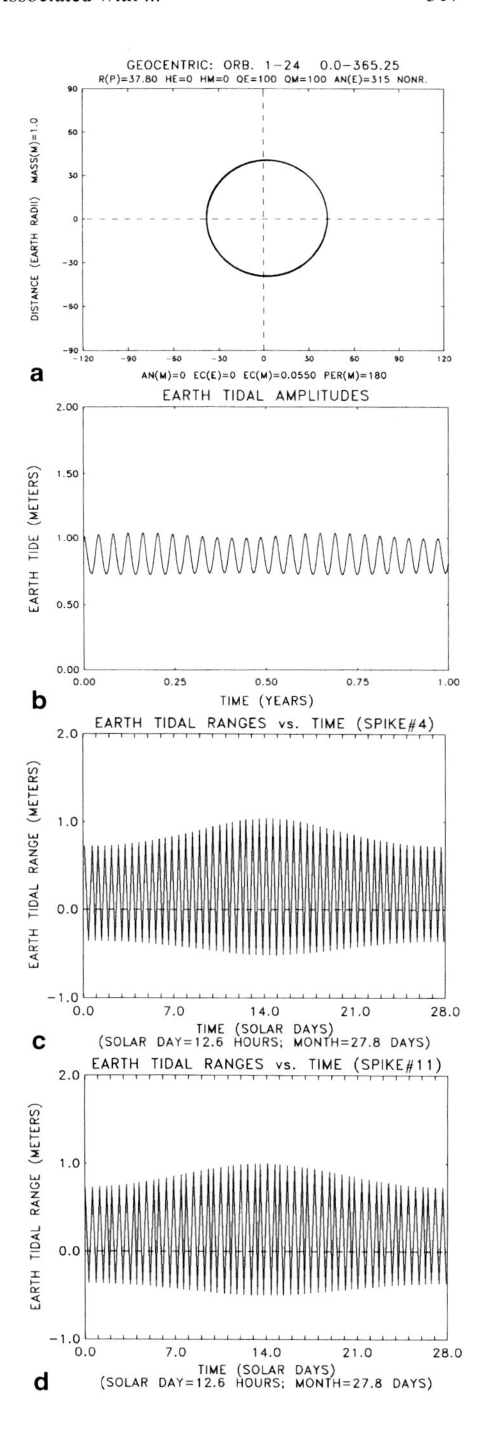

Fig. 8.18 Orbit and tidal setting before the resonance takes effect. **a** One year of orbits for a near circular orbit with ecc. = 0.0550 and a semi-major axis of 40 earth radii. **b** One year of earth (rock) tidal amplitudes for this 24-month year. **c** One month of earth tidal ranges (month # 4) for this near circular orbit. Note that there is not much variation in tidal ranges over the month. Note also that there are two tidal cycles/day. The era day is ~ 13.0 h and there are ~ 28 era days/sidereal month. **d** Another month of earth tides (month # 11). Note that tidal range diagrams for each month of the year for a near circular orbit look very similar. (Note: Solar gravitational tides are not included in any of the tidal amplitude and range diagrams in this section of this chapter)

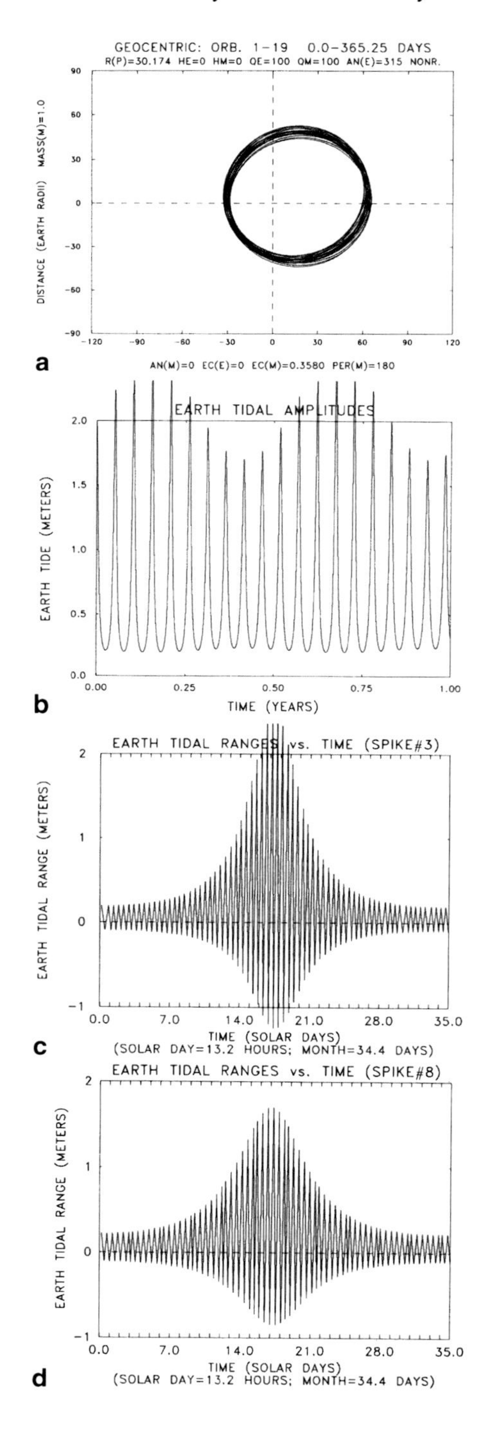

Fig. 8.19 Orbit and tidal setting for the purported maximum eccentricity state for this orbital resonance. **a** One year of orbits for an orbit with ecc. = 0.3580 and semi-major axis of 47 earth radii. **b** One year of earth amplitudes for this 19 month year. Note that the tidal amplitudes are alternatively much higher and much lower than in Fig. 8.18. **c** One month of earth tidal ranges for the highest value for the year. Note that there are fairly high tidal ranges for only about 12 days (~24 tidal cycles) of the 35-day month. **d** one month of earth tidal ranges for the lowest tidal range values for a typical year. Again note that there are fairly high tidal ranges for only about 12 days (~24 tidal cycles) during the 35-day month

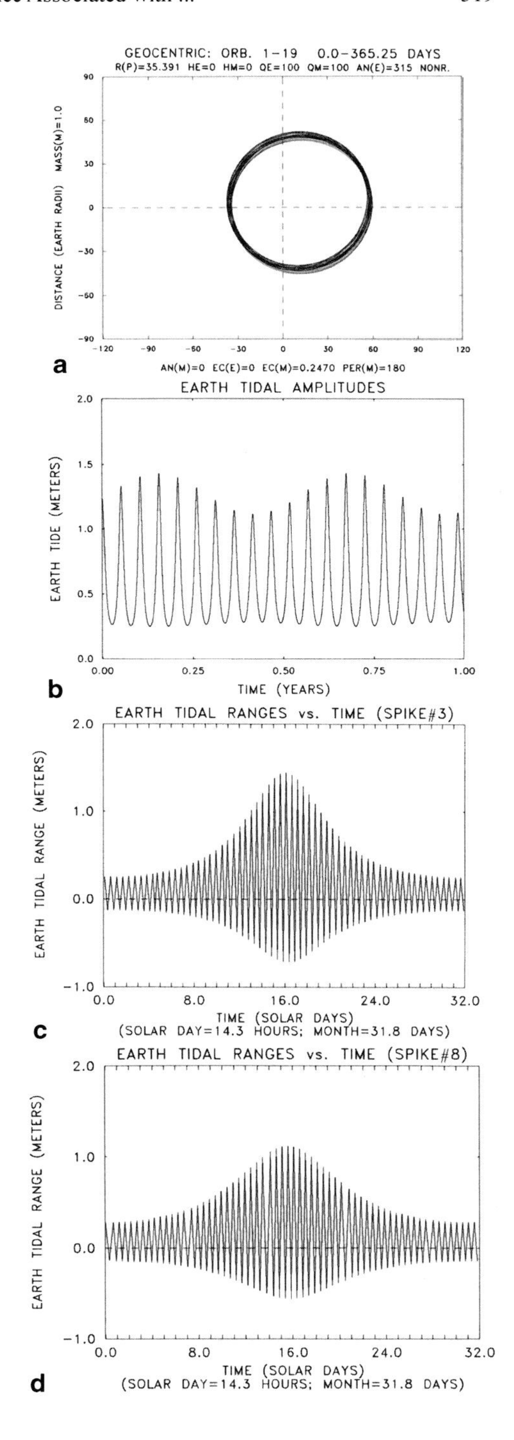

Fig. 8.20 Orbit and tidal setting during an intermediate state in the history of the resonance. **a** One year of orbits for an orbit with ecc. = 0.2470 and semi-major axis of 47 earth radii. **b** One year of earth tidal amplitudes for this 19-month year. **c** One month of earth tidal ranges for the highest tidal month of the scenario. Note that the month now has ~32 days because the rotation rate of the planet has decreased. **d** One month of earth tidal ranges for the lowest tidal month of this year-long scenario

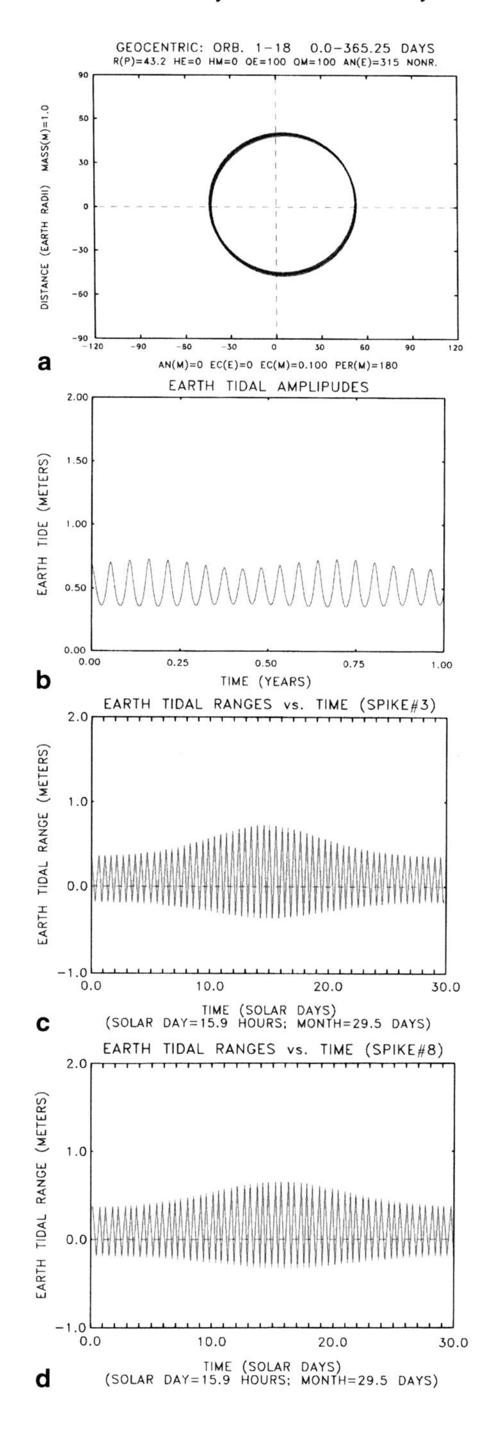

Fig. 8.21 Orbit and tidal setting after the resonance has become ineffective. **a** One year of orbits for a near circular orbit with ecc. = 0.0550 and a semi-major axis of 48 earth radii. **b** One year of earth tidal amplitudes for this 18-month year. **c** One month of earth tides (month # 3) for this near circular orbit. Note that the monthly tidal range pattern looks very similar to that of Fig. 8.18 only the values are a bit lower. **d** One month of earth tides (month # 8) which has values very similar to the month # 3

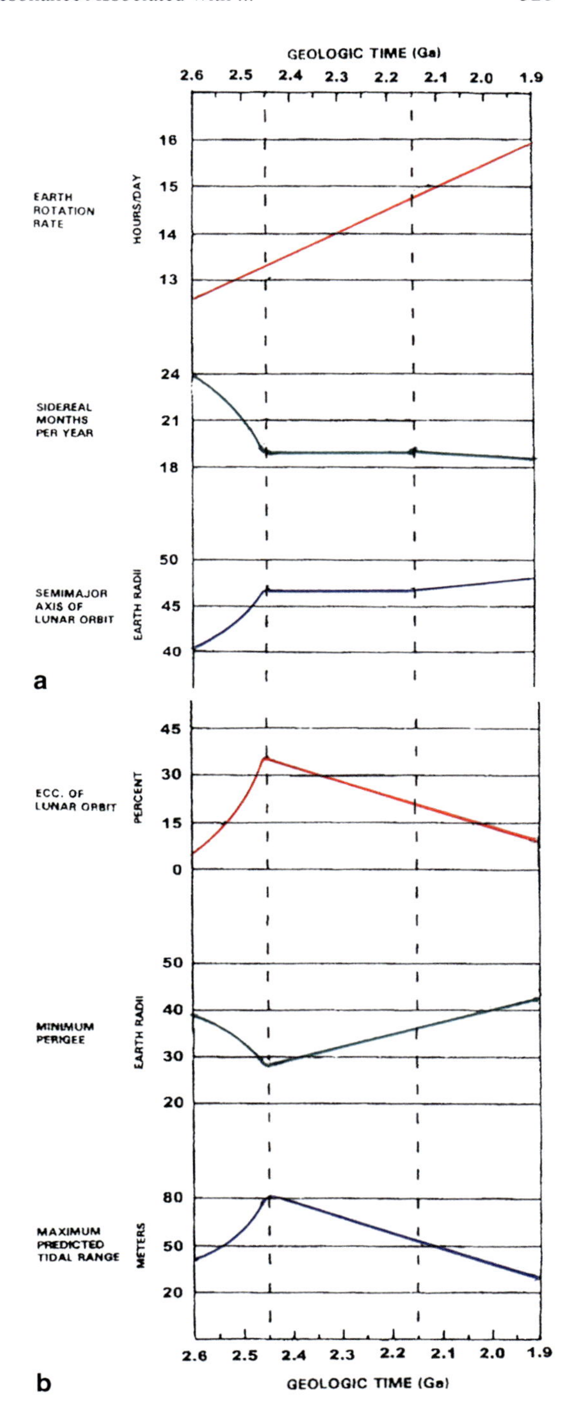

Fig. 8.22 Summary diagrams for some orbit and tidal parameters for this forced eccentricity scenario. Earth rotation rate simply decreases as angular momentum of rotation is transferred to the lunar orbit. The number of months/year changes over 100 Ma or so to the resonant value of 19 sidereal months/year and remains there until the lunar orbit decreases in eccentricity and drifts beyond the orbital resonant value of 47 earth radii. When the sidereal months/year = 19, the semi-major axis remains constant at 47 earth radii. **b** The eccentricity of the lunar orbit increases fairly rapidly to the maximum value while the lunar body is "cool". The Q value of the Moon controls the maximum value of the eccentricity. After the maximum value of eccentricity is attained, then the eccentricity decreases systematically while the semi-major axis remains at the resonant value and as angular momentum is transferred systematically from the rotating Earth to the lunar orbit. The minimum perigee controls the height of the equilibrium tidal amplitudes and ranges. The effective maximum equilibrium rock and ocean tides would be characteristic of a circular orbit of ~28 earth radii rather than 47 earth radii. Likewise, the maximum predicted tidal range, when using the Bay of Fundy tidal amplification factor of ~20, yields a maximum predicted tidal range of about 80 m in favored tidal embayments compared to only ~40 m for the near circular orbits on both sides of the forced eccentricity scenario

8.4.4 Summary and Discussion

Although the perturbation potential of planet Venus is much weaker than that of Jupiter (about 3 months/year for Jupiter and about 5 months/15 years for Venus), once the elliptical orbit is established much geological work can be done. Since the perigee passages are much closer to the Earth for this scenario, the ocean and rock tidal amplitudes and ranges would be somewhat higher and more frequent.

Prediction 1 (above) needs some explanation. There are siltstone-sandstone sequences in the Huronian sequence that have the hallmarks of tidal rhythmites. One extensive sequence of such rocks is at Mirror Lake (just south of the village of Whitefish Falls, Ontario; see Young and Nesbitt 1985, for location). Another sandstone-siltstone sequence that could be a sequence of tidal rhythmites is at the entrance of the "Big Nickel Museum" at Sudbury (Ontario). The Lorrain Quartzite, as well as other quartzite units in the Huronian Supergroup has a number of "packets" of laminated siltsones that appear to the author to be abbreviated sequences of tidal rhythmites. Such abbreviated sequences could be caused by lack of accommodation space at the time of deposition (Davis 2012). My prediction is that as the Huronian Supergroup is studied in more detail, we will find much more evidence of tidal deposition.

Prediction 2 (above) needs additional discussion. The prediction of enhanced tectonic-magmatic activity on Earth is consistent with the evidence for relative movement of cratonic elements of North America (Whitmeyer and Karlstrom 2007) as well as the molybdenum isotope record for the Early Proterozoic relative to the Middle Proterozoic (Reinhard et al. 2009). There is abundant evidence of plate tectonic-like features in the Early Proterozoic, but the Earth's mantle was probably too warm to permit "slab-pull" tectonics (Davies 1992; Stern 2005). Perhaps these plate-tectonic-like features could be caused, in part, by the unidirectional processes of the rock tides during this forced-eccentricity scenario, a process described by Scoppola et al. (2006).

Prediction 3 needs some additional discussion. Since the lunar orbit in a gravitational capture scenario could be circularized to ~20% eccentricity by 3.6 Ga and to ~10% by 3.0 Ga, perhaps the relatively sharp spike in the eccentricity of the lunar orbit at 2.45 Ga would cause enough energy dissipation to cause lunar basalts to be extruded to the lunar surface on the front side of the Moon during this era. Such candidate young flows could be checked by astronauts or robots on future missions to the Moon.

8.5 A Venus Orbit—Lunar Orbit Resonance Associated with a Perigean Cycle of 10 Earth Years (16 Venus Years) (A 10:1 VO-LO Resonance)

Decades ago Stewart (1970) and Lochman-Balk (1970, 1971) concluded that there were many signatures of tidal zone deposition in rocks of Late Precambrian-Early Paleozoic age. But there did not appear to be an apparent cause for significantly higher tides at that time because orbit tracebacks of the Earth-Moon system suggested that the Moon was in a near circular orbit at a distance of 52–55 earth radii.

The main question for this section of the chapter is the following: IS THERE A MECHANISM FOR INDUCING A *SIGNIFICANT ECCENTRICITY* TO THE

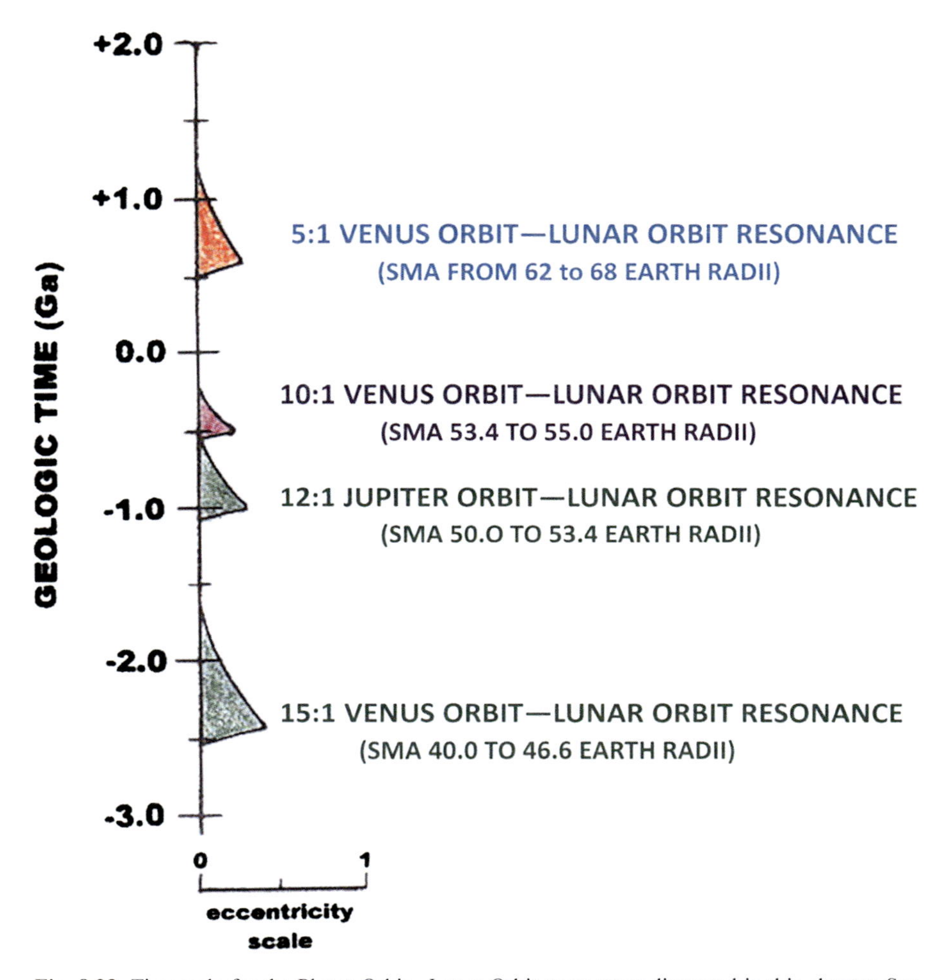

Fig. 8.23 Timescale for the Planet Orbit—Lunar Orbit resonances discussed in this chapter. See Cuk (2007) for some information on VO-LO resonances and Peale and Cassen (1978) for some information on the JO-LO resonance

LUNAR ORBIT DURING THAT ERA THAT WOULD CAUSE *SIGNIFICANTLY HIGHER OCEAN TIDES AT THAT TIME*?

A tentative answer is: YES, MAYBE, if the Perigean Cycle of the lunar orbit is in resonance with the orbit of planet Venus (i.e., A VENUS ORBIT—LUNAR ORBIT RESONANCE).

8.5.1 A Note on the Proposed Time Scale for Planet Orbit—Lunar Orbit Resonances

Figure 8.23 shows the author's view of the distribution of the three VO-LO resonances and the JO-LO resonance. Note that the 10:1 VO-LO resonance is located

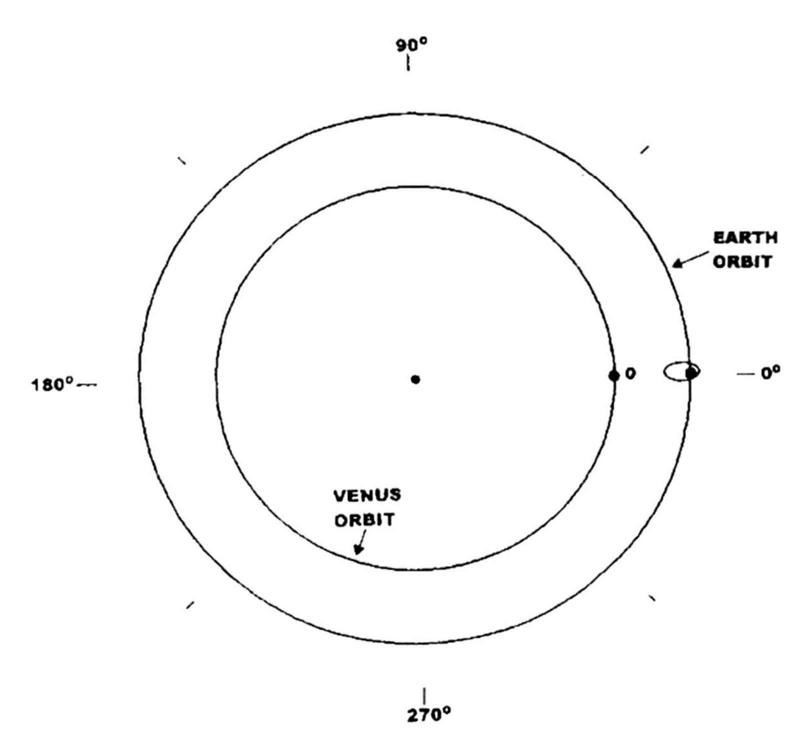

Fig. 8.24 Initial conditions for a VO-LO resonance when the perigean cycle is at 10 earth years (16 venus years). Venus is at the inferior conjunction position and the apogee of the lunar orbit is pointing directly toward planet Venus. Ten years later the apogee of the lunar orbit will have the same orientation. In five years there will be an inferior conjunction at this location (see Fig. 8.25c) on the diagram but the apogee of the lunar orbit is pointing directly away from planet Venus

very close in time to the more powerful JO-LO resonance. The 10:1 VO-LO resonance may seem like a small episode but it may be the most testable of all of the resonances because more of the rock sequences that were deposited during that time are extant. It is also a time that is very important to the evolution of metazoans. The diagram also shows the author's concept that the eccentricity and major axis for all of the Planet Orbit—Lunar Orbit resonances are pumped up to the resonance state over a short period of geologic time and that the eccentricity for each of the orbital resonance scenarios then decreases slowly as angular momentum is transferred from the rotating Earth to the lunar orbit.

8.5.2 Geometry of a Venus Orbit—Lunar Orbit Resonance when the Perigean Cycle is at 10 Earth Years (16 Venus Years) (a 10:1 VO-LO resonance)

Figure. 8.24 shows the initial conditions for an orbital resonance in which the lunar orbit is pointing towards planet Venus during one-half of the Inferior Conjunctions

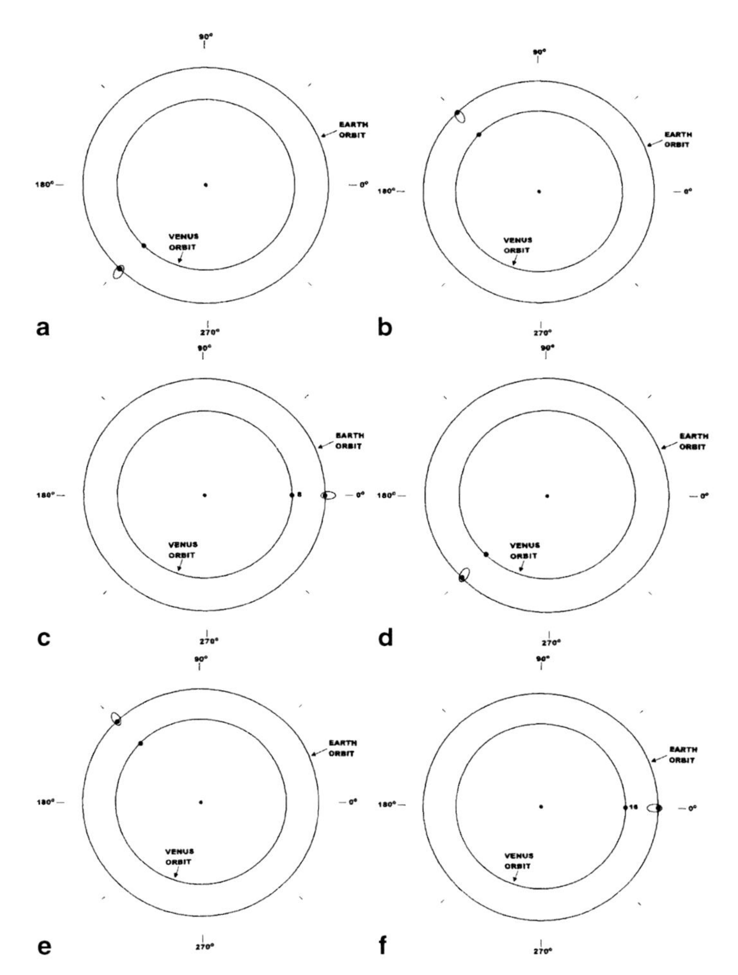

Fig. 8.25 Series of diagrams showing the inferior conjunction positions for one perigean cycle for the lunar orbit. Lunar orbital eccentricity is greatly exaggerated for clarity. **a** First IC (inferior conjunction). **b** Second IC. **c** Third IC (note that the Earth-Moon system is back to the starting position but that the perigean cycle is only one-half completed). **d** Fourth IC. **e** Fifth IC. **f** Sixth IC (note that the Earth-Moon system has returned to the starting position and that the perigean cycle has been completed)

relative to Earth. Figure 8.25 shows the orbital configurations for each of the inferior conjunction positions. It takes 10 earth years (16 venus years) for the lunar orbit to complete a perigean cycle. The lunar orbit for this resonance is ~55 earth radii.

8.5.3 Snapshots of Four Orbit States and the Associated Tidal Regimes

Figure 8.26 shows four orbit states illustrating the general changes in the characteristics of the orbit for this VO-LO resonance. The view is from the north pole of the Solar System; Earth is rotating in the prograde direction and the Moon is revolving in the prograde direction. Figure 8.27 to 8.30 show the orbit and tidal regime for each of these orbital states. Figure 8.31 is a summary of some of the orbital and tidal parameters for this 10:1 VO-LO resonance.

8.5.4 Some Testable Predictions from this Model

The main documents for testing the orbital and tidal features of this orbital resonance model are tidal rhythmite sequences of Early Paleozoic time ranging in age from ~550 to 300 Ma. From the diagrams in Fig. 8.31, the two predictable constants are: (1) the semimajor axis of the lunar orbit should be near 55 earth radii and (2) the number of sidereal months/year should be about 15 throughout this orbital resonance era. Assuming that the rotation rate at the beginning of the resonance is 17.9 h/day, the number of era days/year would change from ~464 to ~438 era days/year (rotation rate ~20 h/day) at the termination of the orbital resonance. The other predictable feature is that there should be an enhancement of tidal zone deposition in the sedimentary rock record of that era (Fig. 8.29 and 8.30).

8.5.5 Summary and Discussion

Cursory examination of the essential features of this VO-LO resonance model suggests a solution for (1) the abundance of tidal zone depositional features for the Lower Paleozoic Cambian-Ordovician quartz-rich sandstone units deposited over the eastern portion of the craton of North America (Lochmann-Balk 1970, 1971; Palmer 1971; Driese 1987; Driese et al. 1981; Hiscott 1982; Tape et al. 2003; (2) the abundance of tidal depositional indicators in the Upper Proterozoic and Lower Paleozoic sandstone dominated sequences of western North America (Stewart 1970), and (3) the abundance of tidal depositional features in the sedimentary sequences of Western Europe, the British Isles, and South America (Holland 1971,1974).

In this orbital resonance model the entire equatorial zone, broadly defined, would be involved to various degrees depending on (a) the shoreline configuration, (b) the shelf width and orientation, and (c) the amphidromic circulation systems operating in the oceans of that era. A certain set of shoreline features would favor amplification of the tidal ranges but others would lead to very little amplification. Nonetheless, the tidal spike nature of the tidal regime of an elliptical orbit could be registered in the sedimentary sequences. More specifically, this model may have a reasonable explanation for specific features reported by Hiscott (1982) from the

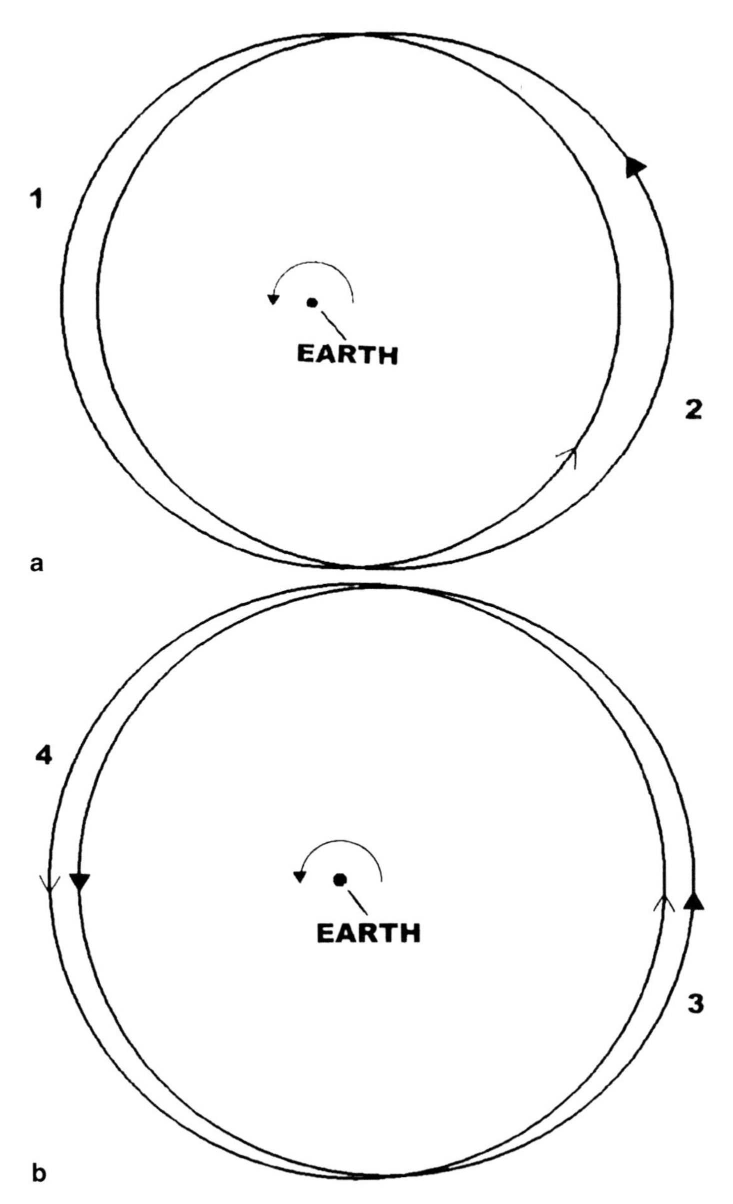

Fig. 8.26 Four orbit states in a typical orbital resonance scenario. **a** Orbit State 1 (about 53.4 earth radii = semi-major axis; orbit ecc. = 0.100); Orbit State 2 (about 55.0 earth radii = semi-major axis; ecc. = 0.250; angular momentum of the lunar orbit is only slightly higher than that of Orbit State 1). **b** Orbit State 3 (semi-major axis remains at the resonant value of 55.0 earth radii; eccentricity = 0.150 and is decreasing as the orbit increases in angular momentum because of the very slow transfer of angular momentum by the rotating Earth to the lunar orbit). Orbit State 4 (ecc. is slowly returning to the ground state of 0.0550; it is at 0.100 in this illustration). The resonance is broken when the transfer of angular momentum increases the semi-major axis of the lunar orbit to beyond the resonant value of 55.0 earth radii)

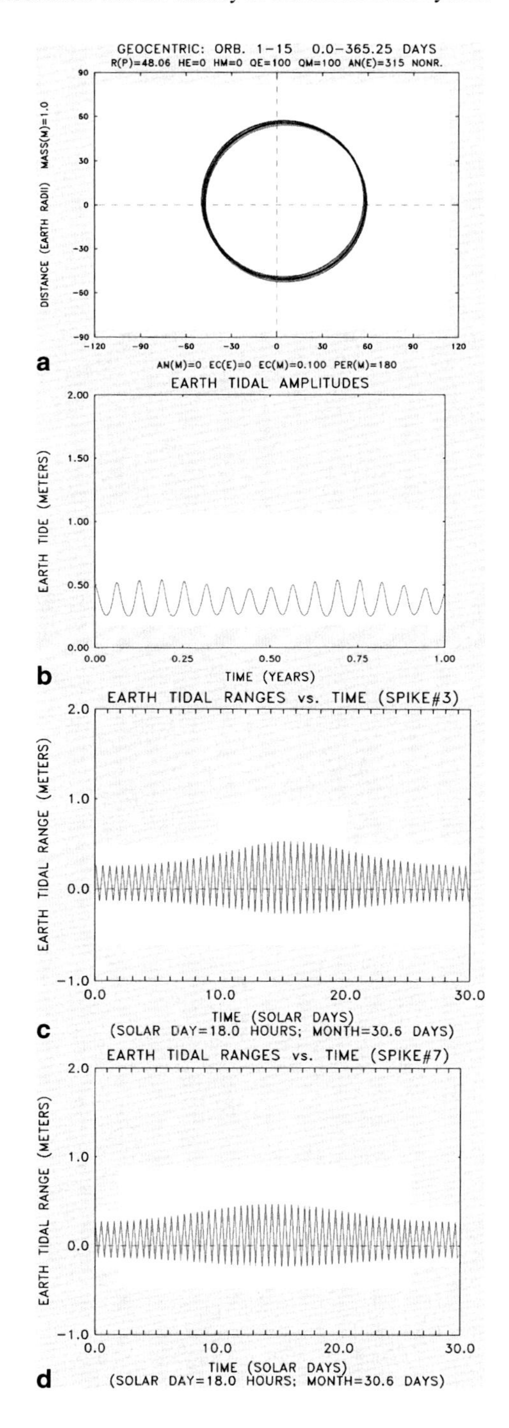

Fig. 8.27 Orbit and tidal setting just before the VO-LO resonance takes effect. The orbit has an eccentricity of 0.100 and a semi-major axis of 53.4 earth radii because the orbit is on the trailing end of the JO-LO orbital resonance. **a** One year of orbits for an elliptical orbit of 0.100 and a semi-major axis of 53.4 earth radii. **b** One year of earth tidal amplitudes for the 15-month year. **c** One month of the highest earth tidal ranges (month # 4) for this elliptical orbit. **d** One month of the lowest earth tidal ranges (month # 7) for this year of tides. (Note: Solar gravitational tides are not included in any of the tidal amplitude and range diagrams in this section of this chapter)

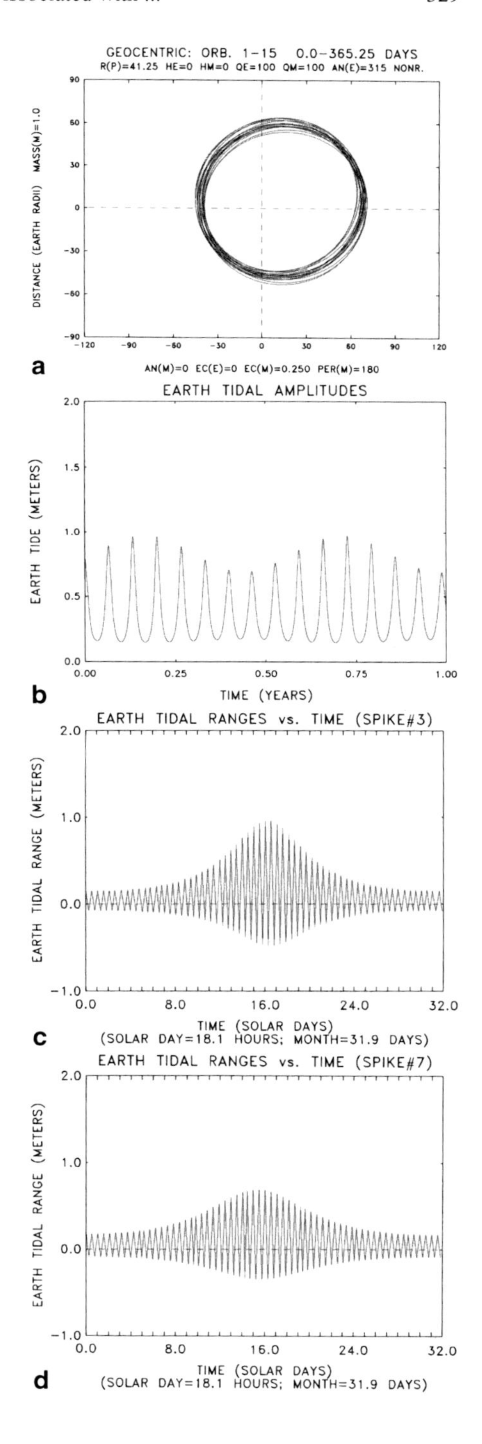

Fig. 8.28 Orbit and tidal setting for the maximum eccentricity state for this orbital resonance. **a** One year of orbits for an orbit with ecc. = 0.250 and semi-major axis of 55 earth raii. **b** One year of earth tidal amplitudes for this 15-month year. **c** One month of the highest earth tidal ranges of the year (month # 3). **d** One month of the lowest earth tidal ranges for the year (month # 7)

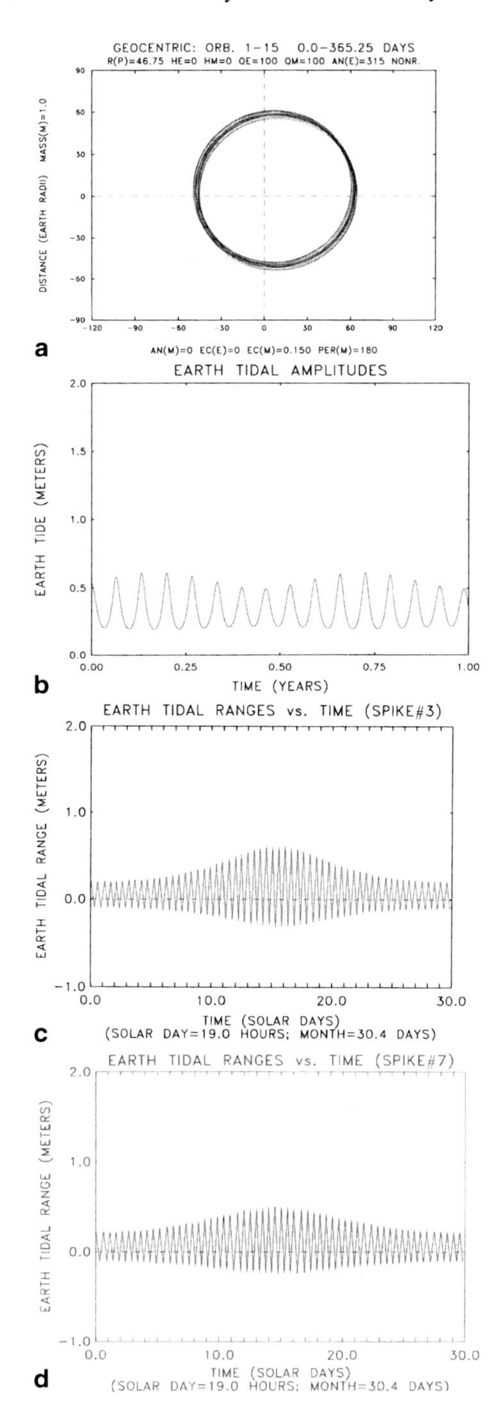

Fig. 8.29 Orbit and tidal setting during an intermediate state in the history of the resonance. **a** One year of orbits for an orbit with ecc. = 0.150 and semi-major axis of 55 earth radii. **b** One year of earth tidal amplitudes for this 15-month year. **c** One month of the highest earth tidal ranges for the year (month # 3). **d** One month of the lowest earth tidal ranges for the year (month # 7)

Fig. 8.30 Orbit and tidal settings after the resonance bas become ineffective. **a** One year of orbits for a near circular orbit with ecc. 0.1000 and a semi-major axis of 55 earth radii. **b** One year of earth tidal amplitudes for this 15-month year for this near circular orbit. **c** One month of earth tidal ranges (month # 3). **d** One month of earth tidal ranges (month # 7) which has values very similar to month # 3

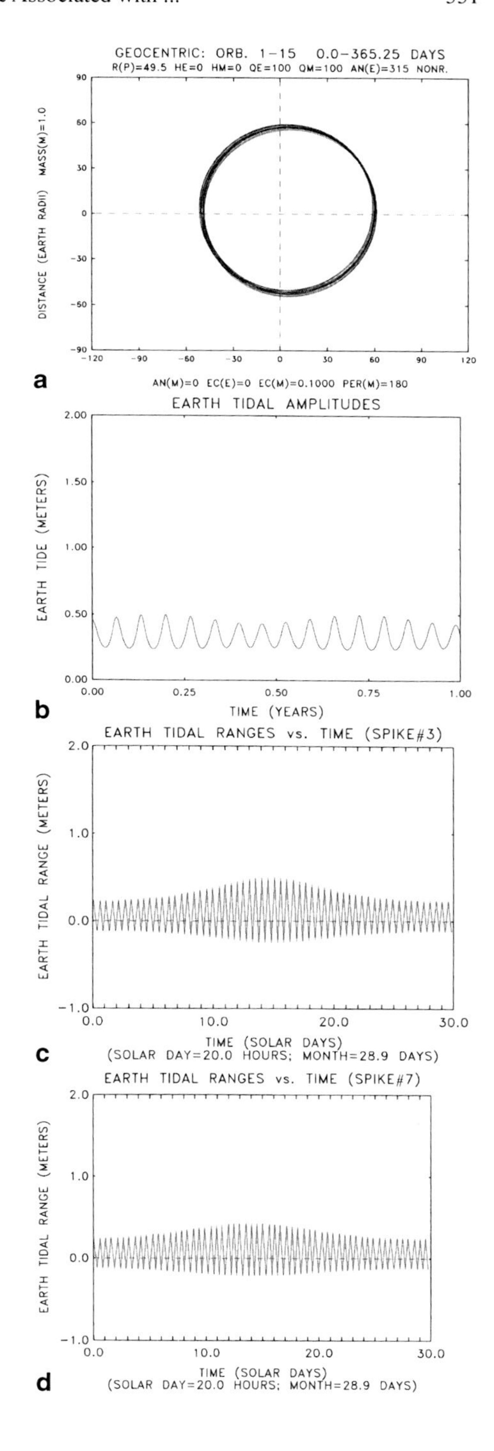

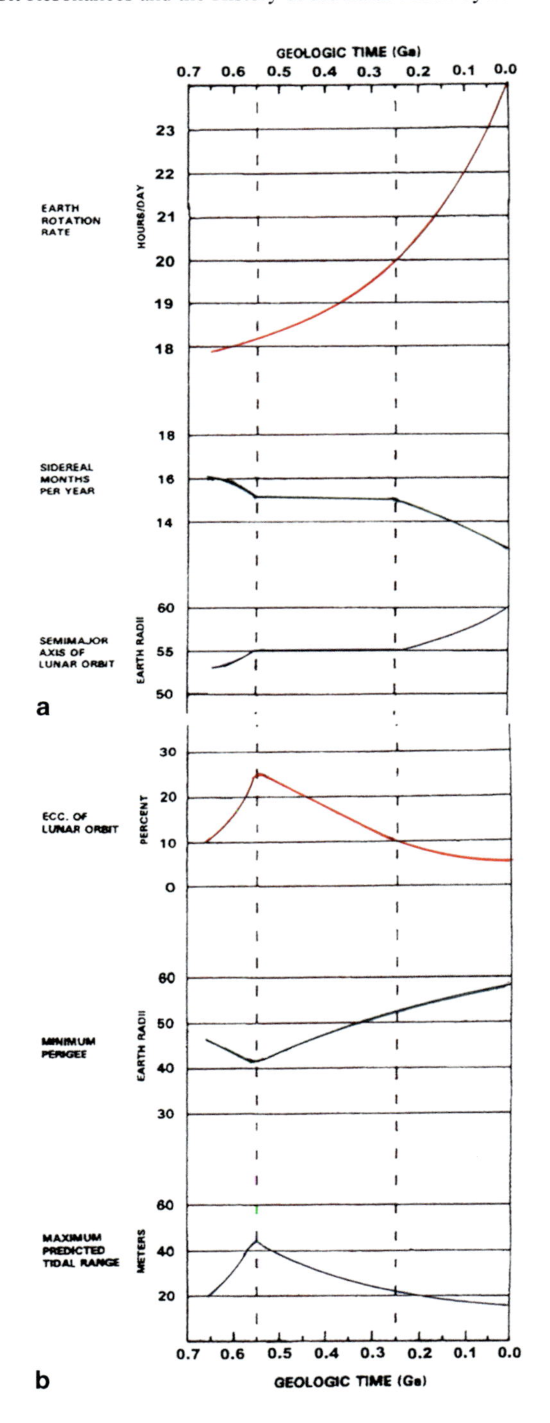

Fig. 8.31 Summary diagrams for some orbit and tidal parameters for this forced eccentricity scenario. **a** Earth rotation rate simply decreases as angular momentum of rotation is transferred to the lunar orbit. The number of sidereal months/year changes from ~16 to over 15 and remains there until the resonance becomes ineffective. The semi-major axis goes from 53.4 earth radii to the resonance value of 55 earth radii. **b** The eccentricity of the lunar orbit increases from 10 % to the maximum value while the lunar body is "cool". The Q value of the Moon controls the maximum value of the eccentricity. After the maximum value of eccentricity is attained, then the eccentricity decreases systematically while the semi-major axis remains at the resonant value and as angular momentum is transferred systematically from the rotating Earth to the lunar orbit. The minimum perigee controls the height of the equilibrium tidal amplitudes and ranges. The maximum equilibrium rock and ocean tides would be characteristic of a circular orbit of 42 earth radii rather than 55 earth radii. Likewise, the maximum predicted tidal range, when using the Bay of Fundy tidal amplification factor of ~20, yields a maximum predicted tidal range of about 42 m in favored tidal embayments compared to only 18 m for a near circular orbit of 55 earth radii

Lower Cambrian Random Formation of the Avalon Peninsula of Newfoundland. He suggested that some of the mud-cracked areas were wetted only occasionally and had high salinity; he thought that they may have been exposed for long periods of time before being wetted again. From the tidal range diagrams of Figs. 8.27 and 8.28, one can visualize a mudflat being flooded every one-half year (i.e., when the Sun and Moon are in alignment to cause the higher tidal range spikes during perigee passages). Of course movement of shallow ocean water by strong winds can also cause periodic wetting of mud-cracked terrane but the flooding by wind action would be much more irregular than that caused by the tidal regime of an elliptical orbit.

8.6 A Venus Orbit—Lunar Orbit Resonance Associated with a Perigean Cycle of 5 Earth Years (8 Venus Years)

There is one more Venus Orbit—Lunar Orbit resonance state to consider and this one is in the future, really the NEAR FUTURE when dealing with the long-term history of the Planet Earth. Practical planetary scientists can "blow this one off" but it is of some importance to those scientists interested in theoretical biology and philosophy in that this VO-LO resonance may extend the habitability of the planet for several 100 Ma.

8.6.1 Geometry and Tidal Regime of a Venus Orbit—Lunar Orbit Resonance when the Perigean Cycle is at 5 Earth Years (8 Venus Years) (A 5:1 VO-LO resonance)

Figure 8.32 shows the initial conditions for an orbital resonance in which the lunar orbit is pointing towards planet Venus during one-third of the Inferior Conjunctions relative to Earth. It will take only 5 earth years (8 Venus years) to complete a perigean cycle. The lunar orbit for this resonance is ~68 earth radii. The positions of the three inferior conjunctions for this resonance are shown in Fig. 8.33. The four orbital states are shown in Fig. 8.34. The orbit and tidal regimes for these orbital states are shown in Figs. 8.35, 8.36, 8.37, and 8.38. Summary diagrams for orbit and tidal parameters are in Fig. 8.39.

8.6.2 Summary and Discussion for this Section

The main effects of a 5:1 VO-LO resonance would be tectonic but there could be some enhancement of tidal deposition along the shorelines of that era. The rock tidal

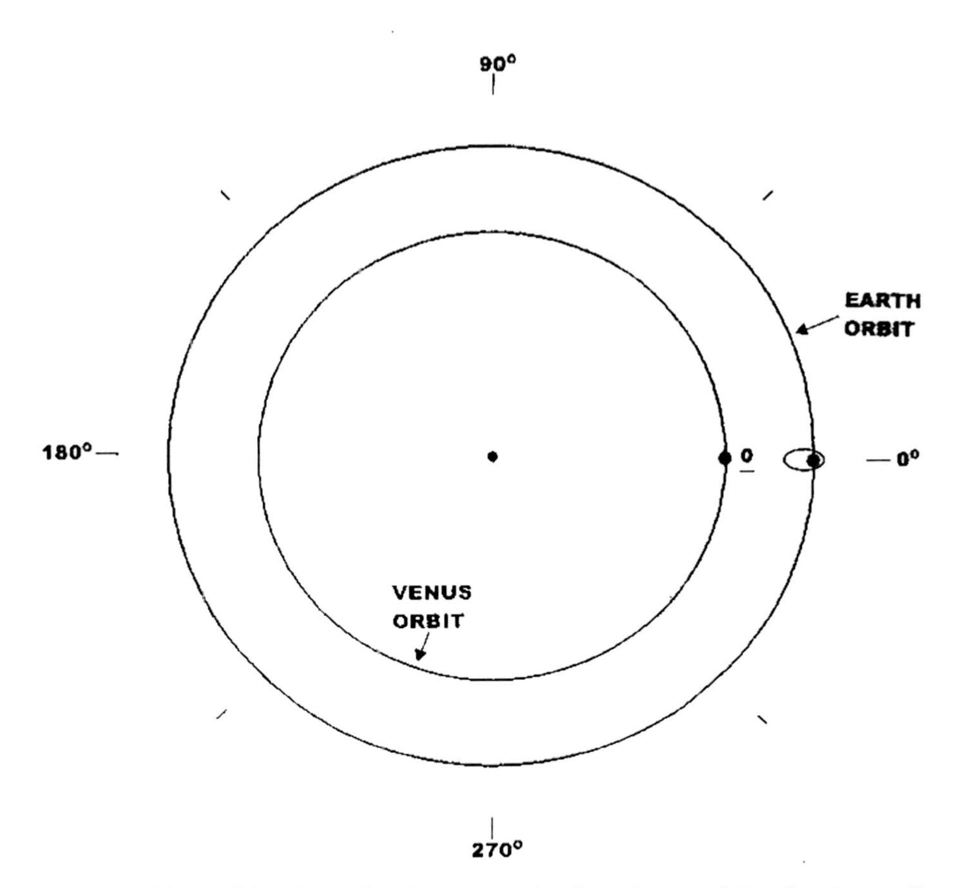

Fig. 8.32 Initial conditions for a VO–LO resonance when the perigean cycle is at 5 earth years (8 venus years). Venus is at the inferior conjunction position and the apogee of the lunar orbit is pointing directly toward planet Venus. Five years later the apogee of the lunar orbit will have the same orientation. Every five years there will be an inferior conjunction at this location on the diagram and the apogee of the lunar orbit *will be* pointing directly at planet Venus

action can be an ancillary force to keep the slab-pull plate-tectonics system operating for a somewhat longer time on a cooling planet. If the Moon was in a circular orbit at that time, then the rock tide regime would not be very helpful for the plate-tectonics operation but an ellipitical orbit with an eccentricity of 0.25 would have a good chance of aiding plate movement and subduction via a process of differential rotation between the lithosphere and mantle suggested by Ricard et al. (1991) and Scoppola et al. (2006). Any volcanism associated with plate subduction would enrich the atmosphere with carbon dioxide which would help the biological system on the planet to function for a few additional million years before ultimate extinction.

Fig. 8.33 Series of diagrams showing the inferior conjunction positions for one perigean cycle of the lunar orbit. Lunar orbital eccentricity is greatly exaggerated for clarity. **a** First IC (inferior conjunction). **b** Second IC. **c** Third IC (note that the Earth-Moon system is back to the starting position and that the perigean cycle *is completed*)

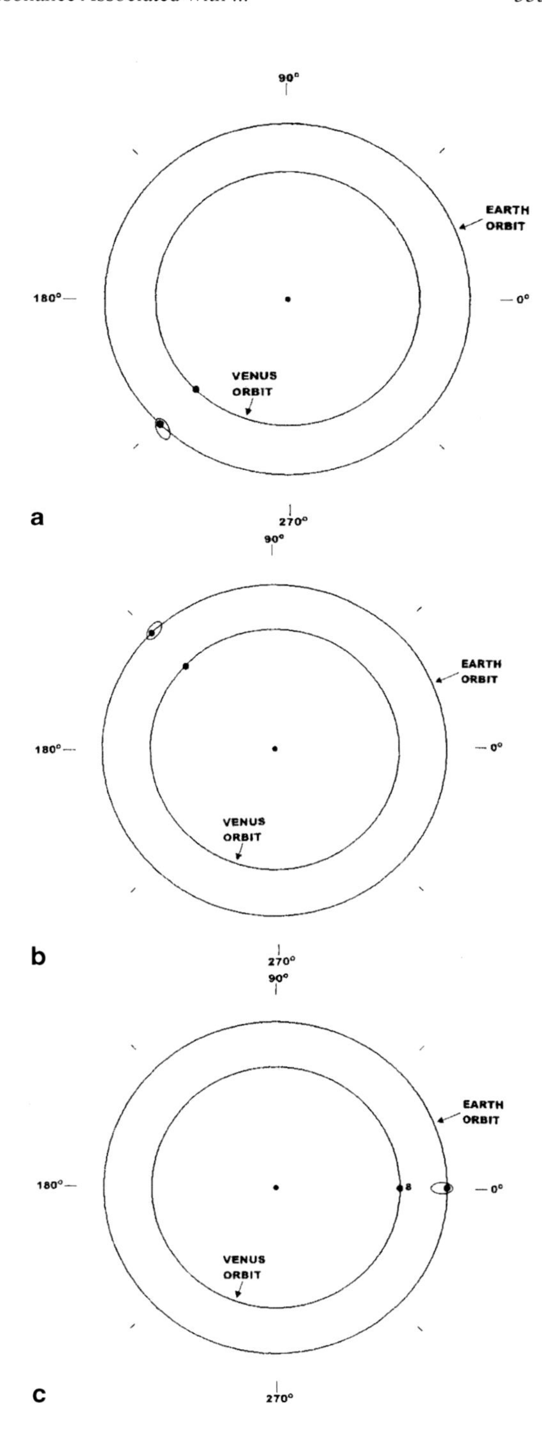

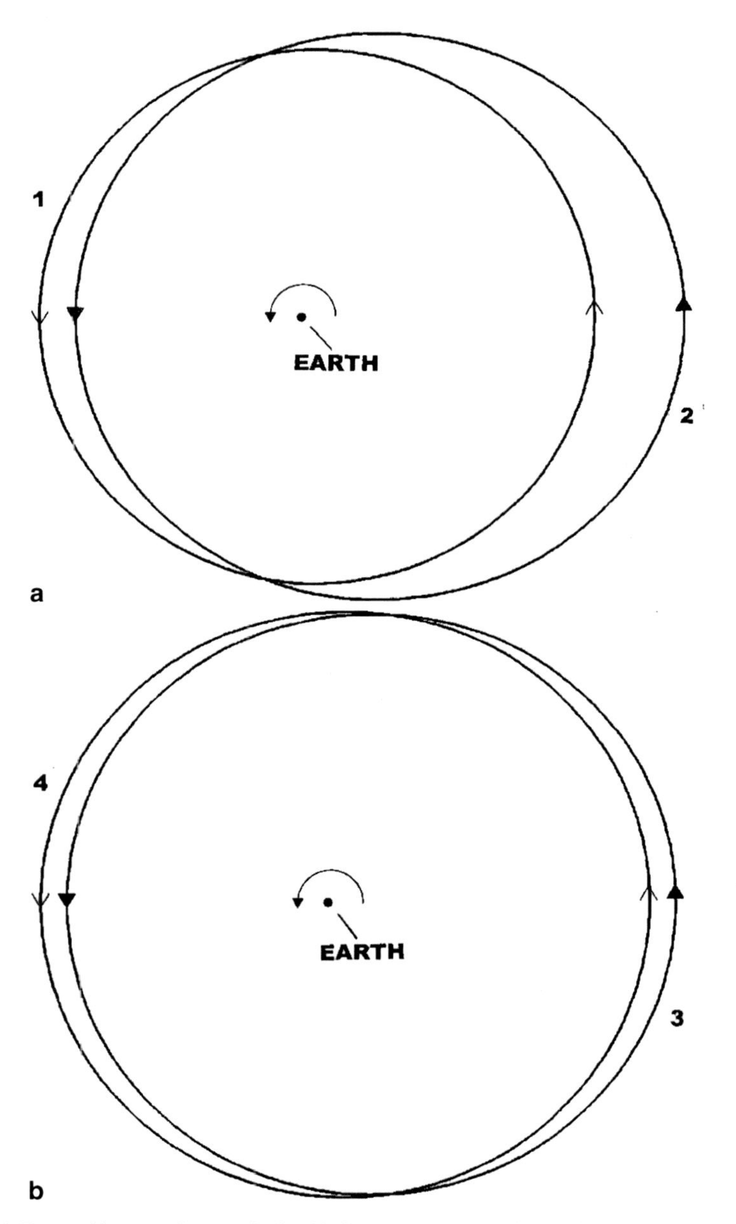

Fig. 8.34 Four orbit states in a typical orbital resonance scenario. Orbit State 1 (about 62 earth radii = semi-major axis; orbit ecc. = 0.0550); Orbit State 2 (about 68 earth radii = semi-major axis; ecc. = 0.257; angular momentum of the lunar orbit is only slightly higher than that of Orbit State 1). **b** Orbit State 3 (semi-major axis remains at the resonant value of 68 earth radii; eccentricity = 0.141 is decreasing as the orbit increases in angular momentum because of the very slow transfer of angular momentum by the rotating Earth to the lunar orbit). Orbit State 4 (eccentricity is back to the ground state of 0.0550; the resonance is broken because the transfer of angular momentum has increased the semi-major axis of the lunar orbit to beyond the resonant value of 68 earth radii)

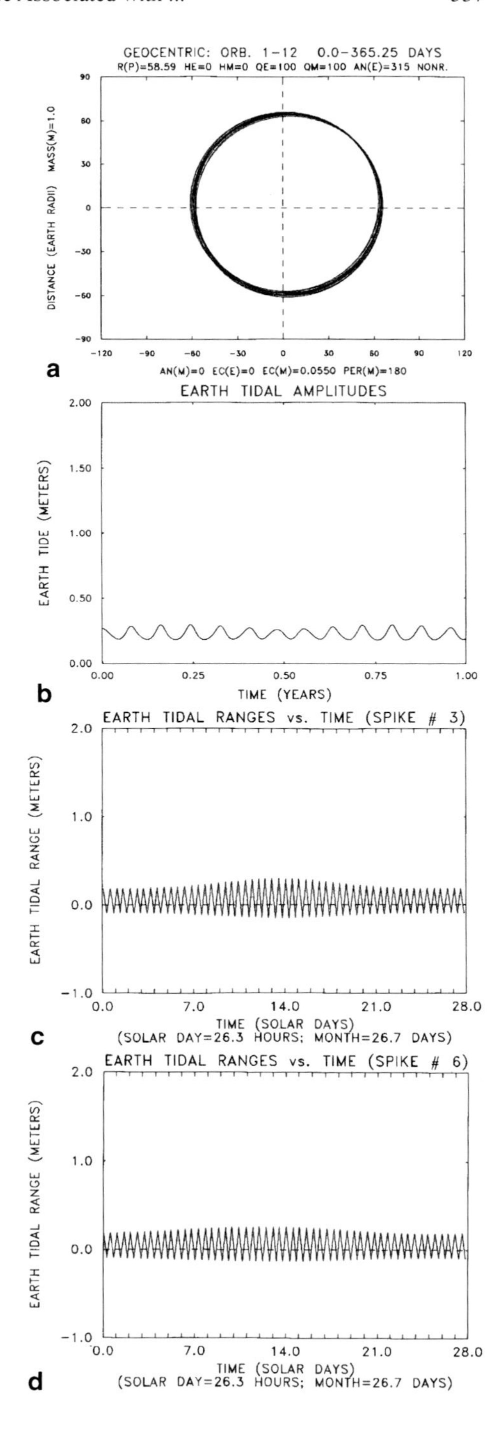

Fig. 8.35 Orbit and tidal setting just before the VO-LO resonance takes effect. The orbit has an eccentricity of 0.0550 and a semi-major axis of 62 earth radii. **a** One year of orbits for a circular orbit of 12 sidereal months. **b** One year of rock tidal amplitudes for this 12-month year. **c** One month of the highest rock tidal ranges (month # 3). **d** One month of the lowest rock tidal ranges (month # 6)

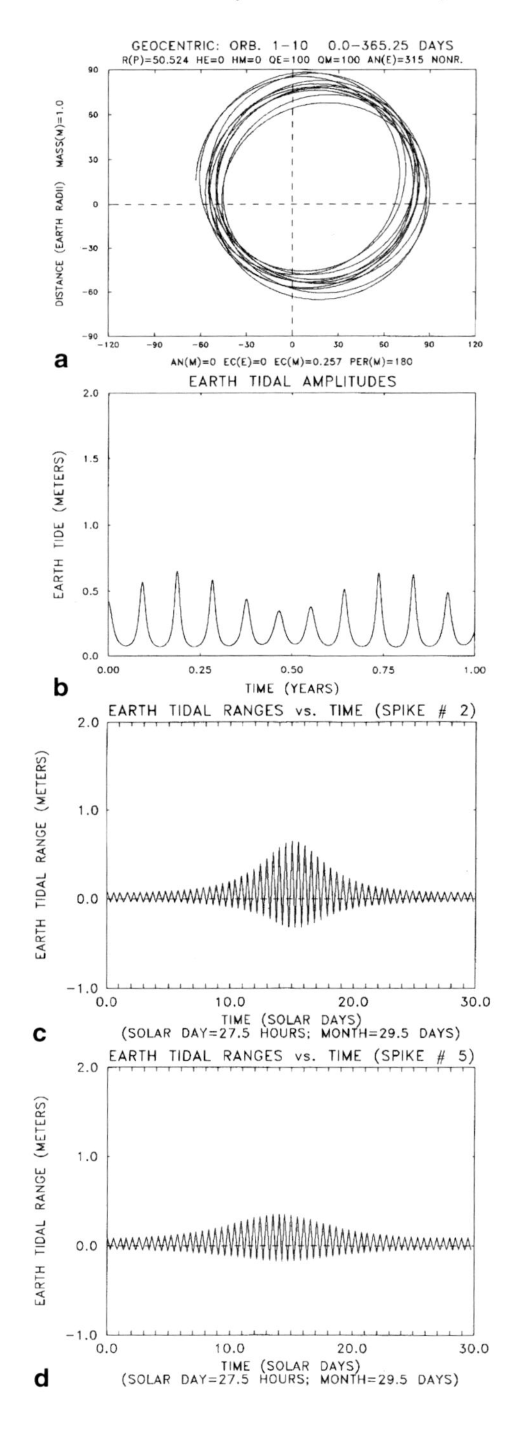

Fig. 8.36 Orbit and tidal setting for the maximum eccentricity state for this orbital resonance. **a** One year of orbits for an orbit with ecc. = 0.257 and semi-major axis of 68 earth radii. **b** One year of rock tidal amplitudes for this 10-month year. **c** One month of the highest rock tidal ranges of the year. **d** One month of the lowest rock tidal ranges for the year

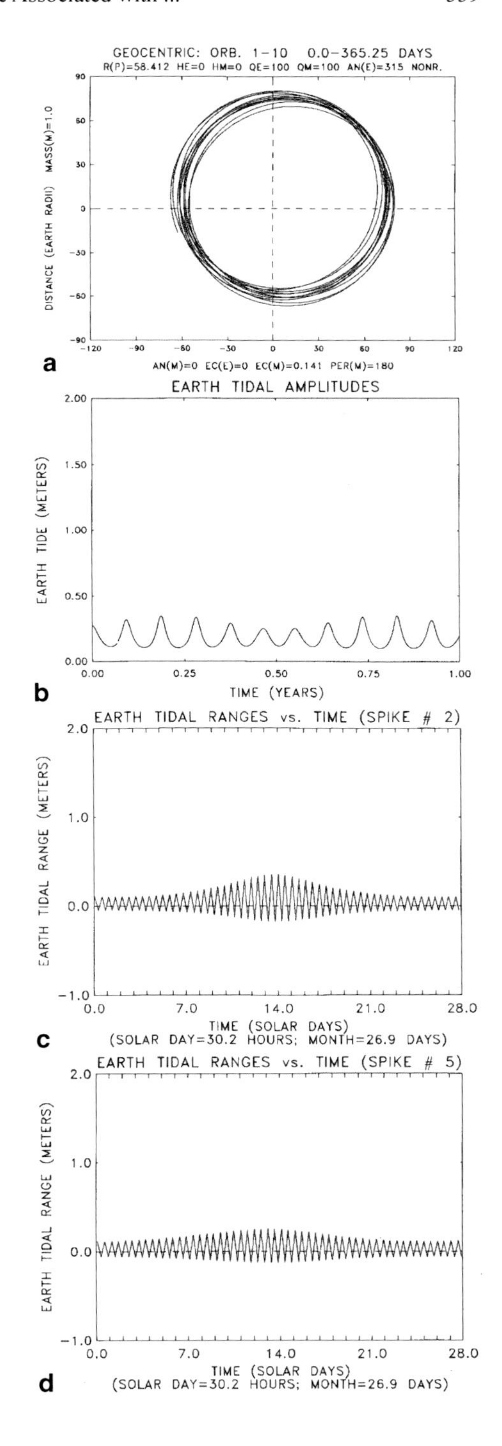

Fig. 8.37 Orbit and tidal setting during an intermediate state in the history of the resonance. **a** One year of orbits for an orbit with ecc. = 0.141 and semi-major axis of 68 earth radii. **b** One year of rock tidal amplitudes for this 10-month year. **c** One month of the highest rock tidal ranges for the year (month # 2). **d** One month of the lowest rock tidal ranges for the year (month # 5)

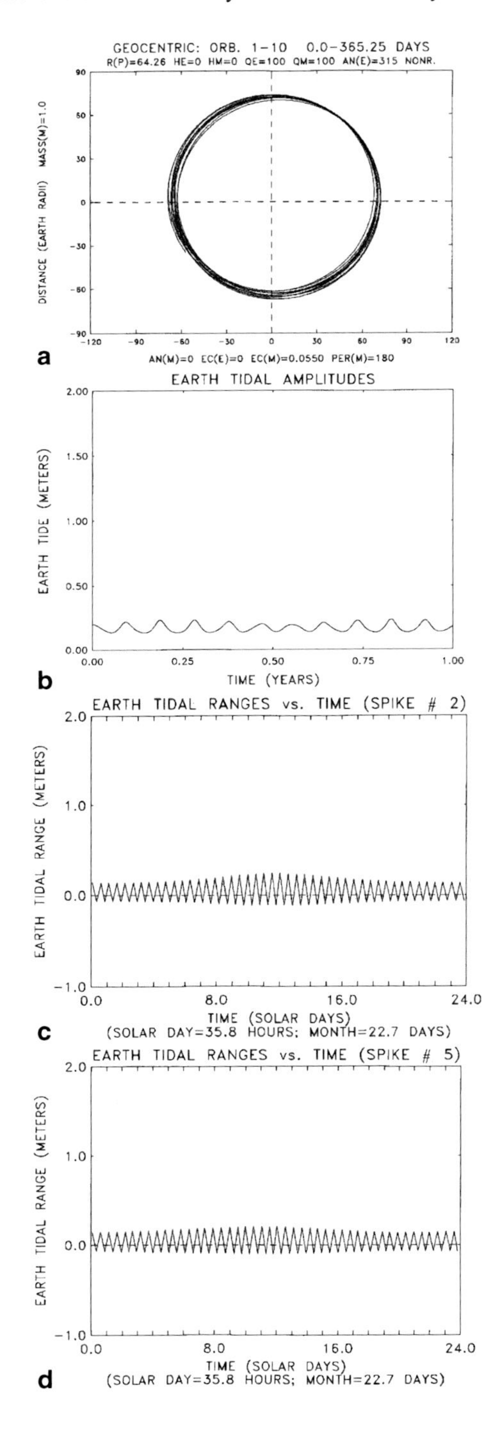

Fig. 8.38 Orbit and tidal setting after the resonance has become ineffective. **a** One year of orbits for a near circular orbit with ecc. = 0.0550 and a semi-major axis of 68 earth radii. **b** One year of rock tidal amplitudes for this 10-month year for this near circular orbit. **c** One month of rock tides (month # 2). **d** One month of rock tides (month # 5). Note that the tidal range values for all of the tidal months are very similar

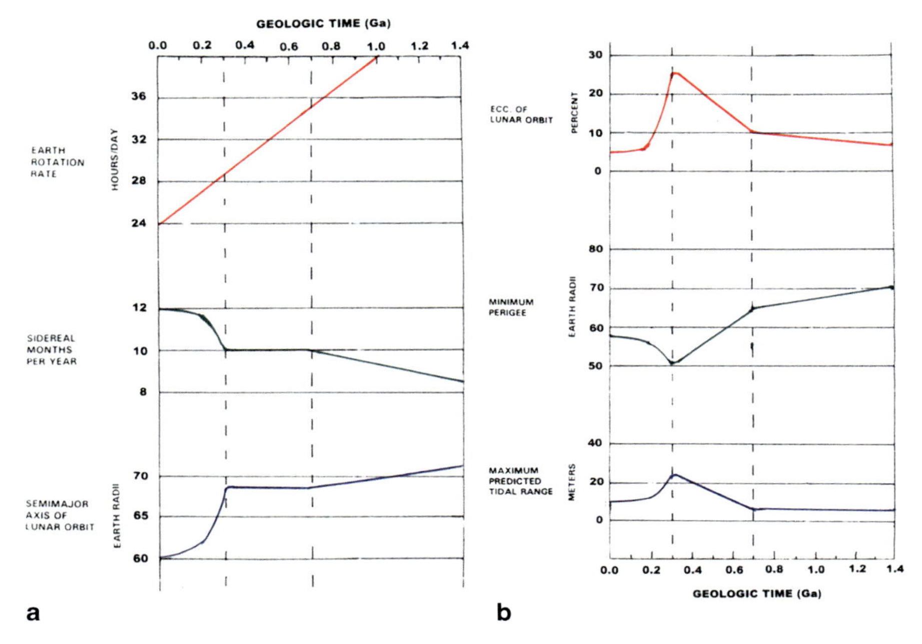

Fig. 8.39 Summary diagrams for some orbit and tidal parameters for this forced eccentricity scenario. **a** Earth rotation rate simply decreases as angular momentum of rotation is transferred to the lunar orbit. The number of sidereal months/year changes from ~12 to ~10 and remains there until the resonance becomes ineffective. The semi-major axis goes from ~62 earth radii to the resonance value of 68 earth radii. **b** The eccentricity of the lunar orbit increases to the maximum value while the lunar body is "cool". The Q value of the Moon controls the maximum value of the eccentricity. After the maximum value of eccentricity is attained, then the eccentricity decreases systematically while the semi-major axis remains at the resonant value and as angular momentum is transferred systematically from the rotating Earth to the lunar orbit. The minimum perigee controls the height of the equilibrium tidal amplitudes and ranges. The maximum equilibrium rock and ocean tides would be characteristic of a circular orbit of 50 earth radii rather than 68 earth radii. Likewise, the maximum predicted tidal range, when using the Bay of Fundy tidal amplification factor of ~20, yields a maximum predicted tidal range of about 22 m in favored tidal embayments compared to only 5 m for a near circular orbit of 68 earth radii

8.7 Summary and Conclusions for this Chapter

Planet Orbit—Lunar Orbit resonances can have a major effect on the tectonic and surface process regimes of planet Earth. These are unusual influences on the planet but the effects are akin to the effects of the Milankovitch cycles. In fact, any detailed numerical calculation of a PO-LO resonance has to incorporate some features of the Milankovitch model. For example, it must be demonstrated that the orbit resonance is stable through the eccentricity cycle of planet Earth in which the planetary orbit oscillates between a near circular orbit to an orbit with about 6% eccentricity over a period of 100 Ka as well as 400 Ka (Imbrie and Imbrie 1979).

The episodes of the PO-LO resonances appear to relate to milestone markers in the evolution of the planet. The one associated with a perigean cycle of 15 years may be the causative agent for the GREAT OXIDATION EVENT as well as an episode of plate movement in Early Proterozoic time. The one associated with the perigean cycle of 12 years may be the causative agent for the beginning of the modern system of plate tectonics. The one associated with a perigean cycle of 10 years may be a causative agent for a MAJOR CHANGE in the biological system; it moved fairly rapidly from a bacteria-algae stage of development which had persisted for hundreds of millions of years to the METAZOAN STAGE of cell metabolism which gave rise to the higher forms of life. And finally, in the FUTUROZOIC, the orbital resonance associated with a perigean cycle of 5 years may keep the biosphere of the planet operating for an additional few 10's of millions of years.

8.8 A Soliloquy on this Chapter

This chapter could be titled "The Contribution of Big Brother and Twin Sister to the Habitability of Planet Earth".

If these PLANET ORBIT—LUNAR ORBIT resonances are stable (and there is a good chance that they are), then many aspects of the development of the biological system as well as the physical-chemical-geological systems would have been affected in either minor, major, or critical ways. Davies (1992) and Stern (2005) have presented suggestive evidence that the modern style of plate tectonics did not start until about 1 Ga ago. Well, the Big Brother resonance action could have played a critical role in that process. Without it, the planet may still be in a quasi-stagnant-lid operation. Such conditions would affect not only the lithosphere but also the atmospheric gas mixture as well as the chemical composition and circulation patterns of the oceans. Big Brother is also the main cause of the heliocentric orbit eccentricity component of the Milankovitch Model, a cycle that manifests itself on the Earth as warm and cold climate cycles. So it looks like Big Brother may have had a critical effect in the development of the metazoan stage of life. Perhaps the plate tectonics step is the critical event that Gould (1994) implied for causing the major step from a bacteria and algae world to a biological system with metazoans.

What is there for our Twin Sister to do? There are three possible resonance states for affecting the lunar orbit and the tidal regime of the Earth. In my view, the events associated with the perigean cycle of 15 years are the most obvious. A few investigators have suggested that something like the modern style of plate tectonics occurred in the 2.5–2.0 Ga era or earlier (Condie and Pease 2008). This plate movement and mantle degassing could have been a partial cause of the regional or global glaciations and perhaps the Great Oxidation Event (Holland 2002, 2009). Then we have an unusual time interval from about 2.0 Ga to 1.0 Ga. Some scientists speculate that these "biological doldrums" or "boring billion" (Bekker and Holland 2012) were caused by a lack of significant plate movement and essentially "stagnant-lid" conditions on the planet. This "stagnant-lid" condition is predictable if the VO-LO resonance was the driver of a short-lived plate tectonics operation in Early Proterozoic time. Elevated rock and ocean tides would have occurred during that span of time (a few 100 Ma) and some of the testing of the model could be done via tidal rhythmite sequences of the Huronian Supergroup and time equivalent rock units in southern Africa and Australia.

Twin Sister exerts a positive influence via the 10 to 1 resonance which may also be involved in the major change from bacteria-algae regime to the METAZOAN stage of evolution. This resonance action immediately follows the JO-LO resonance but nonetheless may have been *critical* for the evolution of metazoans. The *above normal* ocean tidal activity as documented in the sedimentary rock sequences of Early Paleozoic Time (e.g., Stewart 1970; Lockmann-Balk 1970, 1971; Driese et al. 1981; Tape et al. 2003) can be explained via the 10 to 1 resonance. This resonance then fades away in the middle Paleozoic.

Twin Sister can also help out at one time in the FUTUROZOIC, and this event could be a boost, perhaps the final boost, for the higher forms of life on Earth. Without that boost, life would probably become extinct in less than 1 Ga in the future (Kasting 2010).

The main overarching question is: HOW DO THESE POTENTIAL PO–LO RESONANCE ACTIONS FACTOR INTO THE EQUATION FOR FINDING HABITABLE PLANETS WITH HIGHER FORMS OF LIFE? Are these orbital resonances just some major and/or minor additions to the already "LONG CHAIN OF COMPLICATIONS" as suggested by Alfven (1969)?

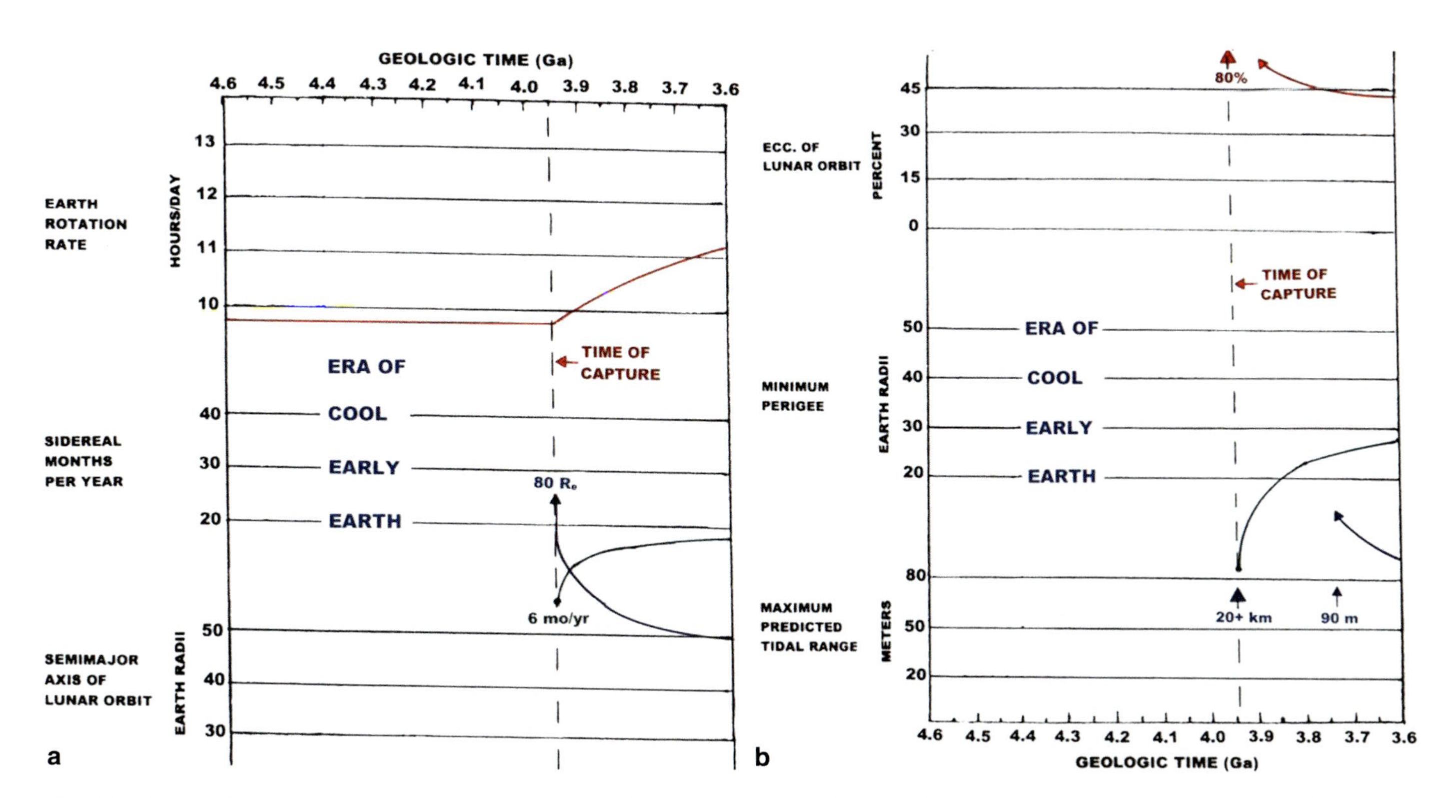

Fig. 8.40 Summary diagrams for some orbit and tidal parameters for the Cool Early Earth era and for Early Archean time (4.6–3.6 Ga). **a** Set 1 parameters. **b** Set 2 parameters

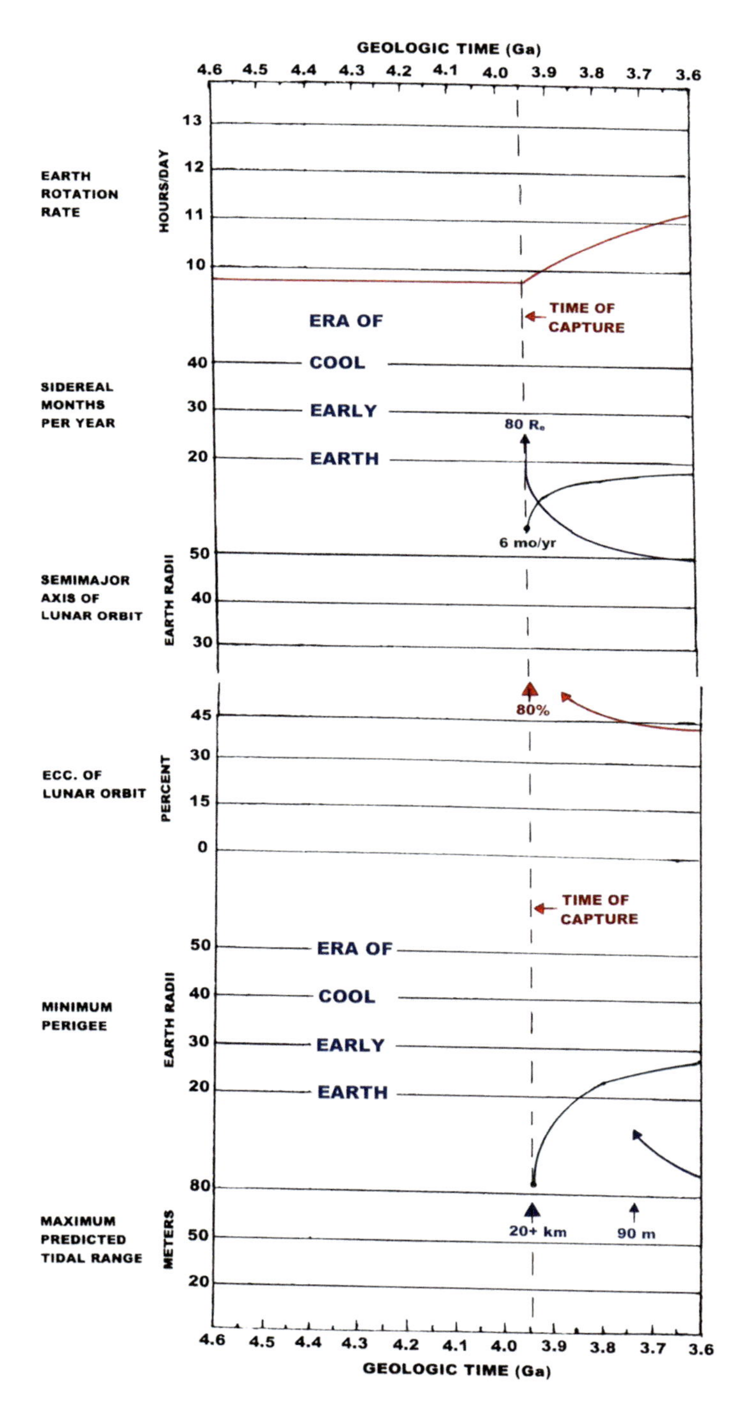

Fig. 8.41 Composite diagram for some orbit and tidal parameters for the Cool Early Earth era and for Early Archean time (4.6–3.6 Ga)

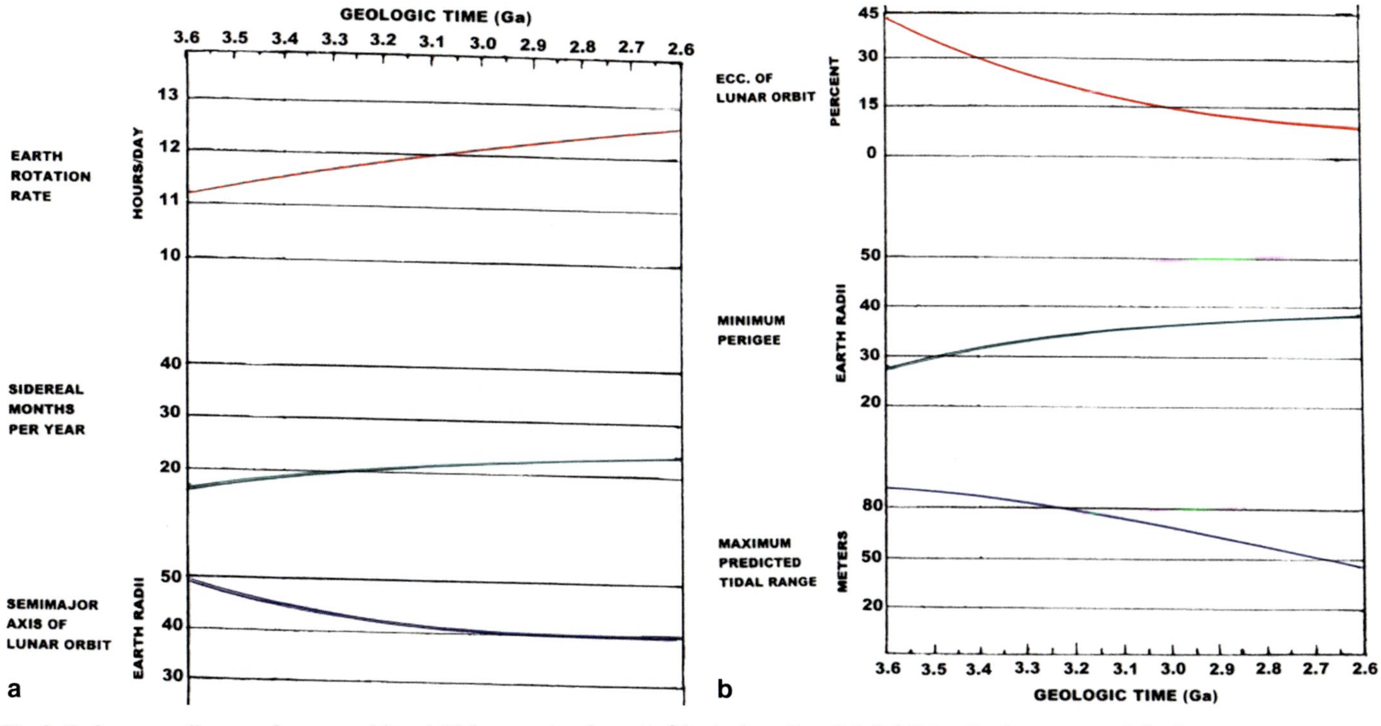

Fig. 8.42 Summary diagrams for some orbit and tidal parameters for part of the Archean Eon (3.6–2.6 Ga). **a** Set 1 parameters. **b** Set 2 parameters

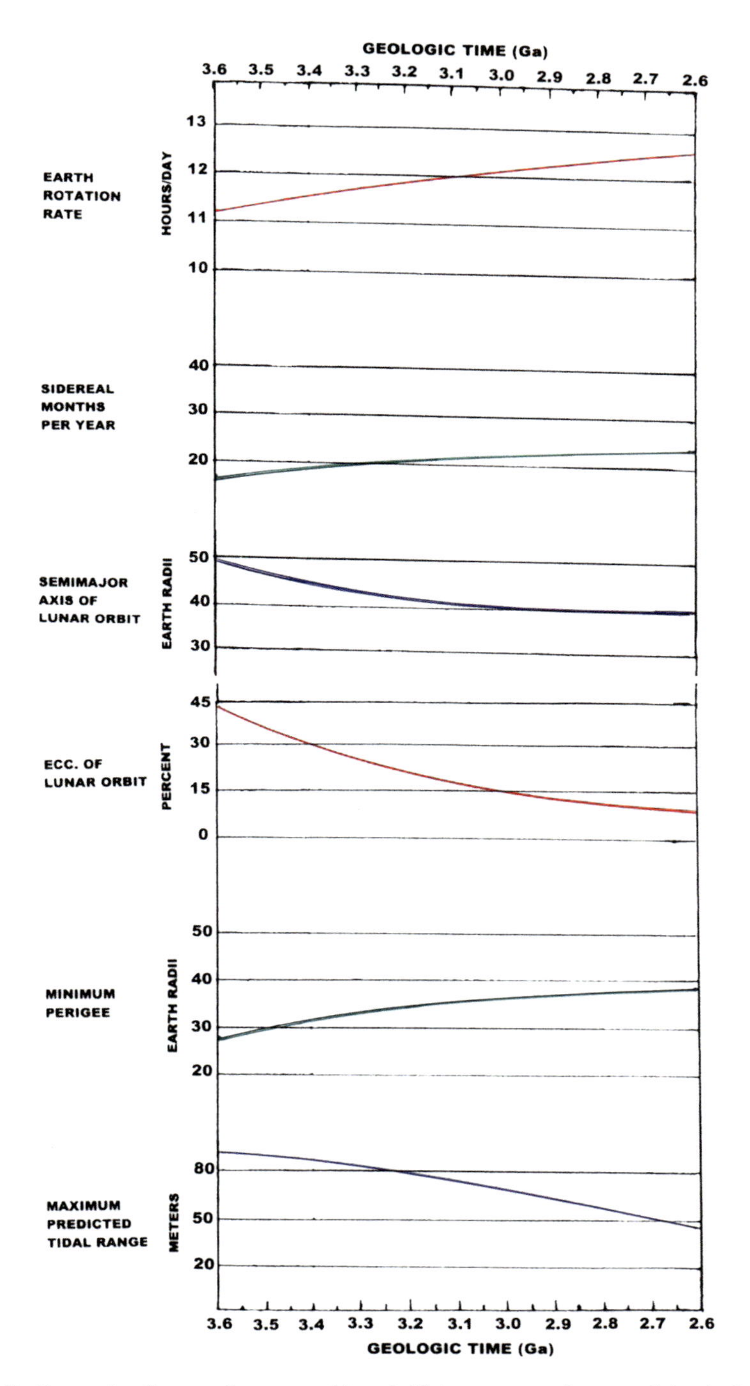

Fig. 8.43 Composite diagram for some orbit and tidal parameters for part of the Archean Eon (3.6–2.6 Ga)

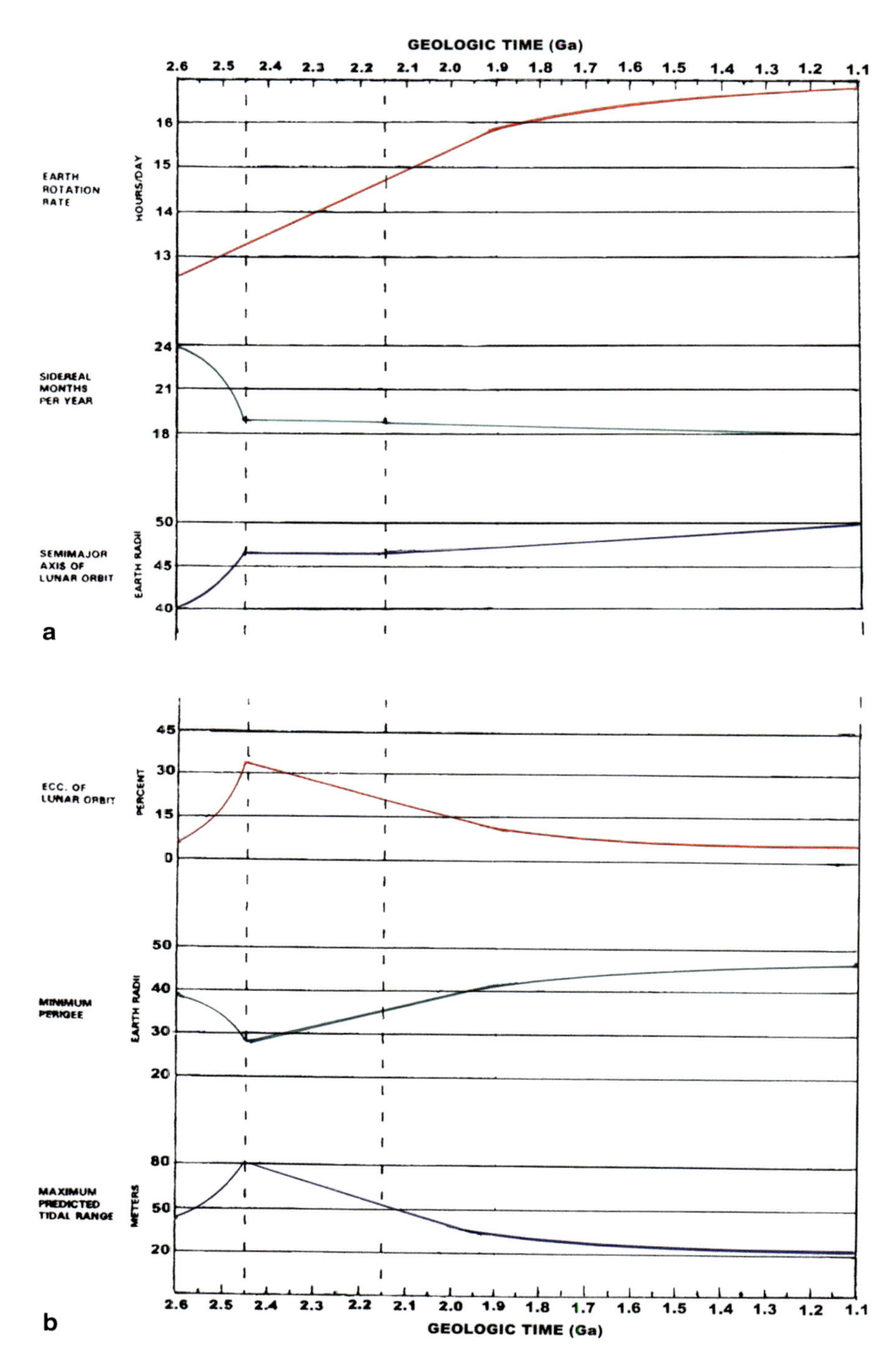

Fig. 8.44 Summary diagrams for some orbit and tidal parameters for the Early Proterozoic to Late Proterozoic eras (2.6–1.1 Ga). **a** Set 1 parameters. **b** Set 2 parameters

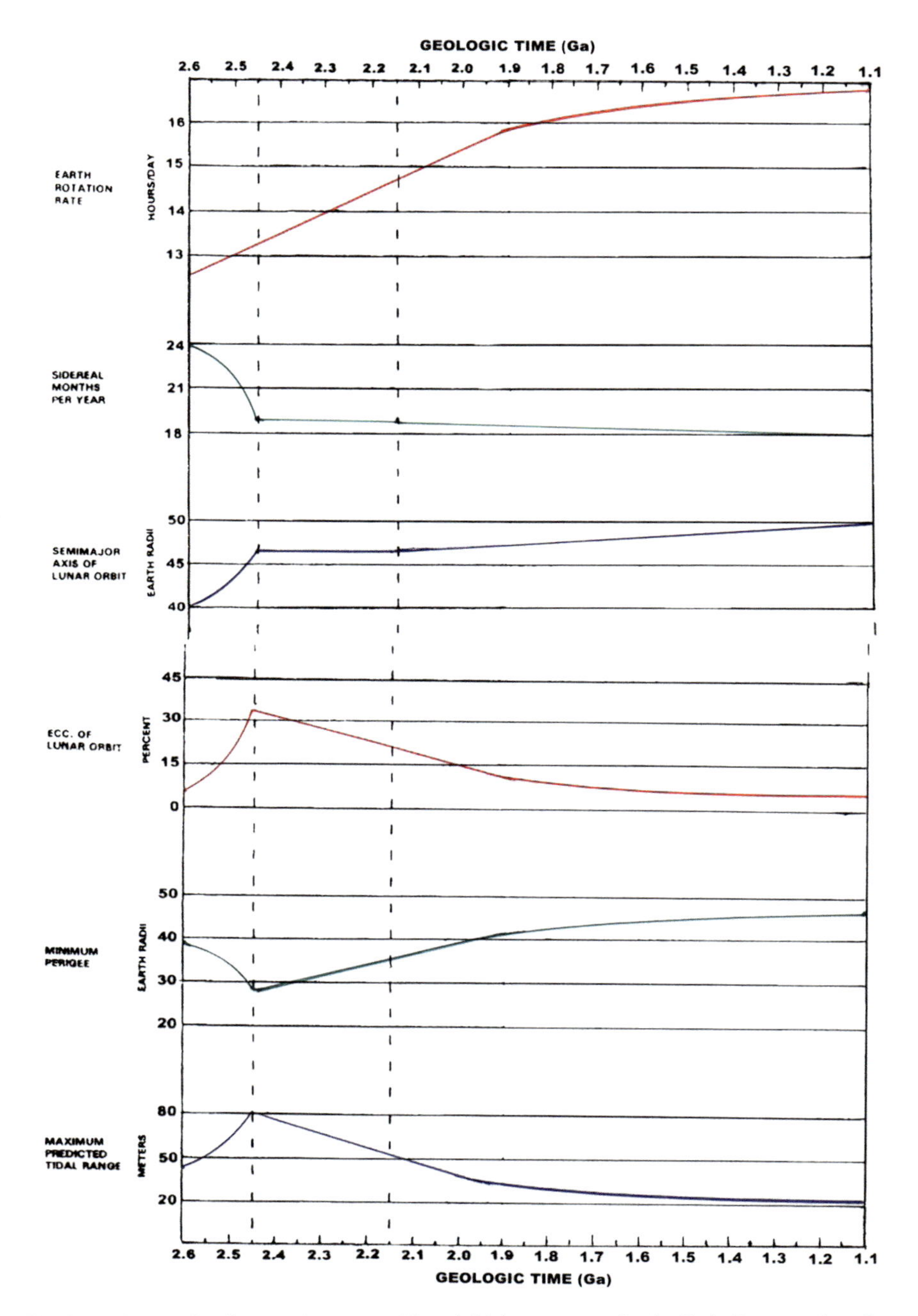

Fig. 8.45 Composite diagram for some orbit and tidal parameters for the Early Proterozoic to Late Proterozoic eras (2.6–1.1 Ga)

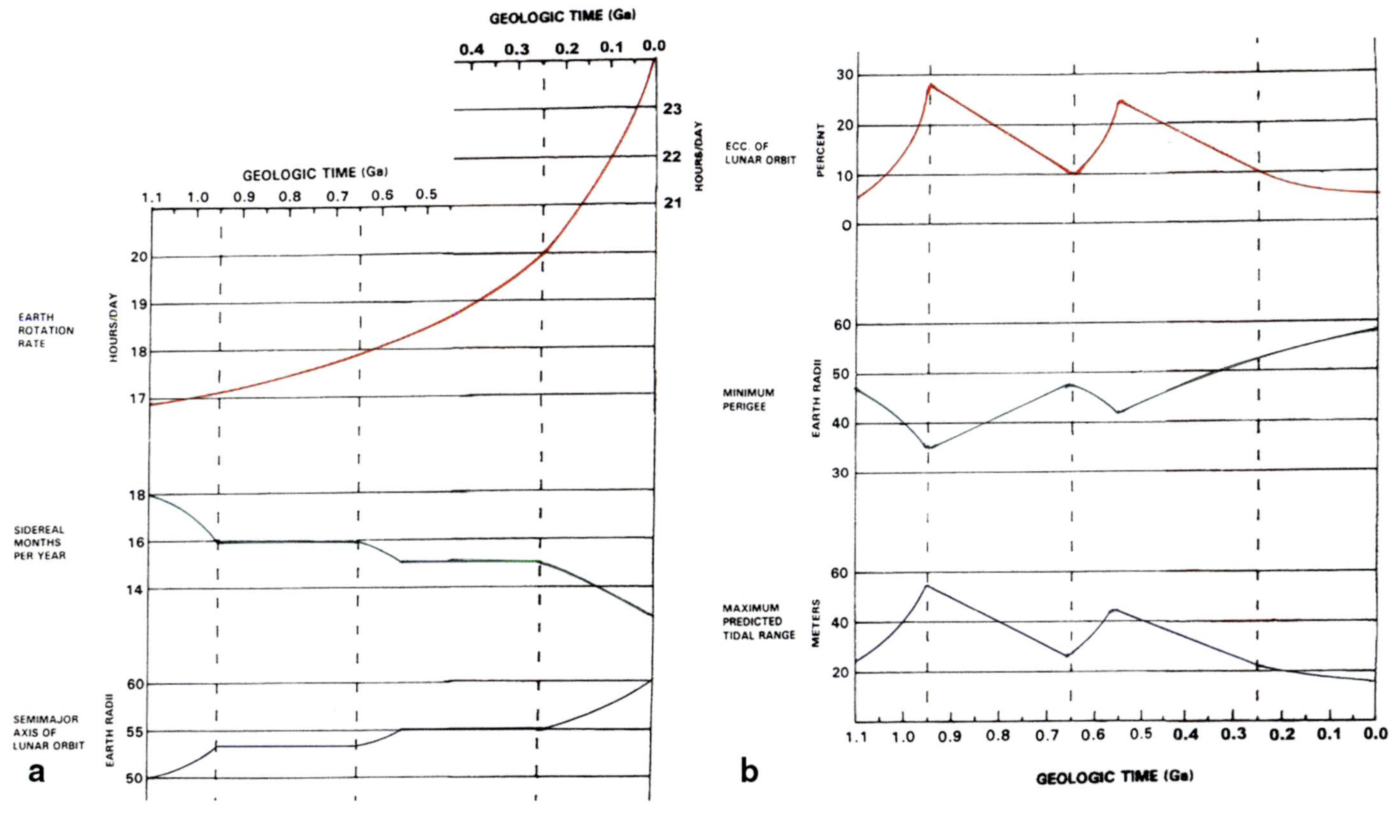

Fig. 8.46 Summary diagrams for some orbit and tidal parameters for the Late Proterozoic and Early Paleozoic eras (1.1 Ga to Present). **a** Set 1 parameters. **b** Set 2 parameters

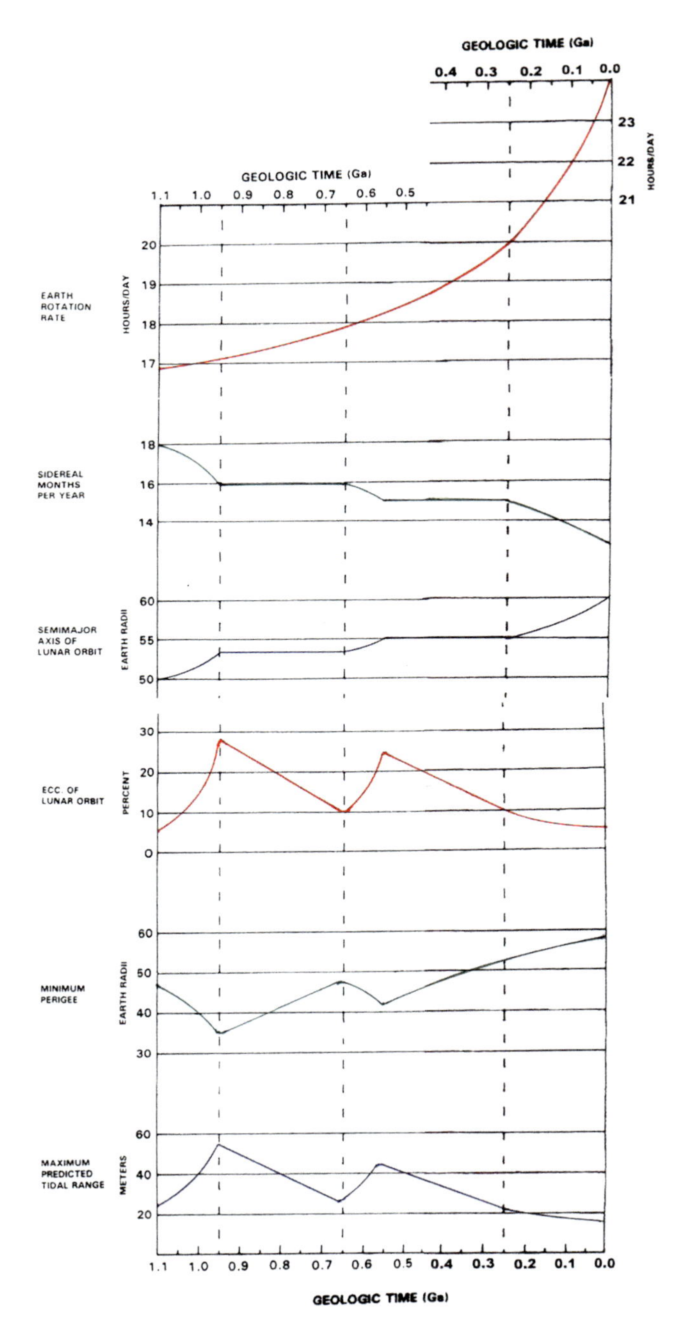

Fig. 8.47 Composite summary diagram for some orbit and tidal parameters for the Late Proterozoic and Early Paleozoic eras (1.1 Ga to Present)

References

Alfven H (1969) Atom, man, and the Universe. The long chain of complications. W. H. Freeman, San Francisco, p 110
Bailey CM, Peters SE, Morton J, Shotwell NL (2007) The Mechum river formation, Virginia Blue Ridge: a record of neoproterozoic and paleozoic tectonics in southeastern Laurentia. Am J Sci 307:1–22
Banks NL (1973) Tide-dominated offshore sedimentation, Lower Cambrian, north Norway. Sedimentology 20:213–228
Bazso A, Dvorak R, Pilat-Lohinger E, Eybl V, Lhotka Ch (2010) A survey of near-mean-motion resonances between Venus and Earth. Celest Mech Dyn Astr 107:63–76
Bostrom RC (2000) Tectonic consequences of Earth's rotation. Oxford University Press, London, p 266
Bekker A, Holland HD (2012) Oxygen overshoot and recovery during the early Paleoproterozoic. Earth Planet Sci Lett 317-318:295–304
Canup RN (2013) Lunar conspiracies. Nature 504:27–29
Chapman DMF (1986) Recurrent phenomena of Venus and the Venus/Earth orbital resonance. J R Astron Soc Canada 80:336–343
Condie KC, Pease V (2008) When did plate tectonics begin on planet Earth? Geological Society of America Special Paper 440, pp 294
Conway-Morris S (1990) Late Precambrian-Early Cambrian metazoan diversification. In: Briggs DEG, Crowther PR (eds) Paleobiology: a synthesis. Oxford Press, Oxford, pp 30–36
Cuk M (2007) Excitation of lunar eccentricity by planetary resonances. Science 318:244
Cuk M, Stewart ST (2012) Making the Moon from a fast-spinning Earth: A giant impact followed by resonant despinning. Science 338:1047–1052
Davies GF (1992) On the emergence of plate tectonics. Geology 20:963–966
Deynoux M, Duringer P, Khatib R, Villeneuve M (1993) Laterally and vertically accreted tidal deposits in the Upper Proterozoic Madina-Kauto Basin, southeastern Senegal, West Africa. Sedimen Geol 84:179–188
Driese SG (1987) An analysis of large-scale ebb-dominated tidal bedforms: evidence for tidal bundles in the Lower Silurian Clinch Sandstone of east Tennessee. Southeast Geol 27:121–140
Driese SG, Byers CW, Dott RH Jr (1981) Tidal deposition in the basal Upper Cambrian Mt. Simon Formation in Wisconsin. J Sediment Petrol 51:367–381
Ehlers TA, Chan MA (1999) Tidal sedimentology and estuarine deposition of the Proterozoic Big Cottonwood Formation, Utah. J Sediment Res 69:1169–1180
Eyles N, Januszczak N (2004) 'Zipper-rift': a tectonic model for neoproterozoic glaciations during the breakup of Rodinia after 750 Ma. Earth Sci Rev 65:1–73
Eyles N, Young GM (1994) Geodynamic controls on glaciation in Earth history. In: Deynoux M, Miller JMG, Domack EW, Eyles N, Fairchild IJ, Young GM (eds) Earth's glacial record. Cambridge University Press, Cambridge, pp 1–28
Gould SJ (1994) The evolution of life on the Earth. Sci Am 271(4):85–91
Grinspoon DH (1997) Venus revealed: a new look below the clouds of our mysterious twin planet. Addison-Wesley, Reading, p 355
Halverson GP, Shields-Zhou G (2011) Chemostratigraphy and the neoproterozoic glaciations. In: Arnaud E, Halverson GP, Shields-Zhou G (eds) The geological record of neoproterozoic glaciations, vol 36. The Geological Society of London, Memoir, London, pp 51–66
Hansen KS (1982) Secular effects of oceanic tidal dissipation on the moon's orbit and earth's rotation. Rev Geophys Space Phys 20:457–480
Hiscott RN (1982) Tidal deposits of the Lower Cambrian Random Formation, eastern Newfoundland: facies and paleoenvironments. Canadian J Earth Sci 19:2028–2042
Hoffman PF, Kaufman AJ, Halverson GP, Schrag DP (1998) A neoproterozoic s curves from marin nowball earth. Science 281:1342–1346
Holland CH (ed) (1971) Cambrian of the new world. Wiley, New York, p 456
Holland CH (ed) (1974) Cambrian of the British Isles, Norden, and Spitsbergen. Wiley, New York, p 300

Holland HD (2002) Volcanic gases, black smokers, and the great oxidation event. Geochim Cosmochim Acta 66:3811–3826
Holland HD (2009) Why the atmosphere became oxygenated: a proposal. Geochim Cosmochim Acta 73:5241–5255
Imbrie J, Imbrie KP (1979) Ice ages; solving the mystery. Enslow Publishers, Short Hills (NJ) pp 224
Jordan TH (1974) Some comments on tidal drag as a mechanism for driving plate motion. J Geophy Res 79:2141–2142
Kasting J (2010) How to find a habitable planet. Princeton University Press, New Jersey, p 326
Kessler LG, Gallop IG (1988) Inner shelf/shoreface intertidal transition, upper Precambrian, Port Askaig Tillite, Isle of Islay, Argyll. In: de Boer PL, van Gelder A, Nio SD (eds) Tide-influenced sedimentary environments and facies. D. Reidell, Dordrecht, pp 341–358
Knoll AD (1991) End of the Proterozoic eon. Sci Am 265(4):64–73
Laskar J (1994) Large-scale chaos in the solar system. Astron Astrophys 287:L9–L12
Laskar J (1995) Large scale chaos and marginal stability in the solar system. In: Proceedings Volume, XIth International Congress of Mathematical Physics, pp 75–120
Laskar J (1996) Large scale chaos and marginal stability in the solar system. Celest Mech Dyn Astron 64:115–162
Link PK, Miller JMG, Christie-Blick N (1994) Glacial-marine facies in a continental rift environment: neoproterozoic rocks of the western United States Cordillera. In: Deynoux M, Miller JMG, Domack EW, Eyles N, Fairchild IJ, Young GM (eds) Earth's glacial record. University of Cambridge Press, Cambridge, pp 29–46
Lochman-Balk C (1970) Upper Cambrian faunal patterns on the craton. Geol Soc Am Bull 81:3197–3224
Lochman-Balk C (1971) The Cambrian of the craton of the United States. In: Holland CH (ed) Cambrian of the new world. Wiley, New York pp 79–167
Macdonald JGF (1963) The internal constitutions of the inner planets and the Moon. Sp Sci Rev 2:473–557
Miller JMG (1994) The neoproterozoic Konnarock formation, southwestern Virginia, USA: glaciolacustrine facies in a continental rift. In: Deynoux M, Miller JMG, Domack EW, Eyles N, Fairchild IJ, Young GM (eds) Earth's glacial record. University of Cambridge Press, Cambridge, pp 47–59
Miller DJ, Eriksson KA (1997) Late Mississippian prodeltaic rhythmites in the Appalachian Basin: a hierarchical record of tidal and climatic periodicities. J Sedi Res 67:653–660
Nelson TH, Temple PG (1972) Mainstream mantle convection: a geologic analysis of plate motion. Am Assoc Petrol Geol Bull 56:226–246
Palmer AR (1971) The cambrian of the appalachian and eastern New England regions, eastern United States. In: Holland CH (ed) Cambrian of the new world. Wiley, New York, pp 169–217
Peale SJ, Cassen P (1978) Contribution of tidal dissipation to lunar thermal history. Icarus 36:245–269
Powell CM, Preiss WV, Gatehouse CG, Krapaz B, Li ZX (1994) South Australian record of a Rodinian epicontinental basin and its mid-neoproterozoic breakup (~ 700 Ma) to form the Palaeo-Pacific ocean. Tectonophysics 237:113–140
Priess WV (ed) (1987) The adelaide geosyncline-late proterozoic stratigraphy: sedimentatikon, palaeontology and tectonics. Geol Surv S Aust Bull 53:438
Rankin DW (1975) The continental margin of eastern North America in the southern Appalachians: the opening and closing of the Proto-Atlantic ocean. Am J Sci 275-A:298–336
Rankin DW (1993) The volcanogenic Mount Rogers formation and the overlying glaciogenic Konnarock formation—two late proterozoic units in southwestern Virginia. U. S. Geol Surv Bull 2029:26
Reinhard CT, Ralswell R, Scott C, Anbar AD, Lyons TW (2009) A late archean sulfidic sea simulated by early oxidative weathering of the continents. Science 326:713–716
Ricard Y, Doglioni C, Sabadina R (1991) Differential rotation between lithosphere and mantle: a consequence of lateral mantle viscosity variations. J Geophys Res 96:8407–8415
Ross MN, Schubert G (1989) Evolution of the lunar orbit with temperature and frequency-dependent dissipation. J Geophys Res 94(B7):9533–9544
Saeed A, Evans JE (1999) Subsurface facies analysis of the late cambrian Mt. Simon Sandstone in western Ohio (midcontinent North America). Open J Geol 2:35–47

Saunders RS (1999) Venus. In: Beatty JK, Petersen CC, Chaikin A (eds) The new solar system, 4th edn. Sky Publishing Company and Cambridge University Press, Cambridge, pp 97–110
Schultz PH, Spudis PD (1983) Beginning and end of lunar mare volcanism. Nature 302:233–236
Scoppola B, Boccaletti D, Bevis M, Carminati E, Doglioni C (2006) The westward drift of the lithosphere: a rotational drag? Geol Soc Am Bull 118:199–209
Shields-Zhou G, Och L (2011) The case for a neoproterozoic oxygenation event: geochemical evidence and biological consequences. GSA Today 21(3):4–11
Sonett CP, Chan MA (1998) Meoproterozoic Earth-Moon dynamics: rework of the 900 Ma big Cottonwood Canyon tidal laminae. Geophys Res Lett 25:539–542
Sonett CP, Kvale EP, Zakharian A, Chan MA, Demko TM (1996) Late proterozoic and paleozoic tides, retreat of the moon, and rotation of the earth. Science 273:100–104
Stern RJ (2005) Evidence from ophiolites, blueschists, and ultrahigh-pressure metamorphic terranes that the modern episode of subduction tectonics began in neoproterozoic time. Geology 33:557–560
Stewart JH (1970) Upper precambrian and lower Cambrian strata in the Southern Great Basin, California and Nevada. United States Geological Survey Professional Paper 620, p 206
Tape CH, Cowan CA, Runkel AC (2003) Tidal-bundle sequences in the jordan sandstone (Upper Cambrian), southeastern Minnesota, U. S. A.: evidence for tides along inboard shorelines of the Sauk epicontinental sea. J Sediment Res 73:354–366
Touma J, Wisdon J (1994) Evolution of the Earth-Moon system. Astron Jour 108:1943–1961
Touma J, Wisdom J (1998) Resonances in the early evolution of the Earth-Moon system. Astron Jour 115:1653–1663
Uyeda S, Kanamori H (1979) Back-arc opening and the mode of subduction. J Geophys Res 84:1049–1061
Webb DH (1982) Tides and the evolution of the Earth-Moon system. Geophys J R Astron Soc 70:261–271
Whitmeyer SJ, Karlstrom KE (2007) Tectonic model for the proterozoic growth of North America. Geosphere 3:220–259. doi:10-1130/GES00055.1
Williams GE (1989a) Tidal rhythmites: geochronometers for the ancient Earth-Moon system. Episodes 12(3):162–171
Williams GE (1989b) Precambrian tidal sedimentary cycles and earth's palorotation. EOS (Am Geophys Union) 70(33):40–41
Williams GE (1997) Precambrian length of day and the validity of tidal rhythmite paleotidal values. Geophys Res Lett 24:421–424
Windley BR (1995) The evolving continents, 3rd edn. Wiley, New York, p 526
Young GM, Nesbitt HW (1985) The Gowganda formation in the sourhern part of the Huronian outcrop bets, Ontario, Canada: statigraphy, depositional environments and the regional tectonic significance. Precambrian Res 29:265–301

Chapter 9
Discussion of the Probability of Finding Habitable Planets for Humans Orbiting Sun-Like Stars

"Planets and stars are supremely important to every member of the human race. Although most people never give it much thought, we live on a planet and are warmed and nourished by a star, not just any planet and not just any star, but a particular kind of planet with certain essential characteristics, and a particular kind of star. The central purpose of this book is to spell out the necessary requirements of planets on which human beings as a biological species (Homo sapiens) can live, and the essential properties required of the stars that provide heat and light to such planets." From Dole (1964, p. vii).

"It is almost certain that no other planet in our solar system now supports the phenomenon of life. The question still remains, however, as a persistent field of speculation, as to how many other stars in the galaxy or the universe as a whole may support a planet like Earth. In science fiction and fantasies like Star Wars, such stars are numerous. Even in the scientific community, a blue-ribbon panel established by the National Aeronautics and Space Administration has designated a major international program to search for extra-terrestrial intelligence (SETI) (Morrison et al. 1977), and a branch of science (so far without content) called exobiology has been designated and widely accepted. In view of such a great interest it seems desirable to seek an estimate of the fraction of stars that might have associated with them an Earthlike planet and the probability that such a planet would be capable of supporting a prolonged evolutionary history of organisms." From Pollard (1979, p. 653).

"My own guess is that, just as we learned that the Sun is an ordinary star, we will find that Earth is an ordinary planet and that life itself is a commonplace phenomenon that exists on most, or all, such planets." From Kasting (2010, p. 298).

Definition of a HABITABLE ZONE in a planetary system Ollivier et al. (2009, pp. 319–320):

> The habitable zone in a planetary system is generally defined as the region in which a terrestrial-type planet may have liquid water at its surface during part of the year (the period of revolution around its parent star).... The notion of the habitable zone should not be considered as a very rigid one.

Robert J. Malcuit, *The Twin Sister Planets Venus and Earth*,
DOI 10.1007/978-3-319-11388-3_9

Using the definition just given, the habitable zone in the Solar System extends approximately from 0.9 to 1.3 AU.

Most scientists that are interested in the habitability of planets can agree on the definition of a habitable zone. (e.g., Hart 1978, 1979; Ward and Brownlee 2002). But from the quotes shown above, it is clear that there is very little agreement for how prevalent life forms are beyond the Earth. We geoscientists know that the Earth has had a complex evolution and that is a large part of the problem. The metazoan life forms on Earth appear to be a result of this complex evolution of the planet. How about the origin of the simpler forms of life? Could a less complex planetary evolution result in less complex life forms? And does this less complex planet evolution scheme make it more probable to find habitable planets in other star-planet systems?

The purpose of this chapter is to demonstrate that when we search for habitable planets *we should be searching for terrestrial planets that have somewhat massive satellites*. The details of the mass and radius of the satellite are not as important as its presence. The direction of revolution of the satellite relative to the direction of rotation of the planet, although important, is not as important *as the stage of development of the planet-satellite system*.

Although the presence of the Moon in orbit about the Earth is mentioned in some treatments of the evolution of habitable planets (e.g., Kasting and Catling 2003; Kasting 2010), the Moon is discussed only as a major stabilizer of the obliquity of the planet under certain conditions of rotation rate for the planet. In their treatment, the effects of the evolution of the Earth-Moon system are absent. In the "Rare Earth" book of Ward and Brownlee (2000) the process of plate tectonics is presented as a process for keeping the planet alive but there is not much attention given to the long term evolution of the Earth-Moon system on the development of the biological system. In contrast, Dole (1964) devotes several pages to the importance of a large satellite to the development of a habitable planet.

9.1 How Simple or How Complicated is the System of Biology on Earth?

Ollivier et al. (2009) give a very good treatment of the essential strategic functions of "life forms" as we understand them on Earth: (1) auto-reproduction (reproduction of identical organisms), (2) evolution by mutations (corresponding to accidental changes in the mechanism of reproduction), and (3) self-regulation with respect to its surroundings (which insures the growth and survival of the individual life form). The special role of carbon and liquid water as well as the chirality is explained in significant detail. They point out that only amino acids with left-handed chirality are used by living organisms on Earth while in abiotic organic molecules there tends to be an equal number of right-handed and left-handed amino acids. Ollivier et al. (2009) also point out that the amino acids found in carbonaceous chondrites have an excess of left-handed forms. This observation suggests, or is consistent with, the concept that prebiotic molecules arrived on Earth from outside (i.e., from the Asteroid Belt). This concept is consistent with these amino acids being delivered to Earth by the water-rich Themis-type asteroids as featured in the Albarede (2009) model.

9.2 My Suggested List of Additional Factors (Models) That Should be Considered in the Development of Higher Forms of Life on Earth

In addition to the numerous conditions listed for the origin and development of a biological system on Earth by a combination of Dole (1964), Ward and Brownlee (2000), and Ollivier et al. (2009), I have four more items for consideration as additions to the list for the development of life forms on our planet.

1. *A COOL EARLY EARTH ERA*: Using an initial rotation rate of about 10 h/day suggested by MacDonald (1963) for the Earth following the accretion and core formation process, a thick basaltic crust forms over a global magma ocean. After an initial period of instability, the basaltic crust becomes resistant to subduction into a less turbulent magma ocean. I think that if the Earth were a stand-alone planet (i.e., no moon or a small moon or two), some portion, perhaps a large portion, of that primitive crust would still be at the surface of the planet. The "aquarioid" asteroids would bring in the water for Earth's oceans along with other volatile elements and the left-handed amino acids and deposit them as a "late veneer" on, in, or through this primitive basaltic crust.
2. *THE FORMATION AND DISTRIBUTION OF THE VULCANOID PLANETOIDS*: *Where did the Moon come from* if it is not the result of giant impact? The Giant Impact Model appears to be incompatible with the Cool Early Earth Model described above. *Thus, we must make a choice here*, it is one or the other and *not* both! A Vulcanoid source for the Moon and sibling planetoids offers a solution for many geological/planetological problems. (a) It explains lunar composition and structure as well as the primitive crustal magnetization. (b) It gives Earth a chance to cool down for about 600 million years before Luna becomes a candidate for capture. (c) Once the paradoxes of the capture process are understood, we know that stable gravitational capture can occur. In addition, a Vulcanoid source yields a planetoid for capture by Venus and may be a source for perhaps all "volcanic" asteroids. Greenwood et al. (2005) suggested that all V-type asteroids had a magma ocean development early in their history. I contend that if they all started out as Vulcanoid planetoids, that condition can be explained.
3. *GRAVITATIONAL CAPTURE OF A SIZEABLE SATELLITE*: Collision of some sort is much more probable. Also, *retrograde* capture is more probable than *prograde* capture. The geometry of Stable Capture Zones, however, can be used to estimate the probability of stable capture for both prograde and retrograde encounter scenarios. The important point is that once a successful (stable)

capture occurs, then a very predictable sequence of events follows. Some of these events are associated with the early post-capture orbit circularization process. After a few *post-capture* close encounters occur then (a) primitive crust in the equatorial zone of the planet (broadly defined) commences to undergo a forced subduction but primitive crustal masses in the polar zones can survive perhaps to become "embryos" of future continents, (b) *Tidal Vorticity Induction* circulation cells (Bostrom 2000) are formed in the upper mantle of the planet via the unidirectional tangential tidal activity associated with capture and these cells, in turn, generate alternate belts of basaltic ocean floor and volcanic arc complexes in the equatorial zone of the planet, (c) *Tidal Vorticity Induction* cells are also set up in the upper mantle of the lunar body and lead to the generation of an exponentially decreasing internal magnetic field within the satellite, (d) eventually the basaltic crust on Earth becomes stabilized and the era of tidally-induced subduction comes to an end because the upper mantle of Earth is too warm to permit "slab-pull" tectonics, and (e) within about a billion years after capture, Earth has returned to a one-plate planet status and would remain so *UNLESS, OR UNTIL, SOMETHING ELSE HAPPENS.*

4. *PLANET ORBIT—LUNAR ORBIT RESONANCES*: These resonances can happen in the solar system but it seems a bit unusual that planet Earth is positioned very favorably to benefit from this activity These "something else" episodes, in my view, are a result of having Twin Sister, Venus, and Big Brother, Jupiter, in the neighborhood. These episodes and their possible results are the following. A Venus Orbit—Lunar Orbit resonance episode when the perigean cycle of the lunar orbit approaches 15 earth years could result in an increase in lunar orbit eccentricity and elevated tides on both Earth and Moon. The sequence of events is the following: (1) a gradual increase in lunar orbit eccentricity over 100 Ma or less; (2) tidally-induced plate tectonics is initiated as the rock tides reach a maximum value (the weak zones in the all-enclosing crust become the subduction zones and spreading centers form wherever the older crust pulls apart); (3) carbon dioxide degassing at volcanic arcs in the presence of cyanobacteria leads to molecular oxygen generation by photosynthesizers [perhaps there are no higher forms of life at that point in time but eventually eukaryotic cells can evolve via the Margulis (1970) autosymbiosis process]; and (4) after about 300 Ma, this VO-LO resonance action drifts away, the tidally-induced plate tectonic action decreases in intensity and by about 2.0 Ga and the crustal complex slowly returns to a stagnant-lid condition.

Then Big Brother, Jupiter, becomes an important character when the perigean cycle approaches 12 earth years. This orbital resonance action is stronger than any of the others because there is a significant gravitational tug on the lunar orbit at apogee for about 4 months every year compared to a significant 3 or 4 month interaction every 5 years with planet Venus during VO-LO resonances. This JO-LO resonance begins at about 1.1 Ga and the lunar orbit is at maximum eccentricity by about 1.0 Ga. The Earth is cool enough for the slab-pull mechanism to work at this point in the history of the planet (Davies 1992; Stern 2005). This is where the rock tides become important. The elevated rock tides can cause a weak version of TVI cell circularization to be established in the upper mantle and the downward limb of these cells aids in the initiation of subduction. Once the process of subduction is initiated on the stagnant-lid planet, the subduction process is self-accelerating and the result is the subduction of a significant mass of crust into the mantle. The carbon dioxide content of the atmosphere of the planet is increased by island arc volcanism and the photosynthesizers that were fairly dormant for some time, will be energized to cause the Second Oxidation Event (Holland 2009).

Now we return to our Twin Sister, Venus, about 300 Ma after the peak eccentricity of the Big Brother episode. Even before the Jupiter resonance action is completed and as the perigean cycle approaches 10 years, the eccentricity of the lunar orbit is increased rapidly and the semi-major axis of the lunar orbit is forced to the resonant value of about 55 earth radii. This is at about 550 Ma ago *and this action could trigger the "Cambrian Explosion"* (Gould 1994) of the metazoan stage of organic evolution. There is a time of elevated rock and ocean tides for a few 100 Ma before the slow-working tidal friction process increases the angular momentum of the lunar orbit to that of a circular orbit of 55 earth radii. During this era of elevated rock and ocean tides, plate tectonics gets well established and the mantle becomes cool enough to perpetuate the process of slab-pull subduction. The rock tides then continue to provide a weak, unidirectional mechanism to keep the plate tectonic process operating (Nelson and Temple 1972; Ricard et al. 1991; Scoppola et al. 2006).

Twin Sister, Venus, has one more chance to aid her "womb-mate", Earth, when the perigean cycle approaches 5 years. This value of the perigean cycle is characteristic of a lunar orbit at about 68 earth radii and this orbital radius is attained under normal tidal friction processes in 800–1000 Ma in the Future. Using an optimistic approach to the lifetime of the biosphere, Kasting (2010) estimated that conditions for life forms would gradually deteriorate to an intolerable thermal temperature level by about 1.0 Ga in the Future. As outlined in Chap. 8, the increase in orbit eccentricity can begin as early as 62 earth radii. Although the tidal amplitudes for this resonance episode are not as high as those in Earth's past, these higher rock and ocean tides can be a "life saver" for a while. Plate Tectonic action is reinitiated, or enhanced, by rock tidal action. This then causes an influx of carbon dioxide into the atmosphere and the photosynthesizers are in business again. Then, as the resonance gradually loses it effectiveness and as the lunar orbit evolves to a circular orbit of about 68 earth radii, the intolerable, torrid thermal conditions eventually return. The orbital resonance episode, however, gives the biosphere an extension of a few 100 Ma. Then as Earth's surface temperature increases because the solar energy is forever increasing because of the increasing solar constant, the ocean water evaporates and plate tectonics and the associated volcanism will gradually cease. The Earth-Moon system, however, persists for several billion years into the future with only fossils of the life forms that populated planet Earth in past geological eras.

Figure 9.1 shows a diagram of a fairly standard version of the evolution of the Earth-Moon system. The only unusual event on this diagram is lunar capture at about 3.95 Ga and the subsequent lunar orbit circularization sequence. Figure 9.2 shows the same orbit evolution scheme with the Planet Orbit—Lunar Orbit resonances

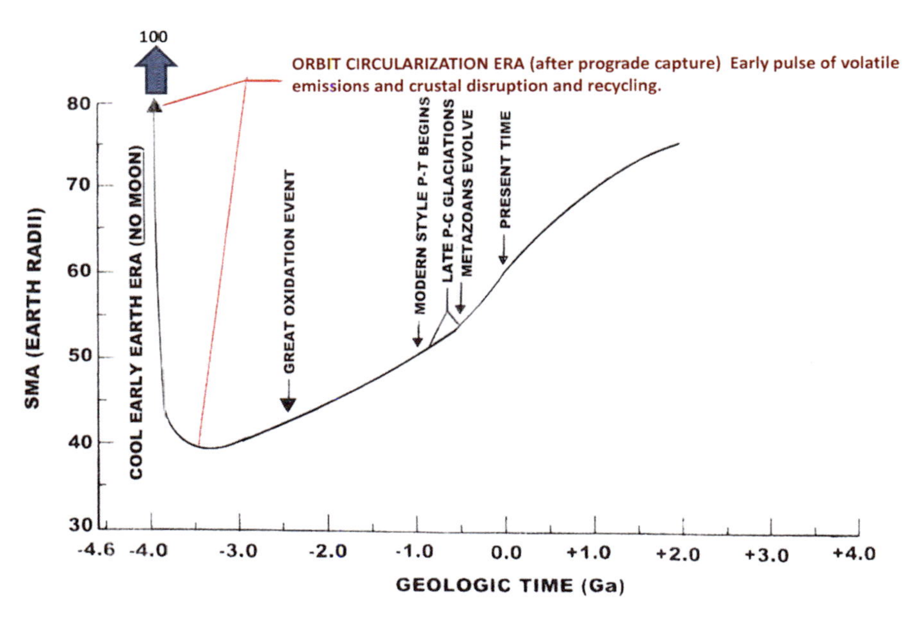

Fig. 9.1 Plot of semi-major axis of the lunar orbit vs. geological time for a "standard version of the evolution of the lunar orbit. The assumption for this plot is that once the lunar orbit circularized after the capture episode, the lunar orbit does not deviate much from a circular orbit and maintains the present eccentricity of 0.0550

shown as "irregularities in the orbit evolution scheme. These PO-LO resonances episodes appear to relate to some of the major "crisis" points in the history of the Earth as well as the history of the biological system of our planet. Perhaps these are major links in "THE LONG CHAIN OF COMPLICATIONS" in the history of our unusual planet.

9.3 The Long Chain of Complications For Explaining Our Existence on the Third Planet From the Sun

Alfven (1969) used the expression that our existence on the third planet from the Sun was the result of "a long chain of complications." Taylor (1998) and Ward and Brownlee (2000) in particular have also emphasized this possibility. These PO-LO resonances are additional complications associated with our existence on the Third Planet from the Sun (Alfven and Alfven 1972). Although these Planet Orbit—Lunar Orbit resonances draw planets Jupiter and Venus into special focus, they are already on board for being factors in the climate history of Earth. Berger et al. (1992) and others demonstrated that Jupiter, with help from Saturn, is the main driver of the heliocentric eccentricity cycle for the Milankovitch Model of climate fluctuations. These changes in orbit eccentricity have been happening from the beginning of the

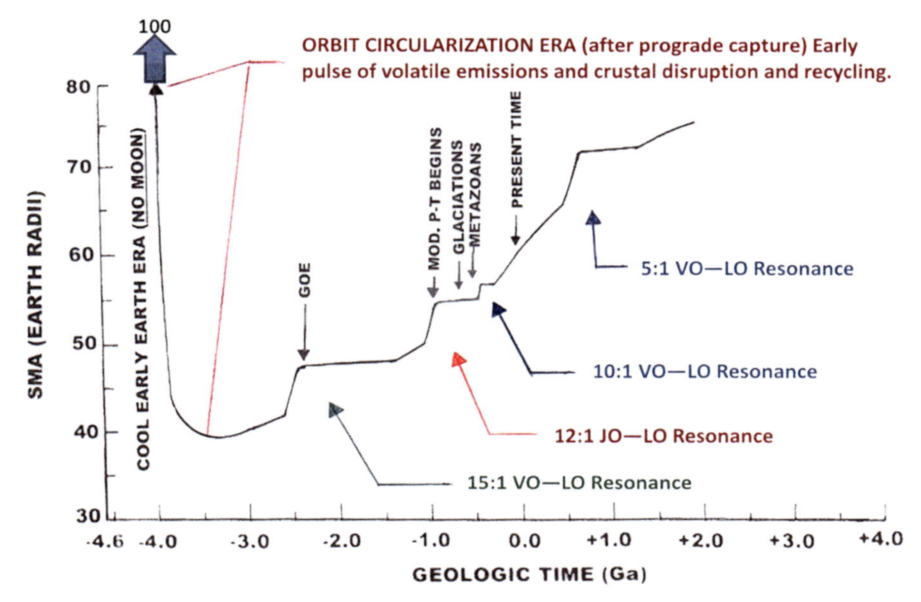

Fig. 9.2 Plot of semi-major axis vs. geologic time for the history of the lunar orbit in which the PO-LO episodes are represented. Note that the lunar orbit eccentricity increases rapidly when each of the resonances is initiated and then simply fades away as the lunar orbit decreases in eccentricity as angular momentum is transferred from the rotating Earth to the lunar orbit. Such rock and ocean tidal episodes could cause significant "punctuation marks" in the rock records of both the Earth and Moon

Solar System and will continue into the distant future. The JO-LO resonance model simply makes Jupiter doubly important to the evolution of planet Earth and its complex biological system. After this consideration it is difficult to believe that Earth is a normal planet and its biological system is to be expected on such a normal planet (Kasting 2010; Broecker 1985; Langmuir and Broecker 2012).

Now let us focus on our Twin Sister, Venus. From the treatment in Chap. 8, Venus has had some effects on planet Earth by way of the VO-LO resonances and these possible effects have been explained in some detail. But the question remains: HAS VENUS AFFECTED EARTH IN OTHER WAYS? If we were attempting to answer this question before 1986, we might say "YES" mainly because of the influence on Mayan, Babylonian, and other cultures (Grinspoon 1997; Swerdlow 1998). Apparently these ancient astronomers were on to something. Chapman (1986) speculated that there must be some dynamical reason for the great regularity of the motion of Venus. Jacques Laskar (1994, 1995, 1996) did extensive calculations including all the planets of the Solar System for 10 Ga into the past and 15 Ga into the future. He found that the orbits of the Jovian planets were stable over this time interval for both eccentricity and inclination. For the terrestrial planets the only two stable planetary orbits for both eccentricity and inclination were Earth and Venus and the orbits of these two planets seemed to co-vary with each other. Planets Mercury and Mars varied considerably in both eccentricity and inclination.

The question is: WHY ARE THE ORBITS OF PLANETS VENUS AND EARTH SO STABLE AND WHY ARE THEY SO STABLE RELATIVE TO ONE ANOTHER? Laskar (1995) stated that there seemed to be a "linear coupling" or an "angular momentum conservation constraint" between Earth and Venus and that this mechanism apparently has been in operation all through geologic time. More recently Bazso et al. (2010) did extensive calculations on the orbital history of the planets in the Solar System and they concentrated their attention on the Earth-Venus relationship. They found that there are three fairly stable orbit-orbit resonance states with the central one being the 8 to 5 resonance (8 Venus years to 5 Earth years). They checked the effects on the Earth's orbit by varying the eccentricity and inclination of the Venus orbit and they found that there were some effects. But to my knowledge, they did not vary the eccentricity and inclination of Earth's orbit to see how that affected the Venus orbit. Maybe that detail will be in future publications. However, Grinspoon (1997, p. 11) states that "computer models of alternate-reality solar systems have shown that the orbit of Venus becomes unstable if you remove Earth, so it seems that these two worlds may have been codependent from birth". The bottom line for this discussion is that the co-variation of the orbits of Venus and Earth is no longer just a "strange coincidence". This relationship apparently has been there from the beginning of the Solar System whether it is from a "linear coupling" or "angular momentum conservation constraint" or some other mechanism. These resulting orbital resonance states affecting the eccentricity and major axis of the lunar orbit are additional favorable episodes in "THE LONG CHAIN OF COMPLICATIONS" that apparently led to the evolution of metazoan life forms on planet Earth (Alfven 1969, 1972). Gould (1994) also raised the question about how and why metazoan life forms originated. He stated that he could understand how bacterial and algal life could evolve under reasonably favorable environmental circumstances. Gould (1994) suggested that there must have been some very unusual events to cause the biological world *to make the giant step to the metazoan stage of life*. I am speculating that perhaps these "unusual" or "special" interactions with Big Brother, JUPITER, and Twin Sister, VENUS, were the "COMPLICATIONS" necessary for the development of the METAZOAN STAGE OF LIFE ON THE PLANET.

9.4 What are the Chances of this Very Long Chain of Complications Happening Elsewhere in a Large Region of Space?

To make a long story much shorter, I come down on the side of Pollard (1979), Taylor (1998), Ward and Brownlee (2000), Ollivier et al. (2009), and many others. The chances of this "long chain of complications" being duplicated in another star-planet system are not good. If there is water, there could be life. However, it is a "giant step" to get to the metazoan stage of life from the bacterial and algal biological stage.

9.5 Summary

We can conclude that the evolution of the biological system of Earth is a VERY, VERY, VERY LONG CHAIN OF COMPLICATIONS. The optimistic view is that the Earth is a somewhat normal terrestrial planet associated with a somewhat normal star (Shklovskii and Sagan 1966; Kasting 2010) and that there are a few earth-like planets out there for discovery, exploration, and development if we can develop the transportation technology to get to them.

The following information from Dole (1964, Table 19) represents an optimistic view of the number of planets. (A light-year is 63,240 astronomical units.)

The Table has the title: "Expected Number of Habitable Planets N within a Sphere of Radius R for from the Sun". The results are the following: radius of 27.2 light-years (1 planet); radius of 34.3 light-years (2 planets); radius of 46.5 light-years (5 planets); radius of 58.5 light-years (10 planets); radius of 100.0 light-years (50 planets).

And some words of wisdom from Dole (1964):

> All in all, the Earth is a wonderful planet to live on, just the way it is. Almost any change in its physical properties, position, or orientation would be for the worse. We are not likely to find a planet that suits us better, although at some future time there may be men who prefer to live on other planets. At the present time, however, the Earth is the only home we have; we would do well to conserve its treasures and to use its resources intelligently.

Pollard (1979) presents a more pessimistic view of finding such planets:

> Although a rough estimate can be made of how many Earthlike planets might exist in the universe, the possibility that life has actually evolved on them cannot be calculated even in principle.

The readers have much latitude to voice a choice on this issue!

References

Albarede F (2009) Volatile accretion history of the terrestrial planets and dynamic implications. Nature 461:1227–1233

Alfven H (1969) Atom, man, and the universe: the long chain of complications. Freeman, San Francisco, p 110

Alfven H, Alfven K (1972) Living on the third planet. Freeman, San Francisco, p 187

Bazso A, Dvorak R, Pilat-Lohinger E, Eybl V, Lhotka Ch (2010) A survey of near-mean-motion resonances between Venus and Earth. Celest Mech Dyn Astron 107:63–76

Berger A, Loutre MF, Laskar J (1992) Stability of the astronomical frequencies over the earth's history for paleoclimate studies. Science 255:560–566

Bostrom RC (2000) Tectonic consequences of earth's rotation. Oxford University Press, London, p 266

Broecker WS (1985) How to build a habitable planet. Eldigio Press, New York, p 291

Chapman DMF (1986) Recurrent phenomena of Venus and the Venus/earth orbital resonance. J R Astron Soc Canada 80:336–343

Davies GF (1992) On the emergence of plate tectonics. Geology 20:963–966

Dole SH (1964) Habitable planets for Man. Blaisdell (a division of Ginn and Company), New York, p 158
Gould SJ (1994) The evolution of life on the earth. Sci Am 271:85–91
Grinspoon DH (1997) Venus revealed. Helix Books, Addison-Wesley, New York, p 355
Greenwood RC, Franchi IA, Jambon A, Buchanan PC (2005) Widespread magma oceans on asteroidal bodies in the early solar system. Nature 435:916–918
Hart MH (1978) The evolution of the atmosphere of the earth. Icarus 33:23–39
Hart MH (1979) Habitable zones about main sequence stars. Icarus 37:351–357
Holland HD (2009) Why the atmosphere became oxygenated: a proposal. Geochim Cosmochim Acta 73:5241–5255
Kasting J (2010) How to find a habitable planet. Princeton University Press, New Jersey, p 326
Kasting J, Catling D (2003) Evolution of a habitable planet. Annu Rev Astron Astrophys 41:429–463
Langmuir CH, Broecker W (2012) How to build a habitable planet. Princeton University Press, New Jersey, p 718
Laskar J (1994) Large-scale chaos in the solar system. Astron Astrophys 287:L9–L12
Laskar J (1995) Large scale chaos and marginal stability in the solar system. In: Proceedings volume, XIth International Congress of Mathematical Physics, pp 75–120
Laskar J (1996) Large scale chaos and marginal stability in the solar system. Celestl Mech Dyn Astron 64:115–162
MacDonald JGF (1963) The internal constitutions of the inner planets and the Moon. Sp Sci Rev 2:473–557
Margulis L (1970) Origin of eukaryotic cells; evidence and research implications for a theory of the origin and evolution of microbial, plant, and animal cells on the Precambrian Earth. Yale University Press, New Haven, p 349
Morrison PJ, Billingham J, Wolfe J (eds) (1977) The search for extraterrestrial intelligence: SETI. NASA Sp Publ 419, Washington, DC, p 276
Nelson TH, Temple PG (1972) Mainstream mantle convection: a geologic analysis of plate motion. Am Assoc Petrol Geol Bull 81:3197–3224
Ollivier M, Encrenaz T, Roques F, Selsis F, Casoli F (2009) Planetary systems: detection, formation and habitability of extrasolar planets. Springer, Berlin, p 340
Pollard WG (1979) The prevalence of earthlike planets. Am Sci 67:653–659
Ricard Y, Doglioni C, Sabadina R (1991) Differential rotation between lithosphere and mantle: a consequence of lateral mantle viscosity variations. J Geophys Res 96:8407–8415
Scoppola B, Boccaletti D, Bevis M, Carminati E, Doglioni C (2006) The westward drift of the lithosphere: a rotational drag? Geol Soc Am Bull 118:199–209
Shklovskii IS, Sagan C (1966) Intelligent life in the universe. Holden-Day, San Francisco, p 509
Stern RJ (2005) Evidence from ophiolites, blueschists, and ultrahigh-pressure metamorphic terranes that the modern episode of subduction tectonics began in Neoproterozoic time. Geology 33:557–560
Swerdlow NM (1998) The Babylonian theory of the planets. Princeton University Press, New Jersey, p 246
Taylor SR (1998) Destiny or chance: our solar system and its place in the Cosmos. Cambridge University Press, Cambridge, p 229
Ward PD, Brownlee D (2000) Rare earth: why complex life is uncommon in the universe. Copernicus (An Imprint of Springer), New York, p 333
Ward PD, Brownlee D (2002) The life and death of planet earth: how the new science of astrobiology charts the ultimate fate of our world. Times Books, New York, p 240

Chapter 10
Summary and Conclusions

From our discussions in this book we can conclude that the terrestrial planets, Venus and Earth, are very similar in physical characteristics but are "worlds apart" on surface and atmospheric conditions. Planet Earth has climate features that resemble "Paradise" and in direct contrast Venus has climate features that resemble "Hades". In this book I am presenting the case that these great differences can be explained by the mode of capture of large satellites: Venus captured a large satellite in *retrograde* orbit and Earth captured a large satellite in *prograde* orbit.

The X-wind model of Shu et al. (2001) yields a suitable environment for the formation of Calcium-Aluminum Inclusions and the CAIs then accumulate to form Vulcanoid planetoids in orbits between the Sun and the orbit of planet Mercury. Gravitational perturbations by Mercury then cause the Vulcanoid planetoids to eventually migrate outward from their place of origin to the vicinity of the terrestrial planets.

Retrograde gravitational capture of a one-half moon-mass planetoid explains the angular momentum regime of the Venus-Adonis system fairly well. Prograde capture of a moon-mass body by Earth explains many features of the Earth-Moon system fairly well. The directional properties of features on the lunar surface, the mascons associated with the circular maria, and the distribution of most of the circular maria along two great-circle patterns can be explained via a tidal disruption model associated with gravitational capture. An Intact Capture Model is compatible with the Cool Early Earth Model because the Earth would be moon-less until the time of capture.

During the early part of the Cool Early Earth Era, water is delivered to both Earth and Venus via impacts of Themis-type asteroids during the first 100 ± 50 Ma (Albarede 2009).

The Twin Sister planets could, in theory, trade places with each other for capture scenarios. Examination of the geometry of Stable Capture Zones (SCZ) suggests that *retrograde* capture of a lunar-mass body by Earth is even more probable than *prograde* capture. The SCZ geometry for Venus, however, strongly favors *retrograde* capture although prograde capture is physically possible (it simply has a low probability of being successful). The bottom line is that planet Earth could have ended up as a "Hades" planet and Venus as a "Paradise" planet. A more likely

Robert J. Malcuit, *The Twin Sister Planets Venus and Earth*,
DOI 10.1007/978-3-319-11388-3_10

(i.e., predictable) outcome is for both planets to end up as "Hades" planets (and we would not be here and the ETs would pass our planet by without much serious consideration).

Nonetheless, *retrograde* capture of a large satellite by either planet could result in habitable conditions for a few billion years. These habitable conditions certainly would be conducive to the evolution of a bacterial-algal biological system but maybe not for a metazoan stage system. In a retrograde capture scenario, the satellite orbit, after an orbit circularization sequence, evolves ever closer to the planet because the angular momentum of the prograde rotation of the planet is in the opposite direction to the retrograde motion of the satellite. The satellite eventually enters with the Roche limit for a solid body, breaks up in orbit, and subsequently the parts of the satellite, large and small, impact onto the surface of the planet after passing through a very dense atmosphere that resulted from mantle degassing that was caused by all the rock tide action before the breakup of the satellite.

A Prograde Capture Scenario will also eventually result in a "Hades" situation because the solar energy output (i.e., the "Solar Constant") is forever increasing as the Sun evolves toward a "red-giant" stage. A reasonable estimate is that some life forms could survive on an Earth-like planet for about one billion years into the future (Kasting 2010) before the surface gets so hot globally that the ocean water begins to evaporate, the water molecules dissociate in the upper atmosphere and the molecular hydrogen escapes from the planet. Algae, bacteria, and associated viruses would be the last biological entities to survive such an scenario.

Finally, there are special perks for Earth by having Big Brother, Jupiter, and Twin Sister, Venus, in the neighborhood because special orbital resonance situations can lead to special rock and ocean tidal episodes on Earth. These tidal episodes are caused by the decreasing period of the perigean cycle as the lunar orbit increases in major axis due to the transfer of angular momentum from the rotating planet to the orbit of the satellite. There are four Planet Orbit—Lunar Orbit resonance states of interest. (1) A resonance state associated with the 15 year period of the perigean cycle, occurring at about 2.5–2.0 Ga, shows promise for explaining the Great Oxidation Event (Holland 2002, 2009; Session et al. 2009). (2) A resonance state associated with the 12 year period of the perigean cycle, occurring at about 1.1–0.5 Ga, shows promise for explaining the Second Oxidation Event (Kump 2010; Shields-Zhou and Och 2011; Sessions et al. 2009) as well as the initiation of the modern style of plate tectonics (i.e., "slab-pull" for the subduction mechanism) (Davies 1992; Stern 2005). (3) A resonance state associated with the 10 year period of the perigean cycle, occurring at about 0.6–0.4 Ga, shows promise for explaining some unusual tide-related characteristics of Late Precambrian and Early Paleozoic sedimentary rock sequences as well as the "Cambrian Explosion" of classical paleontology (Gould 1994). (4) The final resonance state is associated with the 5 year period of the perigean cycle. This resonance, occurring from 0.5 to 1.0 Ga in the future (in what we can call the Futurozoic), shows promise for extending the limited lifetime of the Earth's biological system for a few 100 Ma. These resonance rock tidal action eras may cause an extension of the plate-tectonics era. Plate tectonic processes, in general, recycle carbonate sediments and rock into the mantle and carbon dioxide is emitted to the atmosphere by the associated volcanic arc activity. Photosynthesizing planets then utilize the carbon dioxide and produce molecular oxygen.

In summary, Big Brother, Jupiter, and Twin Sister, Venus, both have had favorable gravitational interactions with planet Earth because of the ever decreasing period of the perigean cycle of the lunar orbit. Without these PLANET ORBIT—LUNAR ORBIT resonances, in my view, the Earth could be a planet characterized by a bacterial-algal biological system. This could have occurred by eliminating the GREAT OXIDATION EVENT, the SECOND OXIDATION EVENT, MODERN STYLE PLATE TECTONICS, and the "CAMBRIAN EXPLOSION".

The bottom line in the search for habitable planets, in my view, is the following. Look for water first, then look for a large satellite, then look for a Sister Planet with a *heliocentric* orbital resonance configuration like planet Venus relative to the planet of interest, and then look for a Big Brother planet like Jupiter that can enter into an orbital resonance with the planet of interest. A PLANET WITH THESE FEATURES AND ORBITAL ENVIRONMENTAL RELATIONSHIPS MAY BE WORTHY OF DETAILED STUDY BY A "BLUE RIBBON" COMMITTEE!

References

Albarede F (2009) Volatile accretion history of the terrestrial planets and dynamic implications. Nature 461:1227–1233

Davies GF (1992) On the emergence of plate tectonics. Geology 20:963–966

Gould SJ (1994) The evolution of life on the earth. Sci Am 271(4):85–91

Holland HD (2002) Volcanic gasses, black smokers, and the great oxidation event. Geochim Cosmochim Acta 66:3811–3826

Holland HD (2009) Why the atmosphere became oxygenated: a proposal. Geochim Cosmochim Acta 73:5241–5255

Kasting J (2010) How to find a habitable planet. Princeton University Press, New Jersey, p 326

Kump LR (2010) Earth's second wind. Science 330:1490–1491

Sessions AL, Doughty DM, Welander PV, Summons RE, Newman DK (2009) The continuing puzzle of the great oxidation event. Curr Biol 19:R567–R574

Shields-Zhou G, Och L (2011) The case for a neoproterozoic oxygenation event: geochemical evidence biological consequences. GSA Today 21(3):4–11

Shu FH, Shang H, Gounelle M, Glassgold AE, Lee T (2001) The origin of chondrules and refractory inclusions in chondritic meteorites. Astrophys J 548:1029–1050

Stern RJ (2005) Evidence from ophiolites, blueschists, and ultrahigh-pressure metamorphic terranes that the modern episode of subduction tectonics began in neoproterozoic time. Geology 33:576–560

Glossary

A

Absolute age A position of the age of a substance on a numerically calibrated time scale (also see relative age).

Accretion process Accumulation of particles from a cloud of debris to form a planet or planetoid.

Accretion torus A heliocentric doughnut-shaped geometric form from which smaller bodies are gravitationally attracted to participate in the planet-building process.

Adonis Proposed name for a phantom satellite of planet Venus.

Adonis Magnus Proposed name for a phantom satellite that gets gravitationally captured in retrograde orbit by an earth-like planet.

Algae Primitive aquatic single-cell or colonial organisms capable of photosynthesis.

Allende meteorite A carbonaceous chondrite meteorite that fell to Earth in 1969 and contained the first intensely studied calcium-aluminum inclusions (CAIs).

Angra Phantom parent body for the Angrite group of meteorites.

Angra dos Reis One of several of the Angrite group of achondritic meteorites.

Angrite meteorites A group of achondritic meteorites that have very similar characteristics and are thought to be fragments of a parent body (Angra) that has not yet been identified. The chemical composition of the angrite meteorites is very similar to that of the Earth's Moon.

Angular momentum A physical property of a rotating body. Similar to linear momentum but is momentum along an arcuate trajectory.

Animal life Metazoan organisms with considerable mobility.

Anorthosite A rock consisting mainly of plagioclase feldspar.

Robert J. Malcuit, *The Twin Sister Planets Venus and Earth*,
DOI 10.1007/978-3-319-11388-3

Anorthositic crust A mass of rock consisting mainly of plagioclase feldspar. The primitive crust of the Earth's Moon and the primitive crust of planet Mercury are both composed of this type of rock which also contains several accessory metallic and ferromagnesian minerals.

Antipodal area on lunar surface An area that is located on the opposite side of the lunar surface. Example: The Mare Ingenii area of the lunar surface is antipodal to a portion of the surface occupied by Mare Imbrium.

Aphelion The farthest distance from the Sun for a body in a heliocentric orbit.

Aphrodite The greek equivalent of the immortal Roman goddess named Venus.

Aquarioid asteroids Proposed name for water-rich (Themis-type) asteroids.

Archean Eon A geologic time period extending from extending from ~3.95 to 2.5 Ga.

Argument of pericenter for orbit of planet The heliocentric position of the closest distance to the Sun for the planet on a 360 degree scale measured in a counterclockwise direction as viewed the north pole of the Solar System.

Argument of pericenter for orbit of planetoid The heliocentric position of the closest distance to the Sun for the planetoid on a 360 degree scale measured in a counterclockwise direction as viewed from the north pole of the Solar System.

Asteroids Small planetoids in heliocentric orbits.

Asteroid zone A zone of heliocentric space between the orbits of Mars and Jupiter which contains most of the orbits of the asteroids.

Asteroidal source of water Albarede (2009) and others have proposed that most of the water and other volatiles were supplied to the surface of the Earth as a "late veneer" from an asteroidal source. There is a class of asteroids (Themis-type) that contain significant water and related volatiles in clay minerals.

Asthenosphere A zone within the Earth's upper mantle that deforms plastically.

Astronomical unit (AU) The average distance between the center of the Earth and the center of the Sun.

Astrophysics The study of the movement of planets, planetoids, stars, galaxies, etc.

Atmosphere The gaseous envelope surrounding a planetary body.

B

Bacteria Single-celled microscopic biota that are building blocks for more complex life forms.

Basalt Fine-grained igneous rocks that are the main rock type on Earth's ocean floors and the main rock type on the surface of planet Venus. The major minerals in basalt are plagioclase, pyroxene, and olivine.

Basaltic glass shards These commonly occur when basaltic lava contains a significant quantity of gas such as water vapor or carbon dioxide. The shards form as the gas bubbles (vesicules) burst and the material then solidifies rapidly as glass.

Benign Estrangement Scenario The situation when a couple of individuals drift apart after a more fruitful relationship. The Earth-Moon system is an analogue for this in that the Moon (Luna) is slowly orbiting farther away from Earth and the tidal interaction with Earth is getting weaker in time.

Beta Regio region of Venus An area of higher topography on the surface of Venus. It is interpreted by some investigators as equivalent to continental crust of Earth.

Big Brother Jupiter Planet Jupiter has a significant gravitational effect on most bodies in the inner part of the Solar System. The effects are greatest in the asteroid zone but Jupiter is the main cause of the heliocentric eccentricity cycles of the Milankovitch model as well as for a purported Jupiter Orbit—Lunar Orbit resonance. This situation does seem analogous to a Big Brother doing kind positive deeds for a little sister.

Biological system An organized group of one of more organisms on a planet or planetoid.

Bipolar outflow Massive outflow of material (dust and gas) in the polar regions of a nascent sun-like star. The outflow appears to be guided by magnetic flux lines. Much of this material escapes from the star.

Body viscosity and Q factor According to Ross and Schubert (1986) the Q factor is a function of body viscosity and low Q values are associated with intermediate values of planetary viscosity for lunar sized bodies.

Brachinite meteorites A small group of olivine-rich achondrites.

C

CAI Calcium-aluminum inclusions found in chondritic meteorites. A composite of all types of CAIs yields a chemical composition very similar to the Earth's Moon (Gast 1972).

Cambrian explosion A term for the "sudden" appearance of the metazoan stage of life forms in the earliest Cambrian Period.

Capture encounter A close gravitational encounter that results in stable capture of a planetoid from a heliocentric orbit into a planetocentric orbit. The planetoid must be sufficiently deformable (1) to store the energy for stable capture and (2) to dissipate a large percentage of that stored energy within the body of the planetoid during the time frame of the close encounter.

Carbon dioxide An oxide of carbon that is an important constituent in the atmospheres of Venus, Earth, and Mars.

Carbonate rocks Rocks composed of calcium carbonate and related carbonate minerals.

Catalytic delamination-driven tectonomagmatic model A petrologic model proposed by Bedard (2006) to explain how granitic crustal material can be formed by reworking a mainly basaltic crust via vertical tectonic processes.

Celestial mechanics The study of the movement and interactions of bodies in space.

Ceres The largest and most massive asteroid (~914 km in diameter).

Chalcophile elements A group of chemical elements that tend to bond with sulfur to form sulfide minerals.

Chemical condensation model A calculation of the sequence of crystallization of minerals from a gas of solar composition as the gas cools under nebular conditions.

Chondritic meteorites Meteorites composed mainly of chondrules.

Chrondules Millimeter-sized spherical particles that were at one time glassy beads in space before they got accreted into chondritic meteorites.

Co-formation model A model for formation of the Earth's Moon by accretion of material from a geocentric orbit.

Collision encounter An encounter between a planet and a planetoid that results in a collision of some sort. Some are solid collisions in which the material of the impactor is consumed by the planet. Others are grazing encounters in which some of the material of the impactor is consumed by the planet and other material either escapes or goes into some sort of planetocentric orbit.

Cometary source of water Martian rocks have a deuterium to hydrogen ratio very similar to that of comets. A reasonable conclusion is that Mars got some, or most, of its water from either a collision with a comet or a collision with a Jupiter-captured comet (i. e., a comet in an asteroidal orbit).

Compressed density Density of a body as it exists in nature.

Continental crust A mass of crystalline material composed mainly of granitic mineralogy.

Continental drift A concept proposed by Alfred Wegener that the continents were moving relative to one another. This idea was the precursor to the Plate Tectonics model.

Continental platform A mass of continental crust the top of which is located above sea level.

Continental shelf A mass of continental crust that is flooded with ocean water.

Cool Early Earth model A model proposed by John Valley and others (2002) that the Earth was cool enough to have liquid water on the surface as early as 4.4 Ga ago. Much of the early evidence for this condition was gained by the study of zircon crystals from the Jack Hills Formation in Western Australia.

Crater density studies Determination of the number and size of craters in a given area of a planetary surface.

Crystallites Very small or poorly formed crystals usually in a glassy matrix.

Crystals A solid formed by the orderly arrangement of ions or atoms.

D

Deuterium A hydrogen atom with a proton and a neutron in its nucleus (i. e., heavy hydrogen).

Deuterium to hydrogen ratio A ratio of heavy hydrogen to normal hydrogen that is characteristic of certain materials.

Devolatilization A process, usually thermal in nature, in which volatile material is expelled for a substance.

Directional properties Features of a portion of a planetary surface that suggest movement in a particular direction. The usual indicators of directionality are elongation and/or asymmetry of the feature.

Disintegrative capture model A model for the origin of the Moon in which material that was tidally disrupted during close encounters of planetoids to the planet is later accumulated to form the Moon. The main objective of the model was to explain how an iron-poor Moon could be formed.

Disk-Wind Model A model for the early phase of solar accretion from a protosolar cloud that is characterized by much turbulence and solar magnetic field activity. The Disk-Wind model would theoretically follow the X-Wind model and looks promising for explaining the origin of chrondrules.

Displacement Love number A deformation parameter for a planetary body that is related to its deformability due to tidal stresses.

E

Early Proterozoic "plate tectonics" Many investigators recognize features in the Early Proterozoic rock record that are very similar to signatures of Modern Plate Tectonics. Many of these signatures are lacking in the Middle Proterozoic rock record. The suggestion is that maybe something like Plate Tectonics operated in Early Proterozoic time and then shut down until Late Proterozoic time (i. e., a discontinuous process).

Earth-Adonis Magnus system An alternate reality scenario in which a lunar-mass planetoid is gravitationally captured into a retrograde geocentric orbit by an earth-like planet. The planet- satellite system then evolves in a similar manner to the Venus-Adonis system.

Earth-tide recycling mechanism for primitive crust A model in which the unidirectional, high and irregular earth tides associated with the capture process and the subsequent orbit circularization era are the proposed mechanism for recycling the Hadean-age primitive crust of planet Earth into the Earth's mantle in the broadly defined equatorial zone over a period of about 300 Ma.

Eccentricity of the orbit of a planet A physical term for the state of ellipticity of the orbit of a planet.

Eccentricity of the orbit of a planetoid A physical term for the state of ellipticity of the orbit of a planetoid.

Electromagnetic induction heating A mechanism for melting the lunar upper mantle by way of electromagnetic waves in the microwave range generated by the Sun.

Elliptical orbits Orbits that are not perfectly circular. The degree of ellipticity is the eccentricity of the orbit.

Energy dissipation The fraction of the energy of tidal distortion that is subsequently deposited in the body. This fraction of energy is determined by the Q of the body.

Eon The longest subdivision of time on the geologic time scale.

Era The next longest subdivision of time on the geologic time scale.

Evection resonances For the purposes of this project an evection resonance is a gravitational interaction between a planet's orbit around the sun and a prograde satellite's orbit around the planet. Given sufficient time in this state of resonance, the eccentricity of the prograde orbit of the satellite can be sequentially increased.

F

Faint Young Sun concept A general concept in astrophysics is that when a star begins its main sequence of "burning" it is in its coolest state. From then on, the star increases in thermal output until it reaches the red-giant stage. Most investigators agree that the Sun about 4.0 billion years ago gave out about 20 % less thermal energy than it does today.

Fall-back process This is a term associated with a gravitational tidal disruption process. After the material is tidally disrupted, then a portion of the material will fall back onto the body that was tidally disrupted. The resulting fall-back pattern depends on the direction of rotation of the body after disruption and on the angular velocity of rotation of the body.

Fatal Attraction Scenario The situation when a couple of individuals are attracted to each other so much as to eventually destroy one or both of the individuals. The Venus-Adonis system is an analogue for this condition in that Adonis is forever getting closer to Venus and the result is the complete destruction of the planetoid Adonis and a transformation of the surface of Venus to an "inferno".

Fission model A model for the origin of the Earth-Moon system in which a rapidly rotating proto-Earth spins off a lunar-sized planetoid which goes into orbit to form the Moon. The mechanism of separation of the proto-Earth and the body of the Moon has never been demonstrated.

Forced eccentricity scenario A process associated with planet orbit—lunar orbit resonances. The lead- in phase of this rhythmic orbital scenario causes the eccentricity and major axis of the lunar orbit to increase but does not increase the angular momentum of the orbit. Once the eccentricity and major axis of the orbit are forced to a maximum value, then angular momentum is very slowly transferred from the rotating planet to the satellite orbit until the satellite orbit is nearly circular again. This process takes hundreds of millions of years.

Frost line This concept is associated with a possible explanation for the size and composition of planet Jupiter. The general idea is that the dust and other particles in the inner part of the solar nebula are progressively devolatilized by a combination of processes like the X-Wind and Disk -Winds. Many of these volatiles are pushed out to the various zones of the asteroid belt and the bulk of the volatiles end up the vicinity of the orbit of planet Jupiter and eventually accete to form planet Jupiter.

FU Orionis Model A era of star evolution characterized by periodic energetic outbursts in the microwave range of radiation.

Futurozoic eon An unofficial major subdivision of geologic time for an interval from the Present to the Distant Future.

G

Ga Billion years.

Gabbro The coarse-grained chemical equivalent of basalt.

Gargantuan Basin A term proposed by Peter Cadogan (1974) for the area of Oceanus Procellarum. His interpretation is that Oceanus Procellarum is a large impact structure. If it is an impact structure most of the typical impact features must be concealed by lava flows.

Gaseous planets Planets with no obvious solid surface and with atmospheres composed mainly of hydrogen and helium.

Geocentric orbit An orbit of a particle orbiting Earth.

Giant Impact Model A model for the origin and evolution of the Earth-Moon system that features a Mars-mass body colliding with Earth and the Moon forms from the debris of the collision.

Global resurfacing event A model proposed to explain the apparent youthful surface of Venus. The main feature for the interpretation is a fairly uniform distribution of impact craters on the surface of the planet.

Granite An igneous rock composed mainly of feldspar, quartz and minor quantities of iron-magnesium bearing minerals.

Granophyre A small mass (pod) of granitic mineralogy in a mass of mafic rock like basalt or gabbro.

Gravitational perturbations Systematic small gravitational tugs on a planetary body.

Grazing encounter An encounter between a planetoid and a planet in which there is contact between the two bodies.

Great circle pattern A pattern that can result from the intersection of an encounter plane on a spherical surface. An example is a great-circle pattern of large circular maria on the lunar surface.

Great Oxidation Event A global event proposed by Heinrich Holland that led to a continuous presence of oxygen in the atmosphere.

H

Habitable zone A heliocentric zone for planets which is suitable for a biological system to exist.

Hadean Eon A major subdivision of geologic time covering the time from the final assembly of the accreted particles to form Earth to the beginning of the Archean Eon (~4.5 to 3.95 Ga ago).

HED meteorites A group of meteorites that are thought to be derived from the asteroid 4 Vesta.

Heliocentric distance Distance measured from the center of the Sun to center of the body of interest.

Heliocentric orbit The path of a planet around the Sun; the orbit can be circular or elliptical.

Hill sphere A spherical boundary around a planet that defines the gravitational sphere of influence of the body.

Hydrogen sulfide A gas composed of two atoms of hydrogen and one of sulfur; this gas is characteristically found in chemically reducing environments.

Hydrothermal metamorphism A process of recrystallization in which some combination heat, pressure, and hot water, steam, or some other volatile substance are involved.

Hypervelocity impact The impact of a body with a velocity that is significantly greater than the escape velocity for that planet or planetoid.

Hypothesis An essentially untested statement or partially tested for explanation of a natural phenomenon.

I

Imbrium sculpture A radial "grooving" pattern associated with the Mare Imbrium. Some investigators suggest that the radial pattern is associated with basin excavation. My interpretation is that it is associated with the deposition of the mare basalts via the directional impact of a "soft" body of mare basalt from a lunar westerly direction.

Inferior conjunction A situation in which a planet in an orbit nearer the Sun moves between an outer planet and the Sun (e. g., Venus passes between Earth and the Sun).

Inorganic processes A process or reaction in which no organisms are involved.

Intact capture model A model in which the encountering planetoid maintains its integrity during the capture process.

Iron line A heliocentric zone within the proto-solar midplane that gets enriched in metallic minerals because of repeated episodes of vaporization of metallic ions due to heat pulses and condensation of metallic ions into metallic mineral structures upon cooling.

Iron meteorites classified as IIIAB A group of meteorites thought to represent the core material of a volcanic asteroid. John Wasson (2013) suggests that this group of iron meteorites may be related to the HED meteorites.

Iron meteorites classified as IVB A group of metallic meteorites that underwent fragmentation and ended up in the Asteroid Belt. Campbell and Humayun (2005) suggest that the parent melt of IVB irons appears to be very similar to that of the Angrite parent body.

Iron-nickel metal An intergrowth of crystals of iron and nickel that commonly occur in certain types of meteorites. Usually a homogeneous crystal forms at high temperature and the Fe and Ni phases form the intergrowth as the temperature of the crystal decreases.

Isostatic equilibrium model for mascons Isostatic equilibrium is the balance between sections of crust and the underlying mantle on a planetary body. Assuming equal thicknesses of crust, the denser the section of crust, the lower that section of crust will be topographically. For example, mare basalts have a higher density than the lunar highland rocks. Therefore, the mare areas should be topographically lower than the adjacent lunar highland rocks.

J

Jack Hills Formation An Archean-age quartz-rich meta-sedimentary rock in Western Australia that has yielded many of the zircon crystals that resulted in the interpretation of the Cool Early Earth model.

Joule heating Energy that results in raising the temperature of a substance.

Jupiter The largest planet in the Solar System.

Jupiter Orbit—Lunar Orbit resonance A resonance caused by a lengthy series of gravitational tugs that cause the major axis and eccentricity of the lunar orbit to increase.

Jupiter-captured comets Comets that have been captured into heliocentric asteroid-like orbits after a close encounter with planet Jupiter

K

Ka Thousand years.

Kona conference A conference on the "Origin of the Moon" that was held in Kona (HI) in October 1984. The Giant Impact Model emerged as the favorite model from this meeting.

Kuiper Belt A heliocentric belt that extends from the orbit of planetoid Pluto out to at least 500 AU and is populated by icy bodies of various masses.

L

Late Heavy Bombardment A purported massive bombardment of the Moon by external bodies between about 4.0 Ga and 3.8 Ga. The source of these impactors is still unknown.

Late Proterozoic continental rift zones There are many continental rift zones identified for this Era. Several of these rift zones also contain sedimentary rocks with features of glacial deposition. The suggestion is that the earth's mantle must have been actively involved in the development of the rift zones.

Late Proterozoic—Early Phanerozoic tidal rhythmites Many sedimentary rock sequences of this age contain features of tidal zone deposition including sequences of tidal rhythmites. Tidal rhythmites are defined as sedimentary sequences that feature cyclic variations in thicknesses of successive laminae in response to changing current velocities associated with lunar cycles.

Late Proterozoic "global" glaciations During this Era of geologic time there are many sedimentary rock sequences that have glacial deposition signatures and paleomagnetic data suggest that many of these deposits formed in very low paleolatitude settings. Analysis of this information has led to the development of the concept of a "snowball earth".

Late veneer of volatiles The Potassium Index of solar system bodies suggests that the activity of the proto-sun must have devolatized the proto-solar cloud in a systematic manner. Albarede (2009) suggests that these volatiles were propelled to the outer asteroid zone and to the vicinity of Jupiter's orbit. Then about 100 Ma plus or minus 50 Ma some of these volatiles were brought back into the realm of the terrestrial planets in the form of water-rich and volatile-rich asteroids. These bodies then impacted onto the surfaces of the rocky planets as a "late veneer".

Length of day The rotation rate of a planet or planetoid.

Lithophile elements A group of chemical elements that tend to bond with silica and/or oxygen and commonly form crustal rock complexes.

Long chain of complications An expression by Hannes Alfven (1969) when assessing the complex nature of the biological system of Earth.

Luna A name for the Moon before it was captured by Earth.

Lunar ash-flow deposits After examining photographs from the lunar orbital missions, J. Hoover Mackin (1969) thought that some of these lunar surface units had characteristics of basaltic ignimbrites and basaltic ash-flow deposits.

Lunar basalt viscosity The measured viscosity of molten lunar basalt is much lower than that of normal molten terrestrial basalts. Thus molten lunar basalts flow much more easily and rapidly than basalts on Earth.

Lunar basaltic spheroids Features that could result from a necking-off process from a lava column on the Moon due to the strong gravitational influence of Earth during a very close encounter between Earth and Moon.

Lunar Capture Model A model that features gravitational capture of Luna from a heliocentric orbit into a geocentric orbit.

Lunar embryo The favored planetesimal for the formation of the Moon in the co-formation model.

Lunar geologic maps Maps showing the spatial distribution of lunar surface rock units. The relative ages are then determined by way of crater density studies of the units.

Lunar KREEP basalts Basalts rich in potassium, rare earth elements, and phosphorous. This type of basalt is thought to be the last magma to crystallize from a lunar magma ocean and also the first to melt when the region of the lunar magma ocean is reheated to above the melting point.

Lunar Magma Ocean A zone about 600 km thick that was apparently melted at the time of lunar formation or remelted after lunar formation. Differentiation of this zone resulted in the development of an all-enclosing anorthositic crust.

Lunar mare Large areas of lunar basalt that can have either circular or irregular outlines.

Lunar nuee ardente-like deposits A good possibility for these is in the Mare Ingenii area which is anti-podal to the Mare Imbrium area.

Lunar satellites A suggestion by Keith Runcorn (1983) and Bruce Conway (1986) was to have a number of lunar satellites impact over a period of time onto the lunar surface to form a great-circle pattern of circular mare basins. Conway (1986) did calculations on the orbital stability of such a group of lunar satellites.

Lunar stratigraphy A stratigraphy developed that was based mainly on a combination of unit overlapping conditions and density of primary craters on the surfaces of the units.

M

Ma Million years.

Magellan Mission A space-craft mission to Venus in the early 1990s that mapped most of the planetary surface via radar imagery. These radar photos were the basis for many of the geologic mapping projects for the surface of Venus.

Magma ocean A zone of molten rock that can form in the outer portion of a planetoid or planet via joule heating.

Magnetic flux lines Lines of a magnetic force field.

Mainstream Mantle Convection model A model developed by Nelson and Temple (1972) to explain the directional features and asymmetry of subduction zones as well as other directional features associated with the plate tectonics model.

Major axis of an orbit The maximum length of an elliptical orbit.

Mantle A zone of the Earth's interior that is beneath the crust and above the metallic core.

Mantle uplift model for mascons A model that features solid body impact to form a basin. This is followed by a mantle upwelling caused by the heat generated by the impact of the solid body.

Mare Ingenii region A region of the Moon that is antipodal to Mare Imbrium and an area characterized by ridges and swirls of unknown origin. A reasonable interpretation is that they are volcanic in origin.

Mare Orientale region A "bull's-eye" structure on the western side of the Moon. This feature has notable asymmetry. The conventional interpretation is that it is an impact structure. In my view, it looks much different than a normal impact structure.

Mare Orientale tidal disruption center This is the initial tidal disruption center in the close encounter model proposed by Malcuit et al. (1975).

Marius Hills volcanic field This volcanic complex is located within a narrow elongate zone in Oceanus Procellarum that is interpreted by Malcuit et al. (1975) as the second tidal disruption zone. After the mare-producing encounter, there would be much tidal pumping action during the orbit circularization sequence. Local volcanism in the Marius Hills area could be expected to continue, on an irregular periodic basis, for a few hundreds of millions of years mainly because there would be no anorthositic crust under this tidal disruption zone and molten lunar basalt with a significant KRREP component could ascend directly from the lunar magma ocean.

Mars The fourth planet from the Sun (Mars is about 0.1 as massive as Earth).

Mass concentrations Areas of excess mass that were mapped by orbiting spacecraft. Most, if not all, of the mass concentrations are associated with circular maria.

Matrix material Fine-grained material binding larger particles together.

Megascopic Large enough to see with the unaided eye.

Mercury accretion torus A heliocentric doughnut-shaped feature that would contain the debris that accreted to form planet Mercury.

Mercury eccentricity belt Heliocentric zone of possible occupancy of planet Mercury when in an elliptical orbit.

Mesosiderite meteorites Meteorites composed of roughly equal quantities of metallic and silicate crystals.

Metallic meteorites Meteorites composed mainly of metallic minerals.

Metal-sulfide core Composition that has been proposed for a core of the Moon. There is really no credible physical evidence for such a lunar core.

Metamorphism A change in form process featuring deformation and/or recrystallization.

Metazoan A multi-cellular biological form whose cells perform different tasks to support the life of the organism.

Microscopic A feature large enough to see with the aid of a microscope.

Microwave wind A "solar wind" in the microwave range of the electromagnetic spectrum.

Milankovitch cycles Cycles of heliocentric orbital eccentricity, planetary obliquity (tilt), and planetary precession as proposed and calculated by Mulitin Milankovitch.

Modern style of plate tectonics Stern (2005) and Davies (1992) and others think that the Earth was too warm to permit slab-pull subduction before about 1.0 Ga ago. The slab-pull subduction mechanism is thought to be a major driving force for the modern style of planet tectonics.

Moment of inertia A property related to the distribution of mass within the body. The coefficient of the moment of inertia is an important factor for this characterization. For example, a homogeneous body has a coefficient of the moment of inertia of 0.4. It is interesting to note that the coefficient of the moment of inertia of the Moon is very near 0.4.

Multi-orbit collision scenario A numerical simulation that results in a collision of some sort after several orbits about the planet.

Multi-orbit escape scenario A three-body numerical simulation in which insufficient energy is dissipated within the interacting bodies for capture and the encountering body escapes back into heliocentric orbit after a number of orbits about the planet.

Multiple-small-moon model A model proposed by MacDonald (1964) to circumvent the apparent problem of a short time scale for lunar orbit evolution. The general idea is that each of the smaller satellites would dissipate much less energy per unit of time than a lunar-mass satellite and the result would be a longer time scale for orbit evolution.

N

Nebular dust and gas Material from previous stellar explosions that gathers together by gravity to form the Sun.

Nebular mid-plane An equatorial belt of infalling dust and gas to form the Sun.

Neptune The eighth planet from the Sun.

Nodal cycle A cycle of the lunar orbit that features the retrograde revolution of the intersection of the lunar orbit and the celestial equatorial plane. The cycle at the present time is 18.6 years, retrograde.

Non-grazing encounter A close encounter between two bodies in which no contact is made.

Non-rotating coordinate system A plotting procedure for planetocentric orbits in which the plot does not rotate as it theoretically revolves about the Sun.

Normal faulting A type of fault commonly caused by crustal extension.

Nullarbor Desert One of the drier deserts in southern South Australia and southern Western Australia. The desert apparently has a lack of trees.

O

Oblique impact An impact that is *not* vertical to the surface (i. e., an impact at an angle).

Obliquity cycle The tilt-angle cycle for the Earth which is ~41 Ka long.

Ocean floor crust Igneous rock that is commonly found under the oceans of Earth; it is mainly basaltic in composition.

Ocean water Water characteristic of Earth's oceans; present salinity is about 3.5 % by weight.

Oceanus Procellarum region A region of very irregular mare shorelines.

Oceanus Procellarum tidal disruption center The proposed site of the second tidal disruption center. Most of the basalt for forming spheroids that are involved in the backfall stage of the tidal disruption model are extracted from this disruption center.

One body isostatic equilibrium model A model in which the only excess density is associated with a basaltic lava lake underlain by a less dense anorthositic crust.

One-orbit escape scenario A three-body numerical simulation in which little or no energy is dissipated within the interacting bodies. The planetoid makes one pass and then escapes back into a heliocentric orbit.

Oort Cloud A spherical zone surrounding the Solar System that is thought to be the source of all comets; the main zone is considered to be between 50,000 and 150,000 AU from the Sun.

Orbit circularization sequence An orbital evolution scheme in which the eccentricity and major axis of the orbit of the planetoid decrease as energy is extracted from the planetoid orbit and dissipated as thermal energy within the interacting bodies by tidal processes.

Orbital traceback A analytical calculation for the evolution of the Earth-Moon system starting with the present conditions and calculating backwards in time. Several of these types of calculations have been done since the first one by Horst Gerstenkorn in 1955.

Orderly component of planetary rotation An explanation of how the rotation rates suggested by MacDonald (1963) could be explained by accumulation of small planetoids from mainly elliptical orbits. These calculations were done by Lissauer and Kary (1991).

Organic processes A process or processes in which organic material is involved.

Oxygen A very reactive gas that is a major constituent of the Earth's atmosphere; it is ~21 volume percent of the atmosphere at the present time.

Oxygen-isotope ratios Ratio of O-16 (the light isotope) relative to either O-18 or O-17 or both; these isotope ratios are sensitive to temperature of formation of certain materials.

P

Palimpsest A surface pattern on which several features are recorded, one superimposed on the other, with some of the older features still visible for possible interpretation.

Pallas Second largest asteroid by diameter (~522 km); third most massive asteroid.

Pallasite meteorites Stony-iron meteorites composed mainly of olivine crystals enclosed by metallic minerals.

Paradigm A commonly accepted concept or model for the explanation of a natural phenomon.

Paradox A statement that seems contradictory, unbelievable, or absurd but may be actually true in fact.

Perigean cycle A cycle in which the position of the perigee moves in a prograde direction around Earth in 8.85 years (at the present time in Earth-Moon system history). The period of the cycle *increases* as the lunar orbital radius *decreases*.

Perigee point The nearest approach of a satellite in earth orbit.

Perihelion The nearest approach distance to the Sun for a body in heliocentric orbit.

Period A subdivision of the geologic time scale that is a subdivision of an era.

Persephone An immortal Greek Goddess who was the daughter of Zeus and Demeter.

Petrographic microscope A microscope that has the capability of using transmitted polarized light to characterize minerals, rocks, and other material.

Phanerozoic Eon A major subdivision of geologic time extending from about 570 Ma to the present.

Phantom body A body that has left a trace of its existence but no longer exists.

Phantom satellite A satellite that has left a physical signature on the characteristics of a planet but no longer exists.

Photo-dissociation Breakdown of an atmospheric gas like water into its hydrogen ions and oxygen ions by way of interaction with ultraviolet (UV) rays from the Sun.

Planet anomaly The position of the planet at the beginning of a three-body numerical simulation calculation.

Planet orbit—lunar orbit resonances A resonance caused by a lengthy series of repeated and systematic gravitational tugs by a planet in an unrelated heliocentric orbit which can increase the eccentricity and major axis of the orbit of the satellite of a planet.

Planetary magnetic field A system of magnetic flux lines generated by an internal magnetic field.

Planetary perturbations A series of repeated and systematic gravitational tugs on a planet by a planet in an unrelated heliocentric orbit.

Planetoid A planet-like unit that is too small to be a planet.

Planetoid anomaly The position of the planetoid relative to the planet at the beginning of a three-body numerical simulation.

Planetology The study of the origin and evolution of planets.

Plate tectonics Formation and movement of a system of rocky plates on a planet; the main processes that result in movement are formation of ocean floor at ocean floor rises and subduction of plates under volcanic arc complexes.

Plant life Metazoan organisms with limited mobility and generally capable of photosynthesis.

Pluto A former planet that is now classified as a planetoid.

Pluto-Charon system Planetoid Pluto has a large satellite (Charon) relative to the size of the planetoid; planetoid Pluto also has several small satellites.

Polar regions sheltered from tidal activity Since most of the tidal activity associated with gravitational capture is confined to the equatorial plane of the Solar System, the most severe rock and ocean tidal activity would be in the broadly defined equatorial zone of the planet. The obliquity (tilt angle) of the planet is a major factor in defining this zone of major tidal activity. Nonetheless, in most cases the polar zones are sheltered from major tidal deformation activity.

Potassium vs. Uranium ratio The K/U fraction is the basis for the Potassium Index for solid bodies in the Solar System; this Potasssium Index appears to be related to the place of origin of planetoids and planets.

Precession cycle This cycle features the retrograde movement of the north pole of the planet relative to the celestial sphere; this cycle is ~20 Ka long.

Primary craters Impact carters that are caused by bodies impacting at a high speed from space.

Primitive crust of Venus A proposed all enclosing crust that formed on Venus during the equivalent of the Cool Early Earth time.

Primordial rotation rate for Earth According to the plot by MacDonald (1963) the primordial rotation rate for Earth should be about 10 h/day prograde.

Primordial rotation rate for Venus According to the plot by MacDonald (1963) the primordial rotation rate for Venus should be about 13.5 h/day prograde

Probability of capture A numerical value that can be obtained from the geometry of an appropriate Stable Capture Zone.

Prograde capture A process in which the captured planetoid is inserted into a geocentric orbit that is in the same direction as the rotation of a planet that is rotating in the prograde direction.

Proterozoic Eon A major subdivision of geologic time extending from 2.5 to 0.57 Ga

Protium The normal atom of hydrogen which consists of only one proton in the nucleus.

Proto-Sun A term for the body of the Sun before the thermonuclear reaction commences.

Q

Q The specific dissipation factor for a body undergoing tidal deformation. 1/Q is the fraction of stored energy that is subsequently dissipated during the deformation cycle.

Quasi-three-body calculation of orbital evolution This is a procedure for making a two-body calculation more realistic by including the effects of close encounters caused by solar gravitational perturbations.

R

Reconnection of magnetic flux lines A major source of energy produced by the X-Wind model; this "reconnection" energy is thought to be the main source of energy for the formation of calcium- aluminum inclusions that are found in chondritic meteorites.

Reconnection ring A shifting belt of intense thermal activity in the proto-solar mid-plane of the X-Wind model. This is the place of formation of CAIs and the belt has a central value ~0.06 AU.

Relative age The age of a rock unit relative to another rock unit as determined by some combination of cross-cutting relations, overlapping ejecta aprons, crater density patterns, etc.

Remanent magnetization The direction and intensity of a magnetic field as recorded by certain iron- bearing rock-forming minerals. This magnetic information is recorded in the rock as it cools through the Curie Point.

Report card for models A grading scheme used by John Wood (1986) to evaluate models of lunar origin presented at the Kona Conference.

Retrograde capture Capture of a satellite from a heliocentric orbit in the opposite direction to the rotation of planet rotating in the prograde direction; the satellite is moving in a clock-wise direction when viewed from the north pole of the Solar System; since the angular momentum vector of the satellite is in the opposite direction to the angular momentum vector of the planet, angular momentum appears to disappear from the system.

Roche limit A geocentric position at which a liquid body becomes unstable and begins to undergo a necking process at the sub-earth point as well as at the anti-sub-earth point. This limit is at ~2.89 earth radii for the Earth-Moon system.

Roche limit for a solid body A geocentric position at which a solid but deformable body becomes unstable and begins to undergo a disintegration or necking process at the sub-earth point as well as at the anti-sub-earth point. This limit is at ~1.6 earth radii.

Rosetta Stone A tablet of black basalt found in 1799 at Rosetta (Egypt) that provided a key for the deciphering of ancient Egyptian writing.

Rotating coordinate system A plotting procedure for planetocentric orbits in which the plot rotates as it theoretically revolves about the Sun.

Rotation rate The rate at which a planet rotates on its axis.

Rules for interpretation of lunar surface features A set of rules (guidelines) proposed by Mutch (1972) and expanded and clarified by Wilhelms (1987, 1993).

Runge-Kutta numerical integration method A fourth-order Runge-Kutta numerical integration scheme is used to step position and velocity coordinates of the bodies forward in time. The precision of the calculations presented in this book are a few parts in 10E5 for energy and a few part in 10E8 for angular momentum.

S

Satellite A body in a planetocentric orbit.

Satellite magnetic field A magnetic field that can get recorded in the rocks of a satellite . The magnetic field can be generated within the body of the satellite or it can be generated by an external source such as a planet or the Sun.

Saturn The sixth planet from the Sun.

Scientific method A procedure for solving problems in the natural sciences.

Second Oxidation Event A second oxididation event during the late Proterozoic time (~ 1.0–0.57 Ga) brought atmospheric oxygen levels up to near modern values (Shields-Zhou and Och, 2011).

Secondary craters Craters formed by impact ejecta from a primary impact crater.

Seed magnetic field A weak magnetic field that can be increased in intensity by a magnetic amplifier in another planetary body.

Semi-major axis One-half the value of the major axis of an elliptical orbit.

Shallow shell magnetic amplifier A mechanism that was proposed for generation of a magnetic field in the Moon by Roman Smoluchowski in 1973. The general principle is that convection cells in the lunar magma ocean region under certain conditions can amplify a weak "seed" magnetic field of Earth origin or of solar origin.

Shorelines of lunar maria These shorelines are at the boundary of the relatively smooth areas of lunar maria and the more rugged topography of the lunar highlands.

Siderophile elements A group of geochemical elements that tend to form metallic minerals and occur mainly in the differentiated core of planets and planetoids.

Sister planet A planet that has very similar physical characteristics to another planet.

SNC meteorites These are the sergottite, nakhlite, and chassignite group of meteorites and they are all thought to be impact debris from planet Mars. For more details see McSween (1999).

Soft-body impact A process associated with the tidal disruption and fallback model of Malcuit et al. (1975).

Solar composition The chemical composition of the Sun.

Solar composition devolatilized This term relates to the chemical composition of chondritic meteorites in general. The chondritic meteorites have very similar abundance ratios to those of the Sun except for volatile elements such as O, C, H, N, Ar, Ne, He and a few others (McSween 1999).

Solar constant This term relates to the thermal output of the Sun. It is not really constant but is thought to be increasing in time. For example, the solar energy output was probably about 20 % less at ~4.0 Ga than it is today.

Solar magnetic field The magnetic field that is generated within the Sun.

Solar nebula A cloud of dust and gas that was inherited from previous stellar explosions in our part of the galaxy. This is the starting composition for the Solar System.

Solar System The system of planets, planetoids, satellites, Kuiper Belt objects, and comets that are associated gravitationally with the Sun.

Solid-body impact The impact of a solid body onto the surface of a planet or planetoid.

Spin-orbit resonance A situation in which the orbital motion of the planet or planetoid controls the spin rate of a planet or planetoid.

Stable capture scenario A three-body capture simulation scenario which results in long-term orbit stability (i. e., there is no fatal collision or an escape).

Stable capture zone A region of parameter space (planet anomaly vs. planetoid eccentricity) in which the orientation of the orbit of the encountering planet is favorable for a stable post-capture orbit if sufficient energy is dissipated for its capture into a planetocentric orbit.

Stable continent model This refers to the paradigm (model) that preceded the Continental Drift/Plate Tectonics model.

Stagnant-lid planet The situation on the surface of a planet or planetoid where there is no organized plate motion but vertical tectonics can occur.

Starting point for numerical simulation The planet anomaly at the beginning of the numerical simulation.

Subduction A process in which one tectonic plate moves beneath another tectonic plate; the process usually results in a linear volcanic arc complex.

Sub-Jupiter point The position at which the object is at the closest position to Jupiter.

Sub-Venus point The position at which the object is at the closest position to Venus.

Sulfide minerals Mainly minerals in which metallic ions bond with sulfur ions; these minerals usually have metallic physical properties such as opaqueness.

Superior conjunction The situation in which a planet is between the Sun and another planet.

T

Terminal lunar cataclysm An early term for what is now called the Late Heavy Bombardment.

Terrestrial planets Planets that have rocky surfaces.

The eighteenth perigee passage scenario A three-body numerical simulation in which there is a very close tidally disruptive encounter on the 18th perigee passage of a stable capture scenario.

Theia A mars-mass phantom planet that undergoes a grazing encounter with the proto-earth to form the Moon via the Giant Impact Model

Theory A well-tested explanation of a natural phenomenon.

Three-body numerical simulation A numerical simulation involving three gravitationally interacting bodies. An example would be a calculation involving the Sun, Earth, and planetoid Luna.

Tidal amplitudes on the planet Equilibrium tidal distortions of the planet.

Tidal amplitudes on the planetoid Equilibrium tidal distortions of the planetoid.

Tidal break-up of Adonis This occurs at the Roche limit for a solid body (at about 1.6 venus radii).

Tidal disruption encounter For tidal disruption to occur on a body like the Earth's Moon the encounter must be very well within the Weightlessness Limit of the Earth-Moon system and there must be a zone of molten rock material in the lunar magma ocean prior to the very close encounter.

Tidal energy dissipation The quantity of tidal energy that is dissipated during a close encounter depends on h (the displacement Love number) and Q (the specific dissipation factor) of the body.

Tidal energy storage The quantity of energy that is stored in a tidally deformed body depends on the h (the displacement Love number) of the body.

Tidal friction This term relates to energy that gets dissipated by tidal activity. At the present time in geologic history most of the tidal energy in earth is dissipated by the action of ocean water. In the early history of the planet the rock tides may have been an important source for dissipation of tidal energy.

Tidal rhythmites Sedimentary rock sequences of that have organized packets of rhythmically-layered sediments. The accepted interpretation is that the couplets (light and dark and/or course and fine) represent diurnal or semi-diurnal cycles and that the larger packets represent fortnightly cycles.

Tidal vorticity induction This results from a process by which rock tidal action causes convection cells to form due to differential rotation between the core and crust of a planet or planetoid. This process is described in more detail in Bostrom (2000).

Time of first perigee passage The time elapsed between the Starting Point and the First Perigee Passage during a numerical simulation of an encounter.

Tsunami-like floods of lunar basalt The rush of a wave of lunar basalt resulting from soft-body impact of a lunar basaltic spheroid to form a circular mare on the lunar surface.

T-Tauri stars This term is used to describe star formation but the meaning of the term has changed somewhat over the past few years. T-Tauri stars were studied by observational astronomers because they were thought to be analogs for the evolution of the Sun. In more recent terminology, the T-Tauri phase of a sun-like star is the fourth in line for energy output. The sequence is: X-Wind, Disk-Wind, FU Orionis, and then T-Tauri phenomena.

Twin sister planets Two planets that have very similar basic body as well as orbital characteristics.

Two body isostatic equilibrium model for mascons A model developed in an attempt to explain mascon development that is consistent with a solid-body impact. The model features a lunar mantle plug that is elevated under the impact site and brings more dense rocks from the lunar mantle to the near surface realm. Only a thin layer of mare basalts is needed on the surface of the mare basin and this mare material comes to the surface at some time after the impact that caused the mare basin.

Two-body analysis A simplified analysis of a more complex problem to check for the feasibility of a reasonable solution. An example is a two-body analysis of the gravitational capture process.

Two-body numerical simulation of orbital evolution An adiabatic calculation for simulating the orbital evolution of a two-body system (Earth and Moon in this case) without the influence of the Sun. The input parameters are the h and Q values of the planet, the h and Q values of the planetoid, and the major axis and eccentricity of the initial elliptical orbit. The time scale for evolution of the system using the method is unrealistically long, especially for the early portion of the calculation.

U

Uncompressed density The uncompressed density of a planetary body is estimated by theoretically decompressing the mineral structures of successive layers within the body.

Uranus The seventh planet from the Sun.

Velocity at infinity The velocity of a body before it comes under the gravitational influence of the planet (this is usually about two times the radius of the Hill Sphere).

V

Venera 8 mission A Soviet space mission to Venus in 1972 that measured some chemical parameters of a few of the surface rocks.

Venocentric orbit Refers to a body in orbit around Venus.

Venus oceans problem The deuterium to hydrogen ratio of water in the outer portion of the atmosphere of Venus suggests that Venus had oceans of water on the surface in earlier geological times.

Venus Orbit—Lunar Orbit resonance in Futurozoic A VO–LO Resonance when the perigean cycle of the Earth-Moon system is at 5 earth years (8 venus years). This will occur about 0.5 to 1.0 Ga in the future.

Venus Orbit—Lunar Orbit resonances There are three possible occurrences in the history and future of the Solar System. Two of them have already left their signatures in the geological record and one is scheduled for the future. All three are caused by a synchronization of the perigean cycle of the Earth-Moon system and the heliocentric orbital motion of Venus.

Venus-Adonis system A planet-satellite pair composed of planet Venus and a phantom satellite called Adonis.

Vesta A basaltic asteroid with a semi-major axis of ~2.36 AU and eccentricity of ~0.09. Vesta is the second most massive asteroid and it is third in diameter at ~500 km.

Vignette A vignette is defined as a short literary composition characterized by compactness, subtlety, and delicacy.

Volcanic arc A linear volcanic complex generally located above a subduction zone; the common rock type is andesite which is intermediate in composition between granite and basalt.

Volcanic-type asteroids These are achondritic meteorites that have all undergone significant heating and melting so that basalt is (or was) present on their surfaces. Basalt is commonly associated with volcanoes on Earth.

Vulcanoid planetoids A name for planetoids formed in the zone between the Sun and Mercury (at about 0.1 to 0.2 AU) as a result of the chemical and physical processing of material by the X-Wind.

Vulcanoid planetoid orbital stability Extensive orbital simulation calculations were done by Evans and Tabachnik (1999, 2002) demonstrating that there are many stable heliocentric orbit located between planet Mercury and the Sun, especially in the 0.1 to 0.2 AU range. They estimate that many of these orbit could be stable for hundreds of millions of years.

Vulcanoid zone A place of origin for Vulcanoid planetoids which would form mainly from CAI material generated in an X-Wind environment.

W

Water-bearing asteroids These are Themis-type asteroids that are known to be enriched with water and other volatiles. These Themis-type asteroids may be the main source for water and some other volatiles on Earth and Venus.

Weightlessness limit The weightlessness limit is the center-to-center distance between the Earth and Moon at which weightlessness occurs at the sub-earth point on the lunar surface (Malcuit et al. 1975). The condition of weightlessness can occur on the lunar back-side as well but it occurs when the center of the Moon is much closer to the Earth.

X

X-Wind model This model was proposed to explain the origin of the very high temperature components found in chondritic meteorites (the CAIs). The very high temperature and very dynamic processes associated with the X-Wind model have been duplicated in part in laboratories but have not been observed in nature because of the opacity of the observable features of nascent stars of roughly solar mass.

Y

Yamato achondrite An achrondrite meteorite that was recovered in the Yamato Mountains area in Antarctica in 1979. This was the first meteorite in terrestrial meteorite collections that was recognized as being derived from the Moon. The number assigned to this meteorite is 791197. The first two numbers are for the year collected and the last four numbers are the collection number (McSween, 1999).

Z

Zircon crystals Crystals composed of zirconium silicate that are very refractory. Zircon crystals commonly contain inclusions of uranium and thorium bearing minerals which contain the age-dating elements for the U-Th-Pb system of radiometric dating.

Author Index

Robert J. Malcuit, *The Twin Sister Planets Venus and Earth*,
DOI 10.1007/978-3-319-11388-3

Subject Index

Robert J. Malcuit, *The Twin Sister Planets Venus and Earth*,
DOI 10.1007/978-3-319-11388-3

d=diagram, t=table, p=photographs